Ⓘnquire

Ⓘnteract

Ⓘnspire

Ⓘnvent

This ⒾScience Interactive Student Textbook Belongs to:

_____
Name

_____
Teacher/Class

Where am I located?

The dot on the map shows where my school is.

Mc Graw Hill Education

PHYSICAL

ⒾSCIENCE

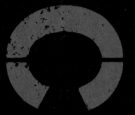

Glencoe

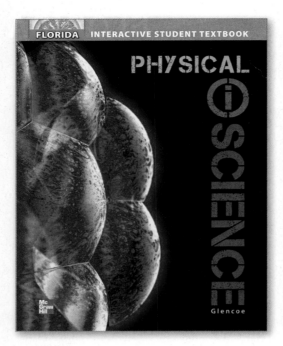

FLORIDA INTERACTIVE STUDENT TEXTBOOK

PHYSICAL iSCIENCE

Glencoe

Bubbles

The iridescent colors of these soap bubbles result from a property called interference. Light waves reflect off both outside and inside surfaces of bubbles. When this happens, the waves interfere with each other and you see different colors. The thickness of the soap film that forms a bubble also affects interference.

The *McGraw-Hill* Companies

 Education

Copyright © 2012 by The McGraw-Hill Companies, Inc. All rights reserved. No part of this publication may be reproduced or distributed in any form or by any means, or stored in a database or retrieval system, without the prior written consent of The McGraw-Hill Companies, Inc., including, but not limited to, network storage or transmission, or broadcast for distance learning.

Send all inquiries to:
McGraw-Hill Education
8787 Orion Place
Columbus, OH 43240-4027

ISBN: 978-0-07-660225-4
MHID: 0-07-660225-7

Printed in the United States of America.

8 9 10 11 12 QVS 18 17 16

The Florida Teacher Advisory Board provided valuable input in the development of the © 2012 Florida student textbooks.

Ray Amil
Union Park Middle School
Orlando, FL

Maria Swain Kearns
Venice Middle School
Venice, FL

Ivette M. Acevedo Santiago, MEd
Resource Teacher
Lake Nona High School
Orlando, FL

Christy Bowman
Montford Middle School
Tallahassee, FL

Susan Leeds
Department Chair
Howard Middle School
Orlando, FL

Rachel Cassandra Scott
Bair Middle School
Sunrise, FL

Authors

American Museum of Natural History
New York, NY

Michelle Anderson, MS
Lecturer
The Ohio State University
Columbus, OH

Juli Berwald, PhD
Science Writer
Austin, TX

John F. Bolzan, PhD
Science Writer
Columbus, OH

Rachel Clark, MS
Science Writer
Moscow, ID

Patricia Craig, MS
Science Writer
Bozeman, MT

Randall Frost, PhD
Science Writer
Pleasanton, CA

Lisa S. Gardiner, PhD
Science Writer
Denver, CO

Jennifer Gonya, PhD
The Ohio State University
Columbus, OH

Mary Ann Grobbel, MD
Science Writer
Grand Rapids, MI

Whitney Crispen Hagins, MA, MAT
Biology Teacher
Lexington High School
Lexington, MA

Carole Holmberg, BS
Planetarium Director
Calusa Nature Center and
Planetarium, Inc.
Fort Myers, FL

Tina C. Hopper
Science Writer
Rockwall, TX

Jonathan D. W. Kahl, PhD
Professor of Atmospheric Science
University of Wisconsin-
Milwaukee
Milwaukee, WI

Nanette Kalis
Science Writer
Athens, OH

S. Page Keeley, MEd
Maine Mathematics and Science
Alliance
Augusta, ME

Cindy Klevickis, PhD
Professor of Integrated Science
and Technology
James Madison University
Harrisonburg, VA

Kimberly Fekany Lee, PhD
Science Writer
La Grange, IL

Michael Manga, PhD
Professor
University of California, Berkeley
Berkeley, CA

Devi Ried Mathieu
Science Writer
Sebastopol, CA

Elizabeth A. Nagy-Shadman, PhD
Geology Professor
Pasadena City College
Pasadena, CA

William D. Rogers, DA
Professor of Biology
Ball State University
Muncie, IN

Donna L. Ross, PhD
Associate Professor
San Diego State University
San Diego, CA

Marion B. Sewer, PhD
Assistant Professor
School of Biology
Georgia Institute of Technology
Atlanta, GA

Julia Meyer Sheets, PhD
Lecturer
School of Earth Sciences
The Ohio State University
Columbus, OH

Michael J. Singer, PhD
Professor of Soil Science
Department of Land, Air and
Water Resources
University of California
Davis, CA

Karen S. Sottosanti, MA
Science Writer
Pickerington, Ohio

Paul K. Strode, PhD
I.B. Biology Teacher
Fairview High School
Boulder, CO

Jan M. Vermilye, PhD
Research Geologist
Seismo-Tectonic Reservoir
Monitoring (STRM)
Boulder, CO

Judith A. Yero, MA
Director
Teacher's Mind Resources
Hamilton, MT

Dinah Zike, MEd
Author, Consultant, Inventor
of Foldables
Dinah Zike Academy; Dinah-
Might Adventures, LP
San Antonio, TX

Margaret Zorn, MS
Science Writer
Yorktown, VA

Consulting Authors

Alton L. Biggs
Biggs Educational Consulting
Commerce, TX

Ralph M. Feather, Jr., PhD
Assistant Professor
Department of Educational
Studies and Secondary Education
Bloomsburg University
Bloomsburg, PA

Douglas Fisher, PhD
Professor of Teacher Education
San Diego State University
San Diego, CA

Edward P. Ortleb
Science/Safety Consultant
St. Louis, MO

Series Consultants

Science

Solomon Bililign, PhD
Professor
Department of Physics
North Carolina Agricultural and
Technical State University
Greensboro, NC

John Choinski
Professor
Department of Biology
University of Central Arkansas
Conway, AR

Anastasia Chopelas, PhD
Research Professor
Department of Earth and Space
Sciences
UCLA
Los Angeles, CA

David T. Crowther, PhD
Professor of Science Education
University of Nevada, Reno
Reno, NV

A. John Gatz
Professor of Zoology
Ohio Wesleyan University
Delaware, OH

Sarah Gille, PhD
Professor
University of California San
Diego
La Jolla, CA

David G. Haase, PhD
Professor of Physics
North Carolina State University
Raleigh, NC

Janet S. Herman, PhD
Professor
Department of Environmental
Sciences
University of Virginia
Charlottesville, VA

David T. Ho, PhD
Associate Professor
Department of Oceanography
University of Hawaii
Honolulu, HI

Ruth Howes, PhD
Professor of Physics
Marquette University
Milwaukee, WI

Jose Miguel Hurtado, Jr., PhD
Associate Professor
Department of Geological
Sciences
University of Texas at El Paso
El Paso, TX

Monika Kress, PhD
Assistant Professor
San Jose State University
San Jose, CA

Mark E. Lee, PhD
Associate Chair & Assistant
Professor
Department of Biology
Spelman College
Atlanta, GA

Linda Lundgren
Science writer
Lakewood, CO

Keith O. Mann, PhD
Ohio Wesleyan University
Delaware, OH

Charles W. McLaughlin, PhD
Adjunct Professor of Chemistry
Montana State University
Bozeman, MT

Katharina Pahnke, PhD
Research Professor
Department of Geology and
Geophysics
University of Hawaii
Honolulu, HI

Jesús Pando, PhD
Associate Professor
DePaul University
Chicago, IL

Hay-Oak Park, PhD
Associate Professor
Department of Molecular
Genetics
Ohio State University
Columbus, OH

David A. Rubin, PhD
Associate Professor of Physiology
School of Biological Sciences
Illinois State University
Normal, IL

Toni D. Sauncy
Assistant Professor of Physics
Department of Physics
Angelo State University
San Angelo, TX

Series Consultants, continued

Malathi Srivatsan, PhD
Associate Professor of
Neurobiology
College of Sciences and
Mathematics
Arkansas State University
Jonesboro, AR

Cheryl Wistrom, PhD
Associate Professor of Chemistry
Saint Joseph's College
Rensselaer, IN

Reading

ReLeah Cossett Lent
Author/Educational Consultant
Blue Ridge, GA

Math

Vik Hovsepian
Professor of Mathematics
Rio Hondo College
Whittier, CA

Series Reviewers

Thad Boggs
Mandarin High School
Jacksonville, FL

Catherine Butcher
Webster Junior High School
Minden, LA

Erin Darichuk
West Frederick Middle School
Frederick, MD

Joanne Hedrick Davis
Murphy High School
Murphy, NC

Anthony J. DiSipio, Jr.
Octorara Middle School
Atglen, PA

Adrienne Elder
Tulsa Public Schools
Tulsa, OK

Carolyn Elliott
Iredell-Statesville Schools
Statesville, NC

Christine M. Jacobs
Ranger Middle School
Murphy, NC

Jason O. L. Johnson
Thurmont Middle School
Thurmont, MD

Felecia Joiner
Stony Point Ninth Grade Center
Round Rock, TX

Joseph L. Kowalski, MS
Lamar Academy
McAllen, TX

Brian McClain
Amos P. Godby High School
Tallahassee, FL

Von W. Mosser
Thurmont Middle School
Thurmont, MD

Ashlea Peterson
Heritage Intermediate Grade
Center
Coweta, OK

Nicole Lenihan Rhoades
Walkersville Middle School
Walkersvillle, MD

Maria A. Rozenberg
Indian Ridge Middle School
Davie, FL

Barb Seymour
Westridge Middle School
Overland Park, KS

Ginger Shirley
Our Lady of Providence Junior-
Senior High School
Clarksville, IN

Curtis Smith
Elmwood Middle School
Rogers, AR

Sheila Smith
Jackson Public School
Jackson, MS

Sabra Soileau
Moss Bluff Middle School
Lake Charles, LA

Tony Spoores
Switzerland County Middle
School
Vevay, IN

Nancy A. Stearns
Switzerland County Middle
School
Vevay, IN

Kari Vogel
Princeton Middle School
Princeton, MN

Alison Welch
Wm. D. Slider Middle School
El Paso, TX

Linda Workman
Parkway Northeast Middle
School
Creve Coeur, MO

With your book!

Answer questions, record data, and interact with images directly in your book!

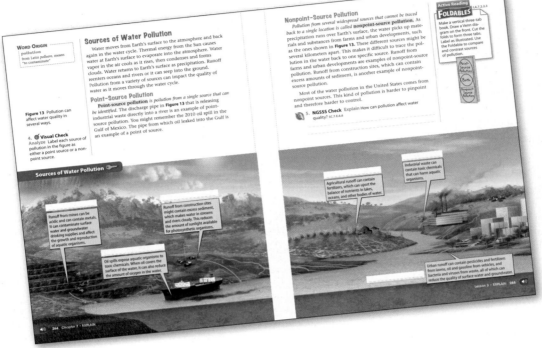

Online!

Log on to **Connect ED** for a digital version of this book that includes

- audio;
- animations;
- virtual labs.

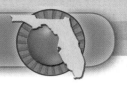

Inquiry
iLAB STATION

Labs, Labs, Labs

Launch Labs at the beginning of every lesson let you be the scientist! The iLAB Station on **ConnectED** has all the labs for each chapter.

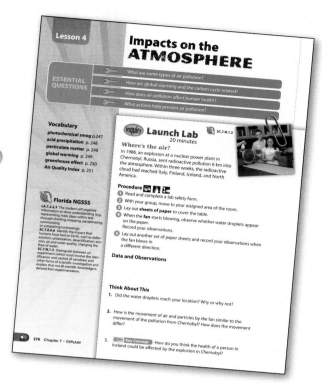

Virtual Labs

Virtual Labs provide a highly interactive lab experience.

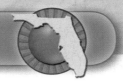

Sequence Words

While you read, watch for words that show the order events happen:

- first
- next
- last
- begins
- second
- later

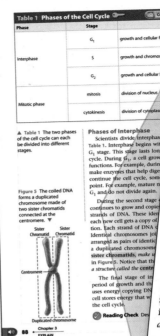

Vocabulary Help

Science terms are highlighted and reviewed to check your understanding.

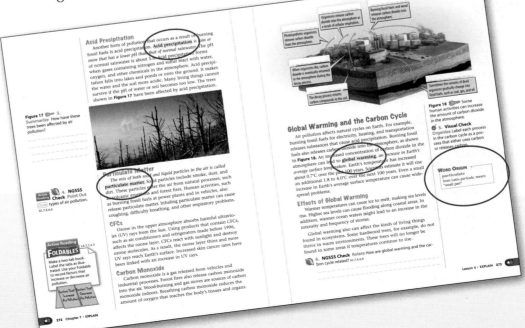

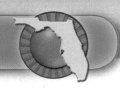

iWrite iScience

Write the answers to questions right in your book!

Concept Map

Each chapter's **Concept Map** gives you a place to show all the science connections you've learned.

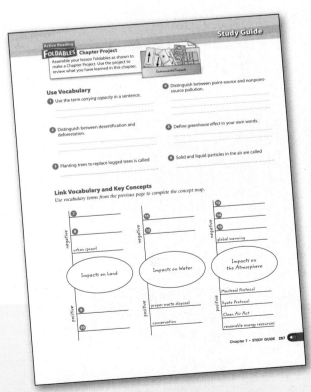

FLORIDA SCIENCE PHYSICAL

TABLE OF CONTENTS

Nature of Science:

Scientific Problem Solving .... NOS 2

Check It! → ☐ Lesson 1   ☐ Lesson 2   ☐ Lesson 3

Skill Practice: How does the strength of geometric shapes differ?

Inquiry Lab: Build and Test a Bridge

Connect ED   **Your online portal to everything you need!**
Video • Audio • Review • ⓘLab Station • WebQuests • Assessment • Concepts in Motion • Personal Tutors • Virtual Labs

Here are some of the exciting digital activities you will find in this chapter!

Virtual Lab: What strategies are involved in solving a science problem?

BrainPOP: Scientific Methods

Page Keeley Science Probe

Unit 1 MOTION AND FORCES

FLORIDA iSCIENCE PHYSICAL

Check It! → ☐ Lesson 1   ☐ Lesson 2   ☐ Lesson 3

Inquiry  iLAB STATION

Try It! then Apply It!

☐ **MiniLabs:** LESSON 1: Why is a reference point useful?

LESSON 2: How can you graph motion?

LESSON 3: How is a change in speed related to acceleration?

Skill Practice: What do you measure to calculate speed?

Inquiry Lab: Calculate Average Speed from a Graph

Connect**ED** **Your online portal to everything you need!**
Video • Audio • Review • iLab Station • WebQuests • Assessment • Concepts in Motion • Personal Tutors • Virtual Labs

Here are some of the exciting digital activities you will find in this chapter!

Virtual Lab: How does horizontal motion affect vertical motion?

Concepts in Motion: Distance vs. Time Graph

BrainPOP: Acceleration

Check It! ☐ Lesson 1 ☐ Lesson 2 ☐ Lesson 3 ☐ Lesson 4

Inquiry iLAB STATION

☐ **MiniLabs:** LESSON 1: How does friction affect motion?

Try It! then Apply It!

LESSON 2: How do forces affect motion?

LESSON 3: How are force and mass related?

LESSON 4: Is momentum conserved during a collision?

Skill Practice: How can you model Newton's first law of motion?

Skill Practice: How does a change in mass or force affect acceleration?

Inquiry Lab: Modeling Newton's Laws of Motion

Here are some of the exciting digital activities you will find in this chapter!

What's Science Got to Do With It?
Tractor Pulls

Concepts in Motion: Satellite Motion

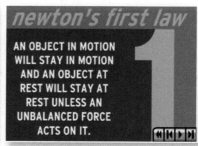

BrainPOP: Newton's Laws of Motion

Check It! → ☐ Lesson 1   ☐ Lesson 2   ☐ Lesson 3

Inquiry iLAB STATION

☐ **MiniLabs:** LESSON 1: How are force and pressure different?

Try It! then Apply It! →

LESSON 2: Can you overcome the buoyant force?

LESSON 3: How does air speed affect air pressure?

Skill Practice: Do heavy objects always sink and light objects always float?

Inquiry Lab: Design a Cargo Ship

ConnectED **Your online portal to everything you need!**
Video • Audio • Review • iLab Station • WebQuests • Assessment • Concepts in Motion • Personal Tutors • Virtual Labs

Here are some of the exciting digital activities you will find in this chapter!

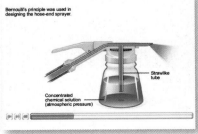

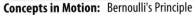

Virtual Lab: Why do things float?   **Concepts in Motion:** Bernoulli's Principle   **Page Keeley Science Probe**

Check It! → □ Lesson 1   □ Lesson 2   □ Lesson 3

Inquiry ①LAB STATION

Try It! then Apply It!

□ **MiniLabs:** LESSON 1: Can a moving object do work?
LESSON 2: How does energy change form?
LESSON 3: How do the particles in a liquid move when heated?

Skill Practice: Can you identify potential and kinetic energy?
Inquiry Lab: Pinwheel Power

ConnectED **Your online portal to everything you need!**
Video • Audio • Review • ①Lab Station • WebQuests • Assessment • Concepts in Motion • Personal Tutors • Virtual Labs

Here are some of the exciting digital activities you will find in this chapter!

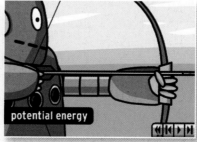

Virtual Lab: What are the relationships between kinetic energy and potential energy?

Concepts in Motion Conduction, Convection and Radiation

BrainPOP: Forms of Energy

(FLORIDA iSCIENCE PHYSICAL logo)

PAGE KEELEY SCIENCE PROBES *Cool It!* 175

LESSON 1 **Thermal Energy, Temperature and Heat** 178

Launch Lab: How can you describe temperature?

LESSON 2 **Thermal Energy Transfers** 186

Launch Lab: How hot is it?

SCIENCE & SOCIETY Insulating the Home 195

LESSON 3 **Using Thermal Energy** 196

Launch Lab: How can you transform energy?

NGSSS for Science Benchmark Practice 206

Mini Benchmark Assessments 208

Check It! → ☐ Lesson 1   ☐ Lesson 2   ☐ Lesson 3

Inquiry ①LAB STATION

Try It! then Apply It!

☐ **MiniLabs:** LESSON 1: How do temperature scales compare?

LESSON 2: How does adding thermal energy affect a wire?

LESSON 3: Can thermal energy be used to do work?

Skill Practice: How do different materials affect thermal energy transfer?

Inquiry Lab: Design an Insulated Container

ConnectED

Your online portal to everything you need!
Video • Audio • Review • ①Lab Station • WebQuests • Assessment • Concepts in Motion • Personal Tutors • Virtual Labs

Here are some of the exciting digital activities you will find in this chapter!

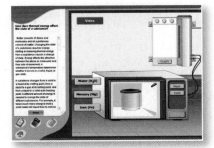

Virtual Lab: How does thermal energy affect the state of a substance?

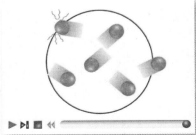

Concepts in Motion: Kinetic Energy

Page Keeley Science Probe

FLORIDA SCIENCE PHYSICAL

TABLE OF CONTENTS

Check It! ☐ Lesson 1 ☐ Lesson 2 ☐ Lesson 3 ☐ Lesson 4

Inquiry LAB STATION

☐ **MiniLabs:** LESSON 1: How can you model an atom?

Try It! then Apply It!

LESSON 2: Can the weight of an object change?

LESSON 3: Can you make ice without a freezer?

LESSON 4: Can you spot the clues for chemical change?

Skill Practice: How can following a procedure help solve a crime?

Skill Practice: How can known substances help you identify unknown substances?

Inquiry Lab: Design an Experiment to Solve a Crime

Here are some of the exciting digital activities you will find in this chapter!

Virtual Lab: What is a balanced chemical equation?

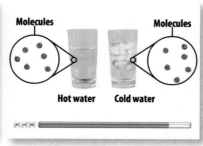

Concepts in Motion Temperature And Molecular Motion

BrainPOP: Measuring Matter

Inquiry iLAB STATION

☐ **MiniLabs:** LESSON 1: How can you make bubble film?

Try It! then Apply It!

LESSON 2: How can you make a water thermometer?

LESSON 3: How does temperature affect the volume?

Skill Practice: How does dissolving substances in water change its freezing point?

Inquiry Lab: Design an Experiment to Collect Data

ConnectED **Your online portal to everything you need!**
Video • Audio • Review • ⓘLab Station • WebQuests • Assessment • Concepts in Motion • Personal Tutors • Virtual Labs

Here are some of the exciting digital activities you will find in this chapter!

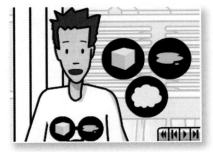

BrainPOP: Matter Changing State

Concepts in Motion States of Matter

Page Keeley Science Probe

FLORIDA SCIENCE PHYSICAL

TABLE OF CONTENTS

Check It! ☐ Lesson 1   ☐ Lesson 2

Inquiry LAB STATION

☐ **MiniLabs:** LESSON 1: How can you gather information about what you can't see?

Try It! then Apply It!

LESSON 2: How many penny isotopes do you have?

Inquiry Lab: Communicate Your Knowledge about the Atom

Connect ED **Your online portal to everything you need!**

Video • Audio • Review • ⓘLab Station • WebQuests • Assessment • Concepts in Motion • Personal Tutors • Virtual Labs

Here are some of the exciting digital activities you will find in this chapter!

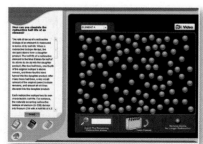

Virtual Lab: How can you simulate the radioactive half-life of an element?

Concepts in Motion: Rutherford's Experiment

BrainPOP: Isotopes

Check It! → ☐ Lesson 1　☐ Lesson 2　☐ Lesson 3

☐ **MiniLabs:**　LESSON 1: How does atom size change across a period?

Try It! then Apply It! ↑

LESSON 2: How well do materials conduct thermal energy?

LESSON 3: Which insulates better?

Skill Practice: How is the periodic table arranged?

Inquiry Lab: Alien Insect Periodic Table

ConnectED　**Your online portal to everything you need!**
Video • Audio • Review • ⓘLab Station • WebQuests • Assessment • Concepts in Motion • Personal Tutors • Virtual Labs

Here are some of the exciting digital activities you will find in this chapter!

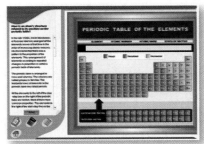

Virtual Lab: How is an atom's structure related to its position on the periodic table?

Concepts in Motion: Atomic Structure

Page Keeley Science Probe

FLORIDA SCIENCE PHYSICAL

TABLE OF CONTENTS

Check It! ☐ Lesson 1   ☐ Lesson 2   ☐ Lesson 3

Inquiry LAB STATION

☐ **MiniLabs:** LESSON 1: How does an electron's energy relate to its position in an atom?

Try It! then Apply It!

 LESSON 2: How do compounds form?

 LESSON 3: How many ionic compounds can you make?

Skill Practice: How can you model compounds?

Inquiry Lab: Ions in Solution

ConnectED **Your online portal to everything you need!**
Video • Audio • Review • ⓘLab Station • WebQuests • Assessment • Concepts in Motion • Personal Tutors • Virtual Labs

Here are some of the exciting digital activities you will find in this chapter!

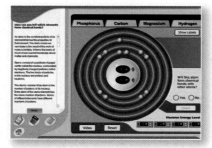

Virtual Lab: How can you tell which elements form chemical bonds?

Concepts in Motion: NaCl Bonding

Concepts in Motion: Ionic Bonding

Check It! → ☐ Lesson 1   ☐ Lesson 2   ☐ Lesson 3

Inquiry iLAB STATION

☐ **MiniLabs:** LESSON 1: How does an equation represent a reaction?

LESSON 2: Can you speed up a reaction?

Try It! then Apply It!

Skill Practice: What can you learn from an experiment?

Inquiry Lab: Design an Experiment to Test Advertising Claims

ConnectED **Your online portal to everything you need!**

Video • Audio • Review • ⓘLab Station • WebQuests • Assessment • Concepts in Motion • Personal Tutors • Virtual Labs

Here are some of the exciting digital activities you will find in this chapter!

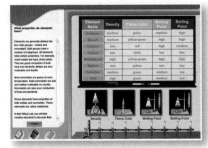

Virtual Lab: What properties do elements have?

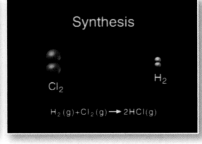

Concepts in Motion: Chemical Reactions

What's Science Got to Do With It? Arson Investigation

FLORIDA SCIENCE PHYSICAL

TABLE OF CONTENTS

Check It! → ☐ Lesson 1   ☐ Lesson 2   ☐ Lesson 3

Inquiry  **⚲ LAB STATION**

☐ **MiniLabs:** LESSON 1: Which one is the mixture?

Try It! then Apply It! ↑ LESSON 2: How much is dissolved?

LESSON 3: Is it an acid or a base?

Skill Practice: How does a solute affect the conductivity of a solution?

Inquiry Lab: Can the pH of a solution be changed?

Connect ED **Your online portal to everything you need!**
Video • Audio • Review • ⓘLab Station • WebQuests • Assessment • Concepts in Motion • Personal Tutors • Virtual Labs

Here are some of the exciting digital activities you will find in this chapter!

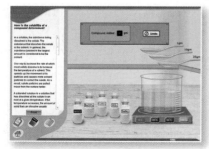

Virtual Lab: How is the solubility of a compound determined?

What's Science Got to Do With It?
Sports Drinks

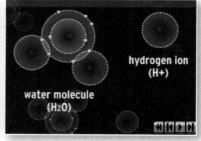

BrainPOP: Acids and Bases

Check It! → ☐ Lesson 1   ☐ Lesson 2   ☐ Lesson 3

Inquiry ①LAB STATION

☐ **MiniLabs:** LESSON 1: How do carbon atoms bond with carbon and hydrogen atoms?

Try It! then Apply It! ↗

LESSON 2: How can you make a polymer?

LESSON 3: How many carbohydrates do you consume?

Skill Practice: How do you test for vitamin C?

Inquiry Lab: Testing for Carbon Compounds

 **ConnectED** **Your online portal to everything you need!**
Video • Audio • Review • ①Lab Station • WebQuests • Assessment • Concepts in Motion • Personal Tutors • Virtual Labs

Here are some of the exciting digital activities you will find in this chapter!

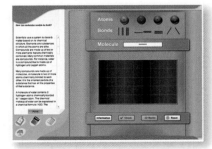

Virtual Lab: How can a molecular model be built?

Concepts in Motion: Polymers

Page Keeley Science Probe

FLORIDA iSCIENCE PHYSICAL

TABLE OF CONTENTS

Unit 5 WAVES, ELECTRICITY, AND MAGNETISM

Check It! → ☐ Lesson 1   ☐ Lesson 2   ☐ Lesson 3

Inquiry iLAB STATION

☐ **MiniLabs:** *Try It! then Apply It!* →

LESSON 1: How do waves travel through matter?

LESSON 2: How are wavelength and frequency related?

LESSON 3: How can reflection be used?

Skill Practice: How are the properties of waves related?

Inquiry Lab: Measuring Wave Speed

Connect ED

Your online portal to everything you need!
Video • Audio • Review • ⓘLab Station • WebQuests • Assessment • Concepts in Motion • Personal Tutors • Virtual Labs

Here are some of the exciting digital activities you will find in this chapter!

Virtual Lab: What are some characteristics of waves?

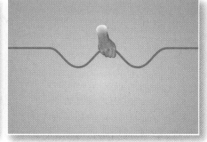

Concepts in Motion: Transverse Waves

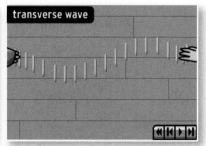

BrainPOP: Waves

Check It! → ☐ Lesson 1   ☐ Lesson 2   ☐ Lesson 3

Inquiry ①LAB STATION   ☐ **MiniLabs:**   LESSON 1: How do you know a sound's direction?

Try It! then Apply It!   LESSON 2: How can you hear beats?

LESSON 3: How fast is sound?

Skill Practice: How can you use a wind instrument to play music?

Inquiry Lab: Make Your Own Musical Instrument

ConnectED **Your online portal to everything you need!**
Video • Audio • Review • ①Lab Station • WebQuests • Assessment • Concepts in Motion • Personal Tutors • Virtual Labs

Here are some of the exciting digital activities you will find in this chapter!

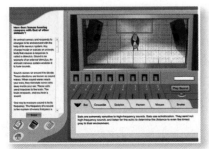

Virtual Lab: How does human hearing compare with that of other animals?

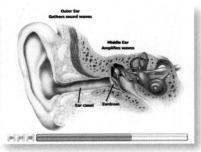

Concepts in Motion: Human Ear

Page Keeley Science Probe

FLORIDA iSCIENCE PHYSICAL

TABLE OF CONTENTS

Check It! → ☐ Lesson 1   ☐ Lesson 2   ☐ Lesson 3

Inquiry iLAB STATION

☐ **MiniLabs:** LESSON 1: How are electric fields and magnetic fields related?

Try It! then Apply It! →

LESSON 3: How does infrared imaging work?

Skill Practice: What's at the edge of a rainbow?

Inquiry Lab: Design an Exhibit for a Science Museum

🖥 **ConnectED**   **Your online portal to everything you need!**
Video • Audio • Review • ⓘLab Station • WebQuests • Assessment • Concepts in Motion • Personal Tutors • Virtual Labs

Here are some of the exciting digital activities you will find in this chapter!

What's Science Got to do With It?
Cell Phones

Concepts in Motion: GPS Satellites

Page Keeley Science Probe

FLORIDA
SCIENCE
PHYSICAL

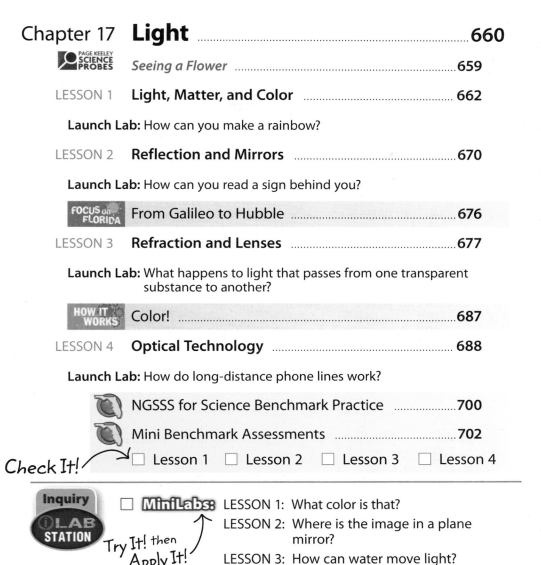
Check It! ☐ Lesson 1 ☐ Lesson 2 ☐ Lesson 3 ☐ Lesson 4

Inquiry **iLAB STATION**

☐ **MiniLabs:** LESSON 1: What color is that?

Try It! then Apply It!

LESSON 2: Where is the image in a plane mirror?

LESSON 3: How can water move light?

LESSON 4: How does a zoom lens work?

Skill Practice: How can you demonstrate the law of reflection?

Skill Practice: How does a lens affect light?

Inquiry Lab: Design Your Own Optical Illusion

Here are some of the exciting digital activities you will find in this chapter!

Virtual Lab: How are colors created?

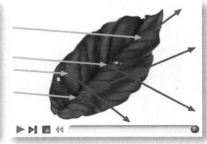

Concepts in Motion: Color and Wavelength

What's Science Got to do With It?
Crime Scene Investigation: Criminals Beware: What you can't see might convict you!

FLORIDA SCIENCE PHYSICAL

TABLE OF CONTENTS

Check It! → ☐ Lesson 1   ☐ Lesson 2   ☐ Lesson 3

Inquiry ⓘLAB STATION

☐ **MiniLabs:** LESSON 1: How can a balloon push or pull?
LESSON 2: When is one more than two?
LESSON 3: What else can a circuit do?

Try It! then Apply It!

Skill Practice: What effect does voltage have on a circuit?
Inquiry Lab: Design an Elevator

🖥 **Connect ED**   **Your online portal to everything you need!**
Video • Audio • Review • ⓘLab Station • WebQuests • Assessment • Concepts in Motion • Personal Tutors • Virtual Labs

Here are some of the exciting digital activities you will find in this chapter!

BrainPOP: Current electricity

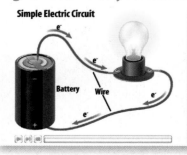

Concepts in Motion: Simple Electric Circuit

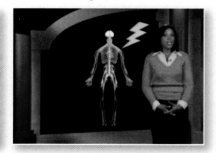

What's Science Got to do With It? Shock Treatment

FLORIDA
SCIENCE
PHYSICAL

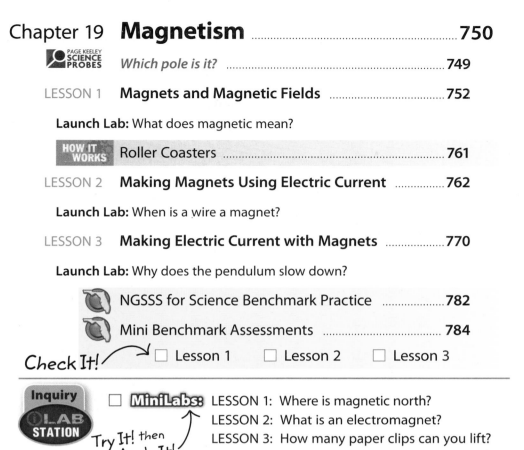

Inquiry **iLAB STATION**

☐ **MiniLabs:** LESSON 1: Where is magnetic north?
Try It! then Apply It! LESSON 2: What is an electromagnet?
LESSON 3: How many paper clips can you lift?

Skill Practice: How can you measure current?
Inquiry Lab: Build a Wind-Powered Generator

ConnectED **Your online portal to everything you need!**
Video • Audio • Review • iLab Station • WebQuests • Assessment • Concepts in Motion • Personal Tutors • Virtual Labs

Here are some of the exciting digital activities you will find in this chapter!

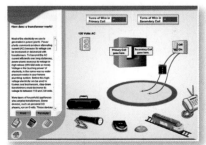

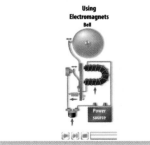

Virtual Lab: How does a transformer work?

Concepts in Motion: Using Electromagnets

BrainPOP: Magnetism

Prove or Disprove?

Two students discussed scientific methods. They disagreed about why scientists test a hypothesis.

Sharla: I think scientists test a hypothesis to disprove it.

Marcos: I think scientists test a hypothesis to prove it.

(Circle) the student you most agree with. Explain why you agree with that student.

FLORIDA
Nature of Science

Scientific Problem SOLVING

Nature of Science

This chapter begins your study of the nature of science, but there is even more information about the nature of science in this book. Each unit begins by exploring an important topic that is fundamental to scientific study. As you read these topics, you will learn even more about the nature of science.

FLORIDA BIG IDEAS

1 **The Practice of Science**
2 **The Characteristics of Scientific Knowledge**
3 **The Role of Theories, Laws, Hypotheses, and Models**

Think About It!

What is scientific inquiry?

This might look like a weird spaceship docking in a science-fiction movie. However, it is actually the back of an airplane engine being tested in a huge wind tunnel. An experiment is an important part of scientific investigations.

1 Why do you think an experiment is important?

_____

_____

_____

2 What is scientific inquiry?

_____

_____

_____

Florida NGSSS

SC.6.N.1.2 Explain why scientific investigations should be replicable.
SC.7.N.1.2 Differentiate replication (by others) from repetition (multiple trials).
SC.8.N.1.3 Use phrases such as "results support" or "fail to support" in science, understanding that science does not offer conclusive 'proof' of a knowledge claim.
SC.6.N.1.4 Discuss, compare, and negotiate methods used, results obtained, and explanations among groups of students conducting the same investigation.
SC.7.N.1.4 Identify test variables (independent variables) and outcome variables (dependent variables) in an experiment.
SC.8.N.1.4 Explain how hypotheses are valuable if they lead to further investigations, even if they turn out not to be supported by the data.
SC.7.N.1.5 Describe the methods used in the pursuit of a scientific explanation as seen in different fields of science such as biology, geology, and physics.
SC.8.N.1.5 Analyze the methods used to develop a scientific explanation as seen in different fields of science.
SC.7.N.1.6 Explain that empirical evidence is the cumulative body of observations of a natural phenomenon on which scientific explanations are based.
SC.8.N.1.6 Understand that scientific investigations involve the collection of relevant empirical evidence, the use of logical reasoning, and the application of imagination in devising hypotheses, predictions, explanations and models to make sense of the collected evidence.
SC.6.N.2.2 Explain that scientific knowledge is durable because it is open to change as new evidence or interpretations are encountered.
SC.6.N.2.3 Recognize that scientists who make contributions to scientific knowledge come from all kinds of backgrounds and possess varied talents, interests, and goals.
SC.6.N.2.1 Distinguish science from other activities involving thought.
SC.8.N.2.1 Distinguish between scientific and pseudoscientific ideas.
SC.8.N.2.2 Discuss what characterizes science and its methods.
SC.6.N.3.1 Recognize and explain that a scientific theory is a well-supported and widely accepted explanation of nature and is not simply a claim posed by an individual. Thus, the use of the term theory in science is very different than how it is used in everyday life.
SC.6.N.3.2 Recognize and explain that a scientific law is a description of a specific relationship under given conditions in the natural world. Thus, scientific laws are different from societal laws.
SC.7.N.3.1 Recognize and explain the difference between theories and laws and give several examples of scientific theories and the evidence that supports them.
SC.8.N.3.1 Select models useful in relating the results of their own investigations.
SC.8.N.3.2 Explain why theories may be modified but are rarely discarded.
SC.6.N.3.3 Give several examples of scientific laws.
SC.8.N.4.1 Explain that science is one of the processes that can be used to inform decision making at the community, state, national, and international levels.

There's More Online!
Video • Audio • Review • ⓘLab Station • WebQuest • Assessment • Concepts in Motion • Multilingual eGlossary

Scientific INQUIRY

ESSENTIAL QUESTIONS

🔑 What are some steps used during scientific inquiry?

🔑 What are the results of scientific inquiry?

🔑 What is critical thinking?

Vocabulary

science p. NOS 4

observation p. NOS 6

inference p. NOS 6

hypothesis p. NOS 6

prediction p. NOS 6

scientific theory p. NOS 8

scientific law p. NOS 8

technology p. NOS 9

critical thinking p. NOS 10

Understanding Science

In a clear night sky, the stars seem to shine like diamonds scattered on black velvet. Why do stars seem to shine more brightly some nights than others?

When you ask questions, such as the one above, you are practicing science. **Science** *is the investigation and exploration of natural events and of the new information that results from those investigations.* You can help shape the future by accumulating knowledge, developing new technologies, and sharing ideas with others.

Throughout history, people of many different backgrounds, interests, and talents have made scientific contributions. Sometimes they overcame a limited educational background and excelled in science. One example is Marie Curie, shown in **Figure 1**. She was a scientist who won two Nobel prizes in the early 1900s for her work with radioactivity. As a young student, Marie was not allowed to study at the University of Warsaw in Poland because she was a woman. Despite this obstacle, she made significant contributions to science.

Active Reading

1. Reflect Infer how people with different backgrounds have contributed to science. How have attitudes changed to include everyone who has an interest in science?

_____

_____

_____

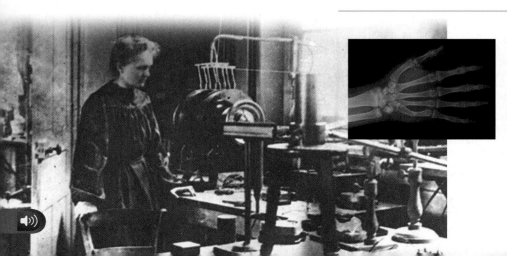

Figure 1 Modern medical procedures such as X-rays, radioactive cancer treatments, and nuclear-power generation are some of the technologies made possible because of the pioneering work of Marie Curie and her associates.

Branches of Science

Scientific study is organized into several branches. The three most common branches are physical science, Earth science, and life science. Each branch focuses on a different part of the natural world.

WORD ORIGIN

science

from Latin *scientia*, means "knowledge" or "to know"

Active Reading **2. Question** In the boxes below, suggest some possible questions people in different branches of science might ask.

Physical Science

Physical science, or physics and chemistry, is the study of matter and energy. The physicist is using an instrument to measure radiation in space.

Physical Science Questions:

- [ ]
- [ ]
- [ ]

Earth Science

Earth scientists study the many processes that occur on Earth, in space, and deep within Earth. This scientist will study a water sample from Mexico.

Earth Science Questions:

- [ ]
- [ ]
- [ ]

Life Science

Life scientists study all organisms and the many processes that occur in them. This life scientist is studying the avian flu virus.

Life Science Questions:

- [ ]
- [ ]
- [ ]

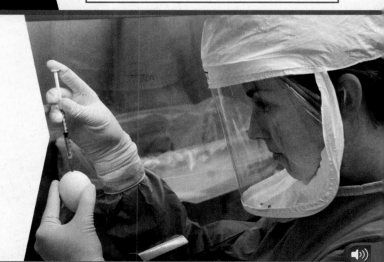

What is Scientific Inquiry?

When scientists conduct investigations, they often want to answer questions about the natural world. They use scientific inquiry—a process that uses a set of skills to answer questions or test ideas. You might have heard these steps called "the scientific method." However, there is no one scientific method. The skills that scientists use to conduct an investigation can be used in any order. One possible sequence is shown in **Figure 2**.

 3. Consider Highlight the term *scientific inquiry* and its definition.

Ask Questions

Imagine warming yourself near a campfire. As you place twigs and logs onto the fire, the fire releases smoke and light. You feel the warmth of the thermal energy being released. These are **observations**—*the results of using one or more of your senses to gather information and taking note of what occurs.* Observations often lead to questions. You ask yourself, "When logs burn, what happens to the wood? Do the logs disappear?"

You might recall that matter can change form, but it cannot be created or destroyed. Therefore, you could infer that the logs do not just disappear. They must undergo some type of change. An **inference** *is a logical explanation of an observation that is drawn from prior knowledge or experience.*

Hypothesize and Predict

You decide to investigate further. You might develop a **hypothesis**—*a possible explanation for an observation that can be tested by scientific investigations.* Your hypothesis about what happens might be: When logs burn, new substances form because matter cannot be destroyed.

When scientists state a hypothesis, they often use it to make predictions. *A* **prediction** *is a statement of what will happen next in a sequence of events.* Predictions based on information might be found when testing the hypothesis. Based on a hypothesis, you might predict that if logs burn, then the substances that make up the logs change into other substances.

Active Reading **4. Analyze** Illustrate the relationship between a hypothesis and a prediction.

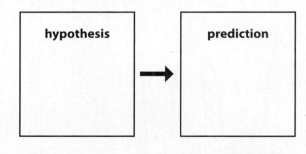

Figure 2 There are many possible steps in the process of scientific inquiry, and they can be performed in a variety of different sequences.

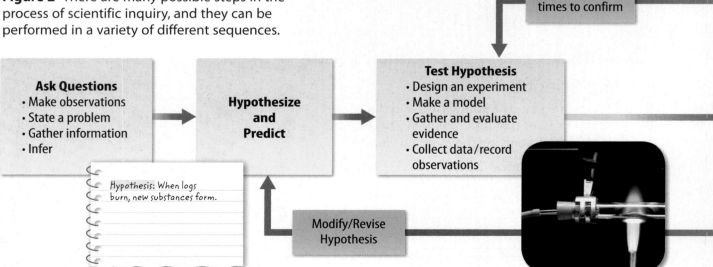

Test Hypothesis and Analyze Results

When you test a hypothesis, you often test your predictions. If a prediction is confirmed, then it supports your hypothesis. If your prediction is not confirmed, you might modify your hypothesis and retest it.

To test your predictions and hypothesis, design an experiment to find out what substances make up wood. Then determine what makes up the ash, the smoke, and other products that form during the burning process. You also could research this topic to find answers to questions.

After doing an experiment or research, analyze your results. You might make additional inferences after reviewing your data. If you find that new substances actually do form when wood burns, your hypothesis is supported. Some methods of testing a hypothesis and analyzing results are shown in **Figure 2**.

Active Reading **5. Suggest** Explain why results might not support a hypothesis.

Discuss what you might do next if your hypothesis is not supported.

Draw Conclusions

After analyzing your results, draw conclusions about your investigation. A conclusion is a summary of the information gained from testing a hypothesis. Like a scientist does, you should test and retest your hypothesis several times to make sure the results are consistent.

Active Reading **6. Define** Highlight the term *conclusion* and its definition.

Communicate Results

Sharing the results of a scientific inquiry is an important part of science. By exchanging information, scientists can evaluate and test others' work, apply new knowledge in their own research, and help keep scientific information accurate. As you do research on the Internet or in science books, you use information that someone else communicated. Scientists exchange information in many ways, as shown below in **Figure 2**.

Active Reading **7. Point out** Underline three reasons scientists exchange experimental results and scientific information.

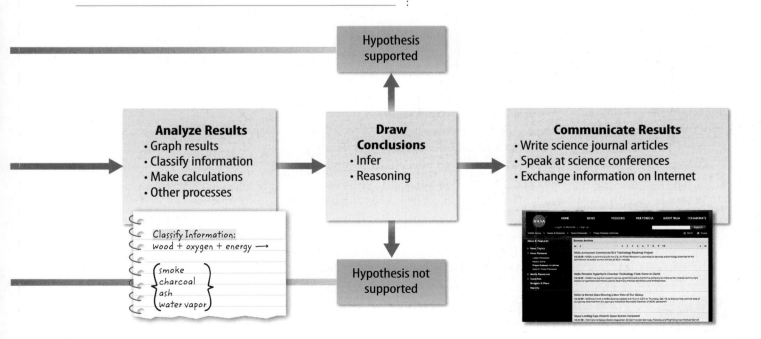

Analyze Results
- Graph results
- Classify information
- Make calculations
- Other processes

Classify Information:
wood + oxygen + energy ⟶

{ smoke
charcoal
ash
water vapor }

Draw Conclusions
- Infer
- Reasoning

Hypothesis supported

Hypothesis not supported

Communicate Results
- Write science journal articles
- Speak at science conferences
- Exchange information on Internet

Unsupported or Supported Hypotheses

Is a scientific investigation a failure and a waste of time if the hypothesis is not supported? Absolutely not! Valuable information can still be gained. The hypothesis can be revised and tested again. Each time a hypothesis is tested, scientists learn more about the topic they are studying.

Scientific Theory

When hypotheses (or a group of closely related hypotheses) are supported through many tests, a scientific theory can develop. A **scientific theory** *is an explanation of observations or events that is based on knowledge gained from many observations and investigations.*

A scientific theory does not develop from just one hypothesis, but from many hypotheses that are connected by a common idea. The kinetic molecular theory described below explains the behavior and energy of particles that make up a gas.

Scientific Law

A scientific law is different from a societal law, which is an agreement on a set of behaviors. A **scientific law** *is a rule that describes a repeatable pattern in nature.* A scientific law does not explain why or how the pattern happens, it only states that it will happen. For example, if you drop a ball, it will fall towards the ground every time. This is a repeated pattern that relates to the law of universal gravitation. The law of conservation of energy, described below, is also a scientific law.

 8. Distinguish <u>Underline</u> the terms scientific law and scientific theory and their definitions.

Kinetic Molecular Theory

The kinetic molecular theory explains how particles that make up a gas move in constant, random motions.

The kinetic molecular theory also assumes that the collisions of particles in a gas are elastic collisions. An elastic collision is a collision in which no kinetic energy is lost. Therefore, kinetic energy among gas particles is conserved.

Law of Conservation of Energy

The law of conservation of energy states that in any chemical reaction or physical change, energy is neither created nor destroyed. The total energy of particles before and after collisions is the same.

Scientific Law v. Scientific Theory

Both are based on repeated observations and can be rejected or modified.

A scientific law states that an event *will* occur. For example, energy will be conserved when particles collide. It does not explain why an event will occur or how it will occur. A law stands true until an observation is made that does not follow the law.

A scientific theory is an explanation of *why* or *how* an event occurred. For example, collisions of particles of a gas are elastic collisions. Therefore, no kinetic energy is lost. A theory can be rejected or modified if someone observes an event that disproves the theory. A theory will never become a law.

Active Reading **9. Point out** Highlight the terms scientific law and scientific theory and their descriptions in the box above.

Results of Scientific Inquiry

Why do you and others ask questions and investigate the natural world? Often, the results of a scientific investigation are new materials and technology, the discovery of new objects, or answers to questions.

Active Reading 10. **Analyze** In the spaces provided below, discuss how technology helps scientists develop new materials, discover new objects or events, and answer questions.

New Materials and Technology

Corporations and governments research and design new materials and technologies. **Technology** *is the practical use of scientific knowledge, especially for industrial or commercial use.* Scientists use technology to design new materials that make bicycles and cycling gear lighter, more durable, safer, and more aerodynamic. Using wind tunnels, scientists test these new materials to see whether they improve the cyclist's performance.

Summarize How might safer, more efficient bike equipment be a product of technology and scientific inquiry?

_____

_____

_____

New Objects or Events

Scientific investigations also lead to newly discovered objects or events. For example, NASA's *Hubble Space Telescope* captured this image of two colliding galaxies. They have been nicknamed the mice, because of their long tails. The tails are composed of gases and young, hot blue stars. If computer models are correct, these galaxies will combine in the future and form one large galaxy.

Explain How did technology aid scientists in the discovery of colliding galaxies?

_____

_____

Answers to Questions

Often scientific investigations are launched to answer *who, what, when, where, why,* or *how* questions. This research chemist investigates new substances found in mushrooms and bacteria. New drug treatments for diseases might be found using new substances. Other scientists look for clues about what causes diseases, whether they can be passed from person to person, and when the disease first appeared.

Infer How is technology used by chemists to develop new knowledge and treatments for the medical field?

_____

_____

_____

Active Reading

FOLDABLES® LA.8.2.2.3

Create a two-tab book and label it as shown. Use it to discuss the importance of evaluating scientific information.

Why is it important to...

...be scientifically literate? | ...use critical thinking?

Evaluating Scientific Information

Are you able to determine if information that claims to be scientifically proven is actually true and scientific instead of pseudoscientific (information incorrectly represented as scientific)? It is important that you are skeptical, identify facts and opinions, and think critically about information. **Critical thinking** *is comparing what you already know with the information you are given in order to decide whether you agree with it.*

Active Reading

11. Differentiate (Circle) the terms *skepticism, critical thinking, opinion,* and *misleading information* and their definitions in both the text above and the text boxes below.

Be A Rock Star!
Do you dream of being a rock star?

Sing, dance, and play guitar like a rock star with the new Rocker-rific Spotlight. A new scientific process developed by Rising Star Laboratories allows you to overcome your lack of musical talent and enables you to perform like a real rock star.

This amazing new light actually changes your voice quality and enhances your brain chemistry so that you can sing, dance, and play a guitar like a professional rock star. Now, there is no need to practice or pay for expensive lessons. The Rocker-rific Spotlight does the work for you.

Dr. Sammy Truelove says, "Before lack of talent stopped someone from achieving his or her dreams of being a rockstar. This scientific breakthrough transforms people with absolutely no talent into amazing rock stars in just minutes. Of the many patients that I have tested with this product, no one has failed to achieve his or her dreams."

Disclaimer: This product was tested on laboratory rats and might not work for everyone.

Skepticism
To be skeptical is to doubt the truthfulness of something. A scientifically literate person can recognize when information misrepresents the facts. Science often is self-correcting because someone usually challenges inaccurate information and tests scientific results for accuracy.

Critical Thinking
Use critical thinking skills to compare what you know with the new information given to you. If the information does not sound reliable, either research and find more information about the topic or dismiss the information as unreliable.

Identifying Facts and Misleading Information
Misleading information often is worded to sound like scientific facts. A scientifically literate person can recognize fake claims and quickly determine when information is false.

Identify Opinions
An opinion is a personal view, feeling, or claim about a topic. Opinions cannot be proven true or false. An opinion might contain inaccurate information.

Science cannot answer all questions.

It might seem that scientific inquiry is the best way to answer all questions. But, science cannot answer all questions. Questions that deal with beliefs, values, personal opinions, and feelings cannot be answered scientifically. It is impossible to collect scientific data on these topics.

Active Reading **12. Apply** Provide two questions that cannot be answered through scientific inquiry.

_____

_____

Safety in Science

Scientists use safe procedures in scientific investigations. During a scientific inquiry, you should always wear protective equipment, as shown in **Figure 3**. You also should learn the meaning of safety symbols, follow your teacher's instructions, and learn to recognize potential hazards.

Figure 3 Always follow safety procedures when doing scientific investigations.

Lesson Review 1

Use Vocabulary

1 **Define** *technology* in your own words.

_____

_____

2 **Use the term** *observation* in a sentence to show its scientific meaning.

_____

_____

Understand Key Concepts 🔑

3 Which action is NOT a way to test a hypothesis? SC.8.N.1.6

(A) analyze results (C) make a model

(B) design an experiment (D) gather and evaluate evidence

4 **Give an example** of a time when you practiced critical thinking.

_____

_____

_____

Interpret Graphics

5 **Complete** the graphic organizer below with some examples of how to communicate the results of scientific inquiry. LA.6.2.2.3

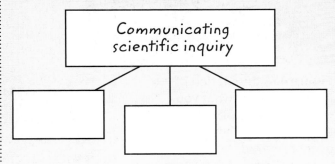

Communicating scientific inquiry

Critical Thinking

6 **Summarize** Your classmate writes the following as a hypothesis:

Red is a beautiful color.

Write a brief explanation to your classmate explaining why this is not a hypothesis.
SC.8.N.1.6

_____

_____

_____

Measurement and SCIENTIFIC TOOLS

ESSENTIAL QUESTIONS

🔑 Why did scientists create the International System of Units (SI)?

🔑 Why is scientific notation a useful tool for scientists?

🔑 How can tools, such as graduated cylinders and triple-beam balances, assist physical scientists?

Vocabulary

description p. NOS 12

explanation p. NOS 12

International System of Units (SI) p. NOS 13

scientific notation p. NOS 15

percent error p. NOS 15

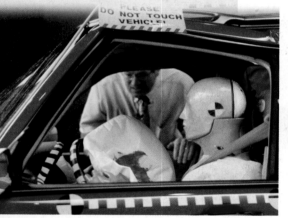

Figure 4 A description of an event details what you observed. An explanation explains why or how the event occurred.

Description and Explanation

Suppose you work for a company to calculate how cars perform during accidents, as shown in **Figure 4**. You measure the acceleration of cars as they crash into other objects.

A **description** *is a spoken or written summary of observations.* Measurements are descriptions of the results of the crash tests. A report discusses the results recorded. An **explanation** *is an interpretation of observations.* You make inferences and explain why the crash damaged the vehicles in specific ways.

A description and an explanation differ. When you describe something, you report your observations. When you explain something, you interpret your observations.

Active Reading 1. **Distinguish** Describe how a description differs from an explanation. Use an example to support your response.

_____

_____

The International System of Units

Different parts of the world use different systems of measurements. This can cause confusion when people communicate their measurements. In 1960, a new system of measurement was adopted. *The internationally accepted system of measurement is the* **International System of Units (SI)**.

Active Reading 2. **Apply** Why did scientists establish the International System of Units?

_____

_____

SI Base Units

When you take measurements during scientific investigations, you will use the SI system, which consists of measurement base units, as shown in **Table 1**. Other units used in the SI system that are not base units are derived from the base units. For example, the liter, used to measure volume, was derived from the base unit for length.

SI Unit Prefixes

The SI system is based on multiples of ten represented by prefixes, as shown in **Table 2**. For example, the prefix *milli-* means 0.001 or 10-3. So, a milliliter is 0.001 L, or 1/1,000 L. Another way to say this is: 1 L is 1,000 times greater than 1 mL.

Converting Among SI Units

To convert from one SI unit to another, you either multiply or divide by a factor of ten. You also can use proportion calculations to make conversions. An example of how to convert between SI units is shown in **Figure 5**.

Active Reading 3. **Summarize** Restate in your own words how to convert 400 mL to liters.

_____

_____

_____

Table 1 SI Base Units

Quantity Measured	Unit (symbol)
Length	meter (m)
Mass	kilogram (kg)
Time	second (s)
Electric current	ampere (A)
Temperature	kelvin (K)
Substance amount	mole (mol)
Light intensity	candela (cd)

Table 2 Prefixes

Prefix	Meaning
Mega- (M)	1,000,000 or (10^6)
Kilo- (k)	1,000 or (10^3)
Hecto- (h)	100 or (10^2)
Deka- (da)	10 or (10^1)
Deci- (d)	0.1 or $\left(\frac{1}{10}\right)$ or (10^{-1})
Centi- (c)	0.01 or $\left(\frac{1}{100}\right)$ or (10^{-2})
Milli- (m)	0.001 or $\left(\frac{1}{1,000}\right)$ or (10^{-3})
Micro- (μ)	0.000001 or $\left(\frac{1}{1,000,000}\right)$ or (10^{-6})

Figure 5 The rock in the photograph has a mass of 17.5 grams. Convert that measurement to kilograms.

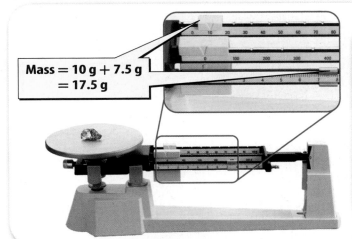

Mass = 10 g + 7.5 g
= 17.5 g

1. Determine the correct relationship between grams and kilograms. There are 1,000 g in 1 kg.

$$\frac{1 \text{ kg}}{1,000 \text{ g}}$$

$$\frac{x}{17.5 \text{ g}} = \frac{1 \text{ kg}}{1,000 \text{ g}}$$

$$x = \frac{(17.5 \text{ g})(1 \text{ kg})}{1,000 \text{ g}} ; x = 0.0175 \text{ kg}$$

2. Check your units. The unit *grams* is canceled out in the equation, so the answer is 0.0175 kg.

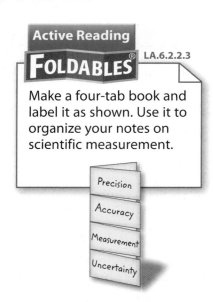

Active Reading

FOLDABLES® LA.6.2.2.3

Make a four-tab book and label it as shown. Use it to organize your notes on scientific measurement.

Precision

Accuracy

Measurement

Uncertainty

WORD ORIGIN

notation

from Latin *notationem*, means "a marking or explanation"

Figure 6 The graduated cylinder is marked in 1-mL increments. The beaker is marked in 50-mL increments. A graduated cylinder provides greater accuracy.

Table 3 Student Density and Error Data
(Accepted value: Density of sodium chloride, 21.7 g/cm³)

	Student A	Student B	Student C
	Density	Density	Density
Trial 1	23.4 g/cm³	18.9 g/cm³	21.9 g/cm³
Trial 2	23.5 g/cm³	27.2 g/cm³	21.4 g/cm³
Trial 3	23.4 g/cm³	29.1 g/cm³	21.3 g/cm³
Mean	23.4 g/cm³	25.1 g/cm³	21.5 g/cm³

Measurement and Uncertainty

The terms *precision* and *accuracy* have specific scientific meanings. Precision is a description of how similar repeated measurements are to each other. Accuracy is a description of how close a measurement is to an accepted value.

The difference between precision and accuracy is illustrated in **Table 3**. Students were asked to find the density of sodium chloride (NaCl). In three trials, students measured the volume and the mass of sodium chloride (NaCl). Then, they calculated the density for each trial and the mean of all three trials. Student A's measurements are the most precise because they are closest to each other. Student C's measurements are the most accurate because they are closest to the scientifically accepted value. Student B's measurements are neither precise nor accurate. They are not close to each other or to the accepted value.

Active Reading **4. Differentiate** Highlight the terms precision and accuracy and their scientific meanings.

Tools and Accuracy

No tool provides a perfect measurement. All measurements have some degree of uncertainty. Some tools produce more accurate measurements than other tools, as shown in **Figure 6**.

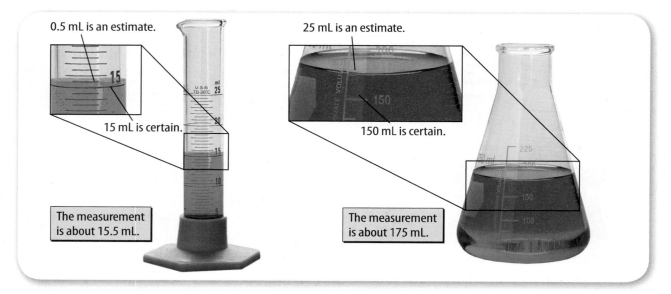

0.5 mL is an estimate.

15 mL is certain.

The measurement is about 15.5 mL.

25 mL is an estimate.

150 mL is certain.

The measurement is about 175 mL.

Scientific Notation

Suppose you are writing a report that includes Earth's distance from the Sun—149,600,000 km—and the density of the Sun's lower atmosphere—0.000000028 g/cm³. These numerals take up too much space and might be difficult to read, so you use **scientific notation**—*a method of writing or displaying very small or very large values in a short form.* To write numerals in scientific notation, use the steps shown to the right.

5. Analyze Why is scientific notation a useful tool?

_____

_____

Percent Error

The densities recorded for NaCl are experimental values because they were calculated during an experiment. Each of these values has some error because the accepted value for table salt density is 21.65 g/cm³. Percent error can help you determine the size of your experimental error. **Percent error** *is the expression of error as a percentage of the accepted value.*

How to Write in Scientific Notation

1 Write the original number.
   A. **149,600,000**
   B. **0.000000028**

2 Move the decimal point to the right or the left to make the number between 1 and 10. Count the number of decimal places moved and note the direction.
   A. **1.49600000** = 8 places to the left
   B. **00000002.8** = 8 places to the right

3 Rewrite the number deleting all extra zeros to the right or to the left of the decimal point.
   A. **1.496**
   B. **2.8**

4 Write a multiplication symbol and the number *10* with an exponent. The exponent should equal the number of places that you moved the decimal point in step 2. If you moved the decimal point to the left, the exponent is positive. If you moved the decimal point to the right, the exponent is negative.
   A. **1.496×10^8**
   B. **2.8×10^{-8}**

Math Skills   MA.6.A.3.6

Solve for Percent Error A student in the laboratory measures the boiling point of water at 97.5°C. If the accepted value for the boiling point of water is 100.0°C, what is the percent error?

1. This is what you know: **experimental value = 97.5°C**
                                   accepted value = 100.0°C

2. This is what you need to find: percent error

3. Use this formula: $\text{percent error} = \dfrac{|\text{experimental value} - \text{accepted value}|}{\text{accepted value}} \times 100\%$

Note that |experimental value – accepted value| refers to the absolute value of this operation.

4. Substitute the known values into the equation and perform the calculations

$$\text{percent error} = \frac{|97.5° - 100.0°|}{100.0°} \times 100\% = 2.50\%$$

Practice

6. Calculate the percent error if the experimental value of the density of gold is 18.7 g/cm³ and the accepted value is 19.3 g/cm³.

Scientific Tools

As you conduct scientific investigations, you will use tools to take measurements. The tools listed here are some of the tools commonly used in science.

Active Reading **7. Apply** Infer a safety rule for each scientific tool discussed.

◀ Science Journal

Use a science journal to record observations, write questions and hypotheses, collect data, and analyze the results of scientific inquiry. All scientists record the information they learn while conducting investigations.

Balances ▶

A balance is used to measure the mass of an object. Units often used for mass are kilograms (kg), grams (g), and milligrams (mg). Two common types of balances are the electronic balance and the triple-beam balance.

Balance safety rule: _____

_____

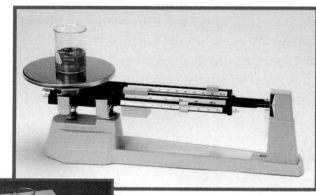

◀ Glassware

Laboratory glassware is used to hold or measure the volume of liquids. Flasks, beakers, test tubes, and graduated cylinders are just some of the different types of glassware available. Volume usually is measured in liters (L) and milliliters (mL).

Glassware safety rule: _____

_____

Thermometers ▶

A thermometer is used to measure the temperature of substances. Although Kelvin is the SI unit of measurement for temperature, in the science classroom, you often measure temperature in degrees Celsius (°C). To avoid breakage, never stir a substance with a thermometer.

Thermometer safety rule: _____

_____

◀ Calculators

A calculator is a scientific tool that you might use in math class. But you also can use it to make quick calculations using your data.

Calculator safety rule: _____

_____

Computers ▼

For today's students and scientists, electronic probes can be attached to computers and handheld calculators to record measurements. There are probes for collecting different kinds of information, such as temperature and the speed of objects. It is difficult to think of a time when computers were not readily available. Scientists can collect, compile, and analyze data more quickly using a computer. Computers are also used to prepare research reports and to share data and ideas with investigators worldwide.

Hardware refers to the physical components of a computer, such as the monitor and the mouse. Computer software refers to the programs that are run on computers, such as word processing, spreadsheet, and presentation programs.

Computer safety rule:

_____

_____

Additional Tools Used by Physical Scientists

pH paper is used to quickly estimate the acidity of a substance. The paper changes color when it comes into contact with an acid or a base.

A hot plate is a small heating device that can be placed on a table or desk to heat substances in the laboratory.

Scientists use stopwatches to measure the time it takes for an event to occur. The SI unit for time is seconds (s).

A spring scale can be used to measure the weight or the amount of force applied to an object. The SI unit for weight is the newton (N).

Active Reading

8. Revisit Briefly recall a time when you used scientific equipment.

_____

_____

Inquiry

LAB STATION

SC.6.N.1.4, SC.6.N.1.5, SC.7.N.1.5, SC.8.N.1.2, SC.8.N.1.6, SC.8.N.3.1

Try It!

Skill Practice *How do geometric shapes differ in strength?* at connectED.mcgraw-hill.com

Lesson Review 2

Use Vocabulary

1. A spoken or written summary of observations is a(n) _____, while a(n) _____ is an interpretation of observations.

Understand Key Concepts 🔑

2. Which type of glassware would you use to measure the volume of a liquid?
   - (A) beaker
   - (B) flask
   - (C) graduated cylinder
   - (D) test tube

3. **Summarize** why recording the diameter of an atom or the distance to the Moon would use scientific notation. MA.6.A.3.6

_____

_____

4. **Explain** why scientists use the International System of Units (SI). SC.8.N.2.2

_____

_____

Interpret Graphics

5. **Identify** List some scientific tools used to collect data. LA.6.2.2.3

Scientific tools

Critical Thinking

6. **Explain** why precision and accuracy should be part of a scientific investigation.

_____

_____

Math Skills   MA.6.A.3.6

7. **Calculate** the percent error if the experimental value for the density of zinc is 9.95 g/cm³, but the accepted value is 7.13 g/cm³.

The Design Process

Create a Solution

Scientists investigate and explore natural events. Then they interpret data and share information learned from those investigations. How do engineers differ from scientists? Engineers construct and maintain the designed world. Look around you and take notice of things that do not occur in nature. Schools, roads, submarines, toys, microscopes, medical equipment, amusement park rides, computer programs, and video games all result from engineering. Science involves the practice of scientific inquiry, but engineering involves The Design Process—a set of methods used to create a solution to a problem or need.

From the first snowshoes to modern ski lifts, people have designed solutions to travel effectively in the snow. Though the term *engineer* did not exist when the first snowshoes were made, the developers of snowshoes used The Design Process just as the engineers of ski lifts do today.

The Design Process

1. Identify a Problem or Need
- Determine a problem or need
- Document all questions, research, and procedures throughout the process

2. Research and Development
- Brainstorm solutions
- Research any existing solutions that address the problem or need
- Suggest limitations

3. Construct a Prototype
- Develop possible solutions
- Estimate materials, costs, resources, and time to develop the solutions
- Select the best possible solution
- Construct a prototype

4. Test and Evaluate Solutions
- Use models to test the solution
- Use graphs, charts, and tables to evaluate results
- Analyze the solution's strengths and weaknesses

5. Communicate Results and Redesign
- Communicate the design process and results to others
- Redesign and modify the solution
- Construct the final solution

It's Your Turn

SC.6.N.1.4,
SC.6.P.12.1,
SC.7.N.1.1,
SC.8.N.1.2,
SC.8.N.4.2

Inquiry
LAB STATION

DESIGN PROCESS LAB Design a Zipline Ride

Case STUDY

 Why are evaluation and testing important in the design process?

 How is scientific inquiry used in a real-life scientific investigation?

Vocabulary

variable p. NOS 21

constant p. NOS 21

independent variable p. NOS 21

dependent variable p. NOS 21

experimental group p. NOS 21

control group p. NOS 21

qualitative data p. NOS 24

quantitative data p. NOS 24

The Minneapolis Bridge Failure

On August 1, 2007, the center section of the Interstate-35W (I-35W) bridge in Minneapolis, Minnesota, suddenly collapsed. A major portion of the bridge fell more than 30 m into the Mississippi River, as shown in **Figure 7**. There were more than 100 cars and trucks on the bridge at the time, including a school bus carrying over 50 students.

The failure of this 8-lane, 581-m long interstate bridge came as a surprise. Drivers do not expect a bridge to drop out from underneath them. The design and engineering processes are supposed to ensure that bridge failures do not happen.

Controlled Experiments

After the bridge collapsed, investigators had to use scientific inquiry to determine why the bridge failed. The investigators designed controlled experiments to help them discover exactly what happened. A controlled experiment is a scientific investigation that answers questions, tests hypotheses, and collects data to determine how one factor affects another.

Active Reading **1. Identify** Highlight the term *controlled experiment* and its description.

Figure 7 At about 6:05 pm, on August 1, 2007, a portion of the Interstate-35W bridge spanning the Mississippi River outside Minneapolis, Minnesota, collapsed. Thirteen people were killed, and many more were injured.

Identifying Variables and Constants

When conducting an experiment, you must identify factors that can affect the experiment's outcome. A **variable** *is any factor that can have more than one value.* In controlled experiments, there are two kinds of variables. The **independent variable** *is the factor that you want to test. It is changed by the investigator to observe how it affects a dependent variable.* The **dependent variable** *is the factor you observe or measure during an experiment.* **Constants** *are the factors in an experiment that do not change.*

Experimental Groups

A controlled experiment usually has at least two groups. The **experimental group** *is used to study how a change in the independent variable changes the dependent variable.* The **control group** *contains the same factors as the experimental group, but the independent variable is not changed.* Without a control, it is impossible to know whether your experimental observations result from the variable you are testing or some other factor.

This case study will explore how the investigators used scientific inquiry to determine why the bridge collapsed. Notebooks in the margin identify what a scientist might write in a science journal.

Simple Beam Bridges

Before you read about the bridge-collapse investigation, think about the structure of bridges. The simplest type of bridge is a beam bridge, as shown in **Figure 8**. This type of bridge has one horizontal beam across two supports. A disadvantage of beam bridges is that they tend to sag in the middle if they are too long or if the load is too heavy.

You can change the independent variable to observe how it affects the dependent variable. Without constants, two independent variables could change at the same time, and you would not know which variable affected the dependent variable.

Active Reading **2. Differentiate** Illustrate the relationship between independent and dependent variables. Then, list two possible constants for a scientific investigation.

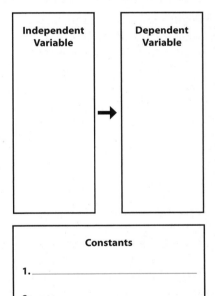

Independent Variable	Dependent Variable

Constants

1._____

2._____

Figure 8 Simple beam bridges are used to effectively span short distances, such as small creeks.

Figure 9 Truss bridges can span long distances and are strengthened by a series of interconnecting triangles called trusses.

Active Reading 3. **Identify** As you read the following paragraph, fill in the tags on the photo above to identify the following parts of a truss bridge: support beam, gusset plate, deck of the bridge, end support, and truss.

Truss Bridges

A truss bridge, shown in **Figure 9**, is supported at its ends, by an assembly of interconnected, triangular trusses to strengthen it. The I-35W bridge, shown in **Figure 10**, was a tress bridge designed with straight beams and trusses connected to vertical supports. The beams in the bridge's deck and the supports came together at structures known as gusset plates, shown below on the right. These steel plates joined the triangular and vertical truss elements to the overhead roadway beams that ran along the deck of the bridge. This area, where the truss structure connects to the roadway portion of the bridge at a gusset plate is called a node.

Figure 10 Trusses were a major structural element of the I-35W bridge. The gusset plates at each node in the bridge, shown on the right, are critical pieces that hold the bridge together.

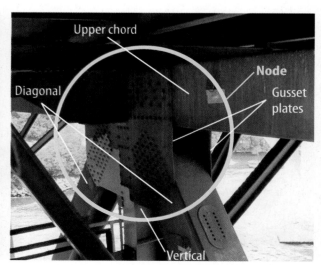

Figure 11 The tangled bridge was further damaged by rescue workers trying to recover victims of the accident.

Bridge Failure Observations

After the bridge collapsed, shown in **Figure 11**, the local sheriff's department handled the initial recovery of the bridge structure. Finding, freeing, and identifying victims was a high priority, and unintentional damage to the fallen bridge occurred. However, investigators eventually recovered the entire structure.

Investigators labeled each part with the location where it was found and the date when it was removed. They then moved the pieces to a nearby park where they placed the pieces in their relative original positions. Examining the reassembled structure, investigators found physical evidence to help determine where the breaks in each section occurred.

The investigators found more clues in a motion-activated security camera recording of the bridge collapse. The video showed about 10 seconds of the collapse, which revealed the sequence of events that destroyed the bridge. Investigators used this video to help pinpoint where the failure began.

Scientists often observe and gather information about an object or an event before proposing a hypothesis. This information is recorded or filed for the investigation.

Observations:
- Recovered parts of the collapsed bridge
- A video showing the sequence of events as the bridge fails and falls into the river

Active Reading

4. Sequence Analyze the paragraphs above. Then, number the steps from 1 to 6 that investigators took in the process of recovery and identification of the destroyed bridge structure.

Asking Questions

What factors caused the bridge to fail? Was the original bridge design faulty? Were bridge maintenance and repair poor or lacking? Was there too much weight on the bridge? Or was it a combination of these factors that caused the bridge to fail? Each of these questions was studied to determine why the bridge collapsed.

Asking questions and seeking answers to those questions is a way that scientists formulate hypotheses.

When gathering information or collecting data, scientists might perform an experiment, create a model, gather and evaluate evidence, or make calculations.

Qualitative data: A thicker layer of concrete was added to the bridge to protect rods.

Quantitative data:
• The concrete increased the load on the bridge by 13.4 percent.

• The modifications in 1998 increased the load on the bridge by 6.1 percent.

• At the time of the collapse in 2007, the load on the bridge increased by another 20 percent.

A hypothesis is a possible explanation for an observation that can be tested by scientific investigations.

Hypothesis: The bridge failed because it was overloaded.

Gathering Information and Data

Investigators reviewed the modifications made to the bridge since it opened in 1967. In 1977, engineers noticed that salt used to deice the bridge during winter weather was causing the reinforcement rods in the roadway to weaken. To protect the rods, engineers applied a thicker layer of concrete to the surface of the bridge roadway. Analysis after the collapse revealed that this extra concrete increased the dead load on the bridge by about 13.4 percent. A load can be a force applied to the structure from the structure itself (dead load) or from temporary loads such as traffic, wind gusts, or earthquakes (live load). Investigators recorded this qualitative and quantitative data. **Qualitative data** *uses words to describe what is observed.* **Quantitative data** *uses numbers to describe what is observed.*

In 1998, modifications were made to the bridge when it was noted that the bridge did not meet current safety standards. Analysis showed that these changes further increased the dead load on the bridge by about 6.1 percent.

An Early Hypothesis

At the time of the collapse, the bridge was undergoing additional renovations. Four piles of sand, four piles of gravel, a water tanker filled with over 11,000 L of water, a cement tanker, a concrete mixer, and other equipment, supplies, and workers were assembled on the bridge. This caused the load on the bridge to increase by 20 percent. In addition, normal vehicle traffic was on the bridge. Did these factors overload the bridge, causing the center section to collapse as shown in **Figure 12?** Only a thorough analysis could answer this question.

Active Reading 5. **Analyze** Highlight the factors that contributed to the load on the bridge at the time of its collapse.

Figure 12 The center section of the bridge broke away and fell into the river.

Figure 13 Engineers used computer models to study the structure and loads on the bridge.

Computer Modeling

The analysis of the bridge was conducted using computer-modeling software, as shown in **Figure 13**. Investigators entered data from the Minnesota bridge into a computer to efficiently perform numerous mathematical calculations. After thorough modeling and analysis, it was determined that the bridge was not overloaded.

Revising the Hypothesis

Analysis of routine assessments conducted in 1999 and 2003 provided additional clues as to why the bridge might have failed. At the time, investigators had taken numerous pictures of the bridge structure. The photos revealed bowing of the gusset plates at node U10. Gusset plates are designed to be stronger than the structural parts they connect. The bowing of the plates possibly indicated a problem with the gusset plate design. Previous inspectors and engineers missed this warning sign.

The accident investigators found that some recovered gusset plates were fractured early in the collapse, while others were not damaged. If the bridge had been properly constructed, none of the plates should have failed.

After evaluating the evidence, investigators formulated the hypothesis that the gusset plates failed, which lead to the bridge collapse. Now investigators had to test this hypothesis.

Hypothesis:
1. ~~The bridge failed because it was overloaded.~~
2. The bridge collapsed because the gusset plates failed.
Prediction:
If a gusset plate is not properly designed, then a heavy load on a bridge will cause the gusset plate to fail.

Active Reading

6. Summarize Based on what investigators discovered about the gusset plates, restate the new hypothesis.

_____

_____

Test the Hypothesis:
• Compare the load on the bridge
when it collapsed with the load limits
of the bridge at each of the main
gusset plates.
• Determine the demand-to-
capacity ratios for the main gusset
plates.
• Calculate the appropriate
thicknesses of the U10 gusset plates.
Independent Variables: actual load
on bridge and load bridge was
designed to handle
Dependent Variable: demand-to-
capacity ratio

Testing the Hypothesis

To calculate the load on the bridge when it collapsed, they estimated the combined weight of the bridge and the traffic on the bridge. The investigators divided the load on the bridge when it collapsed by the known load limits of the bridge to find the demand-to-capacity ratio. The demand-to-capacity ratio provides a measure of a structure's safety.

Analyzing Results

As investigators calculated the demand-to-capacity ratios for each of the main gusset plates, they found that the ratios were particularly high for the U10 node. The U10 plate, shown in **Figure 14,** failed earliest in the bridge collapse. **Table 4** shows the demand-to-capacity ratios for some gusset plates at three nodes. A value greater than 1 means the structure is unsafe. Notice how high the ratios are for the U10 gusset plate compared to the other plates.

Further data showed that the U10 plates were about half the thickness they should have been to support the load they were designed to handle.

Active Reading

7. Infer Why are evaluation and testing an important part of the design process?

_____

_____

Figure 14 The steel plates, or gusset plates, at the U10 node were too thin for the loads the bridge carried.

Table 4 Node-Gusset Plate Analysis							
Gusset Plate	Thickness (cm)	Demand-to-Capacity Ratios for the Upper-Node Gusset Plates					
		Horizontal loads			Vertical loads		
U8	3.5	0.05	0.03	0.07	0.31	0.46	0.20
U10	1.3	1.81	1.54	1.83	1.70	1.46	1.69
U12	2.5	0.11	0.11	0.10	0.71	0.37	1.15

Drawing Conclusions

Over the years, modifications to the I-35W bridge added more load to the bridge. On the day of the accident, traffic and the concentration of construction vehicles and materials added still more load. Investigators concluded that if the U10 gusset plates were properly designed, they would have supported the added load. When the investigators examined the original records for the bridge, they were unable to find any detailed gusset plate specifications. They could not determine whether undersized plates were used because of a mistaken calculation or some other error in the design process. The only thing that they could conclude with certainty was that undersized gusset plates could not reliably hold up the bridge.

The Federal Highway Administration and the National Transportation Safety Board published the results of their investigations. These publications provide valuable information that can be used in future designs. These reports are good examples of why it is important to publish results and to share information.

> **Analyzing Results:** The U10 gusset plates should have been twice as thick as they were to support the bridge.

> **Conclusions:** The bridge failed because the appropriate gusset plates were not installed; therefore, the bridge could not carry the increased load.

Active Reading 8. **Conclude** (Circle) the results and conclusions as to why the bridge collapsed.

Lesson Review 3

Inquiry LAB STATION Try It!

SC.6.N.1.3, SC.7.N.1.6, SC.6.N.1.4, SC.8.N.1.4, SC.7.N.1.3, SC.8.N.3.1 SC.7.N.1.7, SC.8.N.1.2, SC.8.N.1.3, SC.8.N.1.5,

Inquiry Lab *Build and Test a Bridge* at connectED.mcgraw-hill.com

Use Vocabulary

1 **Distinguish** between qualitative data and quantitative data.

Understand Key Concepts 🔑

2 **Give an example** of a situation in your life in which you depend on testing and evaluation in a product design to keep you safe.

Interpret Graphics

3 **Complete** Sequence scientific inquiry steps used in one part of the case study. LA.6.2.2.3

Critical Thinking

4 **Analyze** how the scientific inquiry process differs when engineers design a product, such as a bridge, and when they investigate design failure. SC.8.N.1.5

5 **Evaluate** why the gusset plates were a critical piece in the bridge design. SC.8.N.1.5

Think About It! Scientific inquiry is a multifaceted activity that includes the formulation of investigable questions, the construction of investigations, the collection and evaluation of data, and the communication of results.

Key Concepts Summary

LESSON 1 Scientific Inquiry

- Some steps used during scientific inquiry are making **observations** and **inferences,** developing a **hypothesis,** analyzing results, and drawing conclusions. These steps, among others, can be performed in any order.
- There are many results of scientific inquiry, and a few possible outcomes are the development of new materials and new technology, the discovery of new objects and events, and answers to basic questions.
- **Critical thinking** is comparing what you already know about something to new information and deciding whether or not you agree with the new information.

LESSON 2 Measurement and Scientific Tools

- Scientists developed one universal system of units, the **International System of Units (SI),** to improve communication among scientists.
- **Scientific notation** is a useful tool for writing large and small numbers in a shorter form.
- Tools such as graduated cylinders and triple-beam balances make scientific investigation easier, more accurate, and repeatable.

LESSON 3 Case Study—The Minneapolis Bridge Failure

- Evaluation and testing are important in the design process for the safety of the consumer and to keep costs of building or manufacturing the product at a reasonable level.
- Scientific inquiry was used throughout the process of determining why the bridge collapsed, including hypothesizing potential reasons for the bridge failure and testing those hypotheses.

Vocabulary

science p. NOS 4
observation p. NOS 6
inference p. NOS 6
hypothesis p. NOS 6
prediction p. NOS 6
scientific theory p. NOS 8
scientific law p. NOS 8
technology p. NOS 9
critical thinking p. NOS 10

description p. NOS 12
explanation p. NOS 12
International System of Units (SI) p. NOS 12
scientific notation p. NOS 15
percent error p. NOS 15

variable p. NOS 21
independent variable p. NOS 21
dependent variable p. NOS 21
constants p. NOS 21
qualitative data p. NOS 21
quantitative data p. NOS 21
experimental group p. NOS 24
control group p. NOS 24

Use Vocabulary

1 The _____ contains the same factors as the experimental group, but the independent variable is not changed.

2 The expression of error as a percentage of the accepted value is _____.

3 The process of studying nature at all levels and the collection of information that is accumulated is _____.

4 The _____ are the factors in the experiment that stay the same.

Fill in the correct answer choice.

🔑 Understand Key Concepts

5 Which is NOT an SI base unit? SC.8.N.2.2
- (A) kilogram
- (C) meter
- (B) liter
- (D) second

6 While analyzing results from an investigation, a scientist calculates a very small number that he or she wants to make easier to use. Which does the scientist use to record the number? SC.8.N.2.2
- (A) explanation
- (C) scientific notation
- (B) inference
- (D) scientific theory

7 Which is NOT true of a scientific law? SC.8.N.3.1
- (A) It can be modified or rejected.
- (B) It states that an event will occur.
- (C) It explains why an event will occur.
- (D) It is based on repeated observations.

Critical Thinking

8 **Write** a brief description of the activity shown in the photo. SC.8.N.1.1

_____

_____

_____

Writing in Science

9 **Apply** On a separate piece of paper, write a paragraph that gives examples of how critical thinking, skepticism, and identifying facts and opinions can help you in your everyday life. Be sure to include a topic sentence and concluding sentence in your paragraph. SC.8.N.2.2

Big Idea Review

10 What is scientific inquiry? Explain why it is a constantly changing process. SC.8.N.2.2

_____

_____

_____

Math Skills MA.6.A.3.6

11 The accepted scientific value for the density of sucrose is 1.59 g/cm³. The data from three trials to measure the density of sucrose is shown in the table below. Calculate the percent error for each trial.

Trial	Density	Percent Error
Trial 1	1.55 g/cm³	
Trial 2	1.60 g/cm³	
Trial 3	1.58 g/cm³	

Multiple Choice *Bubble the correct answer.*

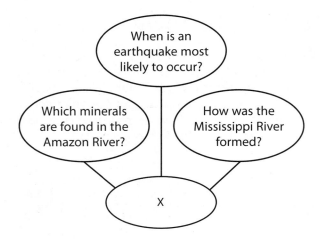

1. The graphic organizer above shows some questions scientists might ask. Which branch of science should go in X? **SC.7.N.1.5**

- (A) chemical science
- (B) Earth science
- (C) life science
- (D) physical science

2. Scientists design experiments in order to test **SC.8.N.2.2**

- (F) conclusions.
- (G) hypotheses.
- (H) observations.
- (I) predictions.

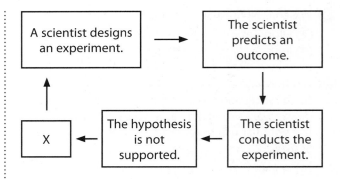

3. The flow chart above shows some of the steps involved in a scientific inquiry. Which would most likely occur at X? **SC.8.N.1.4**

- (A) The scientist analyzes the results.
- (B) The scientist makes a new prediction.
- (C) The scientist modifies or revises the hypothesis.
- (D) The scientist speaks at a science conference.

4. Which statement cannot be tested scientifically? **SC.6.N.1.3**

- (F) Apples are red because they reflect red light waves.
- (G) Apples have more sugar than oranges have.
- (H) Oranges do not grow as well in Canada as they do in Mexico.
- (I) Oranges have a better flavor than apples have.

Copyright © Glencoe/McGraw-Hill, a division of The McGraw-Hill Companies, Inc.

Benchmark Mini-Assessment   Nature of Science • Lesson 2   mini BAT

Multiple Choice *Bubble the correct answer.*

Use the table below to answers questions 1 and 2.

Melting Point of Aspirin (Accepted value: 135°C)			
	Student R	**Student S**	**Student T**	**Student U**
Trial 1	133.7°C	134.6°C	134.8°C	133.9°C
Trial 2	133.4°C	135.5°C	134.5°C	135.0°C
Trial 3	133.9°C	136.1°C	135.3°C	134.5°C
Average	**133.7°C**	**135.4°C**	**134.9°C**	**134.5°C**

1. According to the information in the table, which student's measurements are the most accurate? **MA.6.S.6.2**

(A) Student R

(B) Student S

(C) Student T

(D) Student U

2. According to the information in the table, which student's measurements are the most precise? **SC.8.N.1.6**

(F) Student R

(G) Student S

(H) Student T

(I) Student U

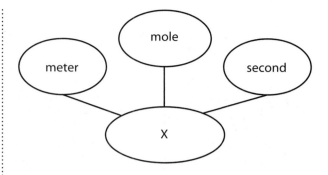

3. The graphic organizer above shows three kinds of measurements. Which description goes in X? **SC.8.N.1.5**

(A) SI base units

(B) SI prefixes

(C) metric units of length

(D) metric units of mass

4. A triple-beam balance is used to measure **SC.7.N.1.5**

(F) density.

(G) mass.

(H) temperature.

(I) volume.

Copyright © Glencoe/McGraw-Hill, a division of The McGraw-Hill Companies, Inc.

Multiple Choice *Bubble the correct answer.*

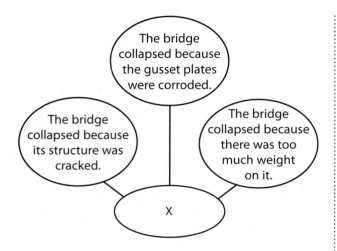

1. In the graphic organizer above, which goes in X? **SC.6.N.1.5**

(A) observations about why the bridge collapsed

(B) possible hypotheses that can be tested on a collapsed bridge

(C) possible reasons why the bridge might have collapsed

(D) predictions about why a bridge might collapse

2. It would be too costly to build actual bridges for testing, so bridge designers use mathematical models instead. These models are **SC.8.N.3.1**

(F) conclusions.

(G) hypotheses.

(H) observations.

(I) predictions.

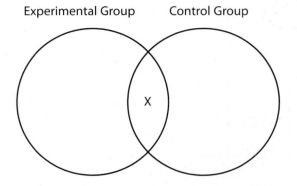

3. In the Venn diagram above, which does NOT go in X? **SC.7.N.1.1**

(A) constants

(B) measurements

(C) dependent variable

(D) independent variable

4. Which statement is true? **SC.8.N.1.4**

(F) A hypothesis becomes a law if several experiments support it.

(G) A hypothesis is true if one experiment supports it.

(H) A hypothesis is untrue if even one experiment does not support it.

(I) A hypothesis might be true even if several experiments do not support it.

Copyright © Glencoe/McGraw-Hill, a division of The McGraw-Hill Companies, Inc.

Notes

Unit 1
Motion & Forces

3500 B.C.
The oldest wheeled vehicle is depicted in Mesopotamia, near the Black Sea.

400 B.C.
The Greeks invent the stone-hurling catapult.

1698
English military engineer Thomas Savery invents the first crude steam engine while trying to solve the problem of pumping water out of coal mines.

1760–1850
The Industrial Revolution results in massive advances in technology and social structure in England.

1769
The first vehicle to move under its own power is designed by Nicholas Joseph Cugnot and constructed by M. Breszin. A second replica is built that weighs 3,629 kg and has a top speed of 3.2 km per hour.

1794
Eli Whitney receives a patent for the mechanical cotton gin.

1817
Baron von Drais invents a machine to help him quickly wander the grounds of his estate. The machine is made of two wheels on a frame with a seat and a pair of pedals. This machine is the beginning design of the modern bicycle.

1903
Wilbur and Orville Wright build their airplane, called the Flyer, and take the first successful, powered, piloted flight.

1976
The first computer for home use is invented by college dropouts Steve Wozniak and Steve Jobs, who go on to found Apple Computer, Inc.

? Inquiry

Visit ConnectED for this unit's **STEM** activity.

Models

SC.6.N.1.3, SC.7.N.1.5, SC.7.N.3.2, SC.6.N.3.4

Have you ridden on an amusement park roller coaster such as the one in **Figure 1?** Did you think to yourself, "I hope I don't fly off this thing"? Before construction begins on a roller coaster, engineers build models of the thrill ride to ensure proper construction and safety. A **model** is a representation of an object, an idea, or a system that is similar to the physical object or idea being studied.

Using Models in Physical Science

Models are used to study things that are too big or too small, happen too quickly or too slowly, or are too dangerous or too expensive to study directly. Different types of models serve different purposes. Roller-coaster engineers build physical models for new, daring coasters. Mathematical and computer models calculate measurements of hills, angles, and loops to ensure a safe ride. Finally, engineers create a blueprint drawing that details the construction of the ride. Studying the various models allows engineers to predict how the actual coaster will behave when it travels through a loop or down a giant hill.

Figure 1 Engineers use various models to design roller coasters.

Active Reading 1. **Point out** Highlight the purpose of using a model.

Types of Models

Physical Model

A physical model is a model that you can see and touch. It shows how parts relate to one another, how something is built, or how complex objects work. Physical models are built to scale. A limitation of a physical model is that it might not reflect the physical behavior of the full-sized object.

Mathematical Model

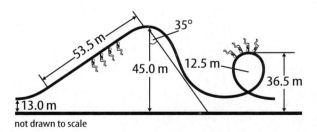

53.5 m   35°

45.0 m   12.5 m

13.0 m   36.5 m

not drawn to scale

A mathematical model uses numerical data and equations to model an event or idea. When designing a thrill ride, engineers use mathematical models to calculate the heights, the angles of loops and turns, and the forces that affect the ride. One limitation of a mathematical model is that it cannot be used to model how different parts are assembled.

Making Models

An important factor in making a model is determining its purpose. You might need a model that physically represents an object, or a model that includes only important elements of an object or a process. When you build a model, first determine the function of the model. What materials should you use? What do you need to communicate to others? **Figure 2** shows two models of a glucose molecule, each with a different purpose.

Limitations of Models

It is impossible to include all the details about an object or an idea into one model. All models have limitations. An engineer must be aware of the information each model does and does not provide. A blueprint of a roller coaster does not show the maximum weight that a car can support. However, a mathematical model would include this information. Scientists and engineers consider the purpose and the limitations of the model they use to ensure they draw accurate conclusions from models.

Active Reading 2. **Determine** Highlight what scientists and engineers must consider when using models in their work.

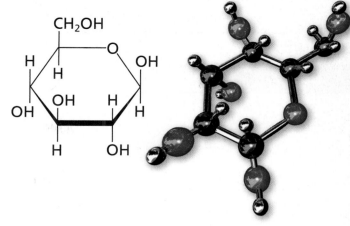

Figure 2 The model on the left is used to represent how the atoms in a glucose molecule bond together. The model on the right is a 3-D representation of the molecule, which shows how atoms might interact.

Inquiry LAB STATION **Try It!**
SC.6.N.3.4, SC.7.N.3.2

MiniLab *Can you model a roller coaster?* at connectED.mcgraw-hill.com

Apply It!
After you complete the lab, answer these questions.

1. **Analyze** How can each type of model be helpful in designing your safe, yet thrilling, coaster ride?

   Physical model:

   Mathematical model:

   Computer model:

2. **Recall** Think of a time you constructed a model.

   What was the purpose of the model?

   What materials did you use?

   How did the model help communicate information to you and your peers?

Computer Simulation

A computer simulation is a model that combines mathematical models with computer graphic and animation programs. Simulations can contain thousands of complex mathematical models. When engineers change variables in mathematical models, they use computer simulation to view the effects of the change.

Name _____ Date _____

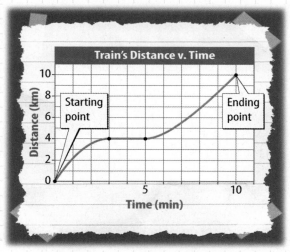

Train Ride?

The graph shows a train traveling between two stations. (Circle) what you think is the best interpretation of the graph.

A: The train sped up during the first 3 minutes, slowed down for several minutes, then sped up again.

B: The train went uphill, traveled on flat ground, then went uphill again.

C: The train sped up for 3 minutes, traveled at a steady speed for 2 minutes, then sped up again.

D: The train sped up then slowed down during the first 3 minutes, stopped for two minutes, then sped up again.

E: The train traveled at a steady speed uphill for most of the way except for in the middle when it stopped for 2 minutes.

Explain the answer you selected to describe the motion of the train.

Describing
MOTION

Results of Scientific Inquiry

Why do you and others ask questions and investigate the natural world? Often, the results of a scientific investigation are new materials and technology, the discovery of new objects, or answers to questions.

Active Reading

10. Analyze In the spaces provided below, discuss how technology helps scientists develop new materials, discover new objects or events, and answer questions.

New Materials and Technology

Corporations and governments research and design new materials and technologies. **Technology** *is the practical use of scientific knowledge, especially for industrial or commercial use.* Scientists use technology to design new materials that make bicycles and cycling gear lighter, more durable, safer, and more aerodynamic. Using wind tunnels, scientists test these new materials to see whether they improve the cyclist's performance.

Summarize How might safer, more efficient bike equipment be a product of technology and scientific inquiry?

_____

_____

_____

New Objects or Events

Scientific investigations also lead to newly discovered objects or events. For example, NASA's *Hubble Space Telescope* captured this image of two colliding galaxies. They have been nicknamed the mice, because of their long tails. The tails are composed of gases and young, hot blue stars. If computer models are correct, these galaxies will combine in the future and form one large galaxy.

Explain How did technology aid scientists in the discovery of colliding galaxies?

_____

_____

_____

Answers to Questions

Often scientific investigations are launched to answer *who, what, when, where, why,* or *how* questions. This research chemist investigates new substances found in mushrooms and bacteria. New drug treatments for diseases might be found using new substances. Other scientists look for clues about what causes diseases, whether they can be passed from person to person, and when the disease first appeared.

Infer How is technology used by chemists to develop new knowledge and treatments for the medical field?

_____

_____

_____

Position and MOTION

ESSENTIAL QUESTIONS

 How does the description of an object's position depend on a reference point?

 How can you describe the position of an object in two dimensions?

 What is the difference between distance and displacement?

Vocabulary

reference point p. 11

position p. 11

motion p. 15

displacement p. 15

 SC.6.N.1.4

Inquiry Launch Lab

10 minutes

How do you get there from here?

How would you give instructions to a friend who was trying to walk from one place to another in your classroom?

Procedure

1. Read and complete a lab safety form.
2. Place a sheet of **paper** labeled *North, East, South,* and *West* on the floor.
3. Walk from the paper to one of the three locations your teacher has labeled in the classroom. Have a partner record the number of steps and the directions of movement.
4. Using these measurements, write instructions other students could follow to move from the paper to the location.
5. Repeat steps 3 and 4 for the other locations.

Data and Observations

Think About This

1. How did your instructions to each location compare to those written by other groups?

2. **Key Concept** How did the description of your movement depend on the point at which you started?

 Florida NGSSS

SC.6.N.1.4 Discuss, compare, and negotiate methods used, results obtained, and explanations among groups of students conducting the same investigation.

LA.6.2.2.3 The student will organize information to show understanding or relationships among facts, ideas, and events (e.g., representing key points within text through charting, mapping, paraphrasing, summarizing, or comparing/contrasting);

Assemble your lesson Foldables as shown to make a Chapter Project. Use the project to review what you have learned in this chapter.

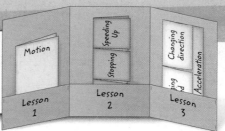

Use Vocabulary

1 A pencil's _____ might be described as 3 cm to the left of the stapler.

2 An object that changes position is in

_____ .

3 If an object is traveling at a _____ , it does not speed up or slow down.

4 An object's _____ includes both its speed and the direction it moves.

5 An object's change in velocity during a time interval, divided by the time interval during which the velocity changed, is its _____ .

6 A truck driver stepped on the brakes to make a quick stop. The truck's _____ is in the opposite direction as its velocity.

Link Vocabulary and Key Concepts

Use vocabulary terms from the previous page to complete the concept map.

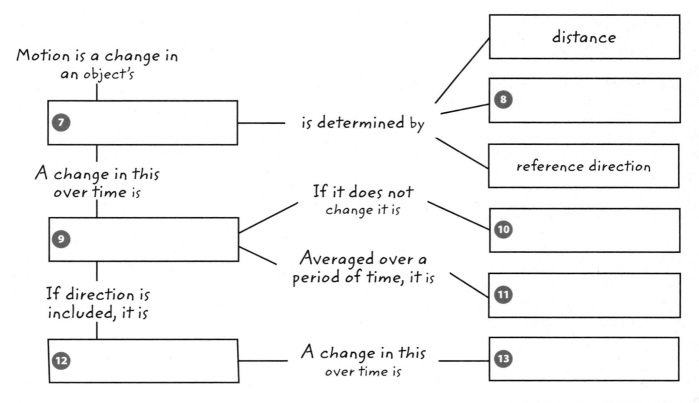

Chapter 1 Review

Fill in the correct answer choice.

🔑 Understand Key Concepts

1 An airplane rolls down the runway. Compared to which reference point is the airplane in motion?
   Ⓐ the cargo the plane carries
   Ⓑ the control tower
   Ⓒ the pilot flying the plane
   Ⓓ the plane's wing

2 Which describes motion in two dimensions?
   Ⓐ a car driving through a city
   Ⓑ a rock dropping off a cliff
   Ⓒ a sprinter on a 100-m track
   Ⓓ a train on a straight track

3 Which line represents the greatest average speed during the 30-s time period? SC.6.P.12.1

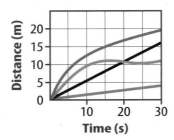

   Ⓐ the blue line
   Ⓑ the black line
   Ⓒ the green line
   Ⓓ the orange line

4 Which describes the greatest displacement?
   Ⓐ walking 3 m east, then 3 m north, then 3 m west
   Ⓑ walking 3 m east, then 3 m south, then 3 m east
   Ⓒ walking 3 m north, then 3 m south, then 3 m north
   Ⓓ walking 3 m north, then 3 m west, then 3 m south

5 Which has the greatest average speed? MA.6.A.3.6
   Ⓐ a boat sailing 80 km in 2 hours
   Ⓑ a car driving 90 km in 3 hours
   Ⓒ a train traveling 120 km in 3 hours
   Ⓓ a truck moving 50 km in 1 hour

Critical Thinking

6 **Describe** A ruler is on the table with the higher numbers to the right. An ant crawls along the ruler from 6 cm to 2 cm in 2 seconds. Describe the ant's distance, displacement, speed, and velocity.

_____

_____

_____

7 **Describe** a theme-park ride that has constant speed but changing velocity. SC.6.P.12.1

_____

_____

_____

8 **Construct** a distance-time graph that shows the following motion: A person leaves a starting point at a constant speed of 4 m/s and walks for 4 s. The person then stops for 2 s. The person then continues walking at a constant speed of 2 m/s for 4 s. SC.6.P.12.1

9 **Calculate** A truck driver travels 55 km in 1 hour. He then drives a speed of 35 km/h for 2 hours. Next, he drives 175 km in 3 hours. What was his average speed? MA.6.A.3.6

10 **Interpret** Keisha measured the distance her friend Morgan ran on a straight track every 2 s. Her measurements are recorded in the table below. What was Morgan's average speed? What was her acceleration? MA.6.A.3.6

Time (s)	Distance (m)
0	0
2	2
4	6
6	8
8	14
10	20

Writing in Science

11 **Write** A friend tells you he is 30 m from the fountain in the middle of the city. Write a short paragraph on a separate sheet oof paper explaining why you cannot identify your friend's position from this description. LA.6.2.2.3

Big Idea Review

12 Nora rides a bicycle for 5 min on a curvy road at a constant speed of 10 m/s. Describe Nora's ride in terms of position, velocity, and acceleration. Compare the distance she rides and her displacement. SC.6.P.12.1

_____

_____

_____

_____

13 What are some ways to describe the motion of the jets in the photograph on page 8?

_____

_____

_____

Math Skills MA.6.A.3.6

14 A model train moves 18.3 m in 122 s. What is the train's average speed?

15 A car travels 45 km in an hour. In each of the next two hours, it travels 78 km. What is the average speed of the car?

16 The speed of a car traveling on a straight road increases from 63 m/s to 75 m/s in 4.2 s. What is the car's acceleration?

17 A girl starts from rest and reaches a walking speed of 1.4 m/s in 3.0 s. She walks at this speed for 6.0 s. The girl then slows down and comes to a stop during a 10.0-s period. What was the girl's acceleration during each of the three time periods? What was her acceleration for the entire trip?

Fill in the correct answer choice.

Multiple Choice

1 Radar tells an air traffic controller that a jet is slowing as it nears the airport. Which might represent the jet's speed? SC.6.P.12.1

Ⓐ 700 h

Ⓑ 700 h/km

Ⓒ 700 km

Ⓓ 700 km/h

Use the diagram below to answer question 2.

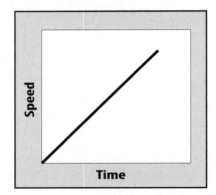

2 What does the graph above illustrate? SC.6.P.12.1

Ⓕ average speed

Ⓖ constant speed

Ⓗ decreasing speed

Ⓘ increasing speed

3 Why is a car accelerating when it is circling at a constant speed? SC.6.P.12.1

Ⓐ It is changing its destination.

Ⓑ It is changing its direction.

Ⓒ It is changing its distance.

Ⓓ It is changing its total mass.

4 Which is defined as the process of changing position? SC.6.P.12.1

Ⓕ displacement

Ⓖ distance

Ⓗ motion

Ⓘ relativity

5 Each diagram below shows two sliding boxes. Which boxes have the same velocity? SC.6.P.12.1

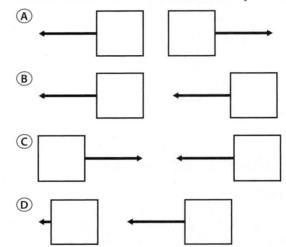

6 In the phrase "two miles southeast of the mall," what is the mall? LA.6.2.2.3

Ⓕ a dimension

Ⓖ a final destination

Ⓗ a position

Ⓘ a reference point

7 The initial speed of a dropped ball is 0 m/s. After 2 seconds, the ball travels at a speed of 20 m/s. What is the acceleration of the ball? MA.6.A.3.6

Ⓐ 5 m/s^2

Ⓑ 10 m/s^2

Ⓒ 20 m/s^2

Ⓓ 40 m/s^2

8 Which could be described by the expression "100 m/s northwest"? SC.6.P.12.1

- Ⓕ acceleration
- Ⓖ distance
- Ⓗ speed
- Ⓘ velocity

Use the diagram below to answer question 9.

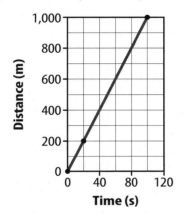

9 In the above graph, what is the average speed of the moving object between 20 and 60 seconds? SC.6.P.12.1

- Ⓐ 5 m/s
- Ⓑ 10 m/s
- Ⓒ 20 m/s
- Ⓓ 40 m/s

10 A car travels 250 km and stops twice along the way. The entire trip takes 5 hours. What is the average speed of the car? SC.6.P.12.1

- Ⓕ 25 km/h
- Ⓖ 40 km/h
- Ⓗ 50 km/h
- Ⓘ 250 km/h

11 Which describes motion in which the person or object is accelerating? SC.6.P.12.1

- Ⓐ A bird flies straight from a tree to the ground without changing speed.
- Ⓑ A dog walks at a constant speed along a straight sidewalk.
- Ⓒ A girl runs along a straight path the same distance each second.
- Ⓓ A truck moves around a curve without changing speed.

12 Richard walks from his home to his school at a constant speed. It takes him 4 min to travel 100 m. Which of the lines in the following distance-time graph could show Richard's motion on the way to school? SC.6.P.12.1

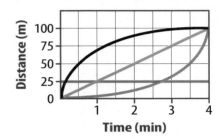

- Ⓕ the black line
- Ⓖ the blue line
- Ⓗ the green line
- Ⓘ the orange line

13 Which is a unit of acceleration? SC.6.P.12.1

- Ⓐ kg/m
- Ⓑ kg·m/s^2
- Ⓒ m/s
- Ⓓ m/s^2

NEED EXTRA HELP?

If You Missed Question...	1	2	3	4	5	6	7	8	9	10	11	12	13
Go to Lesson...	2	3	3	1	2	1	4	2	2	2	3	2	3

Multiple Choice *Bubble the correct answer.*

Use the image below to answer questions 1 and 2.

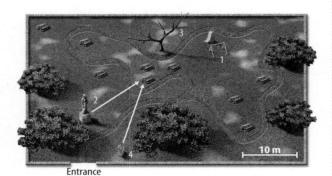

Entrance

10 m

1. Using the park entrance as the reference point, which landmark is the closest?
   LA.6.2.2.3

   (A) 1, the slide

   (B) 2, the statue

   (C) 3, the dead tree

   (D) 4, the drinking fountain

2. To which reference point is a picnic bench closest? **LA.6.2.2.3**

   (F) 1, the slide

   (G) 2, the statue

   (H) 3, the dead tree

   (I) 4, the drinking fountain

Use the image below to answer questions 3 and 4.

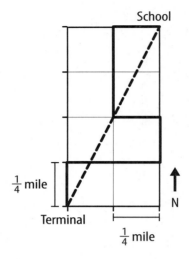

School

$\frac{1}{4}$ mile

Terminal

$\frac{1}{4}$ mile

N

3. Following the bolder line, which distance does the school bus travel on its route each day? **MA.6.A.3.6**

   (A) 0.5 mi

   (B) 1.5 mi

   (C) 2.0 mi

   (D) 2.5 mi

4. The dashed line represents the return route. Which reference directions would you probably use to describe the return route of the bus from the school to its terminal?
   LA.6.2.2.3

   (F) east and west

   (G) north and east

   (H) north and south

   (I) south and west

Copyright © Glencoe/McGraw-Hill, a division of The McGraw-Hill Companies, Inc.

Multiple Choice *Bubble the correct answer.*

Use the table below to answer questions 1 through 4.

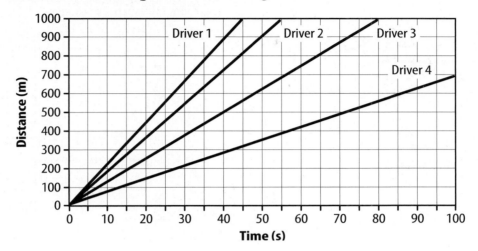

1. The table shows the speeds of four drivers at a superspeedway race. Which driver had the fastest average time? **MA.6.A.3.6**

   Ⓐ Driver 1

   Ⓑ Driver 2

   Ⓒ Driver 3

   Ⓓ Driver 4

2. How much longer did it take Driver 4 than Driver 3 to reach a distance of 700 m? **MA.6.A.3.6**

   Ⓕ 45 seconds

   Ⓖ 50 seconds

   Ⓗ 55 seconds

   Ⓘ 60 seconds

3. What was Driver 4's average speed over the first 700 m of the race? **MA.6.A.3.6**

   Ⓐ 3 m/s

   Ⓑ 6 m/s

   Ⓒ 7 m/s

   Ⓓ 9 m/s

4. If Driver 3 made a pit stop during the race to have the car's tires changed, what would Driver 3's line look like during the stop? **SC.6.P.12.1**

   Ⓕ The line would be horizontal.

   Ⓖ The line would curve down.

   Ⓗ The line would drop straight down.

   Ⓘ The line would slope upward.

Copyright © Glencoe/McGraw-Hill, a division of The McGraw-Hill Companies, Inc.

Benchmark Mini-Assessment   Chapter 1 • Lesson 3

Multiple Choice *Bubble the correct answer.*

1. Which image shows an object that is at rest? **SC.6.P.12.1**

(A)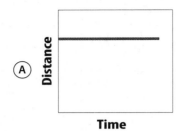

(B)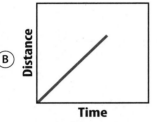

(C)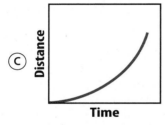

(D)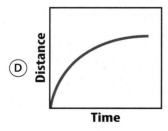

2. Which image shows an object that is speeding up? **SC.6.P.12.1**

(F)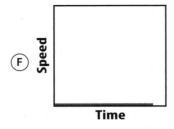

(G)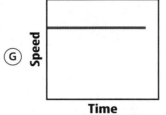

(H)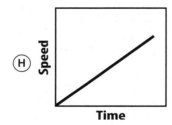

(I)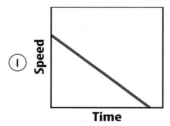

Copyright © Glencoe/McGraw-Hill, a division of The McGraw-Hill Companies, Inc.

Notes

Notes

Ball Toss

Jose tosses a ball high into the air. The ball eventually comes down so his friends can catch it. Jose and his friends have different ideas about why the ball comes back down. This is what they said:

Jose: I think it comes down *because* Earth is pulling on it.

Eddie: I think it comes down *because* it runs out of force.

Lucy: I think it comes down *because* air pressure pushes it down.

Dinah: I think it comes back down *because* no forces are acting on it.

Who do you most agree with? Explain your thinking about why the ball comes back down.

The Laws of MOTION

FLORIDA BIG IDEAS

1 **The Practice of Science**

2 **The Characteristics of Scientific Knowledge**

3 **The Role of Theories, Laws, Hypotheses, and Models**

8 **Properties of Matter**

13 **Forces and Changes in Motion**

Think About It!

How do forces change the motion of objects?

Imagine the sensations these riders experience as they swing around. The force of gravity pulls the riders downward. Instead of falling, however, they move around in circles.

1 What do you think causes the riders to move around?

_____

_____

2 What do you think prevents the riders from falling?

_____

_____

3 How do you think forces change the motion of the riders?

_____

_____

Get Ready to Read

What do you think about the Laws of Motion?

Before you read, decide if you agree or disagree with each of these statements. As you read this chapter, see if you change your mind about any of the statements.

	AGREE	DISAGREE
1 You pull on objects around you with the force of gravity.	☐	☐
2 Friction can act between two unmoving, touching surfaces.	☐	☐
3 Forces acting on an object cannot be added.	☐	☐
4 A moving object will stop if no forces act on it.	☐	☐
5 When an object's speed increases, the object accelerates.	☐	☐
6 If an object's mass increases, its acceleration also increases if the net force acting on the object stays the same.	☐	☐
7 If objects collide, the object with more mass applies more force.	☐	☐
8 Momentum is a measure of how hard it is to stop a moving object.	☐	☐

There's More Online!
Video • Audio • Review • ⓘLab Station • WebQuest • Assessment • Concepts in Motion • Multilingual eGlossary

Gravity and FRICTION

ESSENTIAL QUESTIONS

 What are some contact forces and some noncontact forces?

 What is the law of universal gravitation?

 How does friction affect the motion of two objects sliding past each other?

Vocabulary

force p. 51

contact force p. 51

noncontact force p. 52

gravity p. 53

mass p. 53

weight p. 54

friction p. 55

 Florida NGSSS

SC.6.N.1.1 Define a problem from the sixth grade curriculum, use appropriate reference materials to support scientific understanding, plan and carry out scientific investigation of various types, such as systematic observations or experiments, identify variables, collect and organize data, interpret data in charts, tables, and graphics, analyze information, make predictions, and defend conclusions.

SC.6.N.3.3 Give several examples of scientific laws.

SC.6.P.13.1 Investigate and describe types of forces including contact forces and forces acting at a distance, such as electrical, magnetic, and gravitational.

SC.6.P.13.2 Explore the Law of Gravity by recognizing that every object exerts gravitational force on every other object and that the force depends on how much mass the objects have and how far apart they are.

SC.8.P.8.2 Differentiate between weight and mass recognizing that weight is the amount of gravitational pull on an object and is distinct from, though proportional to, mass.

LA.6.2.2.3 The student will organize information to show understanding or relationships among facts, ideas, and events (e.g., representing key points within text through charting, mapping, paraphrasing, summarizing, or comparing/contrasting);

LA.6.4.2.2 The student will record information (e.g., observations, notes, lists, charts, legends) related to a topic, including visual aids to organize and record information, as appropriate, and attribute sources of information;

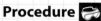 SC.6.P.13.1

(Inquiry) Launch Lab

5 minutes

Can you make a ball move without touching it?

You can make a ball move by kicking it or throwing it. Is it possible to make the ball move even when nothing is touching the ball?

Procedure

1. Read and complete a lab safety form.
2. Roll a **tennis ball** across the floor. Think about what makes the ball move.
3. Toss the ball into the air. Watch as it moves up and then falls back to your hand.
4. Drop the ball onto the floor. Let it bounce once, and then catch it.

Data and Observations

Think About This

1. What made the ball move when you rolled, tossed, and dropped it? What made it stop?

2. Did something that was touching the ball or not touching the ball cause it to move in each case?

Lesson Review 1

Visual Summary

Forces can be either contact, such as a karate chop, or noncontact, such as gravity. Each type is described by its strength and direction.

Gravity is an attractive force that acts between any two objects that have mass. The attraction is stronger for objects with greater mass.

Friction can reduce the speed of objects sliding past each other. Air resistance is a type of fluid friction that slows the speed of a falling object.

Use Vocabulary

1 **Define** *friction* in your own words.

_____

_____

2 **Distinguish** between weight and mass.

_____

_____

Understand Key Concepts 🗝

3 **Explain** the difference between a contact force and a noncontact force. SC.6.P.13.1

4 You push a book sitting on a desk with a force of 5 N, but the book does not move. What is the static friction? SC.6.P.13.3

(A) 0 N   (C) between 0 N and 5 N

(B) 5 N   (D) greater than 5 N

5 **Apply** According to the law of universal gravitation, is there a stronger gravitational force between you and Earth or an elephant and Earth? Why? SC.6.P.13.2

_____

_____

Interpret Graphics

6 **Organize Information** Fill in the table below to describe forces mentioned in this lesson. Add as many rows as you need. LA.6.2.2.3

Force	Description

Critical Thinking

7 **Decide** Is it possible for the gravitational force between two 50-kg objects to be less than the gravitational force between a 50-kg object and a 5-kg object? Explain. SC.6.P.13.1

_____

_____

_____

_____

AMERICAN
MUSEUM OF
NATURAL
HISTORY

Avoiding an Asteroid Collision

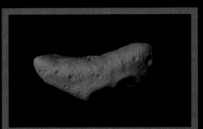

The force of gravity can change the path of an asteroid moving through the solar system.

The Spacewatch telescope in Arizona scans the sky for near-Earth asteroids. Other U.S. telescopes with this mission are in Hawaii, California, and New Mexico.

Meteor Crater in Arizona was created when an asteroid about 50 m wide collided with Earth about 50,000 years ago.

Gravity to the rescue!

Everything in the universe—from asteroids to planets to stars—exerts gravity on every other object. This force keeps the Moon in orbit around Earth and Earth in orbit around the Sun. Gravity can also send objects on a collision course—a problem when those objects are Earth and an asteroid. Asteroids are rocky bodies found mostly in the asteroid belt between Mars and Jupiter. Jupiter's strong gravity can change the orbits of asteroids over time, occasionally sending them dangerously close to Earth.

Astronomers use powerful telescopes to track asteroids near Earth. More than a thousand asteroids are large enough to cause serious damage if they collide with Earth. If an asteroid were heading our way, how could we prevent the collision? One idea is to launch a spacecraft into the asteroid. The impact could slow it down enough to cause it to miss Earth. But if the asteroid broke apart, the pieces could rain down onto Earth!

Scientists have another idea. They propose launching a massive spacecraft into an orbit close to the asteroid. The spacecraft's gravity would exert a small tug on the asteroid. Over time, the asteroid's path would be altered enough to pass by Earth. Astronomers track objects now that are many years away from crossing paths with Earth. This gives them enough time to set a plan in motion if one of the objects appears to be on a collision course with Earth.

It's Your Turn

PROBLEM SOLVING With a group, come up with a plan for avoiding an asteroid's collision with Earth. Present your plan to the class. Include diagrams and details that explain exactly how your plan will work. LA.6.4.2.2

Newton's SECOND LAW

Vocabulary

Newton's second law of motion p. 71

circular motion p. 72

centripetal force p. 72

 Florida NGSSS

SC.6.N.1.1 Define a problem from the sixth grade curriculum, use appropriate reference materials to support scientific understanding, plan and carry out scientific investigation of various types, such as systematic observations or experiments, identify variables, collect and organize data, interpret data in charts, tables, and graphics, analyze information, make predictions, and defend conclusions.

SC.6.N.2.2 Explain that scientific knowledge is durable because it is open to change as new evidence or interpretations are encountered.

SC.6.N.3.3 Give several examples of scientific laws.

SC.6.P.13.3 Investigate and describe that an unbalanced force acting on an object changes its speed, or direction of motion, or both.

SC.7.N.1.4 Identify test variables (independent variables) and outcome variables (dependent variables) in an experiment.

LA.6.2.2.3 The student will organize information to show understanding or relationships among facts, ideas, and events (e.g., representing key points within text through charting, mapping, paraphrasing, summarizing, or comparing/contrasting);

MA.6.A.3.6 Construct and analyze tables, graphs, and equations to describe linear functions and other simple relations using both common language and algebraic notation.

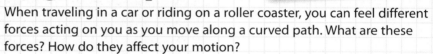

SC.6.P.13.3

 Launch Lab

10 minutes

What forces affect motion along a curved path?

When traveling in a car or riding on a roller coaster, you can feel different forces acting on you as you move along a curved path. What are these forces? How do they affect your motion?

Procedure

1. Read and complete a lab safety form.

2. Attach a piece of **string** about 1 m long to a rolled-up **sock.**

   WARNING: Find a spot away from your classmates for the next steps.

3. While holding the end of the string, swing the sock around in a circle above your head. Notice the force tugging on the string.

4. Repeat step 3 with two socks rolled together. Compare the force of swinging one sock to the force of swinging two socks.

Think About This

1. Describe the forces acting along the string while you were swinging it. Classify each force as balanced or unbalanced.

2. **Key Concept** How does the force from the string seem to affect the sock's motion

1. The archer pulls back the string and takes aim. When she releases the string, the arrow soars through the air. To reach the target, the arrow must quickly reach a high speed. How do you think the arrow is able to move so fast?

_____

_____

_____

_____

_____

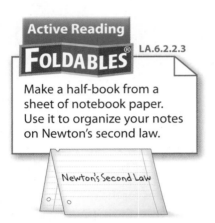

Active Reading

FOLDABLES® LA.6.2.2.3

Make a half-book from a sheet of notebook paper. Use it to organize your notes on Newton's second law.

Newton's Second Law

How do forces change motion?

Picture yourself inline skating along the Pinellas Trail in western Florida. Think about the different ways forces can change your motion. The force applied by your feet to the skates causes the wheels of the skates to turn faster and your speed to increase. When you stop pushing against the skates, your speed slowly decreases because of friction between the skates and the pavement.

Now, suppose a woman pushes a wheelbarrow across a yard. She changes its speed by pushing with more or less force. She changes its direction by pushing the wheelbarrow in the direction she wants to move. Forces change an object's motion by changing its speed, its direction, or both its speed and its direction.

Unbalanced Forces and Velocity

Velocity is speed in a certain direction. Only unbalanced forces change an object's velocity. A bicycle's speed will not increase unless the forces of the person's feet on the pedals is greater than the friction that slows the wheels. A skater's speed will not decrease if the skater pushes back against the ice with a force greater than the friction against the skates. If someone pushes the wheelbarrow with the same force but in the opposite direction that you are pushing, the wheelbarrow's direction will not change.

In the previous lesson, you read about Newton's first law of motion—balanced forces do not change an object's velocity. In this lesson you will read about how unbalanced forces affect the velocity of an object.

Active Reading

2. List Identify two characteristics of motion that can be changed by forces.

_____

_____

Unbalanced Forces on an Object at Rest

An example of how unbalanced forces affect an object at rest is shown in **Figure 16.** At first the ball is not moving. The hand holds the ball up against the downward pull of gravity. Because the forces on the ball are balanced, the ball remains at rest. When the hand moves out of the way, the ball falls downward. You know that the forces on the ball are now unbalanced because the ball's motion changed. The ball moves in the direction of the net force. When unbalanced forces act on an object at rest, the object begins moving in the direction of the net force.

Unbalanced Forces on an Object in Motion

Unbalanced forces change the velocity of a moving object. Recall that one way to change an object's velocity is to change its speed. To understand this, you only need to visit a Florida water park.

Speeding Up At the top of the water slide in **Figure 17,** gravity causes a downward net force on the person as he starts down the slide. Because the direction of the net force is in the direction of the motion, the rider's speed increases.

Slowing Down At the bottom of the water slide in **Figure 17,** friction causes a net force on the rider, opposite to the direction of his motion. Because the net force is in the opposite direction of the motion, the rider's speed decreases until he comes to a stop.

Active Reading **3. Describe** What happens to the speed of a wagon rolling to the right if a net force to the right acts on it?

_____

_____

_____

Balanced Forces

Force exerted by hand

Force due to gravity

Unbalanced Forces

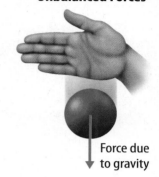

Force due to gravity

Figure 16 When unbalanced forces act on a ball at rest, it moves in the direction of the net force.

Speeding up

Net force

Velocity

Slowing down

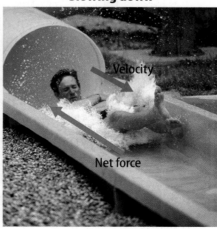

Velocity

Net force

Figure 17 Unbalanced forces can cause an object to speed up or slow down.

Active Reading **4. Predict** How would the net force and velocity arrows in the right photo change if the man grabbed the sides of the slide?

_____

_____

_____

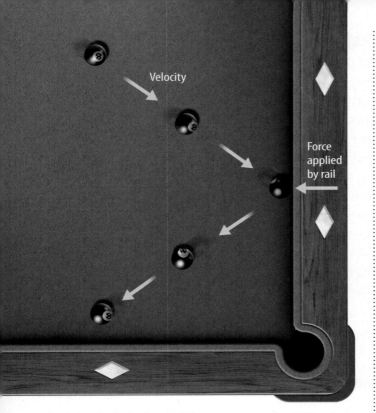

Velocity

Force applied by rail

Figure 18 Unbalanced forces act on the billiard ball, causing its direction to change.

SC.6.P.13.3, SC.6.N.1.1

Inquiry

iLAB STATION **Try It!**

MiniLab *How are force and mass related?* at connectED.mcgraw-hill.com

Apply It!

After you complete the lab, answer the questions below.

1. **Support** Justify the following statement: An object can accelerate in two ways at the same time.

2. **Apply** Think about how you traveled from home to school today. Make a list of all the times during your trip that unbalanced forces act on your body.

Changes in Direction of Motion

Another way that unbalanced forces can change an object's velocity is to change its direction. The ball in **Figure 18** moved at a constant velocity until it hit the rail of the billiard table. The force applied by the rail changed the ball's direction. Likewise, unbalanced forces change the direction of Earth's crust. Recall that the crust is broken into moving pieces called plates. The direction of one plate changes when another plate pushes against it with an unbalanced force.

 Active Reading **5. Identify** List the results of unbalanced forces on the velocity of an object. Give an example of each.

Result of Unbalanced Forces	Example
A.	
B.	
C.	

Unbalanced Forces and Acceleration

You have read how unbalanced forces can change an object's velocity by changing its speed, its direction, or both its speed and its direction. Another name for a change in velocity over time is acceleration. When the man in **Figure 17** began to move down the slide he accelerated because his speed changed. When the billiard ball in **Figure 18** hit the side of the table, the ball accelerated because its direction changed. Unbalanced forces can make an object accelerate by changing its speed, its direction, or both.

Newton's Second Law of Motion

Isaac Newton also described the relationship between an object's acceleration (change in velocity over time) and the net force that acts on an object. *According to* **Newton's second law of motion,** *the acceleration of an object is equal to the net force acting on the object divided by the object's mass.* The direction of acceleration is the same as the direction of the net force.

Notice that the equation for Newton's second law has SI units. Acceleration is expressed in meters per second squared (m/s^2), mass in kilograms (kg), and force in newtons (N). From this equation, it follows that a newton is the same as $kg{\cdot}m/s^2$.

 6. NGSSS Check Explain What is Newton's second law of motion? SC.6.P.13.3

_____

_____

_____

Newton's Second Law Equation

$$\text{acceleration (in } m/s^2) = \frac{\text{net force (in N)}}{\text{mass (in kg)}}$$

$$a = \frac{F}{m}$$

Math Skills   MA.6.A.3.6

Solve for Acceleration For a sudden one-hundredth of a second, a volleyball player strikes a volleyball during her serve. Her fist applies a force of 54 N to the 0.27 kg ball. What is the acceleration of the ball?

1. **This is what you know:**   mass: $m = 0.27$ kg

    force: $F = 54$ N or 54 $kg{\cdot}m/s^2$

2. **This is what you need to find:**   acceleration: a

3. **Use this formula:** $a = \dfrac{F}{m}$

4. **Substitute:** $a = \dfrac{54\text{ N}}{0.27\text{ kg}} = 200\ \dfrac{kg{\cdot}m/s^2}{kg} = 200\ m/s^2$

    the values for **F** and **m**

    into the formula and divide.

Answer: The acceleration of the ball is 200 m/s^2.

Practice

7. A 24-N net force acts on a 8-kg rock. What is the acceleration of the rock?

8. A 30-N net force on a skater produces an acceleration of 0.6 m/s^2. What is the mass of the skater?

9. What net force is acting on a 14-kg wagon that produces an acceleration of 1.5 m/s^2?

Circular Motion

Newton's second law of motion describes the relationship between an object's change in velocity over time, or acceleration, and unbalanced forces acting on the object. You already read how this relationship applies to motion along a line. It also applies to circular motion. **Circular motion** *is any motion in which an object is moving along a curved path.*

Centripetal Force

The ball in **Figure 19** is in circular motion. The velocity arrows show that the ball has a tendency to move along a straight path. Inertia—not a force—causes this motion. The ball's path is curved because the string pulls the ball inward. *In circular motion, a force that acts perpendicular to the direction of motion, toward the center of the curve, is* **centripetal** (sen TRIH puh tuhl) **force.** The figure also shows that the ball accelerates in the direction of the centripetal force.

 10. NGSSS Check Analyze How does centripetal force affect circular motion? SC.6.P.13.3

_____

_____

_____

The Motion of Satellites and Planets

Another object that experiences centripetal force is a satellite. A satellite is any object in space that orbits a larger object. Like the ball in **Figure 19,** a satellite tends to move along a straight path because of inertia. But just as the string pulls the ball inward, gravity pulls a satellite inward. Gravity is the centripetal force that keeps a satellite in orbit by changing its direction. The Moon is a satellite of Earth. As shown in **Figure 19,** Earth's gravity changes the Moon's direction. Similarly, the Sun's gravity changes the direction of its satellites, including Earth.

WORD ORIGIN

centripetal
from latin *centripetus,* means "toward the center"

Figure 19 Inertia of the moving object and the centripetal force acting on the object produce the circular motion of the ball and the Moon.

Active Reading **11. Describe** How does the direction of the velocity of a satellite differ from the direction of its acceleration?

_____

_____

Circular Motion 🔑

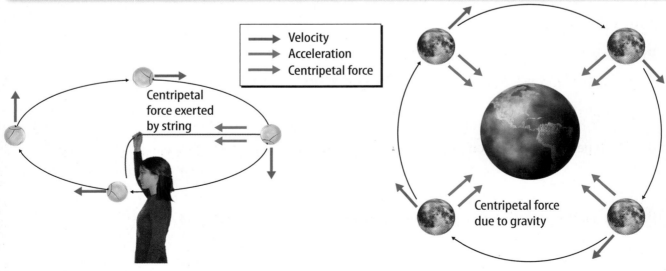

→ Velocity
→ Acceleration
→ Centripetal force

Centripetal force exerted by string

Centripetal force due to gravity

Unbalanced forces cause an object to speed up, slow down, or change direction.

Newton's second law of motion relates an object's acceleration to its mass and the net force on the object.

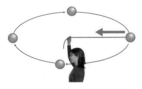

Any motion in which an object is moving along a curved path is circular motion.

Use Vocabulary

1 Explain Newton's second law of motion in your own words.

_____

_____

2 Use the term *circular motion* in a sentence.

_____

_____

Understand Key Concepts 🔑

3 A cat pushes a 0.25-kg toy with a net force of 8 N. According to Newton's second law what is the acceleration of the ball?

Ⓐ 2 m/s²   Ⓒ 16 m/s²

Ⓑ 4 m/s²   Ⓓ 32 m/s²

4 Describe how centripetal force affects circular motion. SC.6.P.13.3

_____

_____

Interpret Graphics

5 Complete each equation according to Newton's second law. LA.6.2.2.3

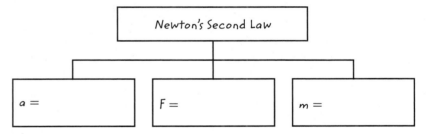

Critical Thinking

6 Design You need to lift an object that weighs 45 N straight up. Draw an illustration that explains the strength and direction of the force you must apply to lift the object. SC.6.P.13.3

Inquiry SC.6.P.13.3, SC.6.N.1.1, SC.7.N.1.4

🔬LAB STATION Try It!

Skill Lab *How does a change in mass or force affect acceleration?* at connectED.mcgraw-hill.com

Math Skills   MA.6.A.3.6

7 The force of Earth's gravity is about 10 N downward. What is the acceleration of a 15-kg backpack if you lift it with a force of 15 N?

Connect Key Concepts 🔑

Create and Review After reading Lesson 4, come back to this page and create your own study guide on Newton's laws of motion using the definition of each law and providing an example of the law in action.

Newton's Laws of Motion

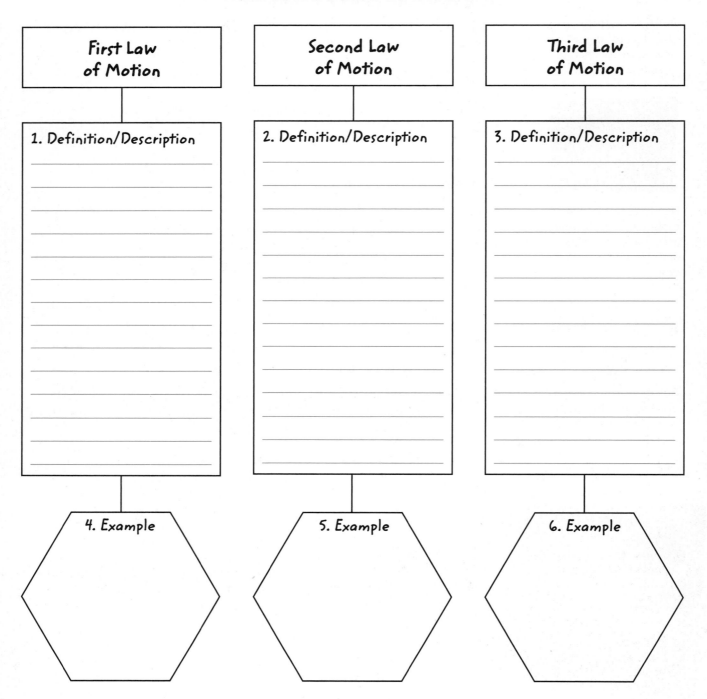

First Law of Motion

1. Definition/Description

Second Law of Motion

2. Definition/Description

Third Law of Motion

3. Definition/Description

4. Example

5. Example

6. Example

Newton's
THIRD LAW

ESSENTIAL QUESTIONS

 What is Newton's third law of motion?

 Why don't the forces in a force pair cancel each other?

 What is the law of conservation of momentum?

Vocabulary

Newton's third law of motion p. 77

force pair p. 77

momentum p. 79

law of conservation of momentum p. 80

 Florida NGSSS

SC.6.N.1.1 Define a problem from the sixth grade curriculum, use appropriate reference materials to support scientific understanding, plan and carry out scientific investigation of various types, such as systematic observations or experiments, identify variables, collect and organize data, interpret data in charts, tables, and graphics, analyze information, make predictions, and defend conclusions.

SC.6.N.1.4 Discuss, compare, and negotiate methods used, results obtained, and explanations among groups of students conducting the same investigation.

SC.6.N.2.2 Explain that scientific knowledge is durable because it is open to change as new evidence or interpretations are encountered.

SC.6.N.3.3 Give several examples of scientific laws.

SC.6.N.3.4 Identify the role of models in the context of the sixth grade science benchmarks.

SC.6.P.13.1 Investigate and describe types of forces including contact forces and forces acting at a distance, such as electrical, magnetic, and gravitational.

SC.6.P.13.3 Investigate and describe that an unbalanced force acting on an object changes its speed, or direction of motion, or both.

SC.7.N.1.3 Distinguish between an experiment (which must involve the identification and control of variables) and other forms of scientific investigation and explain that not all scientific knowledge is derived from experimentation.

Also covers: SC.7.N.1.4, SC.7.N.3.2, SC.8.N.1.6, LA.6.2.2.3

SC.6.P.13.3

Inquiry Launch Lab
10 minutes

How do opposite forces compare?

If you think about forces you encounter every day, you might notice forces that occur in pairs. For example, if you drop a rubber ball, the falling ball pushes against the floor. The ball bounces because the floor pushes with an opposite force against the ball. How do these opposite forces compare?

Procedure 🥽

1. Read and complete a lab safety form.

2. Stand so that you face your lab partner, about half a meter away. Each of you should hold a **spring scale.**

3. Hook the two scales together, and gently pull them away from each other. Notice the force reading on each scale.

4. Pull harder on the scales, and again notice the force readings on the scales.

5. Continue to pull on both scales, but let the scales slowly move toward your lab partner and then toward you at a constant speed.

Think About This

1. Identify the directions of the forces on each scale. Record this information.

2. 🔑 **Key Concept** Describe the relationship you noticed between the force readings on the two scales.

1. To reach the height she needs for her dive, this diver must move up into the air. Does she just jump up? No, she doesn't. She pushes down on the diving board and the diving board propels her into the air. How do you think pushing down causes the diver to move up?

_____

_____

_____

_____

Opposite Forces

Have you ever been on in-line skates and pushed against a wall? When you pushed against the wall, like the boy is doing in **Figure 20,** you started moving away from it. What force caused you to move?

You might think the force of the muscles in your hands moved you away from the wall. But think about the direction of your push. You pushed against the wall in the opposite direction from your movement. It might be hard to imagine, but when you pushed against the wall, the wall pushed back in the opposite direction. The push of the wall caused you to accelerate away from the wall. When an object applies a force on another object, the second object applies a force of the same strength on the first object, but the force is in the opposite direction.

Figure 20 When the skater pushes against the wall, the wall applies a force to the skater that pushes him away from the wall.

Active Reading

2. Illustrate When you are standing, you push on the floor, and the floor pushes on you. Draw a picture of you standing on the floor. Be sure to include the direction and strength of these forces.

Newton's Third Law of Motion

Newton's first two laws of motion describe the effects of balanced and unbalanced forces on one object. Newton's third law relates forces between two objects. *According to **Newton's third law of motion**, when one object exerts a force on a second object, the second object exerts an equal force in the opposite direction on the first object.* An example of forces described by Newton's third law is shown in **Figure 21.** When the gymnast pushes against the vault, the vault pushes back against the gymnast. Notice that the lengths of the force arrows are the same, but the directions are opposite.

 3. NGSSS Check Review Underline Newton's third law of motion. SC.6.P.13.3

Force Pairs

The forces described by Newton's third law depend on each other. *A **force pair** is the forces two objects apply to each other.* Recall that you can add forces to calculate the net force. If the forces of a force pair always act in opposite directions and are always the same strength, why don't they cancel each other? The answer is that each force acts on a different object. In **Figure 22,** the girl's feet act on the boat. The force of the boat acts on the girl's feet. The forces do not result in a net force of zero because they act on different objects. Adding forces can only result in a net force of zero if the forces act on the same object.

Action and Reaction

In a force pair, one force is called the action force. The other force is called the reaction force. The girl in **Figure 22** applies an action force against the boat. The reaction force is the force that the boat applies to the girl. For every action force, there is a reaction force that is equal in strength but opposite in direction.

Newton's Third Law

Force exerted on the vault

Force exerted on the gymnast

Figure 21 The force of the vault propels the gymnast upward.

Force Pairs

Action force exerted by the girl on the boat

Reaction force exerted by the boat on the girl

Figure 22 The force pair is the force the girl applies to the boat and the force that the boat applies to the girl.

Active Reading **4. Describe** How can you tell that the forces don't cancel each other?

_____

_____

_____

_____

Active Reading

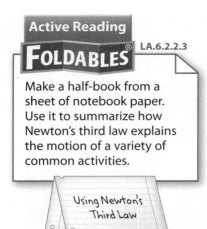

FOLDABLES® LA.6.2.2.3

Make a half-book from a sheet of notebook paper. Use it to summarize how Newton's third law explains the motion of a variety of common activities.

> Using Newton's
> Third Law

Using Newton's Third Law of Motion

When you push against an object, the force you apply is called the action force. The object then pushes back against you. The force applied by the object is called the reaction force. According to Newton's second law of motion, when the reaction force results in an unbalanced force, there is a net force, and the object accelerates. As shown in **Figure 23,** Newton's third law explains how you can swim and jump. It also explains how rockets can be launched into space.

Active Reading 5. **Explain** How does Newton's third law apply to the motion of a bouncing ball?

_____

_____

Action and Reaction Forces 🔑

Figure 23 Every action force has a reaction force in the opposite direction.

◀ **Swimming** When you swim, you push your arms against the water in the pool. The water in the pool pushes back on you in the opposite (forward) direction. If your arms push the water back with enough force, the reaction force of the water on your body is greater than the force of fluid friction. The net force is forward. You accelerate in the direction of the net force and swim forward through the water.

Active Reading 6. **Model** Show Newton's third law of motion in a drawing. Label the terms action force, force, and force pair.

◀ **Rocket Motion** NASA first launched a rocket from Cape Canaveral, FL in 1950. The burning fuel in a rocket engine produces a hot gas. The engine pushes the hot gas out in a downward direction. The gas pushes upward on the engine. When the upward force of the gas pushing on the engine becomes greater than the downward force of gravity on the rocket, the net force is in the upward direction. The rocket then accelerates upward, lifting off the ground.

Momentum

Because action and reaction forces do not cancel each other, they can change the motion of objects. A useful way to describe changes in velocity is by describing momentum. **Momentum** *is a measure of how hard it is to stop a moving object.* It is the product of an object's mass and velocity. An object's momentum is in the same direction as its velocity.

WORD ORIGIN

momentum
from Latin *momentum,* means "movement, impulse"

Momentum Equation

momentum (in kg·m/s) = mass (in kg) × velocity (in m/s)

$$p = m \times v$$

If a large truck and a car move at the same speed, the truck is harder to stop. Because it has more mass, it has more momentum. If cars of equal mass move at different speeds, the faster car has more momentum and is harder to stop.

Newton's first two laws relate to momentum. According to Newton's first law, if the net force on an object is zero, its velocity does not change. This means its momentum does not change. Newton's second law states that the net force on an object is the product of its mass and its change in velocity. Because momentum is the product of mass and velocity, the force on an object equals its change in momentum.

Math Skills   MA.6.A.3.6

Solve for Momentum The large, gentle, slow-moving West Indian manatee can be found concentrated along the Florida coast in winter. An individual manatee can grow to over 3 m long. What is the momentum of a 500-kg manatee foraging for seagrass as it swims along at a speed of 1.5 m/s?

1. **This is what you know:**   mass: $m = 500$ kg
   velocity: $v = 1.5$ m/s

2. **This is what you need to find:**   momentum: p

3. **Use this formula:** $p = m \times v$

4. **Substitute:** $p = 500 \text{ kg} \times 1.5 \text{ m/s} = 750 \text{ kg·m/s}$
   the values for *m* and *v* into the formula and multiply.

Answer: The momentum of the manatee is 750 kg·m/s in the direction of the velocity.

Practice

7. What is the momentum of a 7.25-kg bowling ball rolling at 3.0 m/s?

8. A 55-kg woman has a momentum of 220 kg·m/s. What is her velocity?

Figure 24 The total momentum of all the balls is the same before and after the collision.

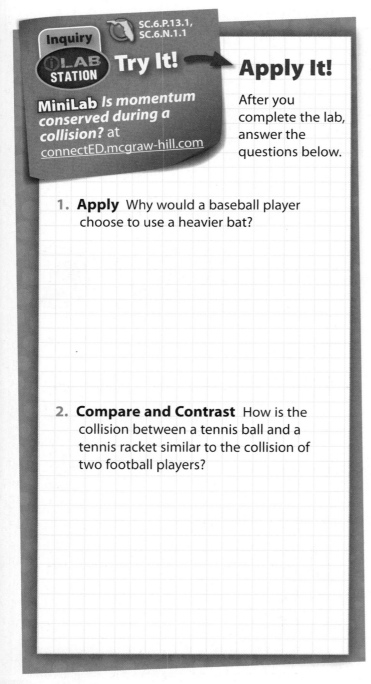

Inquiry

iLAB STATION SC.6.P.13.1, SC.6.N.1.1

Try It!

MiniLab *Is momentum conserved during a collision?* at connectED.mcgraw-hill.com

Apply It!

After you complete the lab, answer the questions below.

1. **Apply** Why would a baseball player choose to use a heavier bat?

2. **Compare and Contrast** How is the collision between a tennis ball and a tennis racket similar to the collision of two football players?

Conservation of Momentum

You might have noticed that if a moving ball hits another ball that is not moving, the motion of both balls changes. The cue ball in **Figure 24** has momentum because it has mass and is moving. When it hits the other balls, the cue ball's velocity and momentum decrease. Now the other balls start moving. Because these balls have mass and velocity, they also have momentum.

The Law of Conservation of Momentum

In any collision, one object transfers momentum to another object. The billiard balls in **Figure 24** gain the momentum lost by the cue ball. The total momentum, however, does not change. *According to the* **law of conservation of momentum,** *the total momentum of a group of objects stays the same unless outside forces act on the objects.* Outside forces include friction. Friction between the balls and the billiard table decreases their velocities, and they lose momentum.

9. **NGSSS Check** Review Highlight the law of conservation of momentum. SC.6.P.13.3

Types of Collisions

Objects collide with each other in different ways. When colliding objects bounce off each other, it is an elastic collision. If objects collide and stick together, such as when one football player tackles another, the collision is inelastic. No matter the type of collision, the total momentum will be the same before and after the collision.

Active Reading 10. **Contrast** How do elastic and inelastic collisions differ?

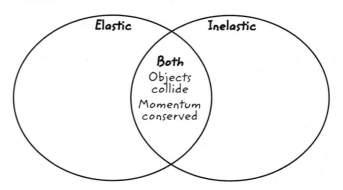

Elastic   Inelastic

Both
Objects collide
Momentum conserved

Newton's third law of motion describes the force pair between two objects.

For every action force, there is a reaction force that is equal in strength but opposite in direction.

In any collision, momentum is transferred from one object to another.

SC.6.P.13.3, SC.6.N.1.1,
SC.6.N.1.4, SC.8.N.1.6,
SC.6.N.3.4, SC.7.N.1.3,
SC.7.N.1.4, SC.7.N.3.2

Inquiry

LAB STATION Try It!

Inquiry Lab *Modeling Newton's Laws of Motion* at connectED.mcgraw-hill.com

Use Vocabulary

1 **Define** *momentum* in your own words.

_____

_____

2 The force of a bat on a ball and the force of a ball on a bat are a(n)

_____ .

Understand Key Concepts

3 **State** Newton's third law of motion. SC.6.N.3.3

_____

_____

4 A ball with momentum 16 kg·m/s strikes a ball at rest. What is the total momentum of both balls after the collision?

- (A) −16 kg·m/s
- (C) 8 kg·m/s
- (B) −8 kg·m/s
- (D) 16 kg·m/s

5 **Identify** A child jumps on a trampoline. The trampoline bounces her up. Why don't the forces cancel each other? SC.6.P.13.3

_____

_____

Interpret Graphics

6 **Organize** Fill in the table. LA.6.2.2.3

Event	Action Force	Reaction Force
A girl kicks a soccer ball.		
A book sits on a table.		

Critical Thinking

7 **Decide** How is it possible for a bicycle to have more momentum than a truck?

_____

_____

_____

Math Skills   MA.6.A.3.6

8 A 2.0-kg ball rolls to the right at 3.0 m/s. A 4.0-kg ball rolls to the left at 2.0 m/s. What is the momentum of the system after a head-on collision of the two balls?

Chapter 2   Study Guide

 Think About It! Contact forces, such as friction, and noncontact forces, such as gravity, act on objects to change their speed, direction of motion, or both.

🔑 Key Concepts Summary

Vocabulary

LESSON 1 Gravity and Friction

- Friction is a **contact force**. Magnetism is a **noncontact force.**
- The law of universal gravitation states that all objects are attracted to each other by **gravity.**
- **Friction** can stop or slow down objects sliding past each other.

force p. 51
contact force p. 51
noncontact force p. 52
gravity p. 53
mass p. 53
weight p. 54
friction p. 55

LESSON 2 Newton's First Law

- An object's motion can only be changed by **unbalanced forces.**
- According to **Newton's first law of motion,** the motion of an object is not changed by **balanced forces** acting on it.
- **Inertia** is the tendency of an object to resist a change in its motion.

net force p. 61
balanced forces p. 62
unbalanced forces p. 62
Newton's first law of motion p. 63
inertia p. 64

LESSON 3 Newton's Second Law

- According to **Newton's second law of motion,** an object's acceleration is the net force on the object divided by its mass.
- In **circular motion,** a **centripetal force** pulls an object toward the center of the curve.

Newton's second law of motion p. 71
circular motion p. 72
centripetal force p. 72

LESSON 4 Newton's Third Law

- **Newton's third law of motion** states that when one object applies a force on another, the second object applies an equal force in the opposite direction on the first object.
- The forces of a **force pair** do not cancel because they act on different objects.
- According to the **law of conservation of momentum,** momentum is conserved during a collision unless an outside force acts on the colliding objects.

Newton's third law of motion p. 77
force pair p. 77
momentum p. 79
law of conservation of momentum p. 80

Active Reading

FOLDABLES® Chapter Project

Assemble your lesson Foldables as shown to make a Chapter Project. Use the project to review what you have learned in this chapter.

Use Vocabulary

1 Kilograms is the SI unit for measuring

_____ .

2 The force of gravity on an object is its

_____ .

3 The sum of all the forces on an object is the

_____ .

4 An object that has _____ acting on it acts as if there were no forces acting on it at all.

5 A car races around a circular track. Friction on the tires is the _____ that acts toward the center of the circle and keeps the car on the circular path.

6 A heavy train requires nearly a mile to come to a complete stop because it has a lot of

_____ .

Link Vocabulary and Key Concepts

Use vocabulary terms from the previous page to complete the concept map.

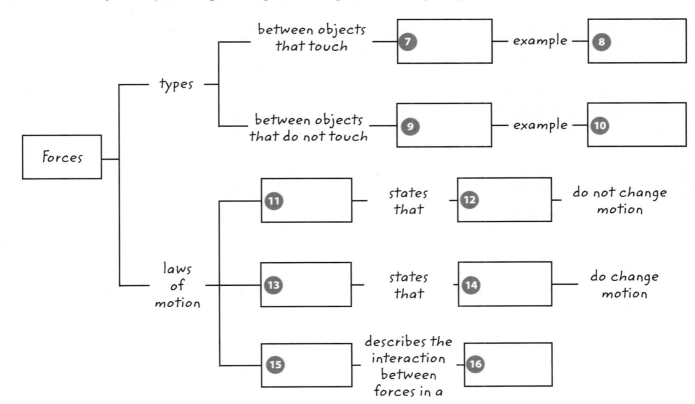

Fill in the correct answer choice.

🔑 Understand Key Concepts

1 The arrows in the figure represent the gravitational force between marbles that have equal mass. SC.6.P.13.2

How should the force arrows look if a marble that has greater mass replaces one of these marbles?
- (A) Both arrows should be drawn longer.
- (B) Both arrows should stay the same length.
- (C) The arrow from the marble with less mass should be longer than the other arrow.
- (D) The arrow from the marble with less mass should be shorter than the other arrow.

2 A person pushes a box across a flat surface with a force less than the weight of the box. Which force is weakest? SC.6.P.13.1
- (A) the force of gravity on the box
- (B) the force of the table on the box
- (C) the applied force against the box
- (D) the sliding friction against the box

3 A train moves at a constant speed on a straight track. Which statement is true? SC.6.P.13.1
- (A) No horizontal forces act on the train as it moves.
- (B) The train moves only because of its inertia.
- (C) The forces of the train's engine balances friction.
- (D) An unbalanced force keeps the train moving.

4 The Moon orbits Earth in a nearly circular orbit. What is the centripetal force? SC.6.P.13.3
- (A) the pull of the Moon on Earth
- (B) the outward force on the Moon
- (C) the Moon's inertia as it orbits Earth
- (D) Earth's gravitational pull on the Moon

Critical Thinking

5 **Predict** If an astronaut moved away from Earth in the direction of the Moon, how would the gravitational force between Earth and the astronaut change? How would the gravitational force between the Moon and the astronaut change? SC.6.P.13.2

_____

_____

_____

_____

_____

6 **Analyze** A box is on a table. Two people push on the box from opposite sides. Which of the labeled forces make up a force pair? Explain your answer. SC.6.P.13.1

_____

_____

_____

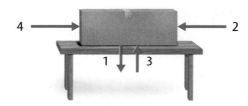

7 **Conclude** A refrigerator has a maximum static friction force of 250 N. Sam can push the refrigerator with a force of 130 N. Amir and André can each push with a force of 65 N. How could they all move the refrigerator? Will the refrigerator move with constant velocity? Why or why not? SC.6.P.13.1

_____

_____

_____

_____

_____

_____

8 **Give an example** of unbalanced forces acting on an object. SC.6.P.13.3

_____

_____

9 **Infer** Two skaters stand on ice. One weighs 250 N, and the other weighs 500 N. They push against each other and move in opposite directions. Which one will travel farther before stopping? Explain your answer. SC.6.P.13.3

_____

_____

_____

Writing in Science

10 Imagine that you are an auto designer. Your job is to design brakes for different automobiles. On a separate sheet of paper, write a four-sentence plan that explains what you need to consider about momentum when designing brakes for a heavy truck, a light truck, a small car, and a van. LA.6.2.2.3

Big Idea Review

11 Explain how balanced and unbalanced forces affect objects that are not moving and those that are moving. SC.6.P.13.3

_____

_____

_____

_____

12 The photo on page 48 shows people on a carnival swing ride. How do forces change the motion of the riders? SC.6.P.13.1

_____

_____

_____

_____

Math Skills   MA.6.A.3.6

Solve One-Step Equations

13 A net force of 17 N is applied to an object, giving it an acceleration of 2.5 m/s^2. What is the mass of the object?

14 A tennis ball's mass is about 0.60 kg. Its velocity is 2.5 m/s. What is the momentum of the ball?

15 A box with a mass of 0.82 kg has these forces acting on it.

9.5 N   6.2 N

8.0 N   8.0 N

What is the strength and direction of the acceleration of the box?

Florida NGSSS · Benchmark Practice

Fill in the correct answer choice.

Multiple Choice

1 A baseball has an approximate mass of 0.15 kg. If a bat strikes the baseball with a force of 6 N, what is the acceleration of the ball? SC.6.P.13.3

Ⓐ 4 m/s^2

Ⓑ 6 m/s^2

Ⓒ 40 m/s^2

Ⓓ 60 m/s^2

Use the diagram below to answer question 2.

2 The person in the diagram above is unable to move the crate from its position. Which is the opposing force? SC.6.P.13.3

Ⓕ gravity

Ⓖ normal force

Ⓗ sliding friction

Ⓘ static friction

3 The mass of a person on Earth is 72 kg. What is the mass of the same person on the Moon where gravity is one-sixth that of Earth?
SC.6.P.13.2

Ⓐ 12 kg

Ⓑ 60 kg

Ⓒ 72 kg

Ⓓ 432 kg

4 A swimmer pushing off from the wall of a pool exerts a force of 1 N on the wall. What is the reaction force of the wall on the swimmer? SC.6.P.13.3

Ⓕ 0 N

Ⓖ 1 N

Ⓗ 2 N

Ⓘ 10 N

Use the diagram below to answer questions 5 and 6.

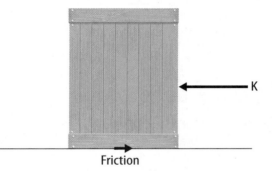

Friction

5 Which term applies to the forces in the diagram above? SC.6.P.13.1

Ⓐ negative

Ⓑ positive

Ⓒ reference

Ⓓ unbalanced

6 In the diagram above, what happens when force K is applied to the crate at rest? SC.6.P.13.3

Ⓕ The crate remains at rest.

Ⓖ The crate moves back and forth.

Ⓗ The crate moves to the left.

Ⓘ The crate moves to the right.

7 What is another term for change in velocity?
SC.6.P.13.3

Ⓐ acceleration

Ⓑ inertia

Ⓒ centripetal force

Ⓓ maximum speed

Use the diagram below to answer question 8.

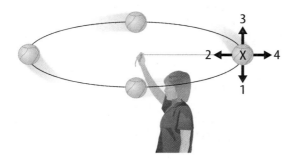

8 The person in the diagram is spinning a ball on a string. When the ball is in position X, what is the direction of the centripetal force? SC.6.P.13.3

(F) 1

(G) 2

(H) 3

(I) 4

9 Which is ALWAYS a contact force? SC.6.P.13.1

(A) electric

(B) friction

(C) gravity

(D) magnetic

10 When two billiard balls collide, which is ALWAYS conserved? SC.6.P.13.3

(F) acceleration

(G) direction

(H) force

(I) momentum

11 A 30-kg television sits on a table. The acceleration due to gravity is 10 m/s². What force does the table exert on the television? SC.6.P.13.1

(A) 0.3 N   (C) 300 N

(B) 3 N   (D) 600 N

12 Which does NOT describe a force pair? SC.6.P.13.3

(F) When you push on a bike's brakes, the friction between the tires and the road increases.

(G) When a diver jumps off a diving board, the board pushes the diver up.

(H) When an ice skater pushes off a wall, the wall pushes the skater away from the wall.

(I) When a boy pulls a toy wagon, the wagon pulls back on the boy.

13 A box on a table has these forces acting on it. SC.6.P.13.3

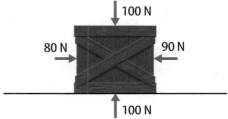

What is the static friction between the box and the table? SC.6.P.13.3

(A) 0 N

(B) 10 N

(C) greater than 10 N

(D) between 0 and 10 N

14 A 4-kg goose swims with a velocity of 1 m/s. What is its momentum? SC.6.P.13.3

(F) 4 N

(G) 4 kg·m/s²

(H) 4 kg·m/s

(I) 4 m/s²

NEED EXTRA HELP?

If You Missed Question...	1	2	3	4	5	6	7	8	9	10	11	12	13	14
Go to Lesson...	3	1	1	4	2	2	3	3	1	4	1	4	1	4

Multiple Choice *Bubble the correct answer.*

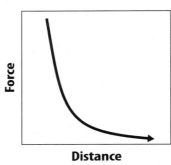

Distance

1. What relationship between gravitational force and distance is shown in the graph above? **SC.6.P.13.2**

   Ⓐ decreases; increases

   Ⓑ decreases; stays the same

   Ⓒ increases; decreases

   Ⓓ increases; stays the same

2. Which sentence below describes a contact force? **SC.6.P.13.1**

   Ⓕ A baseball bat hits a ball.

   Ⓖ A book falls to the floor.

   Ⓗ A leaf floats in the air and falls to the ground.

   Ⓘ A magnetic force pulls a paper clip toward a magnet.

3. Adam and Natalie push on a cart using 150 N of force. The cart does not move. What is the static friction on the cart? **SC.6.P.13.1**

   Ⓐ −150 N

   Ⓑ −100 N

   Ⓒ 0 N

   Ⓓ 150 N

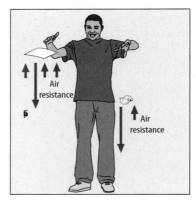

4. What does the image above tell about air resistance on a sheet of paper? **SC.6.P.13.2**

   Ⓕ Air resistance is greater on the crumpled sheet of paper.

   Ⓖ Air resistance is greater on the flat sheet of paper.

   Ⓗ Air resistance is less on the flat sheet of paper.

   Ⓘ Air resistance is the same on both sheets of paper.

Copyright © Glencoe/McGraw-Hill, a division of The McGraw-Hill Companies, Inc.

Benchmark Mini-Assessment   **Chapter 2 • Lesson 2**   mini BAT

Multiple Choice *Bubble the correct answer.*

1. The picture above shows two boys pushing on a dresser. As the boys push, the dresser will **SC.6.P.13.3**

   (A) move to the right with a net force of 100 N.

   (B) move to the right with a net force of 200 N.

   (C) move to the right with a net force of 300 N.

   (D) move to the right with a net force of 400 N.

2. When unbalanced forces are applied to a moving bike, they cause a change in the bike's **SC.6.P.13.3**

   (F) inertia.

   (G) velocity.

   (H) sliding friction.

   (I) static friction.

3. What is the tendency of an object to resist a change in its motion? **SC.6.P.13.3**

   (A) gravity

   (B) inertia

   (C) balanced forces

   (D) unbalanced forces

4. Which pair of arrows represents a pair of balanced forces acting on a box? **SC.6.P.13.3**

   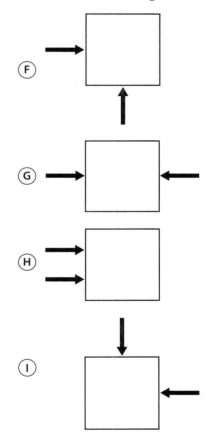

Copyright © Glencoe/McGraw-Hill, a division of The McGraw-Hill Companies, Inc.

Multiple Choice *Bubble the correct answer.*

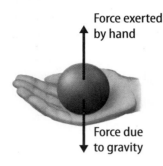

Force exerted by hand

Force due to gravity

1. What does the image above illustrate?
SC.6.P.13.3

- Ⓐ acceleration
- Ⓑ inertia
- Ⓒ balanced forces
- Ⓓ unbalanced forces

2. Which is NOT accelerating? SC.6.P.13.3

- Ⓕ a car that is traveling straight down a country road at 34 km/h
- Ⓖ a marathon runner who speeds up at the last minute to win the race
- Ⓗ a running dog that hears its owner call and suddenly stops
- Ⓘ a volleyball that is spiked over the net by a player

3. A 60 N net force is applied to a 12 kg box. What is the acceleration of the box?
SC.6.P.13.3

- Ⓐ 5 m/s^2
- Ⓑ 50 m/s^2
- Ⓒ 72 m/s^2
- Ⓓ 720 m/s^2

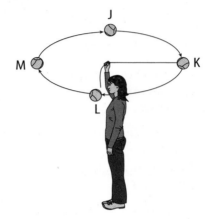

4. The diagram above shows circular motion. In this diagram, the ball at point M is accelerating SC.6.P.13.3

- Ⓕ downward.
- Ⓖ upward.
- Ⓗ to the left.
- Ⓘ to the right.

Copyright © Glencoe/McGraw-Hill, a division of The McGraw-Hill Companies, Inc.

Multiple Choice *Bubble the correct answer.*

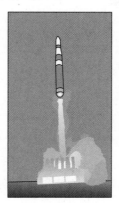

1. What is the reaction force in the image above? **SC.6.N.3.3**

(A) Air pushes down on the rocket as it blasts off.

(B) Gravity pulls down on the rocket as it blasts off.

(C) Hot gases from the rocket engine push down on the ground.

(D) Hot gases push up on the rocket engine.

2. Which describes an action force? **SC.6.P.13.3**

(F) The ground pushes up on a runner's foot during a race.

(G) A gymnast pushes off from a balance beam to perform a back flip.

(H) A table pushes up on a glass that is sitting on top of it.

(I) Water pushes up on a swimmer to keep him afloat.

3. Mai Lin has a mass of 35 kg. What is her momentum if she is running at a velocity of 3 m/s? **MA.6.A.3.6**

(A) 12 kg·m/s

(B) 38 kg·m/s

(C) 70 kg·m/s

(D) 105 kg·m/s

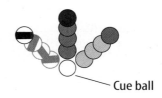
Cue ball

4. Which sentence is true about the image above? **SC.6.N.3.3**

(F) The cue ball has momentum after it hits the other balls.

(G) The cue ball transfers momentum to the other balls.

(H) The momentum of the cue ball increases as it hits the other balls.

(I) The momentum of each billiard ball equals the momentum of the cue ball.

Copyright © Glencoe/McGraw-Hill, a division of The McGraw-Hill Companies, Inc.

Name _____ Date _____

Floating Soap

Imagine you put a bar of Ivory™ soap in a small tub of water. You observe that the bar of soap floats. You wonder what will happen if you carve off a tiny piece of soap the size of a pea and put it in the water. What do you predict will happen? Circle the prediction that best matches your thinking.

A: The tiny piece of soap will float.

B: The tiny piece of soap will sink.

C: The tiny piece of soap will float for a little while, then sink.

D: The tiny piece of soap will sink partly into the water and stay there.

Explain your thinking. What reasoning did you use to make your prediction?

Forces and FLUIDS

FLORIDA BIG IDEAS

1 The Practice of Science

13 Forces and Changes in Motion

Think About It!

In what ways do people use forces in fluids?

Did you know that the helium in parade balloons is considered a fluid? Forces acting on the helium in this balloon keep it in the air. What are those forces?

1 How would you describe a fluid?

_____

_____

_____

2 How do you think forces affect fluids and the objects within them?

_____

_____

_____

3 In what ways do you think people use forces in fluids?

_____

_____

_____

Get Ready to Read

Why do you think about forces?

Before you read, decide if you agree or disagree with each of these statements. As you read this chapter, see if you change your mind about any of the statements.

		AGREE	DISAGREE
1	Air is a fluid.	☐	☐
2	Pressure is a force acting on a fluid.	☐	☐
3	You can lift a rock easily under water because there is a buoyant force on the rock.	☐	☐
4	The buoyant force on an object depends on the object's weight.	☐	☐
5	If you squeeze an unopened plastic ketchup bottle, the pressure on the ketchup changes everywhere in the bottle.	☐	☐
6	Running with an open parachute decreases the drag force on you.	☐	☐

 **Connect ED** **There's More Online!**

Video • Audio • Review • ⓘLab Station • WebQuest • Assessment • Concepts in Motion • Multilingual eGlossary

Pressure and Density of FLUIDS

ESSENTIAL QUESTIONS

 How do force and area affect pressure?

 How does pressure change with depth in the atmosphere and under water?

 What factors affect the density of a fluid?

Vocabulary

fluid p. 97

pressure p. 98

atmospheric pressure p. 100

Florida NGSSS

SC.6.P.13.1 Investigate and describe types of forces including contact forces and forces acting at a distance, such as electrical, magnetic, and gravitational.

SC.7.N.1.6 Explain that empirical evidence is the cumulative body of observations of a natural phenomenon on which scientific explanations are based.

LA.6.2.2.3 The student will organize information to show understanding or relationships among facts, ideas, and events (e.g., representing key points within text through charting, mapping, paraphrasing, summarizing, or comparing/contrasting);

LA.6.4.2.2 The student will record information (e.g., observations, notes, lists, charts, legends) related to a topic, including visual aids to organize and record information, as appropriate, and attribute sources of information;

MA.6.A.3.6 Construct and analyze tables, graphs, and equations to describe linear functions and other simple relations using both common language and algebraic notation.

(Inquiry) Launch Lab

15 minutes

What changes? What doesn't?

What happens to the volume of a liquid as you move the liquid from one container to another?

Procedure

1. Read and complete a lab safety form.
2. As you do each of the following steps, use the markings on the containers to measure the liquid's volume. Record the measurements.
3. Measure 100 mL of water in a **graduated cylinder.**
4. Pour the water into a **200-mL beaker.**
5. Pour the water into a **square plastic container.** Then pour it back into the graduated cylinder.

Data and Observations

Think About This

1. **Describe** What happened to the shape of the water as you moved it from one container to another? What happened to the volume of the water?

2. **Propose** How do you think the results would be different if you had used another liquid, such as honey? What would happen if you did this experiment with a gas?

Inquiry Why Not Straight?

1. Have you ever noticed that dams are thicker at the bottom than they are at the top? Why do you think people build dams this way?

_____

_____

_____

_____

WORD ORIGIN

fluid

from Latin _fluidus,_ means "flowing"

What is a fluid?

When people tell you to drink plenty of fluids when you play a sport, they usually are referring to water or other liquids. It may surprise you to know that gases, such as the oxygen and the nitrogen in the air, are fluids, too. _A_ **fluid** _is any substance that can flow and take the shape of the container that holds it._

Liquids

You probably have poured milk from a carton into a glass or a bowl. Milk, like other liquids, flows and takes the shape of its container. The volume of a liquid remains the same in any container. When you pour all the milk from a carton into a glass, the milk in the glass occupies the same volume it occupied in the carton. It just has a different shape.

Gases

What happens when helium gas fills a balloon? If the balloon is long and thin, then the helium atoms flow into and fill the long, thin shape. If the balloon is spherical, then the helium atoms flow into and fill the spherical shape. Like liquids, helium is a fluid. It takes the shape of its container. However, unlike liquids, helium and other gases do not occupy the same volume in different containers. A gas fills its entire container no matter the size of the container.

Active Reading **2. Organize** Fill in the graphic organizer below to summarize information about fluids.

> Characteristics of a liquid:
>
>
>
> ↑
>
> A fluid is:
>
> ↓
>
> Characteristics of a gas:

Pressure of Fluids

Pressure *is the amount of force per unit area applied to an object's surface.* All fluids, both liquids and gases, apply pressure. The air around you is applying pressure on you right now. Pressure applied on an object by a fluid is related to the weight of the fluid. Like all forces, weight is measured in newtons (N). Pressure can be calculated using the equation below. In the equation, *P* is pressure, *F* is the force applied to a surface, and *A* is the surface area over which the force is applied.

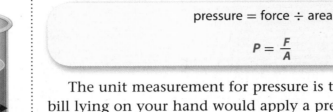

pressure = force ÷ area

$$P = \frac{F}{A}$$

The unit measurement for pressure is the pascal (Pa). A dollar bill lying on your hand would apply a pressure of about 1 Pa. A small carton of milk would apply about 1,000 Pa.

The Direction of Pressure

If the pressure applied by a fluid is related to the weight of the fluid, is the pressure only in the downward direction? No. A fluid applies pressure perpendicular to all sides of an object in contact with the fluid, as shown in **Figure 1.**

Figure 1 The pressure applied by a fluid is perpendicular to the surface of the container in contact with the fluid.

Math Skills   MA.6.A.3.6

Solve for Pressure A surfer on his surfboard applies a force of 645 N on the water. The area of the surfboard is 1.5 m². What is the pressure applied on the water?

1. **This is what you know:**   force: $F = 645$ N
                                    area: $A = 1.5$ m²

2. **This is what you need to find:**   pressure: P

3. **Use this formula:** $P = \frac{F}{A}$

4. **Substitute:** $P = \frac{645 \text{ N}}{1.5 \text{ m}^2}$
   the values for *F* and *A*
   into the formula and divide. $= 430$ Pa

Answer: The pressure is 430 Pa.

Practice

3. The water in a small, plastic swimming pool applies a force of 30,000 N over an area of 75 m². What is the pressure on the pool?

Pressure and Area

You read on the previous page that the amount of pressure on an object depends on the area over which a force is applied. If you tried to push the box of books shown in **Figure 2** with one finger, your finger would probably bend. It might even hurt. Now suppose you use your entire hand to push the box with the same force. Why is it easier?

When you push a box with one finger, the force you apply spreads over a small area—the surface area of your fingertip. The pressure is great, and your finger hurts. If you push the box with your whole hand, you apply the same force, but the force is spread over a larger area. The pressure is less, and it is easier to push the box.

Pressure decreases when the surface area over which a force is applied increases. Pressure increases when the surface area over which a force is applied decreases.

Pressure on box = 500,000 Pa

Pressure on box = 5,000 Pa

Figure 2 🔑 When a force is applied over a small area, the pressure is greater than when the same force is applied over a larger area.

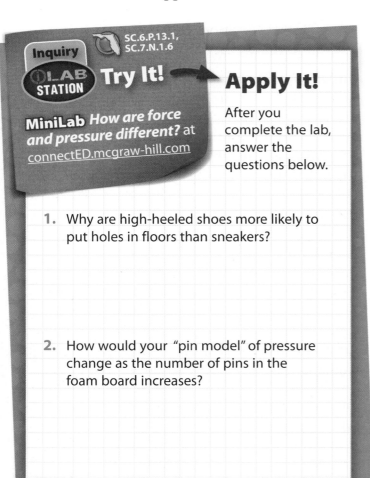

SC.6.P.13.1, SC.7.N.1.6

Inquiry

①LAB STATION **Try It!**

Apply It!

After you complete the lab, answer the questions below.

MiniLab *How are force and pressure different?* at connectED.mcgraw-hill.com

1. Why are high-heeled shoes more likely to put holes in floors than sneakers?

2. How would your "pin model" of pressure change as the number of pins in the foam board increases?

Active Reading **4. Relate** Draw a line from each box in the top row to the correct box in the bottom row to show the relationship between pressure and area. LA.6.2.2.3

Surface area increases		Surface area decreases
Pressure stays the same	Pressure increases	Pressure decreases

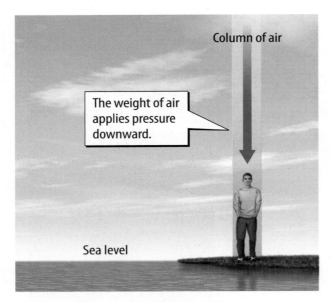

Column of air

The weight of air applies pressure downward.

Sea level

Figure 3 The weight of a column of air applies pressure downward. The greater the height above sea level, the lower the pressure.

5. **NGSSS Check** **Relate** Below is data a scuba diver recorded during a recent dive in the Atlantic Ocean. At various depths, he noted the pressure on his body. Plot the data on the graph provided. Then, answer the question below.

Depth (m)	Pressure (kPa)
0	100
10	200
30	400
60	700
70	800

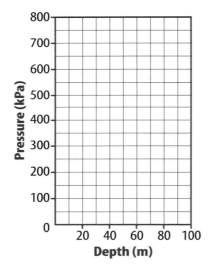

Describe the relationship between the depth of seawater and the pressure it applies. MA.6.A.3.6

_____

_____

_____

Pressure and Depth

Have you ever dived under water and felt pressure in your ears? The deeper you dive, the more pressure you feel because pressure applied by a fluid increases with depth.

Atmospheric Pressure Think of the air above you as a column that extends high into the atmosphere, as shown in **Figure 3.** The weight of the air in this column applies pressure downward. *The ratio of the weight of all the air above you to your surface area is* **atmospheric pressure.** At sea level, the atmospheric pressure on you and everything around you is about 100,000 Pa, or 100 kPa. Now imagine you are on the top of Mount Everest, nearly 9 km above sea level, as shown in **Figure 4.** Atmospheric pressure is only about 33 kPa. The column of air above you is not as tall as it was at sea level. Therefore, the weight of the air applies less pressure downward. Atmospheric pressure increases as you hike down a mountain toward lower elevation. It decreases as you hike up to higher elevation.

Underwater Pressure How does the pressure on the divers in **Figure 4** change as the divers swim deeper? Underwater pressure depends on the sum of the weight of the column of air above an object and the weight of the column of water above the object. As a diver dives farther under water, the air column above him or her stays the same. But the water column increases in height, and it weighs more. Underwater pressure increases with depth. At the deepest part of the ocean, the Mariana Trench, the pressure is 108,600 kPa—more than 1,000 times greater than the pressure at the ocean's surface. Increased pressure with depth explains why the wall of a dam is thicker at the bottom than at the top. The water behind the dam applies more pressure to the bottom of the wall.

Figure 4 On land, atmospheric pressure depends on your elevation. Under water, the water pressure depends on your depth below the water's surface.

Active Reading **6. Infer** What would be the pressure on an airplane flying at an altitude of 5,500 m?

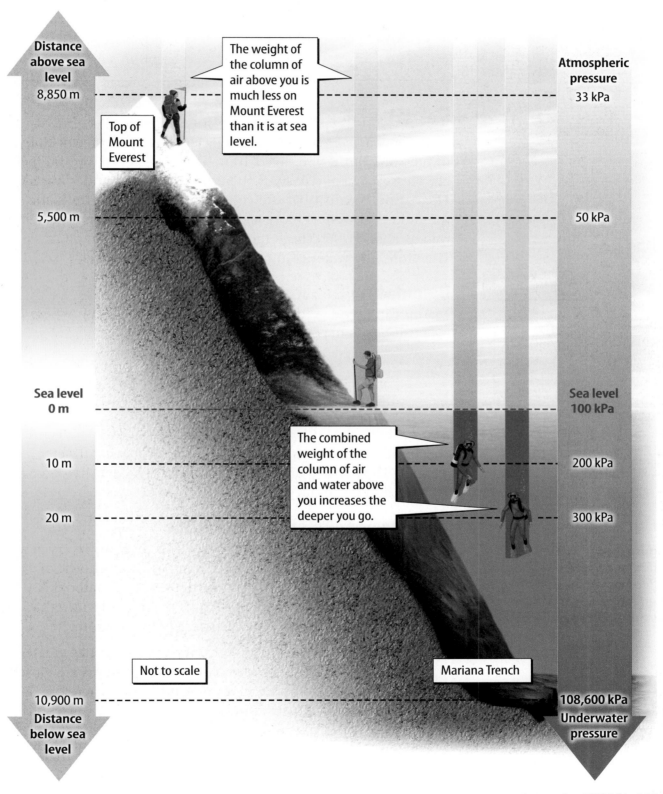

Active Reading

FOLDABLES® LA.6.2.2.3

Make a vertical two-column chart with labels as shown. Use the chart to organize your notes about pressure and density of fluids.

Pressure	Density

Active Reading **7. Characterize** Complete the graphic organizer below to organize information about pressure.

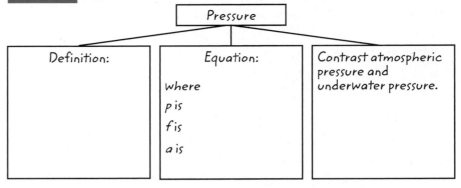

Pressure

Definition:	Equation:	Contrast atmospheric pressure and underwater pressure.
	where	
	p is	
	f is	
	a is	

Density of Fluids

If you walked 10 m down a hill, you would not feel a change in pressure. But if you dived 10 m under water, you would feel a big pressure change. Why is there more pressure under water? The reason is density. If the volume of two fluids is the same, the fluid that weighs more is denser. A 10-m column of water weighs about 1,000 times more than a 10-m column of air. Therefore, the water column is denser.

Calculating Density

You can calculate density with the equation below, where D is density, m is mass, and V is volume.

> Density = mass ÷ volume
>
> $$D = \frac{m}{V}$$

Density is often measured in grams per cubic centimeter (g/cm^3). The density of water is 1.0 g/cm^3. The density of air at 100 kPa is about 0.001 g/cm^3.

Densities of Different Materials

Have you ever noticed that vinegar and oil form layers in a bottle of salad dressing? Each layer, such as those shown in **Figure 5,** has a different density. The density of a material is determined by the masses of the atoms or the molecules that make up the material and the distances between them.

A penny is made mostly of zinc. One zinc atom has greater mass than an entire water molecule. Also, zinc atoms are closer together than the atoms are in a water molecule. Therefore, a penny is denser than water.

Molecules in most solids are closer together than molecules in liquids or gases. Therefore, solids are usually denser than liquids or gases.

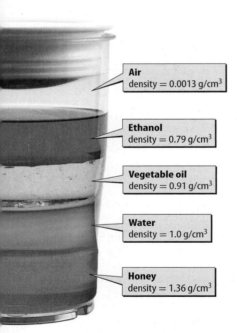

Air
density = 0.0013 g/cm³

Ethanol
density = 0.79 g/cm³

Vegetable oil
density = 0.91 g/cm³

Water
density = 1.0 g/cm³

Honey
density = 1.36 g/cm³

Figure 5 ⚷ Materials have different densities because of differences in the masses of their molecules and in the distances between them.

Active Reading **8. Name** What factors determine the density of fluids? MA.6.A.3.6

Visual Summary

Pressure is high when a force is applied over a small area.

Atmospheric pressure decreases with elevation.

Air
density = 0.0013 g/cm³

Ethanol
density = 0.79 g/cm³

Vegetable oil
density = 0.91 g/cm³

Water
density = 1.0 g/cm³

Honey
density = 1.36 g/cm³

Fluids form layers depending on their densities.

Use Vocabulary

1. **Compare** pressure and atmospheric pressure.

_____

_____

2. Gases and liquids are both _____.

Understand Key Concepts 🔑

3. Ice floats in water. Which must be true?
   - (A) Ice molecules have more mass than water molecules.
   - (B) Ice molecules have less mass than water molecules.
   - (C) A volume of ice has less mass than the same volume of water.
   - (D) One gram of ice has less volume than 1 g of water.

4. **Compare** One person walks 5 m down a hill. Another dives 5 m in a pool. Which feels a greater pressure increase? Explain.

_____

_____

5. **Draw** a beaker filled with two liquids. One liquid is blue with a density of 0.93 g/cm². The other liquid is red with a density of 1.03 g/cm². LA.6.2.2.3

Interpret Graphics

6. **Organize Information** Copy and fill in the following graphic organizer to list two factors that affect density. LA.6.2.2.3

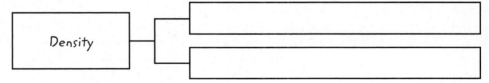

Density

Critical Thinking

7. **Propose** As you walk across a field covered with deep snow, you keep sinking. What could you do to prevent this? Explain your reasoning.

_____

_____

Math Skills

8. What pressure does 50,000 N of water apply to a 0.25-m² surface area of coral?

HOW IT WORKS

Submersibles
Deep-sea craft carry scientists to the depths of the ocean.

Exploring the deep ocean is exciting work. However, the pressure that ocean water applies increases rapidly as depth increases. This pressure limits how far humans can dive without special equipment. One way that humans can explore the deep ocean is to use a submersible.

A submersible is a special watercraft designed to function in deep water. Some submersibles are remote-controlled and do not carry people. Others, such as the *Johnson-Sea-Link* submersible shown below, can carry people. They are designed to protect passengers from pressure in the deep ocean.

A submersible is carried to the dive location on a ship. A lift is used to lower the submersible into the water. At the conclusion of the dive, the submersible is reattached to this lift and hoisted back onto the ship.

HARBOR BRANCH

A steel sphere with an acrylic dome protects the scientists inside from the extreme deep-ocean pressure.

Lights attached to the *JSL* submersible enable scientists to see in the darkness of the deep ocean.

Scientists manipulate robotic arms in order to lift and grab various objects.

This submersible, one of two *Johnson-Sea-Link* submersibles, can transport humans safely to a depth of nearly 1,000 m under water. It carries the oxygen that the pilot and the passengers will need throughout the dive time. The *JSL* submersibles were key to exploring the wreckage of the Civil War ironclad USS *Monitor* off the coast of North Carolina as well as the space shuttle *Challenger* wreckage off the coast of Florida.

It's Your Turn

REPORT Recall that atmospheric pressure decreases with increased elevation. How do airplanes protect people from the decreased air pressure when flying at high elevation? Find out and report what you learn. LA.6.4.2.2

The Buoyant FORCE

ESSENTIAL QUESTIONS

 How are pressure and the buoyant force related?

How does Archimedes' principle describe the buoyant force?

What makes an object sink or float in a fluid?

Vocabulary

buoyant force p. 106

Archimedes' principle p. 108

Florida NGSSS

SC.6.N.1.1 Define a problem from the sixth grade curriculum, use appropriate reference materials to support scientific understanding, plan and carry out scientific investigation of various types, such as systematic observations or experiments, identify variables, collect and organize data, interpret data in charts, tables, and graphics, analyze information, make predictions, and defend conclusions.

SC.6.N.2.2 Explain that scientific knowledge is durable because it is open to change as new evidence or interpretations are encountered.

SC.6.P.13.1 Investigate and describe types of forces including contact forces and forces acting at a distance, such as electrical, magnetic, and gravitational.

SC.7.N.1.6 Explain that empirical evidence is the cumulative body of observations of a natural phenomenon on which scientific explanations are based.

LA.6.2.2.3 The student will organize information to show understanding or relationships among facts, ideas, and events (e.g., representing key points within text through charting, mapping, paraphrasing, summarizing, or comparing/contrasting);

Inquiry Launch Lab

10 minutes

How can objects denser than water float on water?

Many ships are made of aluminum. Aluminum is denser than water. Why do aluminum ships float?

Procedure

1. Read and complete a lab safety form.
2. Use **scissors** to cut three squares (10 cm × 10 cm) of **aluminum foil.**
3. Form a boat shape from one square of foil. Squeeze another square into a tight ball. Fold the third square several times into a square (2 cm × 2 cm). Flatten it completely.
4. Predict whether each object will sink or float. Then, gently place each in a **tub of water.** Record your observations.

Data and Observations

Think About This

1. **Infer** What do you think caused each object to float or sink?

Inquiry Why doesn't it sink?

1. This boat is overloaded. Yet it still floats. What forces are acting on a boat sitting in water? What forces affect whether it floats or sinks?

_____

_____

_____

_____

_____

WORD ORIGIN

buoyant
from Spanish *boyar,* means "to float"

Figure 6 A buoyant force balances the weight of this person and his raft and keeps them afloat.

What is a buoyant force?

Have you ever floated on a raft, like the person in **Figure 6?** What forces are acting on this person and his raft? The gravitational force pulls them down. Recall that the gravitational force on an object is the object's weight. The weight of the person and his raft includes the weight of the column of air above them. Why don't they sink? The total weight of the raft, the person, and the air is balanced by another, upward force. *A* **buoyant** (BOY unt) **force** *is an upward force applied by a fluid on an object in the fluid.* The buoyant force on the raft is equal to the total weight of the raft, so the raft floats.

Buoyant Forces in Liquids

You might ha ve noticed that you can lift someone in water even though you could not lift that person when you are both out of water. This is because of a buoyant force. A buoyant force acts on any object in a liquid. This includes floating objects, such as rafts, ships, and bath toys. It also includes submerged objects, such as rocks at the bottom of a lake.

Buoyant Forces in Air

Objects in a gas also experience a buoyant force. For example, the buoyant force from air keeps a helium balloon up even though a gravitational force pulls downward. An object does not need to be floating in air for a buoyant force to act on it. In fact, a buoyant force from air is acting on you right now. However, the buoyant force acting on you is less than the gravitational force acting on you. Therefore, you do not float.

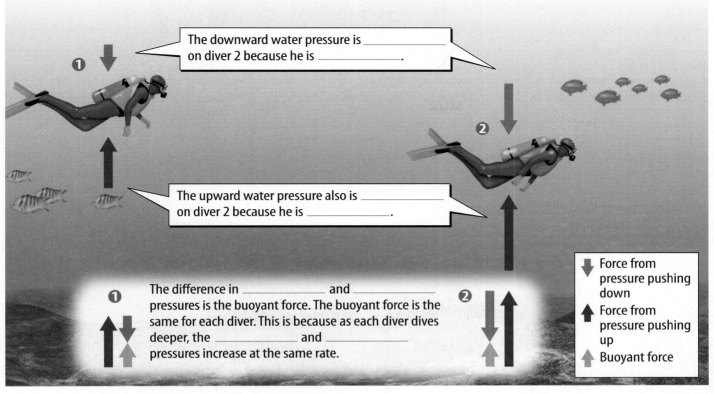

Figure 7 🔑 A diver experiences a buoyant force.

Active Reading 2. **Identify** Fill in the blanks with words from the following list: upward, increase, deeper, greater, downward. Words may be used more than once.

Buoyant Force and Pressure

You read in Lesson 1 that a fluid applies pressure perpendicular to all sides of an object within it. Therefore, the forces from water pressure on the divers in **Figure 7** are in the horizontal and vertical directions. The horizontal forces on the sides of each diver are equal. But, the vertical forces on the top and the bottom of each diver are not equal.

Recall that pressure increases with depth. The pressure at the bottom of each diver is greater than the pressure at the top of each diver. This is illustrated in **Figure 7** by the differences in length of the purple arrows. The dark purple arrow represents the pressure pushing up on each diver. The light purple arrow represents the pressure pushing down on each diver. Notice that the arrows are longer on the diver on the right. That is because this diver is deeper than the other diver. The pressure on both the top and the bottom of this diver is greater.

The difference between the upward force from pressure and the downward force from pressure on each diver is the buoyant force on each diver. The buoyant force is represented by the orange arrows in **Figure 7.** The buoyant force on an object is always in an upward direction because the pressure is always greater at an object's bottom.

Buoyant Force and Depth

Does the buoyant force on a diver change as she dives deeper? Except with extreme changes of depth, the buoyant force does not change significantly. This is because the pressure of the water above the diver and the pressure below the diver increase by about the same amount as the diver goes deeper. The depth of an object submerged in a fluid has little effect on the buoyant force.

Balloon   **Tennis ball**   **Billiard ball**

Figure 8 🔑 The buoyant force is greater on the balloon than on the tennis ball or the billiard ball because the balloon displaces more water.

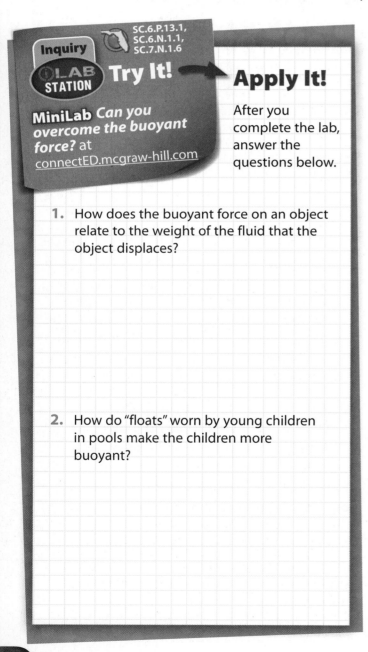

Inquiry

SC.6.P.13.1,
SC.6.N.1.1,
SC.7.N.1.6

🔵LAB STATION Try It!

MiniLab Can you overcome the buoyant force? at
connectED.mcgraw-hill.com

Apply It!

After you complete the lab, answer the questions below.

1. How does the buoyant force on an object relate to the weight of the fluid that the object displaces?

2. How do "floats" worn by young children in pools make the children more buoyant?

Archimedes' Principle

There is another way to think about the buoyant force. What happens to each object in **Figure 8** as it is pushed underwater? As each object is submerged in a large beaker, some water spills out, or is displaced, into a smaller beaker. Notice that the balloon displaces more water than the tennis ball or the billiard ball does. The volume of water displaced by the balloon is equal to the volume of the balloon. Similarly, the volume of water displaced by the tennis ball and the billiard ball is equal to the volume of each ball.

Which small beaker do you think weighs the most? The beaker on the left does because it has the most water in it. According to **Archimedes'** (ar kuh MEE deez) **principle,** *the weight of the fluid that an object displaces is equal to the buoyant force acting on the object.* Because the balloon displaces more water than the billiard ball or the tennis ball does, a greater buoyant force acts on it.

Active Reading 3. **Identify** Underline Archimedes' principle in the text above.

Recall the image of the divers on the previous page. The buoyant force on each diver does not change as the divers go deeper because the volume of each diver does not change. Each displaces the same amount of water at any depth.

Figure 9 If the weight of an object is greater than the buoyant force acting on it, the object sinks.

Sinking and Floating

Now look at the tennis ball and the billiard ball in **Figure 9.** Both have the same volume. But, if you set each on the surface of water and let go, one floats and one sinks.

Buoyant Force and Weight

If you set a tennis ball on the surface of water, it displaces water. With only a little bit of the tennis ball submerged, the tennis ball stops displacing water and floats. Why is this? When the weight of the water displaced by an object equals the weight of the object, the object floats. So, the weight of the water displaced by the tennis ball, or the buoyant force, equals the weight of the tennis ball. The same is true for the leaf in **Figure 9.**

However, when you set a billiard ball on the surface of water, it sinks. Even after the billiard ball is completely submerged and has displaced the maximum volume of water that it can, it still weighs more than the water it displaces. When an object's weight is greater than the buoyant force, the object sinks.

The buoyant force on the billiard ball is greater than the buoyant force on the tennis ball because the billiard ball is completely submerged. The water displaced by the sunken billiard ball weighs more than the water displaced by the floating tennis ball.

Buoyant Force and Density

Because the billiard ball sinks, you know it weighs more than the water it displaces. Therefore, the billiard ball has more mass per volume, or a greater density, than water. If an object is more dense than the fluid in which it is placed, then the buoyant force on that object will be less than the object's weight, and the object will sink.

Active Reading **4. Solve** Read across each row. (Circle) the correct word or phrase shown in blue to summarize the concepts of sinking and floating.

The weight of water displaced by an object equals the weight of the object.	The object will sink/float.	The buoyant force on the object is greater than/less than/equal to the weight of the object.	The density of water is greater than/less than/equal to the density of the object.
The weight of water displaced by an object is less than the weight of the object.	The object will sink/float.	The buoyant force on the object is greater than/less than/equal to the weight of the object.	The density of water is greater than/less than/equal to the density of the object.

Figure 10 🔑 The boat on the left floats because it is filled with air instead of water.

Active Reading

FOLDABLES ® LA.6.2.2.3

Make a vertical two-column chart. Label it as shown. Use the chart to compare and contrast sinking and floating.

Sinking | Floating

Figure 11 🔑 As the balloon loses helium, its density increases and its buoyant force decreases.

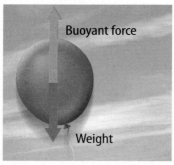

Metal Boats

Aluminum has more than twice the density of water. So, how does the aluminum boat in the left part of **Figure 10** stay afloat?

The boats in **Figure 10** are the same size and made from the same material. The boat on the left is filled with air. The weight of the water displaced equals the weight of the boat plus the air inside. Therefore, the boat floats.

The boat on the right is filled with water. In order for this boat to float, the buoyant force would need to equal the weight of the boat plus the water inside. However, the total weight of the boat and the water inside is greater than the water displaced by the boat, or the buoyant force. So, the boat sinks.

You can also think of this is in terms of density. The density of the boat and air on the left is less than the density of water, so the boat floats. However, the density of the boat and water on the right is greater than the density of water. It sinks. The buoyant force is still greater on the boat on the right because it displaces more water than the boat on the left.

Active Reading **5. Infer** If an object weighing 14 N experiences a 12-N buoyant force, will it sink or float?

Balloons

Why does a helium balloon rise? Helium is less dense than either oxygen or nitrogen in the air. Therefore, the buoyant force acting on a balloon filled with helium is greater than the weight of the balloon, as illustrated in the top of **Figure 11.** The buoyant force pushes the balloon upward. After a day or two, a helium balloon begins to shrink. Helium atoms are so small that they pass between particles that make up the balloon. As the volume of the balloon decreases, the density of the balloon increases. Eventually, the balloon's density is greater than the air's density. The buoyant force on the balloon is less than the balloon's weight, and the balloon falls to the ground.

Visual Summary

A buoyant force results from the difference in pressure between the top and the bottom of an object.

Objects that have the same volume in a fluid experience the same buoyant force.

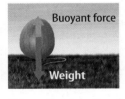

Buoyant force

Weight

When the density of a balloon becomes greater than the density of air, the balloon sinks.

Inquiry   SC.6.P.13.1,
            SC.7.N.1.6
LAB STATION **Try It!**

Skill Lab *Do heavy objects always sink and light objects always float?* at connectED.mcgraw-hill.com

Use Vocabulary

1 State Archimedes' principle in your own words. SC.6.P.13.1

_____

_____

_____

2 Relate buoyant force to pressure.

_____

_____

_____

Understand Key Concepts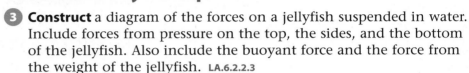

3 Construct a diagram of the forces on a jellyfish suspended in water. Include forces from pressure on the top, the sides, and the bottom of the jellyfish. Also include the buoyant force and the force from the weight of the jellyfish. LA.6.2.2.3

4 The density of water is 1.0 g/cm³. Which of these floats in water?
(A) a ball with a density of 1.7 g/cm³
(B) a ruler with a density of 1.1 g/cm³
(C) a cup that weighs 2 N and displaces 3 N of water when submerged
(D) a toy that weighs 4 N and displaces 3 N of water when submerged

Interpret Graphics

5 Organize Information Complete the graphic organizer stating the relative sizes of forces acting on a sinking object. LA.6.2.2.3

Object sinks

Critical Thinking

6 Use Archimedes' principle to explain why it is easier to lift a rock in water than it is to lift the same rock in air.

_____

_____

_____

_____

Connect Key Concepts 🔑

Complete the graphic organizers below to help you review Lesson 1 and Lesson 2.

1. Order the pressure of fluids at various depths and elevations. (Circle) the examples of atmospheric pressure. Then, describe the pressure in the other examples.

100 feet below the ocean surface	at the top of Mt. Everest
halfway to the top of Mt. Everest	at sea level
20 feet below the ocean surface	at the Mariana Trench

least pressure

greatest pressure

_____

_____

2. Record the equation for density.

density = _____ ÷

_____

3. Express how properties relate to whether an object will sink or float.

Property	Effect
Weight	
Density	

4. Contrast density in the examples of a floating and sunken boat.

Object	Description of Density
Aluminum boat floating in a lake	
Sunken aluminum boat sitting at the bottom of a lake	

5. Compare two factors that affect sinking and floating.

Factor:

Direction of force:
upward

Factor:
weight (gravity)

Direction of force:

Other Effects of
FLUID FORCES

ESSENTIAL QUESTIONS

 How are forces transferred through a fluid?

 How does Bernoulli's principle describe the relationship between pressure and speed?

 What affects drag forces?

Vocabulary

Pascal's principle p. 114
Bernoulli's principle p. 116
drag force p. 118

 Florida NGSSS

SC.6.N.1.1 Define a problem from the sixth grade curriculum, use appropriate reference materials to support scientific understanding, plan and carry out scientific investigation of various types, such as systematic observations or experiments, identify variables, collect and organize data, interpret data in charts, tables, and graphics, analyze information, make predictions, and defend conclusions.

SC.6.P.13.1 Investigate and describe types of forces including contact forces and forces acting at a distance, such as electrical, magnetic, and gravitational.

SC.7.N.1.3 Distinguish between an experiment (which must involve the identification and control of variables) and other forms of scientific investigation and explain that not all scientific knowledge is derived from experimentation.

SC.8.N.1.2 Design and conduct a study using repeated trials and replication.

LA.6.2.2.3 The student will organize information to show understanding or relationships among facts, ideas, and events (e.g., representing key points within text through charting, mapping, paraphrasing, summarizing, or comparing/contrasting);

 Launch Lab
15 minutes

How is force transferred through a fluid?

Fluids apply forces. How are forces transferred through fluids?

Procedure

1. Read and complete a lab safety form.

2. With a partner, use a **nail** to carefully poke three small, identical holes about 1 cm apart along one side of a **drinking straw.** Use **tape** to connect one end of this straw to one end of a **bendable straw.**

3. Tape a **funnel** to the open end of the bendable straw.

4. Keeping the straws and funnel connected, place them in a **tub.** The straw with the holes should be horizontal, with the holes facing upward. Have your partner hold a finger over the open end of the horizontal straw.

5. Gently pour water into the funnel. Record your observations.

Data and Observations

Think About This

1. **Contrast** How did the streams of water pouring out of the holes differ?

2. **Key Concept** **Infer** How do you think force from pressure is transferred through a fluid?

Inquiry Why the Parachute?

1. Have you ever flown in an airplane and noticed that the back edges of the wings move down as the plane prepares to land? These "flaps" increase the drag force on the plane, slowing it down. Similarly, the parachute attached to this race car increases the drag force on the car and slows it down, too. What would happen to the race car if it did not have a parachute?

_____

_____

_____

_____

_____

Fluid Forces Benefits and Challenges

You use forces in fluids to accomplish everyday tasks. You produce a force when you water a garden with a hose or squeeze a plastic ketchup bottle. You make use of a buoyant force when you float on a raft or lift a rock under water.

Forces in fluids also can make tasks difficult or even dangerous. When a car moves at high speeds, air pushes against the car. This increases the amount of fuel the car uses. Fluid forces from floods, tornadoes, and hurricanes can cause damage that costs lives and billions of dollars. How are forces in fluids both useful and dangerous?

Pascal's Principle

Blaise Pascal (pas KAL), a 17th century French physicist, studied the pressures of fluids in closed containers. A fluid cannot flow into or out of a closed container. **Pascal's principle** *states that when pressure is applied to a fluid in a closed container, the pressure increases by the same amount everywhere in the container.* If you push on a closed ketchup bottle, the pressure on the ketchup increases not only under your fingers, but equally throughout the bottle.

Active Reading

Active Reading

FOLDABLES LA.6.2.2.3

Make a horizontal, three-column chart book. Label it as shown. Use it to organize your notes on Pascal's principle, Bernoulli's principle, and drag forces.

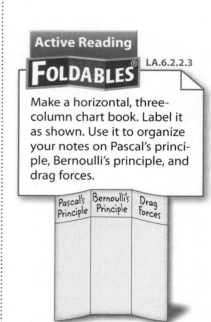

2. Relate Complete the definition of Pascal's principle by filling in the graphic organizer below.

When pressure is applied to a fluid in a closed container,

pressure

_____

_____

_____

_____

_____

Pushing on a Fluid

Pascal's principle applies to fluid power systems. A fluid power system like the one in **Figure 12** uses a pressurized fluid to transfer motion. The figure shows the piston on the left moving down as the piston on the right moves up. In this example, the surface area of the input piston is half that of the output piston. Any input force applied to the input piston results in an output force that is doubled by the output piston.

Why is the output force greater than the input force? Pascal's principle can be written as $F = P \times a$. If pressure (P) is always equal throughout the system, increasing the area of the output piston (a) results in a larger force (F). Does the output piston do more work than the input piston?

No. Recall that *work = force × distance*. In this example, the output force is twice that of the input force. And, if you were to build the fluid power system shown in **Figure 12,** you would find that the input piston is pushed down twice the distance that the output piston moves up. If you use the work equation to solve for the input work and then for the output work of the system, you would find that the two are equal.

Hydraulic Lifts

Automobile mechanics use Pascal's principle to lift a car with a hydraulic lift, as shown in **Figure 13.** A hydraulic lift is an example of a fluid power system that uses a liquid for the fluid. In an oil-filled hydraulic lift, a narrow tube connects to a wider tube under a car. Pushing down on a piston in the narrow tube generates an upward force on a larger piston great enough to lift a car.

Like the fluid power system in **Figure 12,** a hydraulic lift does not reduce the amount of work needed to lift a car. To lift a car, the piston in the narrow tube travels a longer distance than the piston in the wider tube.

Active Reading **4. Identify** On what principle do hydraulic lifts rely?

Pascal's Principle 🔑

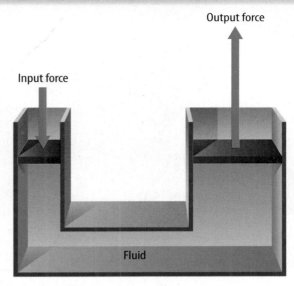
Output force

Input force

Fluid

Figure 12 A fluid power system uses a pressurized fluid to transfer motion.

Active Reading **3. Evaluate** Complete the table below to describe the functions of the input and output pistons of a fluid power system.

Factor	Input piston	Output piston
area		
force		
pressure		
work		
distance		

Figure 13 People rely on Pascal's principle when they use hydraulic lifts. A hydraulic lift increases the force lifting up on a car.

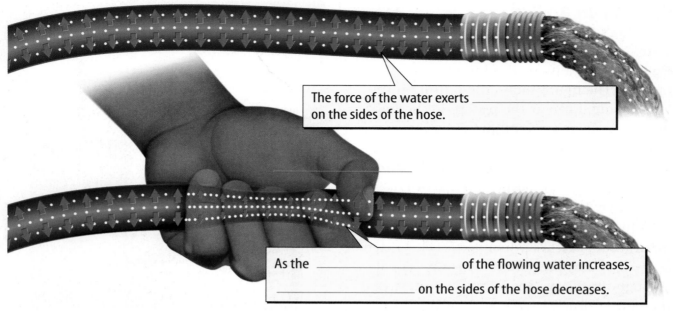

The force of the water exerts _____ on the sides of the hose.

As the _____ of the flowing water increases, _____ on the sides of the hose decreases.

Figure 14 🔑 According to Bernoulli's principle, when water moves faster in the pinched part of a hose, it applies less pressure on the sides of the hose.

 5. Identify Fill in the blanks in the boxes above as you read about Bernoulli's principle.

 6. Point out <u>Underline</u> the relationship between speed and pressure in a fluid.

MiniLab *How does air speed affect air pressure?* at connectED.mcgraw-hill.com

Bernoulli's Principle

Have you ever sprayed a friend with water from a hose? Garden hoses illustrate a principle that describes the relationship between speed and pressure in fluids. **Bernoulli's** (ber NEW leez) **principle** *states that the pressure of a fluid decreases when the speed of that fluid increases.*

Consider the hose shown at the top of **Figure 14.** As the water flows through the hose, it applies a constant pressure on the sides of the hose, and it comes out of the hose at a constant speed.

Suppose you pinch the hose slightly, as shown in the bottom of **Figure 14.** The water speeds up in the pinched part. After it has traveled through the pinched part, it returns to its regular speed. According to Bernoulli's principle, when speed increases, pressure decreases. This means the pressure of the water on the sides of the hose in the pinched part is less than it is elsewhere in the hose.

Damage from High Winds

When strong winds blow over a house, the air outside the house has a high speed. But the air inside the house has almost no speed. According to Bernoulli's principle, increased speed means lowered pressure. Therefore, the pressure outside the house is lower than the pressure inside the house. This is illustrated on the left side of **Figure 15.** The force of the air pressure pushing down on the roof of this house is less than the force of the air pressure pushing up. When the upward force inside the house becomes greater than the downward force outside the house and the force of gravity—the roof begins to rise.

Now think about what happens immediately after the roof lifts off the house, as shown in the right side of **Figure 15.** The wind continues to blow. Its speed is now the same below and above the roof. The air pressure is also the same below and above the roof. The force from the upward air pressure on the roof is balanced by the force from the downward air pressure on the roof. But, the force of gravity adds to the force from the downward air pressure. When the combined downward force is greater than the upward force, the roof crashes back down.

Figure 15 When the upward pressure is greater than the downward pressure, the roof moves up.

Figure 16 Living in a monolithic dome is one way to protect yourself from severe storms.

Storm-Proof Homes in Florida

Engineers have developed a unique wind-resistant building design called monolithic dome construction. As shown in **Figure 16,** a concrete monolithic dome is simple in design and looks as though it is built from a single stone. When finished, the dome is earthquake, tornado, and hurricane resistant. The federal government considers these structures to be near-absolute protection from F5 tornadoes and Category 5 hurricanes. Several monolithic domes in Florida survived direct hits by Hurricane Katrina in 2005. In addition to being safe, sturdy, and weather resistant, monolithic domes are inexpensive, easy to build, and energy efficient.

Active Reading **7. Label** Determine whether pressure is high or low on the left part of the figure.

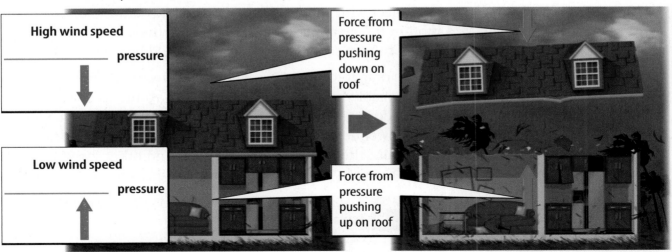

High wind speed

_____ pressure

Low wind speed

_____ pressure

Force from pressure pushing down on roof

Force from pressure pushing up on roof

If wind is moving very quickly outside a house, such as during a tornado, the pressure outside will be less than the pressure inside. This can cause the roof to lift off.

If a roof lifts off a house, the pressure outside the house pushing down on the roof and the pressure inside the house pushing up on the roof are equal.

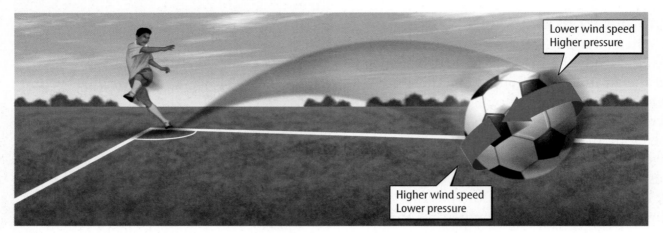

Lower wind speed
Higher pressure

Higher wind speed
Lower pressure

Figure 17 The higher pressure on the left side of the soccer ball causes the ball to curve right.

Active Reading **8. Modify** Which way would the ball curve if it were spinning in the opposite direction?

_____

_____

_____

WORD ORIGIN

drag force

from Old Norse *draga,* means "to draw"; and Latin *fortis,* means "force"

Figure 18 🔑 The parachute increases the drag force on this runner and makes him work harder.

Soccer Kicks

You might have seen a soccer player kick a soccer ball so that it curves around an opposing player. The soccer player puts a spin on the ball as he or she kicks it, as shown in **Figure 17.** This makes the speed of the air on one side of the ball greater than the speed of the air on the other side. According to Bernoulli's principle, the side with the lower speed has greater air pressure acting on it. Because air moves from areas of high to low pressure, the spinning ball curves toward the side with lower pressure.

Drag Forces

Runners often use parachutes to increase the drag force on them and build strength as shown in **Figure 18.** The **drag force** *is a force that opposes the motion of an object through a fluid.* As the speed of an object in a fluid increases, the drag force on that object also increases. The faster the runner runs, the greater the drag force is on him, and the harder he needs to work to resist it.

The drag force on an object also depends on the size and shape of that object. If two objects move in the same direction, the object with the greater surface area toward the direction of motion has a greater drag force on it. If the parachute were larger, the drag force on the runner would be even greater.

The drag force increases when the density of a fluid increases. You probably have felt the pull of water against your legs when you wade in shallow water. When you walk through air, which has a lower density, you barely notice the drag force. Whether you notice them or not, the drag force and other forces in fluids are all around you. They help you work and play.

Active Reading **9. Express** Highlight the definition of *drag force* and explain three factors that affect it.

_____

_____

_____

Visual Summary

People rely on Pascal's principle when they use hydraulic lifts.

The imbalance of pressures in fluids can cause a roof to lift off a house in a severe windstorm.

A soccer player who kicks a curved ball makes use of Bernoulli's principle.

Inquiry

SC.6.P.13.1, SC.6.N.1.1, SC.7.N.1.3

LAB STATION **Try It!**

Inquiry Lab *Design a Cargo Ship* at connectED.mcgraw-hill.com

Use Vocabulary

1 **State** Bernoulli's principle in your own words. SC.6.P.13.1

_____

_____

2 The transfer of forces in fluids in closed containers is explained by

_____ .

Understand Key Concepts 🔑

3 **Identify** Water flows through one pipe section at 4 m/s and then through a second section at 8 m/s. In which section is the pressure greater? Explain why.

_____

_____

_____

4 **Determine** A closed container of liquid soap is 10 cm tall. You squeeze on the container's top with a pressure of 1,000 Pa. What is the pressure at the bottom?

_____

5 Which statement about drag force on an object is true?
- (A) As the speed of the object increases, the drag force increases.
- (B) The drag force decreases when the density of the fluid around the object increases.
- (C) The drag force does not depend on the shape of the object.
- (D) The drag force is applied in the same direction as the motion of the object.

Interpret Graphics

6 **Organize Information** Fill in the graphic organizer below to list four things that affect drag forces. LA.6.2.2.3

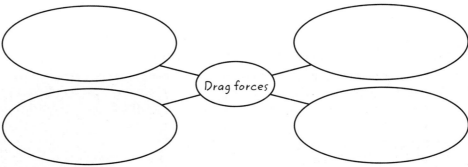

Critical Thinking

7 **Design** a hydraulic lift that could lift 100 N 1 m with a 25-N force.

Chapter 3   Study Guide

 Think About It! People use contact forces through fluids to float objects, to lift objects, and to change the motions of objects.

Key Concepts Summary

LESSON 1 Pressure and Density of Fluids

- **Pressure** is the ratio of force to area.
- **Atmospheric pressure** decreases with elevation. Pressure under water increases with depth.
- The density of a **fluid** depends on the mass of the fluid and its volume.

area = 0.0001 m²

BOOKS

50 N

Vocabulary

fluid p. 97

pressure p. 98

atmospheric pressure p. 100

LESSON 2 The Buoyant Force

- The change in pressure between the top and the bottom of an object results in an upward force called the **buoyant force.**
- **Archimedes' principle** states that the weight of the fluid displaced by an object is equal to the buoyant force on that object.
- An object sinks if its weight is greater than the buoyant force on it. An object does not sink if the buoyant force on it is equal to its weight.

Buoyant force

Weight

buoyant force p. 106

Archimedes' principle p. 108

LESSON 3 Other Effects of Fluid Forces

- **Pascal's principle** states that when pressure is applied to a fluid in a closed container, the pressure increases by the same amount everywhere in the container.
- **Bernoulli's principle** states that when the speed in a fluid increases, the pressure decreases.
- Speed, size, and shape of an object, as well as the density of the fluid in which the object moves, affect the **drag force** on that object.

Pascal's principle p. 114

Bernoulli's principle p. 116

drag force p. 118

FOLDABLES® Chapter Project

Assemble your lesson Foldables as shown to make a Chapter Project. Use the project to review what you have learned in this chapter.

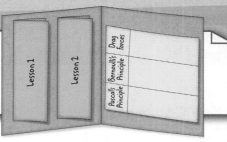

Use Vocabulary

1 Air and water are both _____.

2 A ship does not sink because a(n) _____ acts on it.

3 A column of air exerts _____ on you.

4 According to _____, two objects of equal volume in a fluid experience the same buoyant force.

5 A parachute with a 5-m^2 surface area experiences a much larger _____ than a parachute with a 3-m^2 surface area.

6 Fluid power systems work according to _____.

Link Vocabulary and Key Concepts

Use vocabulary terms from the previous page to complete the concept map.

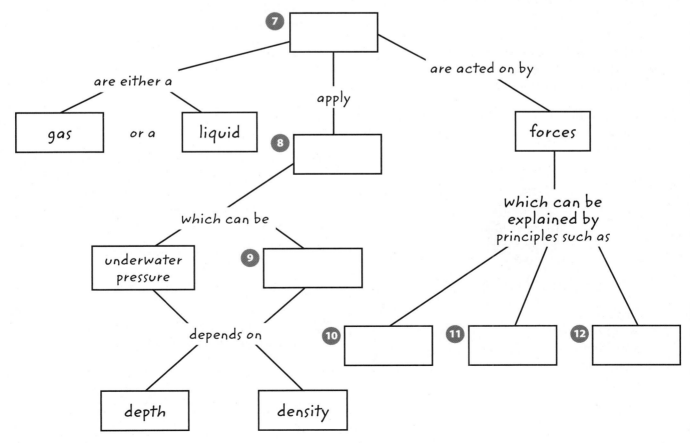

Fill in the correct answer choice.

🔑 Understand Key Concepts

1 Which is NOT a fluid? SC.7.N.1.6

Ⓐ helium
Ⓑ ice
Ⓒ milk
Ⓓ water

2 If you poured the following fluids into a container, which would float on top? SC.6.P.13.1

Ⓐ maple syrup, with a density of 1.33 g/cm³
Ⓑ olive oil, with a density of 0.9 g/m³
Ⓒ seawater, with a density of 1.03 g/cm³
Ⓓ water, with a density of 1.0 g/cm³

3 What pressure does Adam apply to a ball of dough when he pushes on it with a 25-N force? The area of his hand is 0.01 m². MA.6.A.3.6

Ⓐ 0.0004 Pa
Ⓑ 0.5 Pa
Ⓒ 25 Pa
Ⓓ 2,500 Pa

4 Which of these has the greatest pressure applied to it from the surrounding fluid? MA.6.A.3.6

Ⓐ a fish swimming 20 m below the surface
Ⓑ a hawk flying 300 m above sea level
Ⓒ a mountain climber at an altitude of 4,400 m
Ⓓ a person fishing off the coast of California

5 In the diagram of the beach ball floating on water below, what force does the blue arrow represent? SC.6.P.13.1

Ⓐ pressure
Ⓑ weight
Ⓒ buoyant force
Ⓓ drag force

Critical Thinking

6 **Design an Experiment** You are given a material and asked to determine whether it is a fluid. What experiment could you do? SC.6.N.1.1

_____
_____
_____
_____

7 **Propose** The legs of your chair sink into the sand on a beach. What could you do to prevent this? Explain your reasoning. SC.6.N.1.1

_____
_____
_____
_____

8 **Compare** You drop a solid cube into water. The cube is 40 cm on each side. Then you drop another solid object that is 160 cm tall, 20 cm deep, and 20 cm wide and made from the same material as the cube into water. Which object experiences the greater buoyant force? Explain. SC.6.P.13.1

_____
_____
_____

9 **Interpret** A sailor drops an anchor over the side of a ship. When the anchor is 10 m below the ocean's surface, the buoyant force on the anchor is 80 N. What is the buoyant force on the anchor when it sinks to 100 m below the surface? SC.6.P.13.1

_____
_____
_____

10 **Assess** How does opening a parachute change the drag force on a skydiver? SC.6.P.13.1

_____
_____
_____

11 **Explain** Woodchucks live in underground tunnels such as the one shown below. One opening has a dirt mound around it, and air flows across it quickly. The other opening is even with the ground. The air moves across it with less speed. How does this design help ventilate the tunnel? **SC.6.N.1.1**

_____

_____

_____

_____

_____

Air speed

Air speed

Writing in Science

12 **Write** a paragraph on a separate sheet of paper explaining why the *Titanic* sank. Use what you have learned about forces and fluids. **LA.6.2.2.3**

Big Idea Review

13 The braking systems in most automobiles rely on a hydraulic fluid power system. Use Pascal's principle to explain how the pressure of a foot on a car's brake pedal can stop the car. **SC.6.N.1.1**

_____

_____

_____

14 The chapter opener shows a helium balloon in a parade. Imagine you are holding a rope attached to this balloon. What forces do you encounter? Explain at least two ways in which you might encounter or use forces in fluids in your everyday life. **SC.6.N.1.1**

_____

_____

_____

Math Skills **SC.6.N.1.1, SC.6.P.13.1**

Solve a One-Step Equation

15 The buoyant force on an inflatable pool raft with a surface area of 2 m^2 is 36 N. How large is the upward pressure on the raft?

16 A ballerina stands on her toes in a pointe shoe. The pressure on her toes is 454,000 Pa. When she stands on flat feet, the pressure is 22,700 Pa. If the ballerina's weight is 454 N, what is the surface area she stands on when she's on her toes, and what is the surface area she stands on when she stands flat on her foot?

Fill in the correct answer choice.

Multiple Choice

1 The same force is applied over two areas that differ in size. Which is true of the pressure over these areas? SC.6.P.13.1

   (A) The pressure is equal to the force multiplied by the area.

   (B) The pressure on both areas is the same.

   (C) The pressure on the larger area is greater.

   (D) The pressure on the smaller area is greater.

Use the figure to answer question 2.

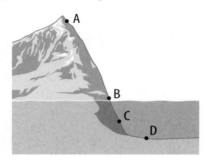

2 The figure shows a side view of a mountain that extends below the water's surface. At which point would the pressure be 100 kPa? SC.6.P.13.1

   (F) A

   (G) B

   (H) C

   (I) D

3 Which characteristics determine the density of a fluid? SC.6.P.13.1

   (A) mass and energy of the particles

   (B) mass of particles and distance between them

   (C) number of particles and distance between them

   (D) shape and energy of the particles

4 Which changes as a solid object moves downward within a fluid? SC.6.N.1.1

   (F) the buoyant force acting on the object

   (G) the mass of the object

   (H) the pressure acting on the object

   (I) the volume of the object

5 Which explains how the buoyant force on an object changes with the weight of the fluid it displaces? SC.6.N.1.1

   (A) Archimedes' principle

   (B) definition of *density*

   (C) definition of *pressure*

   (D) Pascal's principle

Use the figure to answer question 6.

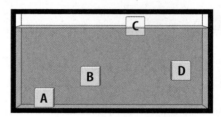

6 The figure shows an aquarium filled with water. Four cubes made out of different materials have been placed in the aquarium. For which object is the buoyant force acting on it equal to the object's weight? SC.6.P.13.1

   (F) A

   (G) B

   (H) C

   (I) D

7 Why are drag forces greater in water than they are in air? SC.6.P.13.1

   (A) Air is denser than water.

   (B) Air is a fluid, but water is not.

   (C) Water is denser than air.

   (D) Water is a fluid, but air is not.

Use the figure to answer questions 8 and 9.

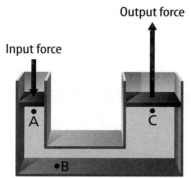

Output force

Input force

A   C

•B

8 When an input force is applied above point A, which is true of the change in fluid pressure at points A, B, and C? SC.6.P.13.1

(F) The fluid pressure increases the most at point A.

(G) The fluid pressure increases the most at point B.

(H) The fluid pressure increases the most at point C.

(I) The fluid pressure increases by the same amount at all three points.

9 Which describes how the input force affects the fluid pressure at different points in the hydraulic lift? SC.6.P.13.1

(A) Bernoulli's principle

(B) definition of buoyant force

(C) equation for density

(D) Pascal's principle

10 Joseph weighs 290 N and displaces 300 N of water as he swims under water in a pool. What is the buoyant force on Joseph? SC.6.P.13.1

(F) 10 N upward

(G) 300 N upward

(H) 290 N downward

(I) 590 N downward

11 Which statement about boats is correct? SC.7.N.1.5

(A) A boat cannot be made from metal because metal has a greater density than water.

(B) A boat floats if its overall density is less than that of water.

(C) A boat floats only if its overall mass per volume is more than water's mass per volume.

(D) A boat floats only if the weight of water it displaces is less than the boat's weight.

12 In the diagram below, how large a force is applied by the piston on the right? SC.6.P.13.1

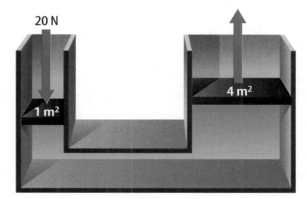

20 N

1 m²   4 m²

(F) 10 N

(G) 20 N

(H) 40 N

(I) 80 N

13 A leaf enters a drainpipe, and the pressure in the water increases from 10 Pa to 30 Pa. What happens to the leaf's speed? SC.7.N.1.5

(A) It decreases.

(B) It increases.

(C) It becomes exactly 30 m/s.

(D) It does not change.

NEED EXTRA HELP?

If You Missed Question...	1	2	3	4	5	6	7	8	9	10	11	12	13
Go to Lesson...	1	1	1	2	2	2	3	3	3	2	2	3	1

Multiple Choice *Bubble the correct answer.*

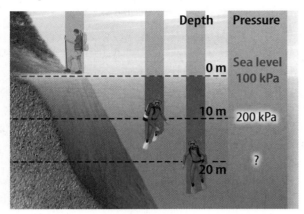

Depth	Pressure
0 m	Sea level 100 kPa
10 m	200 kPa
20 m	?

1. Examine the diagram above. The pressure on the diver at 20 m below sea level is **MA.6.A.3.6**

   (A) 100 kPa.

   (B) 200 kPa.

   (C) 300 kPa.

   (D) 400 kPa.

2. Which statement is true of oxygen gas? **SC.6.P.13.1**

   (F) It applies only downward pressure when in a container.

   (G) It has a given volume that remains the same in any container.

   (H) It is not a fluid.

   (I) It takes the shape of any container that it is in.

3. Water has a density of about 1.0 g/cm^3. Which fluid would sink to the bottom of a beaker of water? **SC.7.N.1.6**

   (A) gasoline (density of 0.66 g/cm^3)

   (B) glycerin (density of 1.26 g/cm^3)

   (C) olive oil (density of 0.9 g/cm^3)

   (D) turpentine (density of 0.85 g/cm^3)

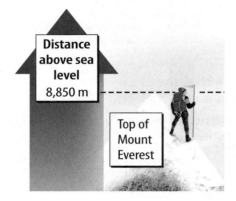

Distance above sea level 8,850 m

Top of Mount Everest

4. Which is true about the situation in the picture? **SC.6.P.13.1**

   (F) Air is so thin at this altitude that it does not apply pressure downward.

   (G) The density of air at this altitude is greater than the density at sea level.

   (H) The hiker experiences less atmospheric pressure here than at sea level.

   (I) The hiker experiences more atmospheric pressure here than at sea level.

Copyright © Glencoe/McGraw-Hill, a division of The McGraw-Hill Companies, Inc.

Multiple Choice *Bubble the correct answer.*

1. Which statement describes the forces that are acting on the billiard ball in the image above? **SC.6.P.13.1**

    (A) The weight force and buoyant force on the ball are equal.

    (B) The weight force on the ball is greater than the buoyant force.

    (C) The weight force on the ball is less than the buoyant force.

    (D) No buoyant force acts on a heavy object such as a billiard ball.

2. Why does a leaf float in water? **SC.6.P.13.1**

    (F) because the buoyant force on the leaf is equal to the weight of the leaf

    (G) because the weight force on the leaf is less than the buoyant force

    (H) because the weight force on the leaf is more than the buoyant force

    (I) because neither buoyant force nor weight force is acting on the leaf

3. In which situation are buoyant forces and weight forces equal? **SC.6.P.13.1**

    (A) when an anchor is dropped over the side of a boat

    (B) when a helicopter takes off

    (C) when a supertanker travels across the ocean

    (D) when a rock is skipped across the water's surface and then sinks

4. The balloon in the picture **SC.6.P.13.1**

    (F) exerts no pressure on the air around it.

    (G) is denser than gases in the air.

    (H) has a greater volume than the air surrounding it.

    (I) weighs less than the buoyant force acting on it.

Copyright © Glencoe/McGraw-Hill, a division of The McGraw-Hill Companies, Inc.

Benchmark Mini-Assessment Chapter 3 • Lesson 3

Multiple Choice *Bubble the correct answer.*

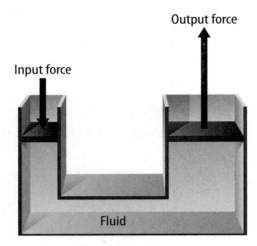

1. In the image above, the surface area of the left piston is half that of the right piston. Which statement is true? **SC.6.P.13.1**

   (A) The output force does more work than the input force.

   (B) The output force is equal to the input force.

   (C) The output force is half as much as the input force.

   (D) The output force is twice as large as the input force.

2. Which statement is true about pressure applied to a fluid in a closed container? **SC.6.P.13.1**

   (F) The pressure has no effect on the fluid in the container.

   (G) The pressure on the fluid inside can damage the container.

   (H) The pressure decreases by the same amount everywhere in the container.

   (I) The pressure increases by the same amount everywhere in the container.

3. Callie wants to be able to spray water across a large garden. What could she do? **SC.7.N.1.3**

   (A) use a longer hose

   (B) use a wider hose

   (C) attach a narrow nozzle to the hose

   (D) attach a wider opening to the hose

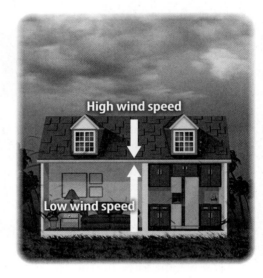

4. Which statement is true about the image above? **SC.6.P.13.1**

   (F) The pressure inside the house decreases when the winds are faster outside the house.

   (G) The pressure inside the house is greater than the pressure outside the house.

   (H) The pressure outside the house is equal to the pressure inside the house.

   (I) The pressure outside the house is greater than the pressure inside the house.

Copyright © Glencoe/McGraw-Hill, a division of The McGraw-Hill Companies, Inc.

Notes

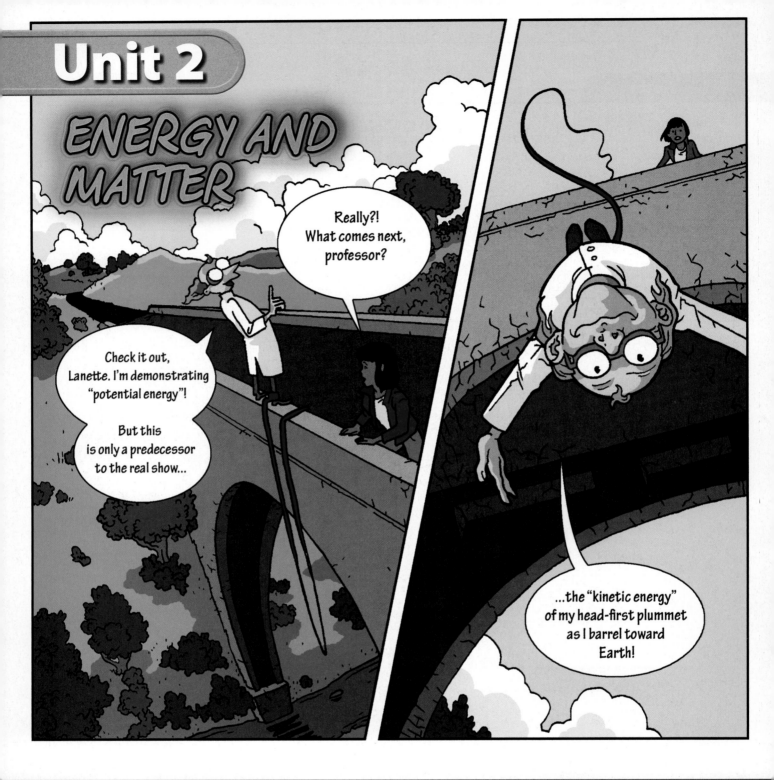

Unit 2

ENERGY AND MATTER

1875 **1900** **1925**

1895
The first X-ray photograph is taken by Wilhelm Konrad Roentgen of his wife's hand. It is now possible to look inside the human body without surgical intervention.

1898
Chemist Marie Curie and her husband Pierre discover radioactivity. They are later awarded the Nobel Prize in Physics for their discovery.

1917
Ernest Rutherford, the "father of nuclear physics," is the first to split atoms.

1934
Nuclear fission is first achieved experimentally in Rome by Enrico Fermi when his team bombards uranium with neutrons.

1939
The Manhattan Project, a code name for a research program to develop the first atomic bomb, begins. The project is directed by American physicist J. Robert Oppenheimer.

1950 1975 2000

1945
American-led atomic bomb attacks on the Japanese cities of Hiroshima and Nagasaki bring about the end of World War II.

1954
Obninsk Nuclear Power Plant, located in the former USSR, begins operating as the world's first nuclear power plant to generate electricity for a power grid. It produces around 5 megawatts of electric power.

2007
Fourteen percent of the world's electricity now comes from nuclear power.

? Inquiry
Visit ConnectED for this unit's STEM activity.

Technology

SC.6.N.1.5, SC.7.N.1.5, SC.8.N.4.2

Scientists use technology to develop materials with desirable properties. **Technology** is the practical use of scientific knowledge, especially for industrial or commercial use. In the late 1800s, scientists developed the first plastic material, called celluloid, from cotton. In the 1900s, scientists developed other plastic materials, such as polystyrene, rayon, and nylon. These new materials were inexpensive, durable, lightweight, and could be molded into any shape.

New technologies often come with problems. For example, many plastics are made from petroleum and contain harmful chemicals. The high pressures and temperatures needed to produce plastics require large amounts of energy. Bacteria and fungi that easily break down natural materials do not easily decompose plastics. Often, plastics accumulate in landfills where they can remain for hundreds, or even thousands, of years, as shown in **Figure 1.**

Figure 1 Nature cannot easily recycle many human-made materials. Much of our trash remains in landfills for years. Scientists are developing materials that degrade quickly. This will help decrease the amount of pollution.

Active Reading

1. Identify What are three negative aspects of producing and using plastics?

_____

_____

_____

Types of Materials

Figure 2 Some organisms produce materials with properties that are useful to people. Scientists are trying to replicate these materials for new technologies.

Most human-made adhesives attach to some surfaces, but not others. Mussels, which are similar to clams, produce a "superglue" that is stronger than anything people can make. It also works on wet or dry surfaces. Chemists are trying to develop a technology that will replicate the mussel glue. This glue would provide solutions to difficult problems. Ships could be repaired under water, chipped teeth could be repaired, and broken bones could be more accurately set in place.

Abalone and other mollusks construct a protective shell from proteins and seawater. The material is four times stronger than human-made metal alloys and ceramics. Using technology, scientists are working to duplicate this material. They hope to use the new product in many ways, including for hip and elbow replacements. Automakers could use these strong, lightweight materials for automobile body panels.

Consider the Possibilities!

Chemists are looking to nature for ideas for new materials. For example, some sea sponges have skeletons that beam light deep inside the animal, similar to the way fiber-optic cables work. A bacterium from a snail-like nudibranch contains compounds that stop other sea creatures from growing on the nudibranch's back. These compounds could be used in paints to stop creatures from forming a harmful crust on submerged parts of boats and docks. Chrysanthemum flowers produce a product that keeps ticks and mosquitoes away. **Figure 2** includes other organisms that produce materials with remarkable properties.

Chemists and biologists are teaming up to understand and replicate the processes that organisms use to survive. Hopefully, these processes will lead to technologies and materials with unique properties that are helpful to people.

Inquiry

⟨i⟩LAB STATION **Try It!**

SC.6.N.1.5, SC.7.N.1.5, SC.8.N.4.2

MiniLab *How would you use it?* at connectED.mcgraw-hill.com

Apply It!

After you complete the lab, answer these questions.

1. **Estimate** How would your invention save money, resources, time, and effort as designed from an organism?

2. **Recognize** What are some problems that might be solved using technology spinoffs from organisms?

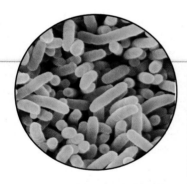

A British company has developed bacteria that produce large amounts of hydrogen gas when fed a diet of sugar. Chemists are working to produce tanks of these microorganisms that produce enough hydrogen to replace other fuels used to heat homes. Bacteria may become the power plants of the future.

Under a microscope, the horn of a rhinoceros looks much like the material used to make the wings of a Stealth aircraft. However, the rhino horn is self-healing. Picture a car with technologically advanced fenders similar to the horn of a rhinoceros; such a car could repair itself if it were in a fender-bender!

Spider silk begins as a liquid inside the spider's body. When ejected through openings, called spinnerets, it becomes similar to a plastic thread. However, its properties include strength five times greater than steel, stretchability greater than nylon, and toughness better than the material in bulletproof vests! Chemists are using technology to make a synthetic spider silk. They hope to someday use the material for cables strong enough to support a bridge or as reinforcing fibers in aircraft bodies.

Notes

Soccer Ball

Five soccer players argued about when a soccer ball has energy. This is what they said:

Jorge: The ball has to be moving to have energy.

Kurt: The ball has energy only at the moment it is kicked.

Amos: The ball has energy only when it is not moving.

Alan: The ball has energy when it is moving and not moving.

Flavio: The ball has no energy. There is no source of energy in the ball.

Who do you agree with the most? Explain your thinking. What rule or reasoning did you use to decide when the soccer ball has energy?

Energy and Energy RESOURCES

FLORIDA BIG IDEAS

1   **The Practice of Science**
3   **The Role of Theories, Laws, Hypotheses, and Models**
4   **Science and Society**
11 **Energy Transfer and Transformations**

Think About It!

What is energy and what are energy resources?

All objects in the photo contain energy. Some objects contain more energy than other objects. The Sun contains so much energy that it is considered an energy source.

1 Where do you think the energy that powers these cars comes from?

_____

_____

2 Do you think the energy in the Sun and the energy in the green plants are related?

_____

_____

_____

3 What do the terms energy and energy resources mean to you?

_____

_____

_____

_____

Get Ready to Read

What do you think about energy?

Before you read, decide if you agree or disagree with each of these statements. As you read this chapter, see if you change your mind about any of the statements.

	AGREE	DISAGREE
1 A fast-moving baseball has more kinetic energy than a slow-moving baseball.	☐	☐
2 A book sitting on a shelf has no energy.	☐	☐
3 Energy can change from one form to another.	☐	☐
4 If you toss a baton straight up, total energy decreases as the baton rises.	☐	☐
5 Nuclear power plants release many dangerous pollutants into the air as they transform nuclear energy into electric energy.	☐	☐
6 Thermal energy from within Earth can be transformed into electric energy at a power plant.	☐	☐

There's More Online!
Video • Audio • Review • ⓘLab Station • WebQuest • Assessment • Concepts in Motion • Multilingual eGlossary

Forms of ENERGY

Vocabulary

energy p. 139

kinetic energy p. 140

potential energy p. 140

work p. 142

mechanical energy p. 143

sound energy p. 143

thermal energy p. 143

electric energy p. 143

radiant energy p. 143

nuclear energy p. 143

 Florida NGSSS

SC.6.N.1.5 Recognize that science involves creativity, not just in designing experiments, but also in creating explanations that fit evidence.

SC.7.N.1.6 Explain that empirical evidence is the cumulative body of observations of a natural phenomenon on which scientific explanations are based.

SC.8.N.4.1 Explain that science is one of the processes that can be used to inform decision making at the community, state, national, and international levels.

LA.6.2.2.3 The student will organize information to show understanding or relationships among facts, ideas, and events (e.g., representing key points within text through charting, mapping, paraphrasing, summarizing, or comparing/contrasting);

SC.7.N.1.6

(Inquiry) **Launch Lab**
20 minutes

Can you change matter?

You observe many things changing. Bubbles form in boiling water. The filament in a lightbulb glows when you turn on a light. How can you cause a change in matter?

Procedure

1 Read and complete the lab safety form.

2 Half-fill a **foam cup** with **sand.** Place the bulb of a **thermometer** about halfway into the sand. *Do not stir.* Record the temperature in the space below.

3 Remove the thermometer and place a lid on the cup. Hold down the lid and shake the cup vigorously for 10 min.

4 Remove the lid. Measure and record the temperature of the sand.

Data and Observations

Think About This

1. What change did you observe in the sand?

2. How could you change your results?

3. **(Key Concept)** What do you think caused the change you observed in the sand?

Fossil Fuels and Rising CO$_2$ Levels

AMERICAN MUSEUM OF NATURAL HISTORY

Investigate the link between energy use and carbon dioxide in the atmosphere.

You use energy every day—when you ride in a car or on a bus, turn on a television or a radio, and even when you send an e-mail.

Much of the energy that produces electricity, heats and cools buildings, and powers engines comes from burning fossil fuels—coal, oil, and natural gas. When fossil fuels burn, the carbon in them combines with oxygen in the atmosphere and forms carbon dioxide gas (CO$_2$).

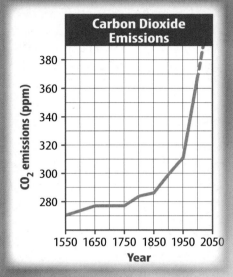

Carbon dioxide is a greenhouse gas. Greenhouse gases absorb energy. This causes the atmosphere and Earth's surface to become warmer. Greenhouse gases make Earth warm enough to support life. Without greenhouse gases, Earth's surface would be frozen.

However, over the past 150 years the amount of CO$_2$ in the atmosphere has increased faster than at any time in the past 800,000 years. Most of this increase is the result of burning fossil fuels. More carbon dioxide in the atmosphere might cause average global temperatures to increase. As temperatures increase, weather patterns worldwide could change. More storms and heavier rainfall could occur in some areas, while other regions could become drier. Increased temperatures could also cause more of the polar ice sheets to melt and raise sea levels. Higher sea levels would cause more flooding in coastal areas.

Developing other energy sources such as geothermal, solar, nuclear, wind, and hydroelectric power would reduce the use of fossil fuels and slow the increase in atmospheric CO$_2$.

It's Your Turn

MAKE A LIST How can CO$_2$ emissions be reduced? Work with a partner. List five ways people in your home, school, or community could reduce their energy consumption. Combine your list with your classmates' lists to make a master list.

LA.6.2.2.3

GREEN SCIENCE

300 Years OF CARBON DIOXIDE

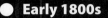

● **1712**

A new invention, the steam engine, is powered by burning coal that heats water to produce steam.

● **Early 1800s**

Coal-fired steam engines, able to pull heavy trains and power steamboats, transform transportation.

● **1882**

Companies make and sell electricity from coal for everyday use. Electricity is used to power the first lightbulbs, which give off 20 times the light of a candle.

● **1908**

The first mass-produced automobiles are made available. By 1915, Ford is selling 500,000 cars a year. Oil becomes the fuel of choice for car engines.

● **Late 1900s**

Electrical appliances transform the way we live, work, and communicate. Most electricity is generated by coal-burning power plants.

● **2007**

There are more than 800 million cars and light trucks on the world's roads.

Energy TRANSFORMATIONS

Vocabulary

law of conservation of energy p. 148

friction p. 149

 Florida NGSSS

SC.6.N.1.1 Define a problem from the sixth grade curriculum, use appropriate reference materials to support scientific understanding, plan and carry out scientific investigation of various types, such as systematic observations or experiments, identify variables, collect and organize data, interpret data in charts, tables, and graphics, analyze information, make predictions, and defend conclusions.

SC.6.N.3.3 Give several examples of scientific laws.

SC.6.P.11.1 Explore the Law of Conservation of Energy by differentiating between potential and kinetic energy. Identify situations where kinetic energy is transformed into potential energy and vice versa.

SC.7.N.1.1 Define a problem from the seventh grade curriculum, use appropriate reference materials to support scientific understanding, plan and carry out scientific investigation of various types, such as systematic observations or experiments, identify variables, collect and organize data, interpret data in charts, tables, and graphics, analyze information, make predictions, and defend conclusions.

SC.7.P.11.2 Investigate and describe the transformation of energy from one form to another.

SC.7.P.11.3 Cite evidence to explain that energy cannot be created nor destroyed, only changed from one form to another.

LA.6.2.2.3 The student will organize information to show understanding or relationships among facts, ideas, and events (e.g., representing key points within text through charting, mapping, paraphrasing, summarizing, or comparing/contrasting);

SC.7.P.11.2

(Inquiry) Launch Lab

15 minutes

Is energy lost when it changes form?

Energy can have different forms. What happens when energy changes from one form to another?

Procedure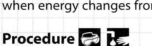

1. Read and complete a lab safety form.

2. Three students should sit in a circle. One student has 30 **buttons,** one has 30 **pennies,** and one has 30 **paper clips.**

3. Each student should exchange 10 items with the student to the right and 10 items with the student to the left.

4. Repeat step 3.

Think About This

1. If the buttons, the pennies, and the paper clips represent different forms of energy, what represents changes from one form of energy to another?

2. [Key Concept] If each button, penny, and paper clip represents one unit of energy, does the total amount of energy increase, decrease, or stay the same? Explain your answer.

Solar Energy in Florida

Using Sunshine to Power Our Lives

How many different ways have you used energy today? You might have ridden in a car or a bus on the way to school, used a hair dryer to get ready, or used a toaster to make breakfast. If you did, you used energy. Furnaces and stoves use thermal energy to heat buildings and to cook food. Cars and other vehicles use mechanical energy to carry people and materials from one part of the country to another.

Energy cannot be made; it must come from the natural world. The surface of Earth receives energy from two sources—the Sun and Earth's interior. Nearly all the energy you used today can be traced to the Sun.

Solar energy is energy from the Sun. Many devices use solar energy for power, including solar-powered calculators, home water-heating tanks, and solar-powered cookers for cooking outdoors. If you look around your city, you might see large, rectangular panels attached to the roofs of buildings or houses. These panels convert radiant energy from the Sun directly into electric energy. These panels are called solar panels.

Solar panels are built from several individual solar cells. The more solar cells in a panel, the more total electric output the panel can produce. Factors that can affect electric output are barriers to direct sunlight, such as buildings, and weather conditions, such as cloudy days.

Solar energy is the reason why the Florida Solar Energy Center (FSEC) exists. In 1975, the Florida legislature created the FSEC after the oil embargo of the 1970s. FSEC is located in the city of Cocoa, near the University of Central Florida. A few of their responsibilities are to conduct research, test and certify solar systems, and develop educational programs. FSEC has worked with the Department of Energy and Habitat for Humanity to construct energy-efficient homes.

This house, located in Tallahasee Florida, is completely solar powered.

It's Your Turn

RESEARCH AND REPORT Find out what solar cells are made of and how they work. Create a model of a solar cell to share with your classmates. LA.6.4.2.2

Energy RESOURCES

ESSENTIAL QUESTIONS

 What are nonrenewable energy resources?

 What are renewable energy resources?

 Why is it important to conserve energy?

Vocabulary

nonrenewable energy resource p. 156

fossil fuel p. 156

renewable energy resource p. 158

inexhaustible energy resource p. 159

Florida NGSSS

SC.6.N.1.5 Recognize that science involves creativity, not just in designing experiments, but also in creating explanations that fit evidence.

SC.7.P.11.3 Cite evidence to explain that energy cannot be created nor destroyed, only changed from one form to another.

SC.8.N.1.6 Understand that scientific investigations involve the collection of relevant empirical evidence, the use of logical reasoning, and the application of imagination in devising hypotheses, predictions, explanations and models to make sense of the collected evidence.

LA.6.2.2.3 The student will organize information to show understanding or relationships among facts, ideas, and events (e.g., representing key points within text through charting, mapping, paraphrasing, summarizing, or comparing/contrasting);

MA.6.A.3.6 Construct and analyze tables, graphs, and equations to describe linear functions and other simple relations using both common language and algebraic notation.

MA.6.S.6.2 Select and analyze the measures of central tendency or variability to represent, describe, analyze, and/or summarize a data set for the purposes of answering questions appropriately.

 Inquiry Launch Lab

SC.6.N.1.5, SC.8.N.1.6

15 minutes

How are energy resources different?

In this activity, the red beans represent an energy resource that is available in limited amounts. The white beans represent an energy resource that is available in unlimited amounts.

Procedure

1. Read and complete a lab safety form.
2. Place **40 red beans** and **40 white beans** in a **paper bag**. Mix the contents of the bag.
3. Each team should remove 20 beans from the bag without looking at the beans. Record the numbers of red and white beans in the space below.
4. Put the red beans aside. They are "used up." Return all the white beans to the bag. Mix the beans in the bag. Repeat steps 3 and 4 three more times.

Data and Observations

Think About This

1. What happened to the number of red beans drawn during each round?

2. What would eventually happen to the red beans in the bag?

3. 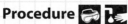 **Key Concept** How would changing the number of beans drawn in each round make the red beans last longer? Explain your answer.

Inquiry **Extracting Energy?**

1. Where does the electric energy come from when you turn on the lights in your home? The answer to that question depends on where you live. Different energy resources are used in different parts of the United States. How does the law of conservation of energy relate to the photograph?

_____

_____

_____

_____

Sources of Energy

Every day you use many forms of energy in many ways. According to the law of conservation of energy, energy cannot be created or destroyed. Energy can only change form. Where does all the energy that you use come from?

Almost all the energy you use can be traced back to the Sun, as shown in **Figure 10**. For example, the chemical energy in the food you eat originally came from the Sun. The energy in fuels, such as gasoline, coal, and wood, also came from the Sun. In addition, a small amount of energy that reaches Earth's surface comes from inside Earth. However, the amount of energy that comes from the Sun each day is about 5,000 times greater than the amount of energy that comes from inside Earth.

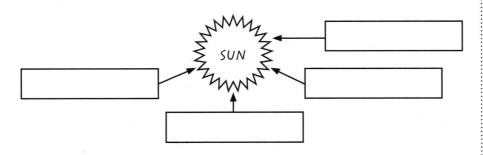

Active Reading

2. Identify Fill in the graphic organizer with sources of energy that can be traced back to the Sun.

Figure 10 These images are three examples of energy that originally came from the Sun.

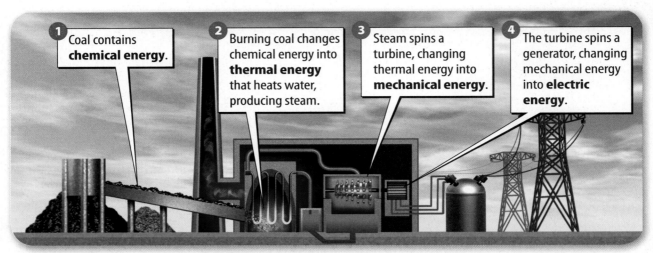

1 Coal contains **chemical energy**.

2 Burning coal changes chemical energy into **thermal energy** that heats water, producing steam.

3 Steam spins a turbine, changing thermal energy into **mechanical energy**.

4 The turbine spins a generator, changing mechanical energy into **electric energy**.

Figure 11 This coal-burning electric power plant transforms chemical energy stored in a fossil fuel into electric energy.

Active Reading

3 Find (Circle) the form of energy that changes water into steam.

Electric Power Plants

Most of the energy you use every day does not come directly from the Sun. Instead, much of the energy is in the form of electric energy. Most of the electric energy comes from electric power plants.

An electric power plant transforms the energy in an energy source into electric energy. There are three main energy sources used in power plants. One source of energy comes from burning fuels, such as coal. The power plant shown in **Figure 11** uses coal as an energy source. Nuclear power plants use the nuclear energy contained in uranium. Hydroelectric power plants convert the kinetic energy in falling water into electric energy.

 4. NGSSS Check Explain What is the energy transformation process to generate electricity? SC.7.P.11.2

_____

_____

_____

Nonrenewable Energy Resources

The coal that a power plant burns is an example of a nonrenewable energy resource. *A **nonrenewable energy resource** is an energy resource that is available in limited amounts or that is used faster than it is replaced in nature.*

Fossil Fuels

The most commonly used nonrenewable energy resources are fossil fuels. **Fossil fuels** *are the remains of ancient organisms that can be burned as an energy source.* Fossil fuels take millions of years to form. They are being used up much faster than they form. Three types of fossil fuels are coal, natural gas, and petroleum.

Active Reading

FOLDABLES LA.6.2.2.3

Make a shutter fold from a sheet of paper. Label it as shown. Use it to organize your notes on nonrenewable and renewable energy resources.

Nonrenewable Energy Resources

Renewable Energy Resources

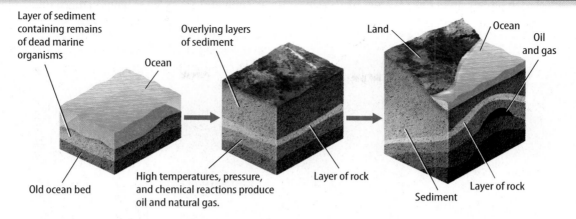

Layer of sediment containing remains of dead marine organisms
Ocean
Overlying layers of sediment
Land
Ocean
Oil and gas
Old ocean bed
High temperatures, pressure, and chemical reactions produce oil and natural gas.
Layer of rock
Sediment
Layer of rock

Figure 12 Petroleum and natural gas formed over millions of years from dead microscopic ocean organisms.

Active Reading 5. **Identify** Highlight the forces that act on the dead organisms in sediment and transforms these materials into oil and gas.

The Formation of Fossil Fuels

The processes that formed fossil fuels began at Earth's surface. Petroleum and natural gas formed from microscopic ocean organisms that died and sank to the ocean floor, as shown in **Figure 12**. These organisms were gradually buried under layers of sediment—sand and mud—and rock.

Over many millions of years, increasing temperature and pressure from the weight of the sediment and rock layers changed the dead organisms into petroleum and natural gas. Coal formed on land from plants that died millions of years ago and were buried under thick layers of sediment and rock.

Using Fossil Fuels

Fossil fuels formed from organisms that changed radiant energy from the Sun to chemical potential energy. The chemical potential energy stored in fossil fuels changes to thermal energy when fossil fuels burn.

Using Petroleum Gasoline, fuel oil, diesel, and kerosene are made from petroleum. These fuels are burned mainly to power cars, trucks, and planes and to heat buildings. Petroleum is also used as a raw material in making plastics and other materials.

Using Coal Electric power plants burn about 90 percent of the coal used in the United States. Coal is also used to directly heat buildings and to produce steel and concrete. Burning coal produces more pollutants than burning other fossil fuels. In some places, these pollutants react with water vapor in the air and create acid rain. Acid rain can harm organisms, such as trees and fish.

Active Reading 6. **Locate** <u>Underline</u> the percentage of coal that is burned in electric power plants in the United States.

Using Natural Gas About half of all homes in the United States use natural gas for heating. Electric power plants burn about 30 percent of the natural gas used in the United States. Burning natural gas produces less pollution than burning other fossil fuels.

Fossil Fuels and Global Warming

Burning fossil fuels releases carbon dioxide gas into Earth's atmosphere. Carbon dioxide is one of the gases that helps keep Earth's surface warm. However, over the past 100 years, Earth's surface has warmed by about 0.7°C. Some of this warming is due to the increasing amount of carbon dioxide produced by burning fossil fuels.

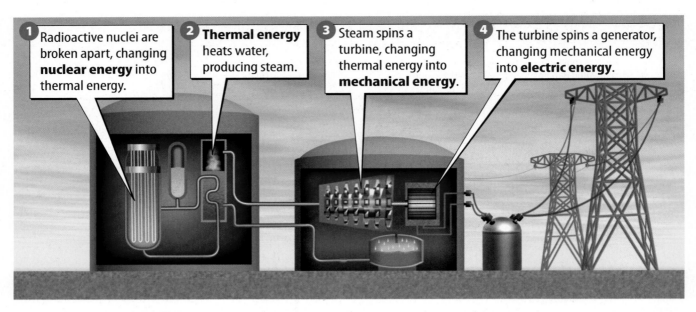

Figure 13 A nuclear power plant transforms nuclear energy into electric energy.

REVIEW VOCABULARY

nuclei
plural form of nucleus; the positively charged center of an atom that contains protons and neutrons.

Math Skills   MA.6.A.3.6

Solve a One-Step Equation
Electric energy is often measured in units called *kilowatt-hours* (kWh). To calculate the electric energy used by an appliance in kWh, use this equation:

$$kWh = (watts/1,000) \times hours$$

Practice

8. A hair dryer is rated at 1,200 watts. If you use the dryer for 0.25 hours, how much electric energy do you use?

Nuclear Energy

Humans can also transform the nuclear energy from uranium **nuclei** into thermal energy. Uranium is found in certain minerals, but significant amounts of uranium are no longer being formed inside Earth. As a result, nuclear energy from uranium is a nonrenewable energy resource.

Active Reading   **7. Determine** <u>Underline</u> why nuclear energy released from uranium nuclei is considered a nonrenewable energy resource.

Nuclear Power Plants In a nuclear power plant, breaking apart uranium nuclei transforms nuclear energy into thermal energy. As shown in **Figure 13**, this thermal energy changes water into the steam that spins the turbine. Unlike fossil fuel power plants, a nuclear power plant does not release pollutants into the air. However, a nuclear power plant does produce harmful nuclear waste.

Storing Nuclear Waste Nuclear waste contains radioactive materials that can damage living things. Some of these materials remain radioactive for thousands of years. Almost all nuclear waste in the United States is currently stored at the nuclear power plants where this waste is produced.

Renewable Energy Resources

Fossil fuels and uranium are being used up faster than they are being replaced. However, there are other energy resources that are not being used up. *A* **renewable energy resource** *is an energy resource that is replaced as fast as, or faster than, it is used.*

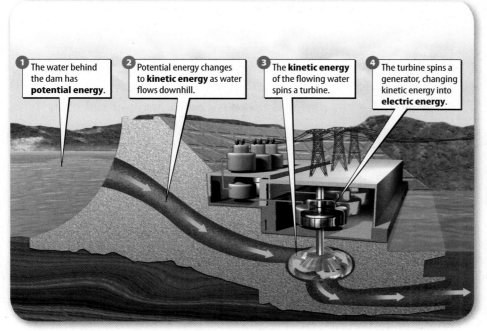

1. The water behind the dam has **potential energy**.

2. Potential energy changes to **kinetic energy** as water flows downhill.

3. The **kinetic energy** of the flowing water spins a turbine.

4. The turbine spins a generator, changing kinetic energy into **electric energy**.

Figure 14 A hydroelectric power plant converts the potential energy of the water stored behind the dam to electric energy.

Hydroelectric Power Plants

The most widely used renewable energy resource is falling water. To generate electric energy from falling water, a dam is built across a river, forming a reservoir (REH zuh vwor). As water falls through tunnels in the dam, the water's potential energy transforms into kinetic energy. **Figure 14** shows how a hydroelectric power plant transforms the kinetic energy in falling water into electric energy. Hydroelectric power plants do not emit pollutants. Florida has two hydroelectric plants, one in Tallahassee and one at the Apalachicola River on the Georgia-Florida border.

Solar Energy

Another renewable energy source is radiant energy from the Sun—solar energy. Because the Sun will produce energy for billions of years, solar energy is also an inexhaustible energy resource. *An* **inexhaustible energy resource** *is an energy resource that cannot be used up.* Less than about 0.1 percent of the energy used in the United States comes directly from the Sun.

Solar energy is converted directly into electric energy by solar cells. Solar cells contain materials that transform radiant energy into electric energy when sunlight strikes the solar cell. Solar cells can be placed on the roof of a building to provide electric energy, as shown in **Figure 15**.

Solar panels are being used in many homes and buildings throughout Florida. The Orange County Convention Center has roof top solar panels, which is the largest collection of solar panels in the southeastern part of the country.

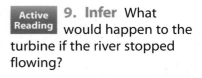

9. Infer What would happen to the turbine if the river stopped flowing?

_____

_____

_____

Figure 15 The panels on the roof of Lyman High School in Longwood, Florida, convert solar energy into electric energy.

Figure 16 Wind turbines convert the kinetic energy in wind into electric energy. Some wind turbines are over 60 m high.

 Active Reading

10. Infer Why are most wind turbines placed on the tops of hills and mountains?

_____

_____

Wind Energy

Wind energy is another inexhaustible energy resource. Modern wind turbines, such as the ones in **Figure 16**, convert the kinetic energy in wind into electric energy. Wind spins a propeller that is connected to an electric generator.

Wind turbines produce no pollution. However, wind turbines are practical only in regions where the average wind speed is more than about 5 m/s. Also, many wind turbines covering a large area are needed to obtain as much electric energy as one fossil fuel–burning power plant.

Biomass

People have often burned materials such as wood, dried peat moss, and manure to stay warm and cook food. These materials come from plants and animals and are called biomass. Because plant and animal materials can be replaced as fast as they are used, biomass is a renewable energy resource.

Some biomass is converted into fuels that can be burned in the engines of cars and other vehicles. Fuels made from biomass are often called biofuels. Using biofuels in vehicles can reduce the use of gasoline and make the supply of petroleum last longer.

Florida accounts for 7 percent of the total U.S. biomass output. Some of Florida's biomass products include sugarcane, citrus, and urban wood waste, as well as forest residues. Many Florida universities conduct biomass research, particularly in ways to use this material for energy.

Geothermal Energy

Thermal energy from inside Earth is called geothermal energy. This energy comes from the decay of radioactive nuclei deep inside Earth. In some places, geothermal energy produces underground pockets of hot water and steam. These pockets are called geothermal reservoirs.

In a few places, wells can be drilled to reach geothermal reservoirs. **Figure 17** shows a geothermal power plant. The hot water and steam in the geothermal reservoir are piped to the surface where they spin a turbine attached to an electric generator.

Figure 17 A geothermal power plant transforms the thermal energy from inside Earth into electric energy.

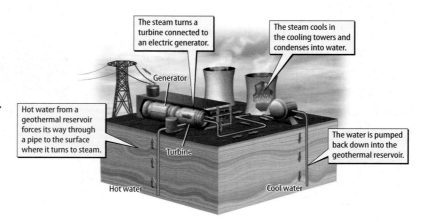

The steam turns a turbine connected to an electric generator.

The steam cools in the cooling towers and condenses into water.

Generator

Hot water from a geothermal reservoir forces its way through a pipe to the surface where it turns to steam.

The water is pumped back down into the geothermal reservoir.

Turbine

Hot water

Cool water

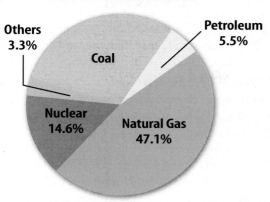

**Sources of Energy
Used in Florida in 2008**

- Others 3.3%
- Coal
- Petroleum 5.5%
- Nuclear 14.6%
- Natural Gas 47.1%

Figure 18 About 96 percent of the energy used in Florida comes from nonrenewable energy resources—fossil fuels and nuclear energy.

Active Reading **11. Calculate** What is the percentage of petroleum used in Florida in 2008?

_____

_____

_____

Conserving Energy Resources

The graph in **Figure 18** shows that fossil fuels provide about 82 percent of the energy used in Florida. Because fossil fuels are a nonrenewable energy resource, the supply decreases as they are used.

Because the supply of fossil fuels is decreasing, there could be shortages of fossil fuels in the future. Conserving energy is one way to reduce the rate at which all energy resources are used. Conserving energy means to avoid wasting energy. For example, turning off the lights when no one is in a room is a way to conserve energy.

In the future, energy resources besides fossil fuels might become more widely used. However, as **Table 2** on the next page shows, all energy resources have advantages and disadvantages. Comparing advantages and disadvantages will help determine which energy resources are used in the future.

Inquiry **LAB STATION** **Try It!**

SC.6.N.1.5, MA.6.A.3.6, LA.6.2.2.3

MiniLab *What energy resources provide our electric energy?* at connectED.mcgraw-hill.com

Apply It! After you complete the lab, answer these questions.

1. Suppose that you have two friends who both want to use an inexhaustible resource to heat their homes. One friend lives in Florida, and the other lives in Michigan. What suggestions could you offer?

2. Create a plan for using renewable or inexhaustible energy sources.

Active Reading **12. Explain** How does conserving energy affect the rate at which energy resources are used?

_____

_____

_____

Table 2 Advantages and Disadvantages of Energy Resources 🔑

Energy Resource	Advantages	Disadvantages
Nonrenewable Energy Resources		
Fossil Fuels	• Easy to transport • Widely available • Relatively inexpensive • Fossil fuel power plants are relatively inexpensive to operate.	• Drilling and surface mining may damage land and wildlife habitats. • Oil spills and leaks can harm wildlife. • Burning fossil fuels can produce air pollution. • Burning fossil fuels produces carbon dioxide that can cause global warming.
Nuclear Energy	• Nuclear power plants are relatively inexpensive to operate. • Does not produce air pollution	• Produces radioactive waste that is difficult to store • Accidents can result in dangerous leaks of radiation. • Nuclear power plants are relatively expensive to build.
Renewable Energy Resources		
Hydroelectric	• Does not pollute the air or water • Hydroelectric power plants are relatively inexpensive to operate.	• Dams damage wildlife habitats. • Dams can affect water quality and reduce the flow of water downstream • Droughts can affect hydroelectric power plants.
Solar	• Does not pollute the air or water • Theoretically inexhaustible supply	• The amount of solar energy that reaches Earth's surface varies, depending on the location, time of day, season, and weather conditions. • A large area is needed to collect enough solar energy for a solar power plant to be viable.
Wind	• Does not pollute the air or water • Can be used in isolated areas where electricity is unavailable	• Wind turbines can be noisy. • Can be disruptive to wildlife • Only generates electricity when the wind is blowing
Geothermal	• Does not pollute the air • Geothermal power plants are relatively inexpensive to operate.	• Geothermal reservoirs are located primarily in the western United States, Alaska, and Hawaii. • Some geothermal plants produce solid wastes that require careful disposal.
Biomass	• Biofuels could replace petroleum fuels in most vehicles.	• Energy from fossil fuels is used to grow biomass. • Some farm land is used to grow biomass instead of food crops. • Burning some biomass produces pollutants, such as smoke.

Active Reading   **13. Point Out** Highlight the energy source that can be noisy and disruptive to wildlife.

Visual Summary

Nonrenewable energy resources, such as fossil fuels, are used faster than they are replaced in nature.

Renewable energy resources, such as wind energy, are replaced in nature as fast as they are used.

Conserving energy, such as driving fuel-efficient cars, is one way to reduce the rate at which energy resources are used.

Inquiry

SC.6.N.1.5,
SC.7.N.1.3,
SC.7.P.11.2

①LAB
STATION **Try It!**

Inquiry Lab *Pinwheel*
Power at
connectED.mcgraw-hill.com

Use Vocabulary

1 **Define** *fossil fuel* in your own words.

_____

_____

Understand Key Concepts 🔑

2 **Compare and contrast** fossil fuels and biofuels.

_____

_____

3 **Explain** Conserving energy makes which type of energy resources last longer?

_____

_____

4 Which of these energy resources is effectively inexhaustible?
- (A) biomass
- (C) nuclear energy
- (B) fossil fuels
- (D) solar energy

Interpret Graphics

5 **Sequence** Sequence the energy transformations that occur in a coal-burning electric power plant. **SC.7.P.11.2**

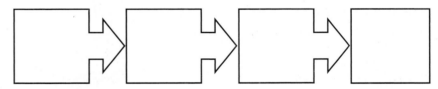

Critical Thinking

6 **Recommend** Fossil fuels form over millions of years. Use this information to explain why fossil fules are a nonrenewable energy resource.

_____

_____

Math Skills MA.6.A.3.6

7 A 22-W compact fluorescent lightbulb (CFL) produces as much light as a 100-W regular lightbulb. How much electric energy in kWh does each bulb use in 10 hours?

Think About It! Energy is the ability to cause change. Energy resources contain energy that can be transferred and transformed into other forms of energy.

Key Concepts Summary

	Vocabulary

LESSON 1 Forms of Energy

- **Energy** is the ability to cause change.
- **Kinetic energy** is the energy an object has because of its motion. **Potential energy** is stored energy.
- Different forms of energy include **thermal energy** and **radiant energy**.

energy p. 139
kinetic energy p. 140
potential energy p. 140
work p. 142
mechanical energy p. 143
sound energy p. 143
thermal energy p. 143
electric energy p. 143
radiant energy p. 143
nuclear energy p. 143

LESSON 2 Energy Transformations

- Any form of energy can be transformed into other forms of energy.
- According to the **law of conservation of energy**, energy can be transformed from one form into another or transferred from one region to another, but energy cannot be created or destroyed.
- **Friction** transforms mechanical energy into thermal energy.

law of conservation of energy p. 148
friction p. 149

LESSON 3 Energy Resources

- A **nonrenewable energy resource** is an energy resource that is available in a limited amount and can be used up.
- A **renewable energy resource** is replaced in nature as fast as, or faster, than it is used.
- Conserving energy, such as turning off lights when they are not needed, is one way to reduce the rate at which energy resources are used.

nonrenewable energy resource p. 178
fossil fuel p. 178
renewable energy resource p. 180
inexhaustible energy resource p. 181

Active Reading
FOLDABLES® Chapter Project

Assemble your Lesson Foldables as shown to make a Chapter Project. Use the project to review what you have learned in this chapter.

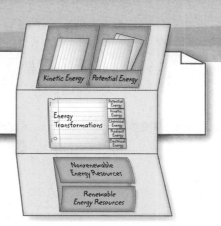

Use Vocabulary

Each of the following sentences is false. Make the sentence true by replacing the italicized word with a vocabulary term.

1 *Thermal energy* is the form of energy carried by an electric current.

_____

2 The *chemical potential energy* of an object depends on its mass and its speed.

_____

3 *Friction* is the transfer of energy that occurs when a force is applied over a distance.

_____

4 Natural gas is considered *an inexhaustible energy resource.*

_____

5 *Radiant energy* is energy that is stored in the nucleus of an atom.

_____

Link Vocabulary and Key Concepts

Use vocabulary terms from the previous page to complete the concept map.

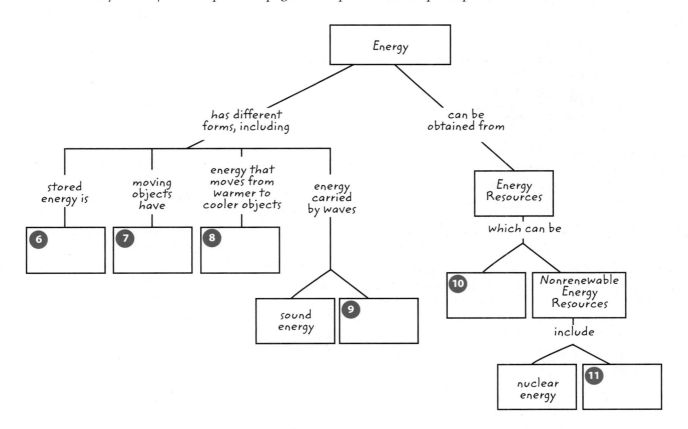

Fill in the correct answer choice.

🔑 Understand Key Concepts

1. What factors determine an object's kinetic energy? SC.6.P.11.1
   - (A) its height and its mass
   - (B) its mass and its speed
   - (C) its size and its weight
   - (D) its speed and its height

2. The gravitational potential energy stored between an object and Earth depends on SC.6.P.11.1
   - (A) the object's height and weight.
   - (B) the object's mass and speed.
   - (C) the object's size and weight.
   - (D) the object's speed and height.

3. When a ball is thrown upward, where does it have the least kinetic energy? SC.6.P.11.1
   - (A) at its highest point
   - (B) at its lowest point when it is moving downward
   - (C) at its lowest point when it is moving upward
   - (D) midway between its highest point and its lowest point

4. What energy transformation occurs in the panels on this roof? SC.7.P.11.2

   - (A) chemical energy to thermal energy
   - (B) nuclear energy to electric energy
   - (C) radiant energy to electric energy
   - (D) sound energy to thermal energy

5. According to the law of conservation of energy, which is always true? SC.6.P.11.1
   - (A) Energy can never be created or destroyed.
   - (B) Energy is always converted to friction in moving objects.
   - (C) The universe is always gaining energy in many different forms.
   - (D) Work is done when a force is exerted on an object.

Critical Thinking

6. **Determine** if work is done on the nail shown below if a person pulls the handle to the left and the handle moves. Explain. SC.6.P.11.1

   _____

   _____

7. **Contrast** the energy transformations that occur in a electrical toaster oven and in an electrical fan. SC.7.P.11.2

   _____

   _____

8. **Infer** A red box and a blue box are on the same shelf. There is more gravitational potential energy between the red box and Earth than between the blue box and Earth. Which box weighs more? Explain your answer. SC.6.P.13.2

   _____

   _____

   _____

9. **Infer** Juanita moves a round box and a square box from a lower shelf to a higher shelf. The gravitational potential energy for the round box increases by 50 J. The gravitational potential energy for the square box increases by 100 J. On which box did Juanita do more work? Explain your reasoning. SC.6.P.11.1

   _____

   _____

10. **Explain** why a skateboard coasting on a flat surface slows down and comes to a stop. SC.6.P.11.1

    _____

    _____

11 **Describe** the difference between the law of conservation of energy and what is meant by conserving energy. SC.6.P.11.1

_____

_____

_____

12 **Decide** Harold stretches a rubber band and lets it go. The rubber band flies across the room. He says this demonstrates the transformation of kinetic energy to elastic potential energy. Is Harold correct? Explain. SC.7.P.11.2

_____

_____

_____

Writing in Science

13 **Write** a short essay on a separate sheet of paper explaining which energy resources you think will be most important in the future. LA.6.2.2.3

Big Idea Review

14 Write an explanation of energy and energy resources for a fourth grader who has never heard of these terms. LA.6.2.2.3

_____

_____

_____

15 Identify five energy transformations in the photo below. SC.7.P.11.2

Math Skills MA.6.A.3.6

16 An electrical water heater is rated at 5,500 W and operates for 106 h per month. How much electric energy in kWh does the water heater use each month?

17 A family uses 1,303 kWh of electric energy in a month. If the power company charges $0.08 cents per kilowatt hour, what is the total electric energy bill for the month?

Fill in the correct answer choice.

Multiple Choice

1 According to the law of conservation of energy, what happens to the total amount of energy in the universe? SC.7.P.11.2

   (A) It remains constant.

   (B) It changes constantly.

   (C) It increases.

   (D) It decreases.

Use the diagram below to answer questions 2 and 3.

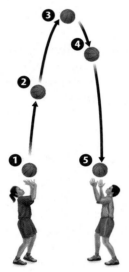

2 At which points is the kinetic energy of the basketball greatest? SC.6.P.11.1

   (F) 1 and 5

   (G) 2 and 3

   (H) 2 and 4

   (I) 3 and 4

3 At which point is the gravitational potential energy at its maximum? SC.6.P.11.1

   (A) 1

   (B) 2

   (C) 3

   (D) 4

Use the diagram below to answer question 4.

Vehicle	Mass	Speed
Car 1	1,200 kg	20 m/s
Car 2	1,500 kg	20 m/s
Truck 1	4,800 kg	20 m/s
Truck 2	6,000 kg	20 m/s

4 Which vehicle has the most kinetic energy? SC.6.P.11.1

   (F) car 1

   (G) car 2

   (H) truck 1

   (I) truck 2

5 In which situation would the gravitational potential energy between you and Earth be greatest? SC.6.P.13.2

   (F) You are running down a hill.

   (G) You are running up a hill.

   (H) You stand at the bottom of a hill.

   (I) You stand at the top of a hill.

6 Which energy source transforms potential energy into electrical energy? SC.7.P.11.2

   (F) geothermal plant

   (G) hydroelectric plant

   (H) nuclear energy plant

   (I) solar energy

7 A bicyclist uses brakes to slow from 3 m/s to a stop. What stops the bike? SC.6.P.11.1

   (A) cohesion

   (B) acceleration

   (C) friction

   (D) gravity

Use the graph below to answer question 8.

8 The work being done in the diagram above transfers energy to SC.7.P.11.2

- Ⓕ the box.
- Ⓖ the floor.
- Ⓗ the girl.
- Ⓘ the shelf.

9 Which is true of energy? SC.6.P.11.1

- Ⓐ It cannot be created or destroyed.
- Ⓑ It cannot change form.
- Ⓒ Most forms cannot be conserved.
- Ⓓ Most forms cannot be traced to a source.

10 Which type of energy is released when a firecracker explodes? SC.6.P.11.1

- Ⓕ chemical potential energy
- Ⓖ elastic potential energy
- Ⓗ electric energy
- Ⓘ nuclear energy

11 Which energy transformation is occurring in the food below? SC.7.P.11.2

- Ⓐ chemical energy to mechanical energy
- Ⓑ electric energy to radiant energy
- Ⓒ nuclear energy to thermal energy
- Ⓓ radiant energy to thermal energy

12 A block of metal is on a table. Which must be true for thermal energy to flow from the table to the metal? SC.7.P.11.2

- Ⓕ Both the table and the metal must be cooler than the surrounding air.
- Ⓖ The metal must be cooler than the table.
- Ⓗ The table and the metal must be in thermal equilibrium.
- Ⓘ The table must be cooler than the metal.

13 Inside the engine of a gasoline-powered car, chemical energy is converted primarily to which kind of energy? SC.7.P.11.2

- Ⓐ electric
- Ⓑ potential
- Ⓒ sound
- Ⓓ waste

14 What is the sequence for the energy transformation that occurs in a hydroelectric plant? SC.7.P.11.2

- Ⓕ chemical energy to kinetic energy to electric energy
- Ⓖ kinetic energy to mechanical energy to electric energy
- Ⓗ mechanical energy to potential energy to electric energy
- Ⓘ potential energy to kinetic energy to electric energy

NEED EXTRA HELP?

If You Missed Question...	1	2	3	4	5	6	7	8	9	10	11	12	13	14
Go to Lesson...	1	2	2	1	1	3	2	1	2	1	2	1	2	3

Multiple Choice *Bubble the correct answer.*

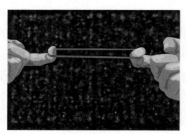

1. Which kind of energy does the image above illustrate? **SC.6.P.11.1**

- (A) active potential energy
- (B) chemical potential energy
- (C) elastic potential energy
- (D) gravitational potential energy

2. Which object will have the MOST kinetic energy? **SC.6.P.11.1**

- (F) a 0.045-kg golf ball traveling at 41 m/s
- (G) a 0.057-kg tennis ball traveling at 20 m/s
- (H) a 0.14-kg baseball traveling at 40 m/s
- (I) a 5-kg bowling ball traveling at 3 m/s

3. Which changes when the kinetic energy of an object changes? **SC.6.P.11.1**

- (A) the object's chemical potential energy
- (B) the object's volume
- (C) the object's direction of motion
- (D) the object's speed

4. Which image below shows the energy that is created by the motion of atoms and molecules within an object? **SC.7.N.1.6**

(F)

(G)

(H)

(I)

Copyright © Glencoe/McGraw-Hill, a division of The McGraw-Hill Companies, Inc.

Multiple Choice *Bubble the correct answer.*

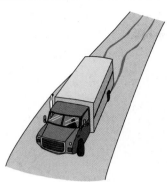

1. Identify the process shown in the image above. **SC.7.P.11.2**

- (A) friction
- (B) work
- (C) chemical energy
- (D) potential energy

2. What kind of device can convert chemical energy to thermal energy? **SC.7.P.11.3**

- (F) bicycle
- (G) microwave
- (H) cellular phone
- (I) gas engine

3. An efficient lightbulb is likely to produce the **SC.7.P.11.1**

- (A) least amount of heat.
- (B) least amount of light.
- (C) most amount of energy.
- (D) most amount of waste energy.

4. Which image represents the energy of a basketball just as it leaves a player's hands? **SC.7.P.11.2**

(F)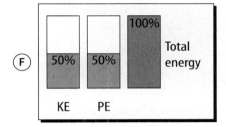

(G)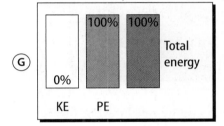

(H)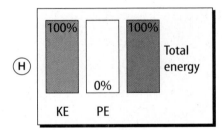

(I)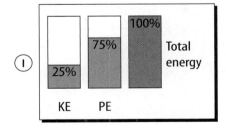

Copyright © Glencoe/McGraw-Hill, a division of The McGraw-Hill Companies, Inc.

Benchmark Mini-Assessment   Chapter 4 • Lesson 3

mini BAT

Multiple Choice *Bubble the correct answer.*

Use the following images to answer questions 1 and 2.

W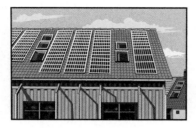

X

Y

Z

1. Which energy resource does NOT cause air pollution but is not practical in all areas? **SC.7.N.1.6**

Ⓐ W

Ⓑ X

Ⓒ Y

Ⓓ Z

2. Which energy resource uses hot water and steam? **SC.7.P.11.2**

Ⓕ W

Ⓖ X

Ⓗ Y

Ⓘ Z

3. Identify the energy resource that is composed of plant materials and is renewable. **SC.7.P.11.2**

Ⓐ coal

Ⓑ biomass

Ⓒ petroleum

Ⓓ natural gas

4. Which type of energy resource produces electricity by converting gravitational potential energy to kinetic energy? **SC.6.P.11.1**

Ⓕ geothermal energy

Ⓖ hydroelectric power

Ⓗ nuclear energy

Ⓘ wind energy

Copyright © Glencoe/McGraw-Hill, a division of The McGraw-Hill Companies, Inc.

Notes

Notes

Cool it!

Sally's mother was about to drink a cup of hot tea. She noticed the tea was too hot to drink. Sally put an ice cube in her mother's tea so that it would be cool enough to drink. Sally and her family each had different ideas about why the tea cooled off after adding an ice cube. This is what they said:

Sally: I think the cold from the ice cube transferred to the hot tea, and this is what cooled it off.

Sally's Mom: I think the cold from the ice and the heat energy from the tea moved back and forth until the tea cooled off.

Sam: I think it cooled the tea because heat energy from the hot tea transferred to the cold ice cube.

Sally's Dad: I don't think the ice cube made any difference. It was the air that cooled it off.

With whom do you most agree? Explain why you agree.

Thermal ENERGY

FLORIDA BIG IDEAS

1 **The Practice of Science**

4 **Science and Society**

11 **Energy Transfer and Transformations**

Think About It!

How can thermal energy be used?

This image shows the thermal energy of cars moving in traffic. The white indicates regions of high thermal energy, and the dark blue indicates regions of low thermal energy.

1 What is thermal energy?

_____

_____

2 How do you think thermal energy is related to temperature and heat?

_____

_____

3 How do you think thermal energy can be used?

_____

_____

_____

Get Ready to Read

What do you think about thermal energy?

Before you read, decide if you agree or disagree with each of these statements. As you read this chapter, see if you change your mind about any of the statements.

	AGREE	DISAGREE
1 Temperature is the same as thermal energy.	☐	☐
2 Heat is the movement of thermal energy from a hotter object to a cooler object.	☐	☐
3 It takes a large amount of energy to significantly change the temperature of an object with a low specific heat.	☐	☐
4 The thermal energy of an object can never be increased or decreased.	☐	☐
5 Car engines create energy.	☐	☐
6 Refrigerators cool food by moving thermal energy from inside the refrigerator to the outside.		

 **There's More Online!**
Video • Audio • Review • ⓘLab Station • WebQuest • Assessment • Concepts in Motion • Multilingual eGlossary

Thermal Energy, Temperature, and HEAT

ESSENTIAL QUESTIONS

 How are temperature and kinetic energy related?

 How do heat and thermal energy differ?

Vocabulary

thermal energy p. 180

temperature p. 181

heat p. 183

 Florida NGSSS

SC.7.N.1.1 Define a problem from the seventh grade curriculum, use appropriate reference materials to support scientific understanding, plan and carry out scientific investigation of various types, such as systematic observations or experiments, identify variables, collect and organize data, interpret data in charts, tables, and graphics, analyze information, make predictions, and defend conclusions.

SC.7.N.1.6 Explain that empirical evidence is the cumulative body of observations of a natural phenomenon on which scientific explanations are based.

SC.7.P.11.4 Observe and describe that heat flows in predictable ways, moving from warmer objects to cooler ones until they reach the same temperature.

LA.6.2.2.3 The student will organize information to show understanding or relationships among facts, ideas, and events (e.g., representing key points within text through charting, mapping, paraphrasing, summarizing, or comparing/contrasting);

MA.6.A.3.6 Construct and analyze tables, graphs, and equations to describe linear functions and other simple relations using both common language and algebraic notation.

MA.6.S.6.2 Select and analyze the measures of central tendency or variability to represent, describe, analyze, and/or summarize a data set for the purposes of answering questions appropriately.

SC.7.P.11.4

 Launch Lab

15 minutes

How can you describe temperature?

Have you ever used Fahrenheit or Celsius to describe the temperature? Why can't you just make up your own temperature scale?

Procedure

1. Read and complete a lab safety form.

2. Use a **ruler** and a **permanent marker** to divide a **clear plastic straw** into equal parts. Number the lines. Give your scale a name.

3. Add a room-temperature **colored alcohol-water mixture** to an **empty plastic water bottle** until it is about $\frac{1}{4}$ full.

4. Place one end of the straw into the bottle with the tip just below the surface of the liquid. Seal the straw onto the bottle top with **clay.**

5. Place the bottle in a **hot water bath**, and observe the liquid in your straw.

Data and Observations

Think About This

1. Why is it important for scientists to use the same scale to measure temperature?

2.  **Key Concept** What are some ways to make the liquid in your thermometer rise or fall?

Insulating the Home

It's what's between the walls that matters.

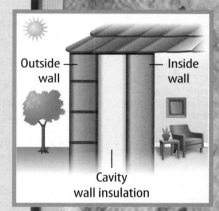

The first requirement of a shelter is to protect you from the weather. If the weather where you live is mild year-round, almost any kind of shelter will do. However, basic shelters, such as huts and tents, are not comfortable during cold winters or hot summers. Over many centuries, societies have experimented with thermal insulators that keep the inside of a shelter warm during winter and cool during summer.

One of the first thermal insulators used in shelters was air. Because air is a poor conductor of thermal energy, using cavity walls became a common form of insulating homes in the United States.

An air gap was not the perfect solution, however. Convection currents in the cavity carried some thermal energy across the gap. At first, no one seemed to mind. But, in the 1970s, the cost of heating and cooling homes suddenly increased. People began looking for a better way to reduce the transfer of thermal energy between the outside and the inside.

Outside wall — Inside wall

Cavity wall insulation

To meet the growing demand for better insulation, scientists began researching to find better insulation materials. If you could stop the convection currents, they reasoned, you could stop the transfer of thermal energy. One way was to use materials such as polymer foam or fiberglass. Each of these trapped air between the walls and held it there.

But how do you install insulation if your house is already built? You poke holes in the walls and blow it in! This process has little effect on your home's structure, and it decreases the cost of heating and cooling the house.

It's Your Turn

MAKE A POSTER A material's insulating ability is rated with an R-value. Find out what an R-value is. Then make a poster showing the ratings of common materials. LA.6.4.2.2

Using Thermal ENERGY

ESSENTIAL QUESTIONS

 How does a thermostat work?

 How does a refrigerator keep food cold?

 What are the energy transformations in a car engine?

Vocabulary

heating appliance p. 197
thermostat p. 198
refrigerator p. 198
heat engine p. 200

 Florida NGSSS

SC.6.N.1.5 Recognize that science involves creativity, not just in designing experiments, but also in creating explanations that fit evidence.

SC.7.N.1.3 Distinguish between an experiment (which must involve the identification and control of variables) and other forms of scientific investigation and explain that not all scientific knowledge is derived from experimentation.

SC.7.N.1.6 Explain that empirical evidence is the cumulative body of observations of a natural phenomenon on which scientific explanations are based.

SC.7.P.11.2 Investigate and describe the transformation of energy from one form to another.

SC.7.P.11.3 Cite evidence to explain that energy cannot be created nor destroyed, only changed from one form to another.

SC.7.P.11.4 Observe and describe that heat flows in predictable ways, moving from warmer objects to cooler ones until they reach the same temperature.

LA.6.2.2.3 The student will organize information to show understanding or relationships among facts, ideas, and events (e.g., representing key points within text through charting, mapping, paraphrasing, summarizing, or comparing/contrasting);

MA.6.S.6.2 Select and analyze the measures of central tendency or variability to represent, describe, analyze, and/or summarize a data set for the purposes of answering questions appropriately.

Also covers: SC.7.N.1.1

 Launch Lab
SC.7.P.11.2
15 minutes

How can you transform energy?

If you rub your hands together very quickly, do they become warm? Where does the thermal energy that raises the temperature of your hands come from?

Procedure 🗒️ 🤚

1. Read and complete a lab safety form.
2. Place a **thermometer strip** on the surface of a **block of wood.** Record the temperature after the thermometer stops changing color.
3. Rub the wood vigorously with **sandpaper** for 30 seconds. Quickly replace the thermometer, and record the temperature.
4. Repeat steps 2 and 3 on another part of the wood. This time, sand the wood for 60 seconds.

Data and Observations

	Starting temp (°C)	Ending temp (°C)
30 s		
60 s		

Think About This

1. Did the temperature of the wood change? Why or why not?

2. When did the wood have the highest temperature? Explain this result.

3. **Key Concept** What energy transformations take place in this activity?

 **Concentrating Energy?**

1. This power plant uses mirrors to focus light toward a tower. The tower then transforms some of the light into thermal energy. What are some advantages of using solar thermal energy?

_____

_____

_____

_____

_____

Thermal Energy Transformations

You can convert other forms of energy into thermal energy. Repeatedly stretching a rubber band makes it hot. Burning wood heats the air. A toaster gets hot when you turn it on.

You also can convert thermal energy into other forms of energy. Burning coal can generate electricity. Thermostats transform thermal energy into mechanical energy to switch heaters on and off. When you convert energy from one form to another, you can use the energy to perform useful tasks.

Remember that energy cannot be created or destroyed. Even though many devices transform energy from one form to another or transfer energy from one place to another, the total amount of energy does not change.

Active Reading **2. Identify** Write a phrase beside each paragraph that summarizes the main point of the paragraph. Use the phrases to review the lesson.

Heating Appliances

A device that converts electric energy into thermal energy is a **heating appliance.** Curling irons, coffeemakers, and clothes irons are some examples of heating appliances.

Other devices, such as computers and cell phones, also become warm when you use them. This is because some electric energy always is converted to thermal energy in an electronic device. However, the thermal energy that most electronic devices generate is not used for any purpose.

 3. NGSSS Check **Write** Complete the concept map of thermal energy transformation. SC.7.P.11.2

Thermal energy can convert to _____ _____	Devices that _____ _____ can perform many useful tasks.	Other forms of energy can convert to _____ _____

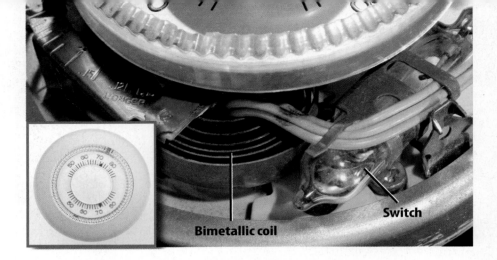

Figure 14 🔑 The coil in a thermostat contains two different metals that expand at two different rates.

Bimetallic coil

Switch

Active Reading **4.** How can using a thermostat save energy in your home?

_____

_____

_____

Active Reading

FOLDABLES® LA.6.2.2.3

Make a vertical four-tab book. Label it as shown. Use it to explain the energy transformation that occurs in each device.

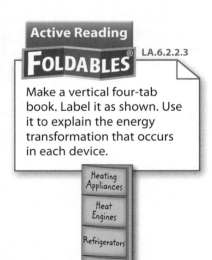

Heating Appliances

Heat Engines

Refrigerators

Thermostats

Thermostats

You might have heard the furnace in your house or in your classroom turn on in the winter. After the room warms, the furnace turns off. _A_ **thermostat** _is a device that regulates the temperature of a system._ Kitchen refrigerators, toasters, and ovens are all equipped with thermostats.

Most thermostats used in home heating systems contain a bimetallic coil. A bimetallic coil is made of two types of metal joined together and bent into a coil, as shown in **Figure 14.** The metal on the inside of the coil expands and contracts more than the metal on the outside of the coil. After the room warms, the thermal energy in the air causes the bimetallic coil to uncurl slightly. This moves a switch that turns off the furnace. As the room cools, the metal on the inside of the coil contracts more than the metal on the outside, curling the coil tighter. This moves the switch in the other direction, turning on the furnace.

Active Reading **5. Explain** How does the bimetallic coil in a thermostat respond to heating and cooling?

_____

_____

_____

Refrigerators

A device that uses electric energy to transfer thermal energy from a cooler location to a warmer location is called a **refrigerator.** Recall that thermal energy naturally flows from a warmer area to a cooler area. The opposite might seem impossible. But, that is exactly how your refrigerator works. So, how does a refrigerator move thermal energy from its cold inside to the warm air outside? Pipes that surround the refrigerator are filled with a fluid, called a coolant, that flows through the pipes. Thermal energy from inside the refrigerator transfers to the coolant, keeping the inside of the refrigerator cold.

Vaporizing the Coolant

A coolant is a substance that evaporates at a low temperature. In a refrigerator, a coolant is pumped through pipes on the inside and the outside of the refrigerator. The coolant, which begins as a liquid, passes through an expansion valve and cools. As the cold gas flows through pipes inside the refrigerator, it absorbs thermal energy from the refrigerator compartment and vaporizes. The coolant gas becomes warmer, and the inside of the refrigerator becomes cooler.

Condensing the Coolant

The coolant flows to an electric compressor at the bottom of the refrigerator. Here, the coolant is compressed, or forced into a smaller space, which increases its thermal energy. Then, the gas is pumped through condenser coils. In the coils, the thermal energy of the gas is greater than that of the surrounding air. This causes thermal energy to flow from the coolant gas to the air behind the refrigerator. As thermal energy is removed from the gas, it condenses, or becomes liquid. Then, the liquid coolant is pumped up through the expansion valve. The cycle repeats.

Active Reading 7. **Name** What type of energy does the coolant in a refrigerator move?

_____

_____

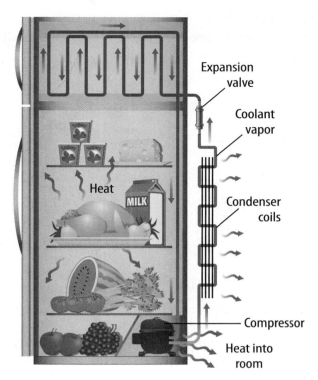

Figure 15 Coolant in a refrigerator moves thermal energy from inside to outside the refrigerator.

Active Reading 6. **Identify** What kind of energy compresses the coolant gas at the bottom of the refrigerator?

_____

_____

Inquiry SC.7.N.1.6

LAB STATION Try It!

MiniLab *Can thermal energy be used to do work?* at connectED.mcgraw-hill.com

Apply It! After you complete the lab, answer this question.

1. **Create** Based on what you have learned from the lab, design a device that can do useful work. Draw your device in the space below.

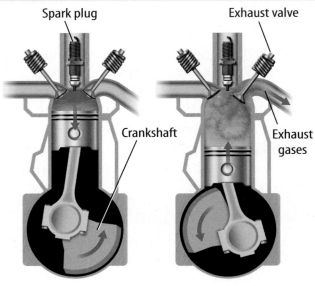

Intake valve | Fuel-air mixture | Spark plug | Exhaust valve

Cylinder | Piston | Crankshaft | Exhaust gases

① The intake valve opens as the piston moves downward, drawing a mixture of gasoline and air into the cylinder.

② The intake valve closes as the piston moves upward, compressing the fuel-air mixture.

③ A spark plug ignites the fuel-air mixture. As the mixture burns, hot gases expand, pushing the piston down.

④ As the piston moves up, the exhaust valve opens, and the hot gases are pushed out of the cylinder.

Figure 16 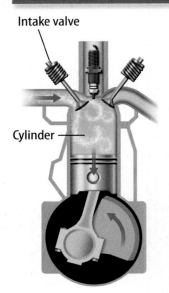 Internal combustion engines transform the chemical energy from fuel to thermal energy, which then produces mechanical energy.

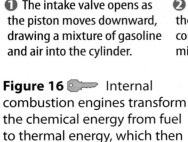

Active Reading **8. Explain** What do the arrows in the pictures represent?

_____

_____

_____

_____

Heat Engines

A typical automobile engine is a heat engine. *A* **heat engine** *is a machine that converts thermal energy into mechanical energy.* When a heat engine converts thermal energy into mechanical energy, the mechanical energy moves the vehicle. Most cars, buses, boats, trucks, and lawn mowers use a type of heat engine called an internal combustion engine. **Figure 16** shows how one type of internal combustion engine converts thermal energy into mechanical energy.

Perhaps you have heard someone refer to a car as having a six-cylinder engine. A cylinder is a tube with a piston that moves up and down. At one end of the cylinder, a spark ignites a fuel-air mixture. The ignited fuel-air mixture expands and pushes the piston down. This action occurs because the fuel's chemical energy converts to thermal energy. Some of the thermal energy immediately converts to mechanical energy.

A heat engine is not efficient. Most automobile engines only convert about 20 percent of the chemical energy in gasoline into mechanical energy. The remaining energy from the gasoline is lost to the environment.

Active Reading **9. Identify** What forms of energy are output from a heat engine?

_____

_____

Visual Summary

A bimetallic coil inside a thermostat controls a switch that turns a heating or cooling device on or off.

A refrigerator keeps food cold by moving thermal energy from the inside of the refrigerator out to the refrigerator's surroundings.

In a car engine, chemical energy in fuel is transformed into thermal energy. Some of this thermal energy is then transformed into mechanical energy.

SC.7.P.11.4,
SC.6.N.1.5,
SC.7.N.1.1,
SC.7.N.1.3,
MA.6.S.6.2

Inquiry

LAB STATION Try It!

Inquiry Lab *Design an Insulated Container* at connectED.mcgraw-hill.com

Use Vocabulary

1. A _____ is a device that converts electric energy into thermal energy.

2. **Explain** how an internal combustion engine works.

_____

_____

Understand Key Concepts 🔑

3. **Describe** the path of thermal energy in a refrigerator.

_____

_____

4. Which sequence describes the energy transformation in an automobile engine? SC.7.P.11.2

   (A) chemical → thermal → mechanical

   (B) thermal → kinetic → potential

   (C) thermal → mechanical → potential

   (D) thermal → chemical → mechanical

5. **Explain** how a thermostat uses electric energy, mechanical energy, and thermal energy.

_____

_____

_____

Interpret Graphics

6. **Sequence** Use the graphic organizer to show the steps involved in one cycle of an internal combustion engine. LA.6.2.2.3

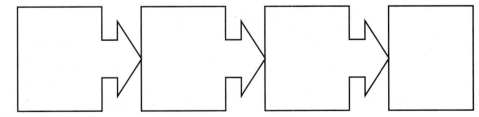

Critical Thinking

7. **Explain** how two of the devices you read about in this chapter could be used in one appliance.

_____

_____

_____

_____

Chapter 5 · Study Guide

Think About It! Thermal energy is conserved as it transfers by conduction, radiation, and convection. Thermal energy is transformed in devices such as thermostats, refrigerators, and automobile engines.

 Key Concepts Summary

Vocabulary

LESSON 1 Thermal Energy, Temperature, and Heat

- The **temperature** of a material is the average kinetic energy of the particles that make up the material.
- **Heat** is the movement of **thermal energy** from a material or an area with a higher temperature to a material or area with a lower temperature.
- When a material is heated, the material's temperature changes.

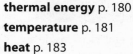

thermal energy p. 180
temperature p. 181
heat p. 183

LESSON 2 Thermal Energy Transfers

- When a material has a low **specific heat,** transferring a small amount of energy to the material increases its temperature significantly.
- When a material is heated, the thermal energy of the material increases and the material expands.
- Thermal energy can be transferred by **conduction, radiation,** or **convection.**

radiation p. 187
conduction p. 188
thermal conductor p. 188
thermal insulator p. 188
specific heat p. 189
thermal contraction p. 190
thermal expansion p. 190
convection p. 192
convection current p. 193

LESSON 3 Using Thermal Energy

- The two different metals in a bimetallic coil inside a **thermostat** expand and contract at different rates. The bimetallic coil curls and uncurls, depending on the thermal energy of the air, pushing a switch that turns a heating or cooling device on or off.
- A **refrigerator** keeps food cold by moving thermal energy from inside the refrigerator out to the refrigerator's surroundings.
- In a car engine, chemical energy in fuel is transformed into thermal energy. Some of this thermal energy is then transformed into mechanical energy.

heating appliance p. 197
thermostat p. 198
refrigerator p. 198
heat engine p. 200

Active Reading

FOLDABLES® Chapter Project

Assemble your lesson Foldables as shown to make a Chapter Project. Use the project to review what you have learned in this chapter.

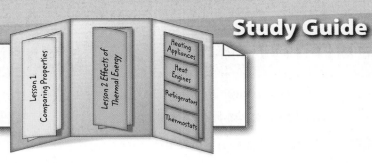

Use Vocabulary

1 When you increase the _____ of a cup of hot cocoa, you increase the average kinetic energy of the particles that make up the hot cocoa.

2 The increase in volume of a material when heated is _____.

3 A(n) _____ is used to control the temperature in a room.

4 Thermal energy is transferred by _____ between two objects that are touching.

5 A fluid moving in a circular pattern because of convection is a _____.

6 Define *heating appliance* in your own words.

_____

_____

_____

Link Vocabulary and Key Concepts

Use vocabulary terms from the previous page to complete the concept map.

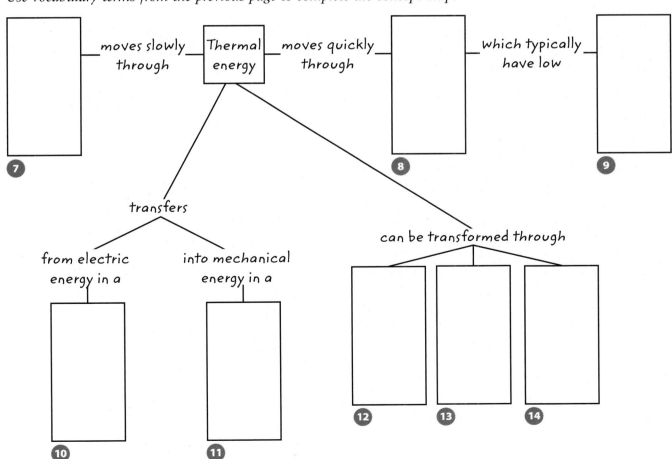

Fill in the correct answer choice.

🔑 Understand Key Concepts

1 Which would decrease a material's thermal energy? SC.7.P.11.4

Ⓐ heating the material

Ⓑ increasing the kinetic energy of the particles that make up the material

Ⓒ increasing the temperature of the material

Ⓓ moving the material to a location where the temperature is lower

2 You put a metal spoon in a bowl of hot soup. Why does the spoon feel hotter than the outside of the bowl? SC.7.P.11.4

Ⓐ The bowl is a better conductor than the spoon.

Ⓑ The bowl has a lower specific heat than the spoon.

Ⓒ The spoon is a good thermal insulator.

Ⓓ The spoon transfers thermal energy better than the bowl does.

3 In the picture to the right, thermal energy moves from the SC.7.P.11.4

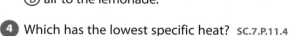

Ⓐ glass to the air.

Ⓑ lemonade to the air.

Ⓒ ice to the lemonade.

Ⓓ air to the lemonade.

4 Which has the lowest specific heat? SC.7.P.11.4

Ⓐ an object that is made out of metal

Ⓑ an object that does not transfer thermal energy easily

Ⓒ an object with electrons that do not move easily

Ⓓ an object that requires a lot of energy to change its temperature

5 Which does NOT occur in an internal combustion engine? SC.7.P.11.2

Ⓐ Most of the thermal energy is wasted.

Ⓑ Thermal energy forces the piston downward.

Ⓒ Thermal energy is converted into chemical energy.

Ⓓ Thermal energy is converted into mechanical energy.

Critical Thinking

6 **Compare** A swimming pool with a temperature of 30°C has more thermal energy than a cup of soup with a temperature of 60°C. Explain why this is so. SC.7.P.11.4

_____

_____

_____

_____

7 **Contrast** A spoon made of aluminum and a spoon made of steel have the same mass. The aluminum spoon has a higher specific heat than the steel spoon. Which spoon becomes hotter when placed in a pan of boiling water? SC.7.P.11.4

_____

_____

_____

_____

8 **Describe** How do convection currents influence Earth's climate? SC.7.P.11.4

_____

_____

_____

_____

_____

9 **Diagram** A room has a heater on one side and an open window letting in cool air on the opposite side. Diagram the convection current in the room. Label the warm air and the cool air. Use the space below to draw your diagram. SC.7.P.11.4

10 **Evaluate** When engineers build bridges, they separate sections of the roadway with expansion joints such as the one below that allow movement between the sections. Why are expansion joints necessary? **SC.7.P.11.4**

_____

_____

11 **Explain** Why is conduction slower in a gas than in a liquid or a solid? **SC.7.P.11.4**

_____

_____

Writing in Science

12 **Research** various types of heat engines that have been developed throughout history. Write 3–5 paragraphs on a separate sheet of paper explaining the energy transformations in one of these engines. **LA.6.2.2.3**

Big Idea Review

13 **Describe** each of the three ways thermal energy can be transferred. Give an example of each. **SC.7.P.11.4**

_____

_____

_____

14 What do the different colors in this photograph indicate? **SC.7.P.11.4**

_____

_____

_____

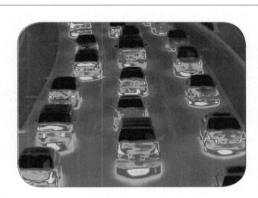

Math Skills   MA.6.A.3.6

Convert Between Temperature Scales

15 If water in a bath is at 104°F, then what is the temperature of the water in degrees Celsius?

16 Convert -40°C to degrees Fahrenheit.

Fill in the correct answer choice.

Multiple Choice

1 Which statement describes the thermal energy of an object? SC.7.P.11.4

(A) kinetic energy of particles + potential energy of particles

(B) kinetic energy of particles ÷ number of particles

(C) potential energy of particles ÷ number of particles

(D) kinetic energy of particles ÷ (kinetic energy of particles + potential energy of particles)

2 Which term describes a transfer of thermal energy? SC.7.P.11.2

(F) heat

(G) specific heat

(H) temperature

(I) thermal energy

Use the figures below to answer question 3.

Sample X   **Sample Y**

3 The figures show two different samples of air. In what way do they differ? SC.7.P.11.4

(A) Sample X is at a higher temperature than sample Y.

(B) Sample X has a higher specific heat than sample Y.

(C) Particles of sample Y have a higher average kinetic energy than those of sample X.

(D) Particles of sample Y have a higher average thermal energy than those of sample X.

Use the table below to answer question 4.

Material	Specific Heat (in J/g·K)
Air	1.0
Copper	0.4
Water	4.2
Wax	2.5

4 The table shows the specific heat of four materials. Which statement can be concluded from the information in the table? SC.7.P.11.4

(F) Copper is a thermal insulator.

(G) Wax is a thermal conductor.

(H) Air takes the most thermal energy to change its temperature.

(I) Water takes the most thermal energy to change its temperature.

5 Which term describes what happens to a cold balloon when placed in a hot car? SC.7.P.11.4

(A) thermal conduction

(B) thermal contraction

(C) thermal expansion

(D) thermal insulation

6 A girl stirs soup with a metal spoon. Which process causes her hand to get warmer? SC.7.P.11.4

(F) conduction

(G) convection

(H) insulation

(I) radiation

7 Which is the name for thermal energy that is transferred only from a high temperature to a lower temperature? SC.7.P.11.4

(A) potential energy

(B) kinetic energy

(C) heat

(D) conduction

Use the figure below to answer questions 8–9.

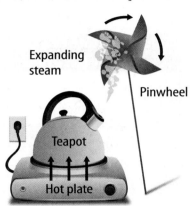

Expanding steam

Pinwheel

Teapot

Hot plate

8 Which term describes the transfer of thermal energy between the hot plate and the teapot? SC.7.P.11.4

Ⓕ conduction

Ⓖ convection

Ⓗ insulation

Ⓘ radiation

9 Which energy transformations are taking place in this system? SC.7.P.11.2

Ⓐ electrical → thermal → chemical

Ⓑ electrical → thermal → mechanical

Ⓒ thermal → electrical → chemical

Ⓓ thermal → electrical → mechanical

10 Which is an example of thermal energy transfer by conduction? SC.7.P.11.4

Ⓕ water moving in a pot of boiling water

Ⓖ warm air rising from hot pavement

Ⓗ the warmth you feel sitting near a fire

Ⓘ the warmth you feel holding a cup of hot tea

11 Which statement about radiation is correct? SC.7.P.11.4

Ⓐ In solids, radiation transfers electromagnetic energy but not thermal energy.

Ⓑ Cooler objects radiate the same amount of thermal energy as warmer objects.

Ⓒ Radiation occurs in fluids such as gas and water but not solids such as metals.

Ⓓ Radiation transfers thermal energy from the Sun to Earth.

12 The device below detects an increase in room temperature as SC.7.P.11.4

Ⓕ an increase in thermal energy causes a bimetallic coil to curl.

Ⓖ an increase in thermal energy causes a bimetallic coil to uncurl.

Ⓗ a switch causes a bimetallic coil to curl.

Ⓘ a switch causes a bimetallic coil to uncurl.

13 Which energy conversion typically occurs in a heating appliance? SC.7.P.11.2

Ⓐ chemical energy to thermal energy

Ⓑ electric energy to thermal energy

Ⓒ thermal energy to chemical energy

Ⓓ thermal energy to mechanical energy

NEED EXTRA HELP?

If You Missed Question...	1	2	3	4	5	6	7	8	9	10	11	12	13
Go to Lesson...	1	1	1	2	2	2	1	2	3	2	2	3	3

Multiple Choice *Bubble the correct answer.*

Use the image below to answer questions 1 and 2.

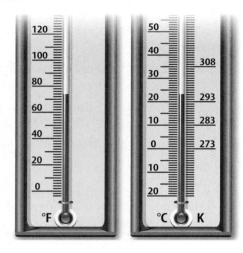

1. Look at the image above. A hot summer day is about 95°F. What is the temperature in Kelvin? **SC.7.P.11.4**

- (A) 35 K
- (B) 95 K
- (C) 308 K
- (D) 368 K

2. At which temperature does an object have the greatest average kinetic energy? **SC.6.P.11.1**

- (F) 0°F
- (G) 30°C
- (H) 40°F
- (I) 273 K

Use the image below to answer questions 3 and 4.

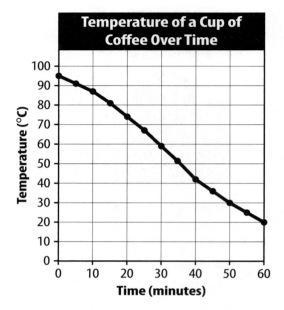

3. Which statement BEST describes the energy of the particles in the cup of coffee? **SC.7.P.11.4**

- (A) Kinetic energy is increasing.
- (B) Mechanical energy is decreasing.
- (C) Potential energy is increasing.
- (D) Thermal energy is decreasing.

4. Which statement correctly describes heat in the graph above? **SC.7.P.11.4**

- (F) Thermal energy from the air heats the coffee.
- (G) Thermal energy from the air heats the cup.
- (H) Thermal energy from the coffee heats the air.
- (I) Thermal energy from the cup heats the coffee.

Copyright © Glencoe/McGraw-Hill, a division of The McGraw-Hill Companies, Inc.

Multiple Choice *Bubble the correct answer.*

Use the image below to answer questions 1 and 2.

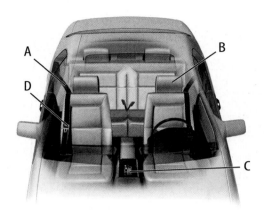

1. Based on the image above, which objects are conductors? **LA.6.2.2.3**

Ⓐ A and B

Ⓑ A and C

Ⓒ B and C

Ⓓ C and D

2. Based on the image above, which objects have a high specific heat? **LA.6.2.2.3**

Ⓕ A and B

Ⓖ A and C

Ⓗ B and C

Ⓘ C and D

3. An internal combustion engine converts thermal energy to which form of energy? **SC.7.P.11.2**

Ⓐ chemical

Ⓑ electrical

Ⓒ mechanical

Ⓓ radiant

4. Which is an example of radiation? **SC.7.P.11.2**

Ⓕ A cold glass of milk warms up while sitting on the table.

Ⓖ A cup of tea gets cooler when an ice cube is added to it.

Ⓗ A metal spoon gets warm when left on a sunny kitchen counter.

Ⓘ A pan of water gets hot after an electric stove is turned on.

Copyright © Glencoe/McGraw-Hill, a division of The McGraw-Hill Companies, Inc.

Benchmark Mini-Assessment Chapter 5 • Lesson 3

Multiple Choice *Bubble the correct answer.*

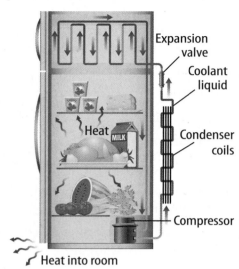

Expansion valve

Coolant liquid

Heat

Condenser coils

Compressor

Heat into room

1. Based on the figure above, how does a refrigerator transfer removed heat to the air? **SC.7.P.11.4**

Ⓐ The compressor increases the temperature of the coolant gas, which heats the air.

Ⓑ The compressor takes in air from outside the refrigerator, leading to a transfer of heat.

Ⓒ The condenser coils remove heat from the coolant liquid and turn it into a gas, which absorbs heat.

Ⓓ Thermal energy moves from inside the refrigerator into the colder coolant gas.

2. In which device is heat released but the heat is not used for any specific purpose? **SC.7.P.11.4**

Ⓕ cell phone

Ⓖ coffee maker

Ⓗ curling iron

Ⓘ electrical oven

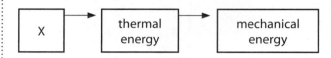

| X | → | thermal energy | → | mechanical energy |

3. The flowchart above shows energy conversions in a typical gas-powered car engine. What should appear in the box marked X? **SC.7.P.11.2**

Ⓐ chemical energy

Ⓑ electrical energy

Ⓒ mechanical energy

Ⓓ thermal energy

4. What kind of energy conversion takes place in a computer? **SC.7.P.11.2**

Ⓕ electrical energy to chemical energy

Ⓖ electrical energy to thermal energy

Ⓗ mechanical energy to thermal energy

Ⓘ thermal energy to mechanical energy

Copyright © Glencoe/McGraw-Hill, a division of The McGraw-Hill Companies, Inc.

Notes

Name _____ Date _____

PAGE KEELEY
**SCIENCE
PROBES**

Does amount matter?

Some properties of matter depend on how much matter there is. Put an *X* next to all of the statements you think are true about properties of matter.

_____ A. The more you have of a substance, the greater its density is.

_____ B. The more you have of a substance, the greater its volume is.

_____ C. The more you have of a substance, the higher the temperature needed to reach its boiling point.

_____ D. The more you have of a substance, the greater its mass is.

_____ E. The more you have of a substance, the lower the temperature needed to freeze it.

_____ F. The more you have of a substance, the less its electrical conductivity.

Explain your thinking. What rule or reasoning did you use to decide whether the amount of matter made a difference in its properties?

Foundations of
CHEMISTRY

FLORIDA BIG IDEAS

1   The Practice of Science
3   The Role of Theories, Laws,
     Hypotheses, and Models
8   Properties of Matter
9   Changes in Matter
11   Energy Transfer and Transformations
13   Forces and Changes in Motion

What is matter, and how does it change?

This siphonophore (si FAW nuh fawr) lives in the Arctic Ocean. Its tentacles have a very powerful sting. However, the most obvious characteristic of this organism is the way it glows.

1 What might cause the siphonophore to glow?

_____

2 How do you think its glow helps the siphonophore survive?

_____

3 What changes happen in the matter that makes up the organism?

_____

_____

Get Ready to Read

What do you think about chemistry?

Before you read, decide if you agree or disagree with each of these statements. As you read this chapter, see if you change your mind about any of the statements.

	AGREE	DISAGREE
1 The atoms in all objects are the same.	☐	☐
2 You cannot always tell by an object's appearance whether it is made of more than one type of atom.	☐	☐
3 The weight of a material never changes, regardless of where it is.	☐	☐
4 Boiling is one method used to separate parts of a mixture.	☐	☐
5 Heating a material decreases the energy of its particles.	☐	☐
6 When you stir sugar into water, the sugar and water evenly mix.	☐	☐
7 When wood burns, new materials form.	☐	☐
8 Temperature can affect the rate at which chemical changes occur.	☐	☐

There's More Online!
Video • Audio • Review • ⓘLab Station • WebQuest • Assessment • Concepts in Motion • Multilingual eGlossary

Classifying MATTER

ESSENTIAL QUESTIONS

🔑 What is a substance?

🔑 How do atoms of different elements differ?

🔑 How do mixtures differ from substances?

🔑 How can you classify matter?

Vocabulary

matter p. 217

atom p. 217

substance p. 219

element p. 219

compound p. 220

mixture p. 221

heterogeneous mixture p. 221

homogeneous mixture p. 221

dissolve p. 221

Florida NGSSS

SC.7.N.3.2 Identify the benefits and limitations of the use of scientific models.

SC.8.N.3.1 Select models useful in relating the results of their own investigations.

SC.8.P.8.5 Recognize that there are a finite number of elements and that their atoms combine in a multitude of ways to produce compounds that make up all of the living and nonliving things that we encounter.

SC.8.P.8.7 Explore the scientific theory of atoms (also known as atomic theory) by recognizing that atoms are the smallest unit of an element and are composed of sub-atomic particles (electrons surrounding a nucleus containing protons and neutrons).

SC.8.P.8.8 Identify basic examples of and compare and classify the properties of compounds, including acids, bases, and salts.

SC.8.P.8.9 Distinguish among mixtures (including solutions) and pure substances.

Also covers: SC.8.N.1.1, LA.6.2.2.3

 Launch Lab

15 minutes

How do you classify matter?

Can you classify an object based on its description?

Procedure

1. Read and complete a lab safety form.
2. Place a selection of **objects** on a table. Discuss how you might separate the objects into the following groups with these characteristics:

   a. Every object is the same and has only one part.

   b. Every object is the same but is made of more than one part.

   c. Individual objects are different. Some have one part, and others have more than one part.

3. Identify the objects that meet the requirements for group *a*, and record them below. Repeat with groups *b* and *c*. Any object can be in more than one group.

Data and Observations

Think About This

1. **Identify** Does any object from the selection belong in all three of the groups (*a*, *b*, and *c*)? Explain.

2. **Apply** What other objects in your classroom would fit into group *b*?

3. 🔑 **Key Concept** **Describe** What descriptions could you use to classify other items around you?

Chemical Properties and CHANGES

ESSENTIAL QUESTIONS

 What is a chemical property?

 What are some signs of chemical change?

 Why are chemical equations useful?

 What are some factors that affect the rate of chemical reactions?

Vocabulary

chemical property p. 242

chemical change p. 243

concentration p. 246

 Florida NGSSS

SC.6.N.1.5 Recognize that science involves creativity, not just in designing experiments, but also in creating explanations that fit evidence.

SC.7.N.1.3 Distinguish between an experiment (which must involve the identification and control of variables) and other forms of scientific investigation and explain that not all scientific knowledge is derived from experimentation.

SC.7.N.1.6 Explain that empirical evidence is the cumulative body of observations of a natural phenomenon on which scientific explanations are based.

SC.7.N.1.7 Explain that scientific knowledge is the result of a great deal of debate and confirmation within the science community.

SC.8.N.1.1 Define a problem from the eighth grade curriculum using appropriate reference materials to support scientific understanding, plan and carry out scientific investigations of various types, such as systematic observations or experiments, identify variables, collect and organize data, interpret data in charts, tables, and graphics, analyze information, make predictions, and defend conclusions.

SC.8.N.1.5 Analyze the methods used to develop a scientific explanation as seen in different fields of science.

SC.8.P.9.2 Differentiate between physical changes and chemical changes.

SC.8.P.9.3 Investigate and describe how temperature influences chemical changes.

LA.6.2.2.3 The student will organize information to show understanding or relationships among facts, ideas, and events (e.g., representing key points within text through charting, mapping, paraphrasing, summarizing, or comparing/contrasting);

 Launch Lab SC.8.N.1.1, SC.8.P.9.2

15 minutes

What can colors tell you?

You mix red and blue paints to get purple paint. Iron changes color when it rusts. Are color changes physical changes?

Procedure

1. Read and complete a lab safety form.

2. Divide a **paper towel** into thirds. Label one section *RCJ*, the second section *A*, and the third section *B*.

3. Dip one end of three **cotton swabs** into **red cabbage juice** (RCJ). Observe the color, and set the swabs on the paper towel, one in each of the three sections.

4. Add one drop of **substance A** to the swab in the *A* section. Observe any changes, and record observations.

5. Repeat step 4 with **substance B** and the swab in the *B* section.

6. Observe **substances C** and **D** in their **test tubes**. Then pour C into D. Rock the tube gently to mix. Record your observations.

Data and Observations

Think About This

1. **Describe** What happened to the color of the red cabbage juice when substances A and B were added?

2. **Key Concept** **Point Out** Which of the changes you observed was a physical change? Explain your reasoning.

1. As this car burns, some materials change to ashes and gases. The metal might change form if the fire is hot enough, but it probably won't burn. Why do fabric, leather, and paint burn but many metals do not? Think of something that burns. What are its properties before and after it burns?

_____

_____

_____

_____

Active Reading

2. Describe
Underline the definition for chemical property and give two examples.

_____

_____

Chemical Properties

Recall that a physical property is a characteristic of matter that you can observe or measure without changing the identity of the matter. However, matter has other properties that can be observed only when the matter changes from one substance to another. *A* **chemical property** *is a characteristic of matter that can be observed as it changes to a different type of matter.* For example, what are some chemical properties of a piece of paper? Can you tell by just looking at it that it will burn easily? The only way to know that paper burns is to bring a flame near the paper and watch it burn. When paper burns, it changes into different types of matter. The ability of a substance to burn is a chemical property. The ability to rust is another chemical property.

Comparing Properties

You now have read about physical properties and chemical properties. All matter can be described using both types of properties. For example, a wood log is solid, rounded, heavy, and rough. These are physical properties that you can observe with your senses. The log also has mass, volume, and density, which are physical properties that can be measured. The ability of wood to burn is a chemical property. This property is obvious only when you burn the wood. It also will rot, another chemical property you can observe when the log decomposes, becoming other substances. When you describe matter, you consider both its physical and its chemical properties.

Chemical Changes

Recall that during a physical change, the identity of matter does not change. However, *a **chemical change** is a change in matter in which the substances that make up the matter change into other substances with new physical and chemical properties.* For example, when iron undergoes a chemical change with oxygen, rust forms. The substances that undergo a change no longer have the same properties because they no longer have the same identity.

Active Reading

3. Differentiate Explain what happens to physical and chemical properties of matter during a physical and chemical changes.

Physical Change:

_____

_____

Chemical Change:

_____

_____

Active Reading

FOLDABLES® LA.6.2.2.3

Use a sheet of paper to make a chart with four columns. Use the chart throughout this lesson to explain how the identity of matter changes during a chemical change.

Action/ Matter	Signs of Chemical Change	Explain the Chemical Reaction	What affects the reaction rate?

WORD ORIGIN

chemical
from Greek *chemeia*, means "cast together"

Signs of Chemical Change

How do you know when a chemical change occurs? What signs show you that new types of matter form? As shown in **Figure 16**, signs of chemical changes include the formation of bubbles or a change in odor, color, or energy.

It is important to remember that these signs do not always mean a chemical change occurred. Think about what happens when you heat water on a stove. Bubbles form as the water boils. In this case, bubbles show that the water is changing state, which is a physical change. The types of evidence shown in **Figure 16** indicates that chemical changes might be occurring. However, the only proof that a chemical change has occurred is the formation of a new substance.

Figure 16 Sometimes you can observe clues that a chemical change has occurred.

Some Signs of Chemical Change 🗝

Bubbles   **Energy change**   **Odor change**   **Color change**

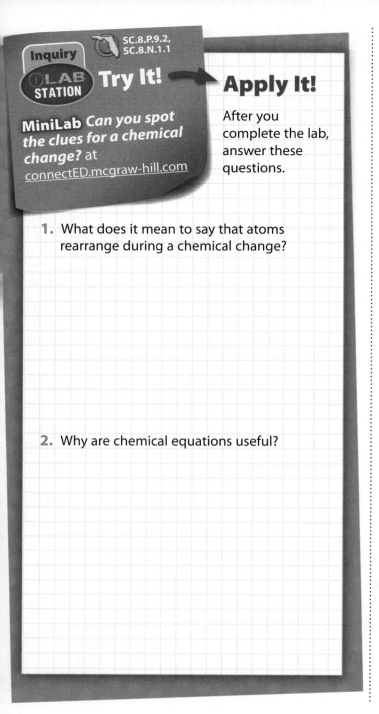

Apply It!

MiniLab *Can you spot the clues for a chemical change?* at connectED.mcgraw-hill.com

After you complete the lab, answer these questions.

1. What does it mean to say that atoms rearrange during a chemical change?

2. Why are chemical equations useful?

Explaining Chemical Reactions

You might wonder why chemical changes produce new substances. Recall that particles in matter are in constant motion. As particles move, they collide with each other. If the particles collide with enough force, the bonded atoms that make up the particles can break apart. These atoms then rearrange and bond with other atoms. When atoms bond in new combinations, new substances form. This process is called a chemical reaction. Chemical changes occur during chemical reactions.

Using Chemical Formulas

A useful way to understand what happens during a chemical reaction is to write a chemical equation. A chemical equation shows the chemical formula of each substance involved in the reaction. The formulas to the left of the arrow represent the reactants. Reactants are the substances present before the reaction takes place. The formulas to the right of the arrow represent the products. Products are the new substances present after the reaction. The arrow indicates that a reaction has taken place.

Active Reading 4. **Describe** Why are chemical equations useful?

_____

_____

_____

Figure 17 Chemical formulas and other symbols are parts of a chemical equation.

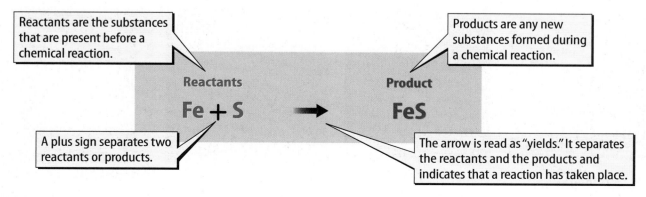

Reactants are the substances that are present before a chemical reaction.

Products are any new substances formed during a chemical reaction.

Reactants

Product

Fe + S ⟶ FeS

A plus sign separates two reactants or products.

The arrow is read as "yields." It separates the reactants and the products and indicates that a reaction has taken place.

Balancing Chemical Equations

Look at the equation in **Figure 17**. There is one iron (Fe) atom on the reactants side and one iron atom on the product side. This is also true for the sulfur (S) atoms. Recall that during both physical and chemical changes, mass is conserved—the total mass before and after a change must be equal. Therefore, in a chemical equation, the number of atoms of each element before a reaction must equal the number of atoms of each element after the reaction. A balanced chemical equation illustrates conservation of mass.

Figure 18 explains how to write and balance a chemical equation. When balancing an equation, you cannot change the chemical formula of any reactants or products.

Changing a formula changes the identity of the substance. Instead, you can place coefficients, or multipliers, in front of formulas. Coefficients change the amount of the reactants and products present. For example, an H_2O molecule has two H atoms and one O atom. Placing the coefficient *2* before H_2O ($2H_2O$) means that you double the number of H atoms and O atoms present:

> 2×2 H atoms = 4 H atoms
>
> 2×1 O atom = 2 O atoms

Note that $2H_2O$ is still water. However, it describes two water particles instead of one.

Figure 18 Equations must be balanced because mass is conserved during a chemical reaction.

Balancing Chemical Equations Example

When methane (CH_4)—a gas burned in furnaces—reacts with oxygen (O_2) in the air, the reaction produces carbon dioxide (CO_2) and water (H_2O). Write and balance a chemical equation for this reaction.

1. Write the equation, and check to see if it is balanced.

a. Write the chemical formulas with the reactants on the left side of the arrow and the products on the right side.	**a.** $CH_4 + O_2 \rightarrow CO_2 + H_2O$

b. Count the atoms of each element in the reactants and in the products.
■ Note which elements have a balanced number of atoms on each side of the equation.
■ If all elements are balanced, the overall equation is balanced. If not, go to step 2.
	b. reactants → products
	  C=1 C=1 **balanced**
	  H=4 H=2 **not balanced**
	  O=2 O=3 **not balanced**

2. Add coefficients to the chemical formulas to balance the equation.

a. Pick an element in the equation whose atoms are not balanced, such as hydrogen. Write a coefficient in front of a reactant or a product that will balance the atoms of the chosen element in the equation.	**a.** $CH_4 + O_2 \rightarrow CO_2 + 2H_2O$

b. Recount the atoms of each element in the reactants and the products, and note which are balanced on each side of the equation.	**b.** C=1 C=1 _____
	  H=4 H=4 _____
	  O=2 O=4 _____

c. Repeat steps 2a and 2b until all atoms of each element in the reactants equal those in the products.	**c.** $CH_4 + 2O_2 \rightarrow CO_2 + 2H_2O$
	  C=1 C=1 _____
	  H=4 H=4 _____
	  O=4 O=4 _____

3. Write the balanced equation that includes the coefficients:

Active Reading **5. Label** In parts b and c of step 2, decide whether the atoms are balanced or unbalanced on each side of the equation. Write your answer on the lines. Then write the balanced equation in step 3.

Factors that Affect the Rate of Chemical Reactions 🔑

Figure 19 The rate of most chemical reactions increases with an increase in temperature, concentration, or surface area.

1 Temperature

Chemical reactions that occur during cooking happen at a faster rate when temperature increases.

2 Concentration

Acid rain contains a higher concentration of acid than normal rain does. As a result, a statue exposed to acid rain is damaged more quickly than a statue exposed to normal rain.

3 Surface Area

When an antacid tablet is broken into pieces, the pieces have more total surface area than the whole tablet does. The pieces react more rapidly with water because more of the broken tablet is in contact with the water.

The Rate of Reactions

Recall that the particles that make up matter are constantly moving and colliding with one another. Different factors can make these particles move faster and collide harder and more frequently. These factors affect the rate of a chemical reaction, as shown in **Figure 19**.

1 A higher **temperature** usually increases the rate of reaction. When the temperature is higher, the particles move faster. Therefore, the particles collide with greater force and more frequently.

2 **Concentration** *is the amount of substance in a certain volume.* A reaction occurs faster if the concentration of at least one reactant increases. When concentration increases, there are more particles available to bump into each other and react.

3 **Surface area** also affects reaction rate if at least one reactant is a solid. If you drop a whole effervescent antacid tablet into water, the tablet reacts with the water. However, if you break the tablet into several pieces and then add them to the water, the reaction occurs more quickly. Smaller pieces have more total surface area, so more space is available for reactants to collide.

Chemistry

To understand chemistry, you need to understand matter. You must comprehend the arrangement of atoms in different types of matter. You also need to be able to distinguish physical properties from chemical properties and describe ways these properties can change. In later chapters and courses, you will examine each of these topics more closely.

Active Reading **6. Recognize** Describe two chemical changes that have happened in your home this week.

_____

_____

_____

A chemical property is observed only as a material undergoes a chemical change and changes identity.

Signs of a possible chemical change include bubbles, energy change, and change in odor or color.

Reactants		Product
Fe + S	→	FeS

Chemical equations show the reactants and products of a chemical reaction and that mass is conserved.

Inquiry

LAB STATION

SC.7.N.1.3,
SC.8.P.9.2,
SC.6.N.1.5,
SC.7.N.1.6,
SC.7.N.1.7,
SC.8.N.1.1,
SC.8.N.1.5

Try It!

Inquiry Lab *Design an Experiment to Solve a Crime* at connectED.mcgraw-hill.com

Use Vocabulary

1 The amount of substance in a certain volume is its _____.

2 **Use the term** *chemical change* in a complete sentence.

_____

_____

Understand Key Concepts

3 **List** some signs of chemical change. SC.8.P.9.2

_____

_____

4 Which property of matter changes during a chemical change but does NOT change during a physical change? SC.8.P.9.2

(A) energy   (C) mass

(B) identity   (D) volume

5 **State** why chemical equations are useful.

_____

_____

6 **Analyze** What affects the rate at which acid rain reacts with a statue?

_____

_____

Interpret Graphics

7 **Examine** Explain how the diagram below shows conservation of mass.

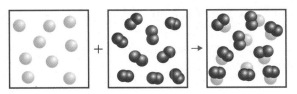

_____

_____

_____

Critical Thinking

8 **Compile** a list of three physical changes and three chemical changes you have observed recently. SC.8.P.9.2

_____

_____

_____

Chapter 6 Study Guide

Think About It! Matter takes up space and has mass, which gives it inertia. Matter can be classified by its physical and chemical properties.

🔑 Key Concepts Summary

| | **Vocabulary** |

LESSON 1 Classifying Matter

- A **substance** is a type of **matter** that always is made of atoms in the same combinations.
- **Atoms** of different elements have different numbers of protons.
- The composition of a substance cannot vary. The composition of a **mixture** can vary.
- Matter can be classified as either a substance or a mixture.

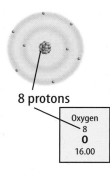

8 protons

Oxygen
8
O
16.00

matter p. 217
atom p. 217
substance p. 219
element p. 219
compound p. 220
mixture p. 221
heterogeneous mixture p. 221
homogeneous mixture p. 221
dissolve p. 221

LESSON 2 Physical Properties

- **Physical properties** of matter include size, shape, texture, and state.
- Physical properties such as **density**, melting point, boiling point, and size can be used to separate mixtures.

physical property p. 226
mass p. 228
density p. 229
solubility p. 230

LESSON 3 Physical Changes

- A change in energy can change the state of matter.
- When something dissolves, it mixes evenly in a substance.
- The masses before and after a change in matter are equal.

physical change p. 235

LESSON 4 Chemical Properties and Changes

- **Chemical properties** include ability to burn, acidity, and ability to rust.
- Some signs that might indicate **chemical changes** are the formation of bubbles and a change in odor, color, or energy.
- Chemical equations are useful because they show what happens during a chemical reaction.
- Some factors that affect the rate of chemical reactions are temperature, **concentration**, and surface area.

chemical property p. 242
chemical change p. 243
concentration p. 246

Active Reading

FOLDABLES® **Chapter Project**

Assemble your lesson Foldables as shown to make a Chapter Project. Use the project to review what you have learned in this chapter. Fasten the Foldable from Lesson 4 on the back of the board.

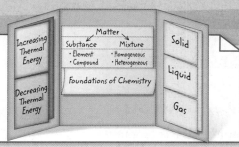

Use Vocabulary

Give two examples of each of the following.

1 element

_____

2 compound

_____

3 homogeneous mixture

_____

4 heterogeneous mixture

_____

5 physical property

_____

6 chemical property

_____

7 physical change

_____

8 chemical change

_____

Link Vocabulary and Key Concepts

Use vocabulary terms from the previous page to complete the concept map.

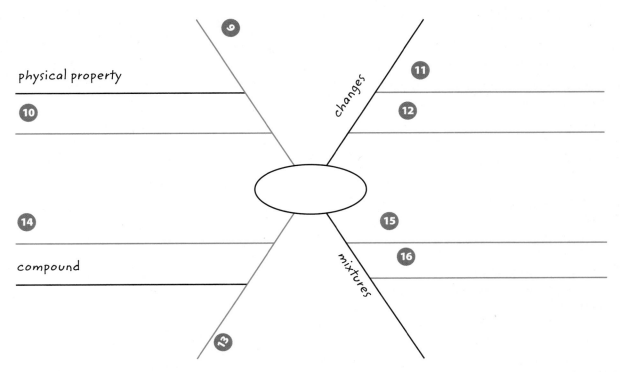

Fill in the correct answer choice.

🔑 Understand Key Concepts

1 The formula $AgNO_3$ represents a compound made of which atoms? SC.8.P.8.8

Ⓐ 1 Ag, 1 N, 1 O
Ⓑ 1 Ag, 1 N, 3 O
Ⓒ 1 Ag, 3 N, 3 O
Ⓓ 3 Ag, 3 N, 3 O

2 Which is an example of an element? SC.8.P.8.7

Ⓐ air
Ⓑ water
Ⓒ sodium
Ⓓ sugar

3 Which property explains why copper often is used in electrical wiring? SC.8.P.8.7

Ⓐ conductivity
Ⓑ density
Ⓒ magnetism
Ⓓ solubility

4 The table below shows densities for different substances.

Substance	Density (g/cm³)
1	1.58
2	0.32
3	1.52
4	1.62

For which substance would a 4.90-g sample have a volume of 3.10 cm³? SC.8.P.8.4

Ⓐ substance 1
Ⓑ substance 2
Ⓒ substance 3
Ⓓ substance 4

5 Which would decrease the rate of a chemical reaction?

Ⓐ increase in concentration
Ⓑ increase in temperature
Ⓒ decrease in surface area
Ⓓ increase in both surface area and concentration

Critical Thinking

6 **Compile** a list of ten materials in your home. Classify each material as an element, a compound, or a mixture. SC.8.P.8.9

_____
_____
_____
_____
_____
_____
_____

7 **Evaluate** Would a periodic table based on the number of electrons in an atom be as effective as the one shown in the back of this book? Why or why not? SC.8.P.8.7

_____
_____
_____

8 **Construct** an explanation for how the temperature and energy of a material changes during the physical changes represented by the diagram below. SC.8.P.8.1

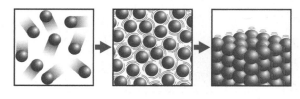

_____
_____
_____
_____

9 **Revise** the definition of physical change given in this chapter so it mentions the type and arrangement of atoms. SC.8.P.9.2

_____
_____
_____
_____
_____

10 **Find an example** of a physical change in your home or school. Describe the changes in physical properties that occur during the change. Then explain how you know the change is not a chemical change. SC.8.P.9.2

_____

_____

_____

_____

11 **Develop** a list of five chemical reactions you observe each day. For each, describe one way that you could either increase or decrease the rate of the reaction. SC.8.P.9.3

_____

_____

_____

_____

_____

_____

_____

_____

_____

Writing in Science

12 **Write** a poem at least five lines long to describe the organization of matter by the arrangement of its atoms. Be sure to include both the names of the different types of matter as well as their meanings. Use a separate sheet of paper LA.6.2.2.3

Big Idea Review

13 How does the chapter opener photo show an example of a physical change, a chemical change, a physical property, and a chemical property? SC.8.P.9.2

Physical Change _____

_____

Chemical Change _____

_____

Physical Property _____

_____

Chemical Property _____

_____

Math Skills   MA.6.A.3.6

Use Ratios

15 A sample of ice at 0°C has a mass of 23 g and a volume of 25 cm^3. Why does ice float on water? (The density of water is 1.00 g/cm^3.)

16 The table below shows the masses and the volumes for samples of two different elements.

Element	Mass (g)	Volume (cm^3)
Gold	386	20
Lead	22.7	2.0

Which element sample in the table has greater density?

Fill in the correct answer choice.

Multiple Choice

1 Which describes how mixtures differ from substances? SC.8.P.8.9

(A) Mixtures are homogeneous.

(B) Mixtures are liquids.

(C) Mixtures can be separated physically.

(D) Mixtures contain only one kind of atom.

Use the figure below to answer question 2.

A   **B**   **C**   **D**

2 Which image in the figure above is a model for a compound? SC.8.P.8.8

(F) A

(G) B

(H) C

(I) D

3 Which is a chemical property? SC.8.P.9.2

(A) the ability to be compressed

(B) the ability to be stretched into thin wire

(C) the ability to melt at low temperature

(D) the ability to react with oxygen

4 You drop a sugar cube into a cup of hot tea. What causes the sugar to disappear in the tea? SC.8.P.8.9

(F) It breaks into elements.

(G) It evaporates.

(H) It melts.

(I) It mixes evenly.

5 Which is an example of a substance? SC.8.P.8.9

(A) air

(B) lemonade

(C) soil

(D) water

Use the figure below to answer question 6.

6 The figure above is a model of atoms in a sample at room temperature. Which physical property does this sample have? SC.8.P.8.4

(F) It can be poured.

(G) It can expand to fill its container.

(H) It has a high density.

(I) It has a low boiling point.

7 Which observation is a sign of a chemical change? SC.8.P.9.2

(A) bubbles escaping from a carbonated drink

(B) iron filings sticking to a magnet

(C) lights flashing from fireworks

(D) water turning to ice in a freezer

8 Zinc, a solid metal, reacts with a hydrochloric acid solution. Which will increase the reaction rate? SC.8.P.9.3

(F) cutting the zinc into smaller pieces

(G) decreasing the concentration of the acid

(H) lowering the temperature of the zinc

(I) pouring the acid into a larger container

Use the figure below to answer question 9.

9 In the figure above, what will be the mass of the final solution if the solid dissolves in the water? SC.8.P.9.1

Ⓐ 5 g

Ⓑ 145 g

Ⓒ 150 g

Ⓓ 155 g

10 Which is NOT represented in a chemical equation? SC.8.P.9.2

Ⓕ chemical formula

Ⓖ product

Ⓗ conservation of mass

Ⓘ reaction rate

11 Which physical change is represented by the diagram below? SC.8.P.8.4

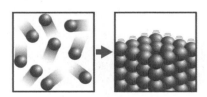

Ⓐ condensation

Ⓑ deposition

Ⓒ evaporation

Ⓓ sublimation

12 Which chemical equation is unbalanced? SC.8.P.8.4

Ⓕ $2KClO_3 \rightarrow 2KCl + 3O_2$

Ⓖ $CH_4 + 2O_2 \rightarrow CO_2 + 2H_2O$

Ⓗ $Fe_2O_3 + CO \rightarrow 2Fe + 2CO_2$

Ⓘ $H_2CO_3 \rightarrow H_2O + CO_2$

13 Which is a size-dependent property? SC.8.P.9.2

Ⓐ boiling point

Ⓑ conductivity

Ⓒ density

Ⓓ mass

14 Why is the following chemical equation said to be balanced? SC.8.P.9.2

$$O_2 + 2PCl_3 \rightarrow 2POCl_3$$

Ⓕ There are more reactants than products.

Ⓖ There are more products than reactants.

Ⓗ The atoms are the same on both sides of the equation.

Ⓘ The coefficients are the same on both sides of the equation.

15 The elements sodium (Na) and chlorine (Cl) react and form the compound sodium chloride (NaCl). Which is true about the properties of these substances? SC.8.P.8.8

Ⓐ Na and Cl have the same properties.

Ⓑ NaCl has the properties of Na and Cl.

Ⓒ All the substances have the same properties.

Ⓓ The properties of NaCl are different from the properties of Na and Cl.

NEED EXTRA HELP?

If You Missed Question...	1	2	3	4	5	6	7	8	9	10	11	12	13	14	15
Go to Lesson...	1	1	4	3	1	2	4	4	3	4	3	2	1	3	1

Multiple Choice *Bubble the correct answer.*

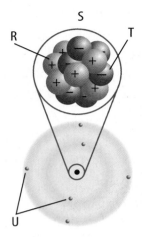

1. Which part of the atom in the diagram above determines the kind of atom it is? **SC.8.P.8.7**

(A) R

(B) S

(C) T

(D) U

2. The three types of elements are metals, nonmetals, and **SC.8.P.8.6**

(F) compounds.

(G) gases.

(H) liquids.

(I) metalloids.

Use the table below to answer questions 3 and 4.

Jar	Contents
J	salt water
K	rusted iron filings (Fe_2O_3)
L	crushed concrete (limestone, sand, and other rocks)
M	gold dust (Au)

3. Which jar contains the compound? **SC.8.P.8.8**

(A) Jar J

(B) Jar K

(C) Jar L

(D) Jar M

4. Which jar contains the heterogeneous mixture? **SC.8.P.8.9**

(F) Jar J

(G) Jar K

(H) Jar L

(I) Jar M

Copyright © Glencoe/McGraw-Hill, a division of The McGraw-Hill Companies, Inc.

Multiple Choice *Bubble the correct answer.*

	Diameter	Length
Rod #1	2 cm	12 cm
Rod #2	4 cm	32 cm
Rod #3	5 cm	40 cm

1. Which physical property of copper rods is being compared in the table above? **SC.8.P.8.2**

    (A) conductivity

    (B) density

    (C) mass

    (D) solubility

2. Object A is dropped into water and sinks. Object B is dropped into water and floats. What does this tell you about Object A compared to Object B? **SC.8.P.8.3**

    (F) Object A has greater volume.

    (G) Object A has more mass.

    (H) Object A is denser.

    (I) Object A is more soluble.

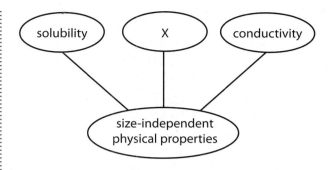

3. The diagram above shows three physical properties that are size-independent. Which property would NOT go in the oval that contains X? **SC.8.P.8.4**

    (A) density

    (B) volume

    (C) boiling point

    (D) state of matter

4. A container has a mixture of iron filings and sand. Which physical property can most likely help separate the substances? **SC.8.P.8.4**

    (F) density

    (G) magnetism

    (H) boiling point

    (I) melting point

Copyright © Glencoe/McGraw-Hill, a division of The McGraw-Hill Companies, Inc.

Multiple Choice *Bubble the correct answer.*

W	Cutting hair
X	Chopping onions
Y	Making lemonade
Z	Sanding wood

1. In the table above, which is NOT a physical change in size or shape? **LA.6.2.2.3**

   Ⓐ W

   Ⓑ X

   Ⓒ Y

   Ⓓ Z

2. Frozen carbon dioxide is better known as "dry ice." As it warms, it changes from a solid directly to a gas. This is an example of **SC.7.P.11.1**

   Ⓕ condensation.

   Ⓖ deposition.

   Ⓗ evaporation.

   Ⓘ sublimation.

3. Which is taking place when dew drops form? **SC.7.P.11.1**

   Ⓐ condensation

   Ⓑ deposition

   Ⓒ melting

   Ⓓ sublimation

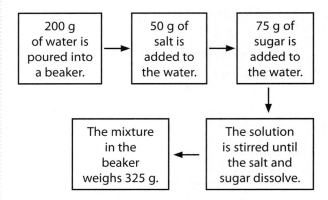

4. Which is demonstrated by the steps shown in the flow chart above? **SC.8.P.9.1**

   Ⓕ addition of thermal energy

   Ⓖ change in state of matter

   Ⓗ conservation of mass

   Ⓘ creation of a compound

Copyright © Glencoe/McGraw-Hill, a division of The McGraw-Hill Companies, Inc.

Multiple Choice *Bubble the correct answer.*

Chemical Change   **Physical Change**

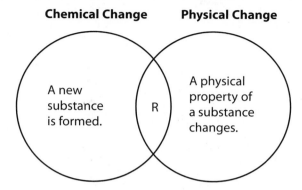

A new substance is formed.   R   A physical property of a substance changes.

1. The Venn diagram above compares chemical changes to physical changes. Which does NOT belong at R? **SC.8.P.9.2**

  (A) change in color

  (B) change in density

  (C) change in shape

  (D) change in state of matter

2. Over time, hydrogen peroxide (H_2O_2) naturally changes into water and oxygen: $2H_2O_2 \longrightarrow 2H_2O + O_2$. How do you know that this is a chemical change? **SC.8.P.9.2**

  (F) The chemical equation is balanced.

  (G) Hydrogen peroxide evaporates.

  (H) Hydrogen peroxide has more energy.

  (I) Two new substances were formed.

3. Which would NOT be a sign that a chemical change has occurred? **SC.8.P.9.2**

  (A) Lithium mixes with oxygen and produces a flame.

  (B) A piece of metal wire left outside turns orange with rust.

  (C) A puddle of water slowly disappears as it evaporates.

  (D) A rotting apple emits a bad odor.

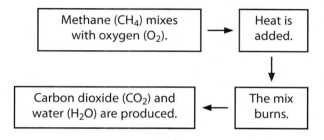

Methane (CH_4) mixes with oxygen (O_2). → Heat is added. ↓ The mix burns. ← Carbon dioxide (CO_2) and water (H_2O) are produced.

4. Which chemical equation correctly shows the chemical reaction from the flow chart above? **SC.8.P.9.2**

  (F) $CH_4 + O_2 \longrightarrow CO_2 + 2H_2O$

  (G) $CH_4 + 2O_2 \longrightarrow CO_2 + H_2O$

  (H) $CH_4 + 2O_2 \longrightarrow CO_2 + 2H_2O$

  (I) $2CH_4 + O_2 \longrightarrow 2CO_2 + H_2O$

Copyright © Glencoe/McGraw-Hill, a division of The McGraw-Hill Companies, Inc.

Name _____ Date _____

Notes

What's the difference?

Five friends were talking about the differences among solids, liquids, and gases. They each agreed that the differences have to do with the particles in each type of matter. However, they disagreed about which differences determine whether the matter is a solid, liquid, or gas. This is what they said:

Gwyneth: I think it has to do with the number of particles.

George: I think it has to do with the shape of the particles.

Hoda: I think it has to do with the size of the particles.

Natalie: I think it has to do with the movement of the particles.

William: I think it has to do with how hard or soft the particles are.

With whom do you agree most? _____ Explain why you agree with that friend.

States of
MATTER

FLORIDA BIG IDEAS

1   **The Practice of Science**
2   **The Characteristics of Scientific Knowledge**
3   **The Role of Theories, Laws, Hypotheses, and Models**
8   **Properties of Matter**
9   **Changes in Matter**
11 **Energy Transfer and Transformations**
13 **Forces and Changes in Motion**

What physical changes and energy changes occur as matter goes from one state to another?

When you look at this blob of molten glass, can you envision it as a beautiful vase? The solid glass was heated in a furnace until it formed a molten liquid. Air is blown through a pipe to make the glass hollow and give it form.

1   Can you identify a solid, a liquid, and a gas in the photo?

_____

_____

_____

2   What physical changes and energy changes do you think occurred when the glass changed state?

_____

_____

_____

What do you think about states of matter?

Before you read, decide if you agree or disagree with each of these statements. As you read this chapter, see if you change your mind about any of the statements.

	AGREE	DISAGREE
1 Particles moving at the same speed make up all matter.	☐	☐
2 The particles in a solid do not move.	☐	☐
3 Particles of matter have both potential energy and kinetic energy.	☐	☐
4 When a solid melts, thermal energy is removed from the solid.	☐	☐
5 Changes in temperature and pressure affect gas behavior.	☐	☐
6 If the pressure on a gas increases, the volume of the gas also increases.	☐	☐

There's More Online!
Video • Audio • Review • ⓘLab Station • WebQuest • Assessment • Concepts in Motion • Multilingual eGlossary

Solids, Liquids, and GASES

 How do particles move in solids, liquids, and gases?

 How are the forces between particles different in solids, liquids, and gases?

Vocabulary

solid p. 265

liquid p. 266

viscosity p. 266

surface tension p. 267

gas p. 268

vapor p. 268

Florida NGSSS

SC.7.N.1.6 Explain that empirical evidence is the cumulative body of observations of a natural phenomenon on which scientific explanations are based.

SC.8.P.8.1 Explore the scientific theory of atoms (also known as atomic theory) by using models to explain the motion of particles in solids, liquids, and gases.

SC.8.N.3.1 Select models useful in relating the results of their own investigations.

LA.6.2.2.3 The student will organize information to show understanding or relationships among facts, ideas, and events (e.g., representing key points within text through charting, mapping, paraphrasing, summarizing, or comparing/contrasting);

SC.8.N.3.1

 **Launch Lab**

10 minutes

How can you see particles in matter?

It's sometimes difficult to picture how tiny objects, such as the particles that make up matter, move. However, you can use other objects to model the movement of these particles.

Procedure

1. Read and complete a lab safety form.

2. Place about 50 **copper pellets** into a **plastic petri dish**. Place the cover on the dish, and secure it with **tape**.

3. Hold the dish by the edges. Gently vibrate the dish from side to side no more than 1–2 mm. Observe the pellets. Record your observations.

4. Repeat step 3, vibrating the dish less than 1 cm from side to side.

5. Repeat step 3, vibrating the dish 3–4 cm from side to side.

Data and Observations

Think About This

1. **Infer** If the pellets represent particles in matter, what do you think the shaking represents?

2. **Assess** In which part of the experiment do you think the pellets were like a liquid? Explain.

3. **Key Concept** **Consider** If the pellets represent molecules of water, what do you think are the main differences among molecules of ice, water, and vapor?

Changes in STATE

ESSENTIAL QUESTIONS

 How is temperature related to particle motion?

How are temperature and thermal energy different?

What happens to thermal energy when matter changes from one state to another?

Vocabulary

kinetic energy p. 272

temperature p. 272

thermal energy p. 273

vaporization p. 275

evaporation p. 276

condensation p. 276

sublimation p. 276

deposition p. 276

 Florida NGSSS

SC.6.P.11.1 Explore the Law of Conservation of Energy by differentiating between potential and kinetic energy. Identify situations where kinetic energy is transformed into potential energy and vice versa.

SC.6.P.13.1 Investigate and describe types of forces including contact forces and forces acting at a distance, such as electrical, magnetic, and gravitational.

SC.7.N.1.6 Explain that empirical evidence is the cumulative body of observations of a natural phenomenon on which scientific explanations are based.

SC.7.P.11.1 Recognize that adding heat to or removing heat from a system may result in a temperature change and possibly a change of state.

SC.8.N.1.3 Use phrases such as "results support" or "fail to support" in science, understanding that science does not offer conclusive 'proof' of a knowledge claim.

SC.8.P.8.1 Explore the scientific theory of atoms (also known as atomic theory) by using models to explain the motion of particles in solids, liquids, and gases.

SC.8.P.9.1 Explore the Law of Conservation of Mass by demonstrating and concluding that mass is conserved when substances undergo physical and chemical changes.

Also covers: SC.8.N.1.1, LA.6.2.2.3, MA.6.A.3.6

(Inquiry) Launch Lab

10 minutes

Do liquid particles move?

If you look at a glass of milk sitting on a table, it appears to have no motion. But appearances can be deceiving!

Procedure

1. Read and complete a lab safety form.
2. Use a **dropper** to place one drop of **2 percent milk** on a **glass slide**. Add a **cover slip**.
3. Place the slide on a **microscope** stage, and focus on low power. Focus on a single globule of fat in the milk. Observe the motion of the globule for several minutes. Record your observations.

Data and Observations

Think About This

1. **Explain** Describe the motion of the fat globule.

2. **Infer** What do you think caused the motion of the globule?

3. **Key Concept** What do you think would happen to the motion of the fat globule if you warmed the milk? Explain.

Steam Heat?

1. When you look at this tropical Florida scene, you probably don't think about states of matter. However, water is one of the few substances that you frequently observe as three states of matter at Earth's temperatures. What energy changes are involved when water changes state?

_____

_____

_____

_____

Kinetic and Potential Energy

When snow begins to melt after a snowstorm, all three states of water are present. The snow is a solid, the melted snow is a liquid, and the air above the snow and ice contains water vapor, a gas. What causes particles to change state?

Kinetic Energy

Recall that the particles that make up matter are in constant motion. These particles have **kinetic energy,** *the energy an object has due to its motion.* The faster particles move, the more kinetic energy they have. Within a given substance, such as water, particles in the solid state have the least amount of kinetic energy. This is because they only vibrate in place. Particles in the liquid state move faster than particles in the solid state. Therefore, they have more kinetic energy. Particles in the gaseous state move very quickly and have the most kinetic energy of all particles of a given substance.

Temperature *is a measure of the average kinetic energy of all the particles in an object.* Within a given substance, a temperature increase means that the particles, on average, are moving at greater speeds, or have a greater average kinetic energy. For example, water molecules at 25°C are generally moving faster and have more kinetic energy than water molecules at 10°C.

Active Reading **2. Relate** How do kinetic energy and temperature relate to particle motion? Draw arrows to show a correlating increase or decrease in each column.

Particle Motion	Kinetic Energy of Particles	Temperature
↑		

Potential Energy

In addition to kinetic energy, particles have potential energy. Potential energy is stored energy due to the interactions between particles or objects. For example, when you pick up a ball and then let it go, the gravitational force between the ball and Earth causes the ball to fall toward Earth. Before you let the ball go, it has potential energy.

Potential energy typically increases when objects get farther apart and decreases when they get closer together. The basketball in the top part of **Figure 8** is farther off the ground than it is in the bottom part of the figure. The farther an object is from Earth's surface, the greater the gravitational potential energy. As the ball gets closer to the ground, the potential energy decreases.

You can think of the potential energy of particles in a similar way. The chemical potential energy is due to the position of the particles relative to other particles. The chemical potential energy of particles increases and decreases as the distances between particles increases or decreases. The particles in the top part of **Figure 8** are farther apart than the particles in the bottom part. The particles that are farther apart have greater chemical potential energy.

Thermal Energy

Thermal energy *is the total potential and kinetic energies of an object.* You can change an object's state of matter by adding or removing thermal energy. When you add thermal energy to an object, the particles either move faster (increased kinetic energy) or get farther apart (increased potential energy) or both. The opposite is true when you remove thermal energy from an object. If enough thermal energy is added or removed, a change of state can occur.

 3. **NGSSS Check** **Explain** How do thermal energy and temperature differ? SC.7.P.11.1

Figure 8 The potential energy of the ball depends on the distance between the ball and Earth. The potential energy of particles in matter depends on the distances between the particles.

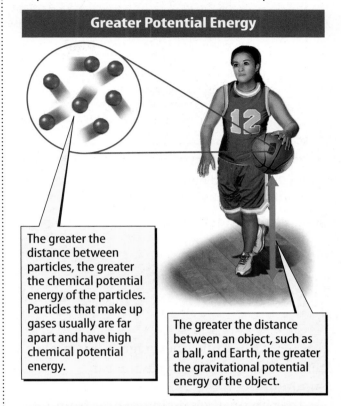

Greater Potential Energy

The greater the distance between particles, the greater the chemical potential energy of the particles. Particles that make up gases usually are far apart and have high chemical potential energy.

The greater the distance between an object, such as a ball, and Earth, the greater the gravitational potential energy of the object.

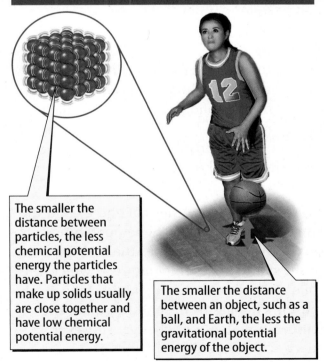

Less Potential Energy

The smaller the distance between particles, the less chemical potential energy the particles have. Particles that make up solids usually are close together and have low chemical potential energy.

The smaller the distance between an object, such as a ball, and Earth, the less the gravitational potential energy of the object.

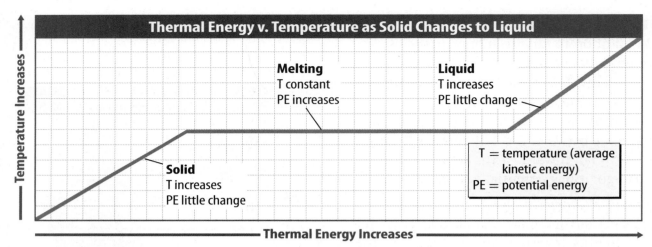

Figure 9 🔑 Adding thermal energy to matter causes the particles that make up the matter to increase in kinetic energy, potential energy, or both.

✔ 4. **Visual Check** **Point Out** During melting, which factor remains constant?

Solid to Liquid or Liquid to Solid

When you drink a beverage from an aluminum can, do you recycle the can? Aluminum recycling is one example of a process that involves changing matter from one state to another by adding or removing thermal energy.

Melting

The first part of the recycling process involves melting aluminum cans. To change matter from a solid to a liquid, thermal energy must be added. The graph in **Figure 9** shows the relationship between increasing temperature and increasing thermal energy (potential energy + kinetic energy).

At first, both the thermal energy and the temperature increase. The temperature stops increasing when it reaches the melting point of the matter, the temperature at which the solid state changes to the liquid state. As aluminum changes from solid to liquid, the temperature does not change. However, energy changes still occur.

Energy Changes

What happens when a solid reaches its melting point? Notice that the line on the graph is horizontal. This means that the temperature, or average kinetic energy, stops increasing. However, the amount of thermal energy continues to increase. How is this possible?

Once a solid reaches its melting point, the average speed of particles does not change, but the distance between the particles does change. The particles move farther apart, and potential energy increases. Once a solid completely melts, the addition of thermal energy will cause the kinetic energy of the particles to increase again, as shown by a temperature increase.

Freezing

After the aluminum melts, it is poured into molds to cool. As the aluminum cools, thermal energy leaves it. Freezing is a process that is the reverse of melting. The temperature at which matter changes from the liquid state to the solid state is its freezing point. To observe the temperature and thermal energy changes that occur to hot aluminum blocks, move from right to left on the graph in **Figure 9.**

During evaporation, a liquid vaporizes only at its surface.

During boiling, a liquid vaporizes at its surface and within the liquid.

Bubbles, or vaporized particles, rise to the top of the liquid and escape from the container.

Liquid to Gas or Gas to Liquid

When you heat water, do you ever notice how bubbles begin to form at the bottom and rise to the surface? The bubbles contain water vapor, a gas. *The change in state of a liquid into a gas is* **vaporization.** **Figure 10** shows two types of vaporization— evaporation and boiling.

Boiling

Vaporization that occurs within a liquid is called boiling. The temperature at which boiling occurs in a liquid is called its boiling point. In **Figure 11,** notice the energy changes that occur during this process. The kinetic energy of particles increases until the liquid reaches its boiling point.

At the boiling point, the potential energy of particles begins increasing. The particles move farther apart until the attractive forces no longer hold them together. At this point, the liquid changes to a gas. When boiling ends, if thermal energy continues to be added, the kinetic energy of the gas particles begins to increase again. Therefore, the temperature begins to increase again as shown on the graph.

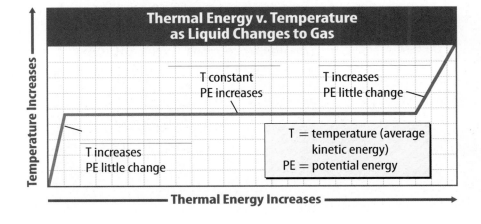

Thermal Energy v. Temperature as Liquid Changes to Gas

Temperature Increases

T constant
PE increases

T increases
PE little change

T increases
PE little change

T = temperature (average kinetic energy)
PE = potential energy

Thermal Energy Increases

Figure 10 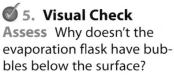 Boiling and evaporation are two kinds of vaporization.

✓ **5. Visual Check**
Assess Why doesn't the evaporation flask have bubbles below the surface?

_____

_____

_____

_____

Figure 11 🔑 When thermal energy is added to a liquid, kinetic energy and potential energy changes occur.

Active Reading **6. Label** Using **Figure 9** as a guide, label the portion of the graph that represents a liquid, a gas, and boiling.

WORD ORIGIN

evaporation

from Latin *evaporare,* means "disperse in steam or vapor"

Evaporation

Unlike boiling, **evaporation** *is vaporization that occurs only at the surface of a liquid.* Liquid in an open container will vaporize, or change to a gas, over time due to evaporation.

Condensation

Boiling and evaporation are processes that change a liquid to a gas. A reverse process also occurs. When a gas loses enough thermal energy, the gas changes to a liquid, or condenses. *The change of state from a gas to a liquid is called* **condensation.**

Solid to Gas or Gas to Solid

Is it possible for a solid to become a gas without turning to a liquid first? Yes, in fact, dry ice does. Dry ice, as shown in **Figure 12,** is solid carbon dioxide. It turns immediately into a gas when thermal energy is added to it. The process is called sublimation. **Sublimation** *is the change of state from a solid to a gas without going through the liquid state.* As dry ice sublimes, it cools and condenses the water vapor in the surrounding air, creating a thick fog.

The opposite of sublimation is deposition. **Deposition** *is the change of state of a gas to a solid without going through the liquid state.* For deposition to occur, thermal energy has to be removed from the gas. You might see deposition in autumn when you wake up and there is frost on the grass. As water vapor loses thermal energy, it changes into a solid known as frost.

Active Reading

7. Generalize Why are sublimation and deposition unusual changes of state?

_____

_____

_____

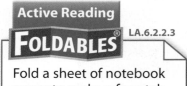

Active Reading

FOLDABLES LA.6.2.2.3

Fold a sheet of notebook paper to make a four-tab Foldable as shown. Label the tabs, define the terms, and record what you learn about each term under the tabs.

Vaporization
Boiling Evaporation
Condensation
Sublimation
Deposition

Figure 12 Dry ice sublimes—goes directly from the solid state to the gas state—when thermal energy is added. Frost is an example of the opposite process—deposition.

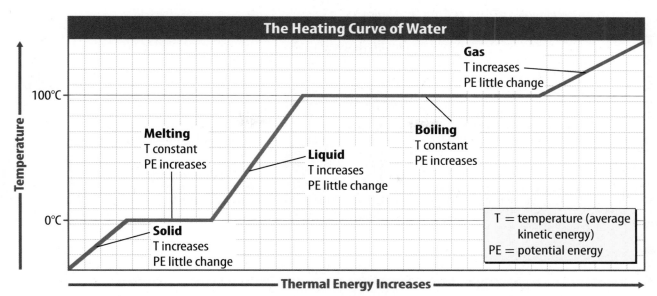

The Heating Curve of Water

Temperature

100°C

0°C

Gas
T increases
PE little change

Melting
T constant
PE increases

Liquid
T increases
PE little change

Boiling
T constant
PE increases

Solid
T increases
PE little change

T = temperature (average kinetic energy)
PE = potential energy

← Thermal Energy Increases →

Figure 13 🔑 Water undergoes energy changes and state changes as thermal energy is added and removed.

States of Water

Water is the only substance that exists naturally as a solid, a liquid, and a gas within Earth's temperature range. To better understand the energy changes during a change in state, it is helpful to study the heating curve of water, as shown in **Figure 13.**

Adding Thermal Energy

Suppose you place a beaker of ice on a hot plate. The hot plate transfers thermal energy to the beaker and then to the ice. The temperature of the ice increases. Recall that this means the average kinetic energy of the water molecules increases.

At 0°C, the melting point of water, the water molecules vibrate so rapidly that they begin to move out of their places. At this point, added thermal energy only increases the distance between particles and decreases attractive forces—melting occurs. Once melting is complete, the average kinetic energy of the particles (temperature) begins to increase again as more thermal energy is added.

When water reaches 100°C, the boiling point, liquid water begins to change to water vapor. Again, kinetic energy is constant as vaporization occurs. When the change of state is complete, the kinetic energy of molecules increases once more, and so does the temperature.

Removing Thermal Energy

The removal of thermal energy is the reverse of the process shown in **Figure 13.** Cooling water vapor changes the gas to a liquid. Cooling the water further changes it to ice.

Active Reading **8. Describe** What are the changes in thermal energy as water goes from a solid to a liquid? When thermal energy is removed from water, where do you think the thermal energy goes?

_____

_____

_____

_____

_____

_____

_____

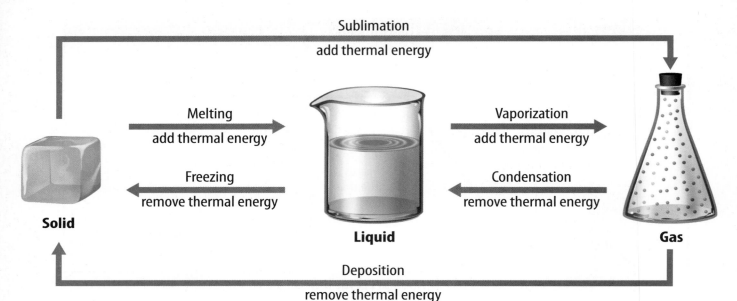

Sublimation
add thermal energy

Melting
add thermal energy

Vaporization
add thermal energy

Freezing
remove thermal energy

Condensation
remove thermal energy

Solid

Liquid

Gas

Deposition
remove thermal energy

Figure 14 🔑 For a change of state to occur, thermal energy must move into or out of matter.

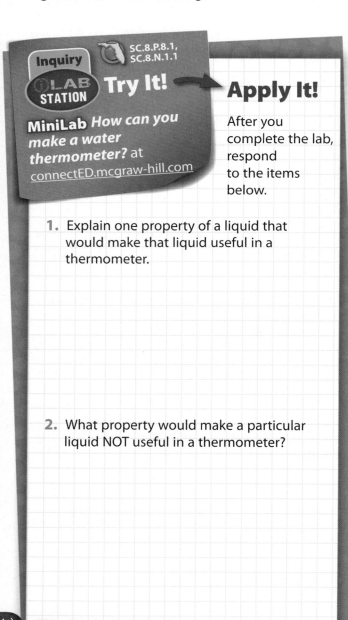

SC.8.P.8.1, SC.8.N.1.1

Inquiry

LAB STATION **Try It!**

MiniLab *How can you make a water thermometer?* at connectED.mcgraw-hill.com

Apply It!

After you complete the lab, respond to the items below.

1. Explain one property of a liquid that would make that liquid useful in a thermometer.

2. What property would make a particular liquid NOT useful in a thermometer?

Conservation of Mass and Energy

The diagram in **Figure 14** illustrates the energy changes that occur as thermal energy is added or removed from matter. Notice that opposite processes, melting and freezing and vaporization and condensation, are shown. When matter changes state, matter and energy are always conserved.

When water vaporizes, it appears to disappear. If the invisible gas were captured and its mass were added to the mass of the remaining liquid, you would see that matter is conserved. This is also true for energy. Surrounding matter, such as air, often absorbs thermal energy. If you measured all the thermal energy, you would find that energy is conserved.

All matter has thermal energy. Thermal energy is the sum of potential and kinetic energy.

When thermal energy is added to a liquid, vaporization can occur.

When enough thermal energy is removed from matter, a change of state can occur.

Inquiry **LAB STATION**
SC.7.P.11.1, SC.8.P.8.1, SC.7.N.1.6, SC.8.N.1.1, SC.8.N.1.3, MA.6.A.3.6

Try It!

Skill Lab *How does dissolving substances in water change its freezing point?* at connectED.mcgraw-hill.com

Use Vocabulary

1 The measure of average kinetic energy of the particles in a material is _____ .

2 **Define** *kinetic energy* and *thermal energy* in your own words. SC.6.P.11.1

_____

_____

_____

3 The change of a liquid into a gas is known as

_____ .

Understand Key Concepts 🔑

4 The process that is opposite of condensation is known as SC.7.P.11.1
   (A) deposition.   (C) melting.
   (B) freezing.   (D) vaporization.

5 **Explain** how temperature and particle motion are related.

_____

_____

6 **Describe** the relationship between temperature and thermal energy.

_____

_____

7 **Generalize** the changes in thermal energy when matter increases in temperature and then changes state. SC.7.P.11.1

_____

_____

_____

Interpret Graphics

8 **Summarize** Fill in the graphic organizer below to identify the two types of vaporization that can occur in matter. LA.6.2.2.3

Critical Thinking

9 **Summarize** the energy and state changes that occur when freezing rain falls and solidifies on a wire fence. SC.7.P.11.1

_____

_____

10 **Compare** the amount of thermal energy needed to melt a solid and the amount of thermal energy needed to freeze the same liquid. SC.7.P.11.1

_____

_____

Connect Key Concepts 🔑

Characterize Fill in the chart below to organize information about solids, liquids, and gases. Describe the energy changes that take place as a solid changes to a liquid and as a liquid changes to a gas.

1. definite _____

2. particle arrangement:

Solids

3. attractive forces:

5. particle motion:

4. Energy Changes:

6. particle motion:

7. attractive forces:

Liquids

8. indefinite _____

10. definite _____

9. Energy Changes:

11. indefinite _____

12. attractive forces:

Gases

13. vapor:

14. particle motion and arrangement:

The Behavior of GASES

ESSENTIAL QUESTIONS

 How does the kinetic molecular theory describe the behavior of a gas?

 How are temperature, pressure, and volume related in Boyle's law?

 How is Boyle's law different from Charles's law?

Vocabulary

kinetic molecular theory p. 282

pressure p. 283

Boyle's law p. 284

Charles's law p. 285

Florida NGSSS

SC.6.N.2.1 Distinguish science from other activities involving thought.

SC.6.N.3.3 Give several examples of scientific laws.

SC.6.P.13.1 Investigate and describe types of forces including contact forces and forces acting at a distance, such as electrical, magnetic, and gravitational.

SC.7.N.1.6 Explain that empirical evidence is the cumulative body of observations of a natural phenomenon on which scientific explanations are based.

SC.7.N.3.1 Recognize and explain the difference between theories and laws and give several examples of scientific theories and the evidence that supports them.

SC.7.P.11.1 Recognize that adding heat to or removing heat from a system may result in a temperature change and possibly a change of state.

SC.8.P.8.1 Explore the scientific theory of atoms (also known as atomic theory) by using models to explain the motion of particles in solids, liquids, and gases.

MA.6.A.3.6 Construct and analyze tables, graphs, and equations to describe linear functions and other simple relations using both common language and algebraic notation.

MA.6.S.6.2 Select and analyze the measures of central tendency or variability to represent, describe, analyze, and/or summarize a data set for the purposes of answering questions appropriately.

Also covers: SC.8.N.1.1, LA.6.2.2.3

Inquiry Launch Lab
SC.7.N.1.6
15 minutes

Are volume and pressure of a gas related?

Pressure affects gases differently than it affects solids and liquids. How do pressure changes affect the volume of a gas?

Procedure

1. Read and complete a lab safety form.
2. Stretch and blow up a **small balloon** several times.
3. Finally, blow up the balloon to a diameter of about 5 cm. Twist the neck, and stretch the mouth of the balloon over the opening of a **plastic bottle.** Tape the neck of the balloon to the bottle.
4. Squeeze and release the bottle several times while observing the balloon. Record your observations.

Data and Observations

Think About This

1. **Infer** Why doesn't the balloon deflate when you attach it to the bottle?

2. **Assess** What caused the balloon to inflate when you squeezed the bottle?

3. **Key Concept** **Evaluate** Using this lab as a reference, do you think pressure and volume of a gas are related? Explain.

Inquiry **Survival Gear?**

1. Why do some pilots wear oxygen masks? Planes fly at high altitudes where the atmosphere has a lower pressure and gas molecules are less concentrated. If the pressure is not adjusted inside the airplane, a pilot must wear an oxygen mask to inhale enough oxygen to keep the body functioning. How do you think the concentration of molecules of gas changes as pressure decreases or increases?

_____

_____

_____

_____

REVIEW VOCABULARY

theory

(noun) an explanation of things or events that is based on knowledge gained from many observations and investigations

FOLDABLES LA.6.2.2.3

Fold a sheet of notebook paper to make a three-tab Foldable and label as shown. Use your Foldable to compare two important gas laws.

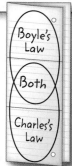

Understanding Gas Behavior

Pilots do not worry as much about solids and liquids at high altitudes as they do gases. That is because gases behave differently than solids and liquids. Changes in temperature, pressure, and volume affect the behavior of gases more than they affect solids and liquids.

The explanation of particle behavior in solids, liquids, and gases is based on the kinetic molecular theory. The **kinetic molecular theory** *is an explanation of how particles in matter behave.* Some basic ideas in this theory are

• small particles make up all matter;

• these particles are in constant, random motion;

• the particles collide with other particles, other objects, and the walls of their container;

• when particles collide, no energy is lost.

You have read about most of these, but the last two statements are very important in explaining how gases behave.

Active Reading **2. Recite** Highlight the main ideas of the kinetic-molecular theory.

| Greatest volume, least pressure | Less volume, more pressure | Least volume, most pressure |

Figure 15 🔑 As pressure increases, the volume of the gas decreases.

What is pressure?

Particles in gases move constantly. As a result of this movement, gas particles constantly collide with other particles and their container. When particles collide with their container, pressure results. **Pressure** *is the amount of force applied per unit of area.* For example, gas in a cylinder, as shown in **Figure 15,** might contain trillions of gas particles. These particles exert forces on the cylinder each time they strike it. When a weight is added to the plunger, the plunger moves down, compressing the gas in the cylinder. With less space to move around, the particles that make up the gas collide with each other more frequently, causing an increase in pressure. The more the particles are compressed, the more often they collide, increasing the pressure.

Pressure and Volume

Figure 15 also shows the relationship between pressure and volume of gas at a constant temperature. What happens to pressure if the volume of a container changes? Notice that when the volume is greater, the particles have more room to move. This additional space results in fewer collisions within the cylinder, and pressure is less. The gas particles in the middle cylinder have even less volume and more pressure. In the cylinder on the right, the pressure is greater because the volume is less. The particles collide with the container more frequently. Because of the greater number of collisions within the container, pressure is greater.

WORD ORIGIN

pressure
from Latin *pressura,* means "to press"

Active Reading **3. Relate** Fill in the graphic organizer to relate the volume to the pressure of a gas at a constant temperature.

Greater volume

Less volume

Solve Equations

Boyle's law can be stated by the equation

$$V_2 = \frac{P_1 V_1}{P_2}$$

P_1 and V_1 represent the pressure and volume before a change. P_2 and V_2 are the pressure and volume after a change. Pressure is often measured in kilopascals (kPa). For example, what is the final volume of a gas with an initial volume of 50.0 mL if the pressure increases from 600.0 kPa to 900.0 kPa?

1. Replace the terms in the equation with the actual values.

$$V_2 = \frac{(600.0\ \text{kPa})(50.0\ \text{mL})}{(900.0\ \text{kPa})}$$

2. Cancel units, multiply, and then divide.

$$V_2 = \frac{(600.0\ \text{kPa})(50.0\ \text{mL})}{(900.0\ \text{kPa})}$$

$$V_2 = 33.3\ \text{mL}$$

Practice

5. What is the final volume of a gas with an initial volume of 100.0 mL if the pressure decreases from 500.0 kPa to 250.0 kPa?

Boyle's Law

You read that the pressure and volume of a gas are related. Robert Boyle (1627–1691), a British scientist, was the first to describe this property of gases. **Boyle's law** *states that pressure of a gas increases if the volume decreases and pressure of a gas decreases if the volume increases, when temperature is constant.* This law can be expressed mathematically as shown to the left.

Active Reading   4. **Identify** <u>Underline</u> the relationship between pressure and volume of a gas if temperature is constant.

Boyle's Law in Action

You have probably felt Boyle's law in action if you have ever traveled in an airplane. While on the ground, the air pressure inside your middle ear and the pressure of the air surrounding you are equal. As the airplane takes off and begins to increase in altitude, the air pressure of the surrounding air decreases. However, the air pressure inside your middle ear does not decrease. The trapped air in your middle ear increases in volume, which can cause pain. These pressure changes also occur when the plane is landing. You can equalize this pressure difference by yawning or chewing gum.

Graphing Boyle's Law

This relationship is shown in the graph in **Figure 16.** Pressure is on the *x*-axis, and volume is on the *y*-axis. Notice that the line decreases in value from left to right. This shows that as the pressure of a gas increases, the volume of the gas decreases.

Figure 16   The graph shows that as pressure increases, volume decreases. This is true only if the temperature of the gas is constant.

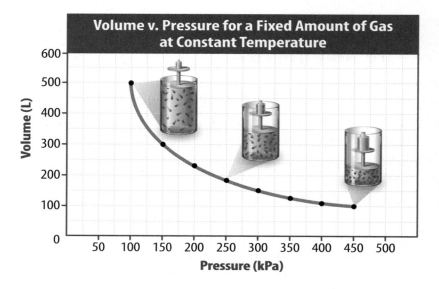

Figure 17 🔑 As the temperature of a gas increases, the kinetic energy of the particles increases. The particles move farther apart, and volume increases.

Lower temperature, less volume

Higher temperature, greater volume

Temperature and Volume

Pressure and volume changes are not the only factors that affect gas behavior. Changing the temperature of a gas also affects its behavior, as shown in **Figure 17.** The gas in the cylinder on the left has a low temperature. The average kinetic energy of the particles is low, and they move closer together. The volume of the gas is less. When thermal energy is added to the cylinder, the gas particles move faster and spread farther apart. This increases the pressure from gas particles, which push up the plunger. This increases the volume of the container.

Charles's Law

Jacque Charles (1746–1823) was a French scientist who described the relationship between temperature and volume of a gas. **Charles's law** *states that the volume of a gas increases with increasing temperature, if the pressure is constant.* Charles's practical experience with gases was most likely the result of his interest in balloons. Charles and his colleague were the first to pilot and fly a hydrogen-filled balloon in 1783.

| **Active Reading** | **6. Identify** <u>Underline</u> the relationship between temperature and volume of a gas if pressure is constant. |

Inquiry SC.8.P.8.1, SC.8.N.1.1

LAB STATION **Try It!**

MiniLab *How does temperature affect the volume?* at connectED.mcgraw-hill.com

Apply It!

After you complete the lab, respond to the items below.

1. After an automobile has been driven a long distance, its tires become very warm from friction with the road. Explain what you would expect to have happened to the pressure of the air in the tires.

2. Most tire companies recommend that air pressure in automobile tires be measured when the tires are cold. Why is this a good idea?

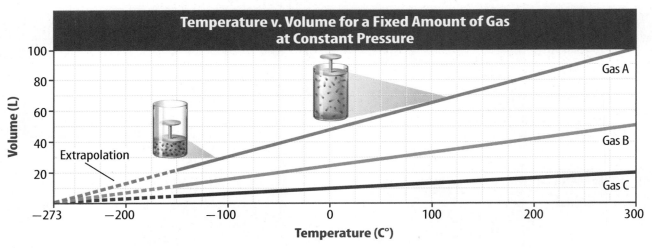

Temperature v. Volume for a Fixed Amount of Gas at Constant Pressure

Volume (L) / Temperature (C°)

Extrapolation

Gas A
Gas B
Gas C

Figure 18 🔑 The volume of a gas increases when the temperature increases at constant pressure.

✔️ **7. Visual Check**
Infer What do the dashed lines mean?

_____

_____

_____

Charles's Law in Action

You have probably seen Charles's law in action if you have ever taken a balloon outside on a cold winter day. Why does a balloon appear slightly deflated when you take it from a warm place to a cold place? When the balloon is in cold air, the temperature of the gas inside the balloon decreases. Recall that a decrease in temperature is a decrease in the average kinetic energy of particles. As a result, the gas particles slow down and begin to get closer together. Fewer particles hit the inside of the balloon. The balloon appears partially deflated. If the balloon is returned to a warm place, the kinetic energy of the particles increases. More particles hit the inside of the balloon and push it out again. The volume increases.

Graphing Charles's Law

The relationship described in Charles's law is shown in the graph of three gases in **Figure 18.** Temperature is on the *x*-axis, and volume is on the *y*-axis. Notice that the lines are straight and represent increasing values. Each line in the graph is extrapolated to −273°C. *Extrapolated* means the graph is extended beyond the observed data points. This temperature also is referred to as 0 K (kelvin), or absolute zero. This temperature is theoretically the lowest possible temperature of matter. At absolute zero, all particles are at the lowest possible energy state and do not move. The particles contain a minimal amount of thermal energy (potential energy + kinetic energy).

Active Reading **8. Label** In the table to the right, state whether temperature, pressure, and volume increase, decrease, or remain constant in each law.

Law	Temperature	Pressure	Volume
Charles's Law			
Boyle's Law			

Visual Summary

The explanation of particle behavior in solids, liquids, and gases is based on the kinetic molecular theory.

As volume of a gas decreases, the pressure increases when at constant temperature.

At constant pressure, as the temperature of a gas increases, the volume also increases.

SC.7.P.11.1,
SC.8.P.8.1,
SC.6.N.2.1,
SC.8.N.1.1,
MA.6.A.3.6,
MA.6.S.6.2

Inquiry

⑪LAB STATION

Try It!

Inquiry Lab *Design an Experiment to Collect Data* at connectED.mcgraw-hill.com

Use Vocabulary

1 **List** the basic ideas of the kinetic molecular theory.

_____

_____

_____

2 _____ is force applied per unit area.

Understand Key Concepts 🔑

3 Which is held constant when a gas obeys Boyle's law? SC.7.N.1.6

  (A) motion   (C) temperature

  (B) pressure   (D) volume

4 **Describe** how the kinetic molecular theory explains the behavior of a gas. SC.7.N.1.6

_____

_____

5 **Contrast** Charles's law with Boyle's law.

_____

_____

Interpret Graphics

6 **Explain** What happens to the particles to the right when more weights are added?

_____

7 **Identify** Fill in the graphic organizer below to list three factors that affect gas behavior. LA.6.2.2.3

Critical Thinking

8 **Describe** what would happen to the pressure of a gas if the volume of the gas doubles while at a constant temperature. MA.6.A.3.6

_____

_____

Math Skills   MA.6.A.3.6

9 **Describe** The pressure on 400 mL of a gas is raised from 20.5 kPa to 80.5 kPa. What is the final volume of the gas?

Chapter 7 Study Guide

Think About It! A change of state occurs as energy transfers to or from a material. Energy is conserved, or the same amount of energy is transferred, as a material changes state and then back to its original form.

 Key Concepts Summary

Vocabulary

solid p. 265
liquid p. 266
viscosity p. 266
surface tension p. 267
gas p. 268
vapor p. 268

LESSON 1 Solids, Liquids, and Gases

- Particles vibrate in **solids**. They move faster in **liquids** and even faster in **gases**.
- The force of attraction among particles decreases as matter goes from a solid, to a liquid, and finally to a gas.

Solid	Liquid	Gas

LESSON 2 Changes in State

- Because **temperature** is defined as the average **kinetic energy** of particles and kinetic energy depends on particle motion, temperature is directly related to particle motion.
- **Thermal energy** includes both the kinetic energy and the potential energy of particles in matter. However, temperature is only the average kinetic energy of particles in matter.
- Thermal energy must be added or removed from matter for a change of state to occur.

kinetic energy p. 272
temperature p. 272
thermal energy p. 273
vaporization p. 275
evaporation p. 276
condensation p. 276
sublimation p. 276
deposition p. 276

LESSON 3 The Behavior of Gases

- The **kinetic molecular theory** states basic assumptions that are used to describe particles and their interactions in gases and other states of matter.
- **Pressure** of a gas increases if the volume decreases, and pressure of a gas decreases if the volume increases, when temperature is constant.
- **Boyle's law** describes the behavior of a gas when pressure and volume change at constant temperature. **Charles's law** describes the behavior of a gas when temperature and volume change, and pressure is constant.

kinetic molecular theory p. 282
pressure p. 283
Boyle's law p. 284
Charles's law p. 285

288 Chapter 7 • STUDY GUIDE

Active Reading
FOLDABLES® Chapter Project

Assemble your lesson Foldables as shown to make a Chapter Project. Use the project to review what you have learned in this chapter.

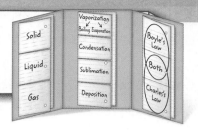

Use Vocabulary

Replace the underlined word with the correct term.

1 Matter with a definite shape and a definite volume is known as a <u>gas</u>.

_____

2 <u>Surface tension</u> is a measure of a liquid's resistance to flow.

_____

3 The gas state of a substance that is normally a solid or a liquid at room temperature is a <u>pressure</u>.

_____

4 <u>Boiling</u> is vaporization that occurs at the surface of a liquid.

_____

5 <u>Boyle's law</u> is an explanation of how particles in matter behave.

_____

6 When graphing a gas obeying <u>Boyle's law</u>, the line will be a straight line with a positive slope.

_____

Link Vocabulary and Key Concepts

Use vocabulary terms from the previous page to complete the concept map.

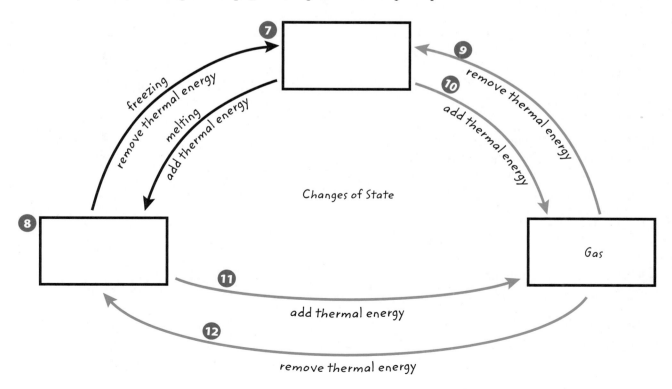

Fill in the correct answer choice.

Understand Key Concepts

1. What would happen if you tried to squeeze a gas into a smaller container? **SC.8.P.8.1**
   - Ⓐ The attractive forces between the particles would increase.
   - Ⓑ The force of the particles would prevent you from doing it.
   - Ⓒ The particles would have fewer collisions with the container.
   - Ⓓ The repulsive forces of the particles would pull on the container.

2. Which type of motion in the figure below best represents the movement of gas particles? **SC.8.N.3.1**

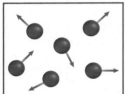
Motion 1

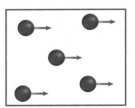
Motion 2

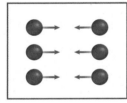
Motion 3

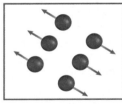
Motion 4

   - Ⓐ motion 1
   - Ⓑ motion 2
   - Ⓒ motion 3
   - Ⓓ motion 4

3. A pile of snow slowly disappears into the air, even though the temperature remains below freezing. Which process explains this? **SC.7.P.11.1**
   - Ⓐ condensation
   - Ⓑ deposition
   - Ⓒ evaporation
   - Ⓓ sublimation

4. Which unit is a density unity?
   - Ⓐ cm^3
   - Ⓑ cm^3/g
   - Ⓒ g
   - Ⓓ g/cm^3

Critical Thinking

5. **Explain** how the distances between particles in a solid, a liquid, and a gas help determine the densities of each. **SC.8.P.8.1**

_____

_____

_____

_____

6. **Describe** what would happen to the volume of a balloon if it were submerged in hot water. **SC.7.P.11.1**

_____

_____

_____

_____

7. **Assess** The particles of an unknown liquid have very weak attractions for other particles in the liquid. Would you expect the liquid to have a high or low viscosity? Explain your answer. **SC.8.P.8.1**

_____

_____

_____

_____

8. **Rank** these liquids from highest to lowest viscosity: honey, rubbing alcohol, and ketchup. **SC.8.P.8.1**

_____

_____

_____

9 **Evaluate** Each beaker below contains the same amount of water. The thermometers show the temperature in each beaker. Explain the kinetic energy differences in each beaker. SC.6.P.11.1

_____

_____

_____

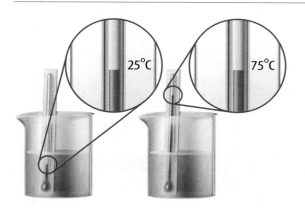

25°C   75°C

10 **Summarize** A glass with a few milliliters of water is placed on a counter. No one touches the glass. Explain what happens to the water after a few days. SC.7.P.11.1

_____

_____

_____

Writing in Science

11 **Write** a paragraph that describes how you could determine the melting point of a substance from its heating or cooling curve. Use a separate sheet of paper. LA.6.2.2.3

Big Idea Review

12 During springtime in Alaska, frozen rivers thaw and boats can navigate the rivers again. What physical changes and energy changes occur to the ice molecules when ice changes to water? Explain the process in which water in the river changes to water vapor. SC.7.P.11.1

_____

_____

_____

_____

13 Look at the chapter opener photo, and explain how the average kinetic energy of the particles changes as the molten glass cools. What instrument could you use to verify the change in the average kinetic energy of the particles? SC.6.P.11.1

_____

_____

_____

_____

Math Skills   MA.6.A.3.6

Solve Equations

14 The pressure on 1 L of a gas at a pressure of 600 kPa is lowered to 200 kPa. What is the final volume of the gas?

15 A gas has a volume of 30 mL at a pressure of 5000 kPa. What is the volume of the gas if the pressure is lowered to 1,250 kPa?

Florida NGSSS

Benchmark Practice

Fill in the correct answer choice.

Multiple Choice

1 Which property applies to matter that consists of particles vibrating in place? SC.8.P.8.1

   Ⓐ has a definite shape

   Ⓑ takes the shape of the container

   Ⓒ flows easily at room temperature

   Ⓓ Particles are far apart.

Use the figure below to answer questions 2 and 3.

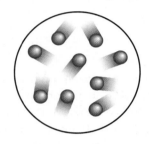

2 Which state of matter is represented above? SC.8.N.8.1

   Ⓕ amorphous solid

   Ⓖ crystalline solid

   Ⓗ gas

   Ⓘ liquid

3 Which best describes the attractive forces between particles shown in the figure? SC.8.P.8.1

   Ⓐ The attractive forces keep the particles vibrating in place.

   Ⓑ The particles hardly are affected by the attractive forces.

   Ⓒ The attractive forces keep the particles close together but still allow movement.

   Ⓓ The particles are locked in their positions because of the attractive forces between them.

4 What happens to matter as its temperature increases? SC.7.P.11.1

   Ⓕ The average kinetic energy of its particles decreases.

   Ⓖ The average thermal energy of its particles decreases.

   Ⓗ The particles gain kinetic energy.

   Ⓘ The particles lose potential energy.

Use the figure to answer question 5.

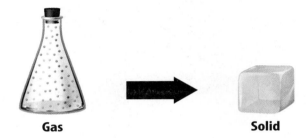

Gas   Solid

5 Which process is represented in the figure? SC.7.P.11.1

   Ⓐ deposition

   Ⓑ freezing

   Ⓒ sublimation

   Ⓓ vaporization

6 Which is a fundamental assumption of the kinetic molecular theory? SC.8.P.8.1

   Ⓕ All atoms are composed of subatomic particles.

   Ⓖ The particles of matter move in predictable paths.

   Ⓗ No energy is lost when particles collide with one another.

   Ⓘ Particles of matter never come into contact with one another.

7 Which is true of the thermal energy of particles? SC.7.P.11.1

Ⓐ Thermal energy includes the potential and the kinetic energy of the particles.

Ⓑ Thermal energy is the same as the average kinetic energy of the particles.

Ⓒ Thermal energy is the same as the potential energy of the particles.

Ⓓ Thermal energy is the same as the temperature of the particles.

Use the graph below to answer question 8.

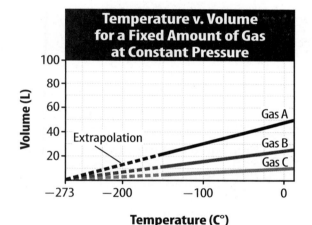

8 Which relationship is shown in the graph?
SC.7.N.1.6

Ⓕ Boyle's law

Ⓖ Charles's law

Ⓗ kinetic molecular theory

Ⓘ definition of thermal energy

9 Which is a form of vaporization? SC.7.P.11.1

Ⓐ condensation

Ⓑ evaporation

Ⓒ freezing

Ⓓ melting

10 When a needle is placed on the surface of water, it floats. Which idea best explains why this happens? SC.8.N.3.1

Ⓕ Boyle's law

Ⓖ molecular theory

Ⓗ surface tension

Ⓘ viscosity theory

11 In which material would the particles be most closely spaced? SC.8.P.8.1

Ⓐ air

Ⓑ brick

Ⓒ syrup

Ⓓ water

Use the graph below to answer questions 12 and 13.

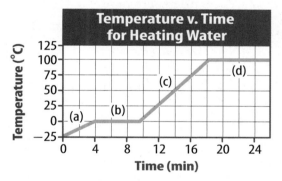

12 Which area of the graph above shows melting of a solid? MA.6.A.3.6

Ⓕ a   Ⓗ c

Ⓖ b   Ⓘ d

13 Which area or areas of the graph above show a change in the potential energy of the particles? SC.6.P.11.1

Ⓐ a   Ⓒ b and d

Ⓑ a and c   Ⓓ c

NEED EXTRA HELP?

If You Missed Question...	1	2	3	4	5	6	7	8	9	10	11	12	13
Go to Lesson...	1	1	1	2	2	3	2	3	1	2	2	3	2

Multiple Choice *Bubble the correct answer.*

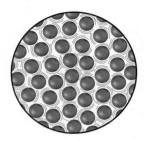

1. The image above could represent the structure of molecules in **SC.8.P.8.1**

   (A) diamonds.

   (B) milk.

   (C) oxygen.

   (D) wood.

2. Which is NOT true about the particles in a plastic spoon? **SC.8.P.8.5**

   (F) The particles are tightly packed.

   (G) The particles have strong attractive forces.

   (H) The particles slip past each other.

   (I) The particles vibrate in place.

3. Sami pops a helium balloon at a birthday party. What will happen to the particles of helium that were in the balloon? **SC.8.P.8.1**

   (A) The particles will disappear.

   (B) The particles will move together.

   (C) The particles will spread out evenly within the room.

   (D) The particles will stick together and sink to the floor.

4. The image above illustrates **SC.8.P.8.1**

   (F) how liquid particles differ from gas particles.

   (G) how liquid particles slip past each other.

   (H) how particles attract each other at a liquid's surface.

   (I) how particles in a liquid attract each other in all directions.

Copyright © Glencoe/McGraw-Hill, a division of The McGraw-Hill Companies, Inc.

Multiple Choice *Bubble the correct answer.*

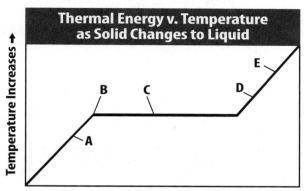

Thermal Energy v. Temperature as Solid Changes to Liquid

1. At which section on the graph above is there a change in thermal energy without a temperature change? **SC.8.N.3.1**

   (A) Point A to Point B

   (B) Point B to Point C

   (C) Point C to Point D

   (D) Point D to Point E

2. When water reaches its boiling point, **SC.6.P.11.1**

   (F) it begins to condense.

   (G) it loses thermal energy.

   (H) its kinetic energy decreases.

   (I) its potential energy increases.

3. David watches as water droplets appear on the outside of his glass of lemonade. The process he is watching is called **SC.7.N.1.6**

   (A) condensation.

   (B) deposition.

   (C) evaporation.

   (D) sublimation.

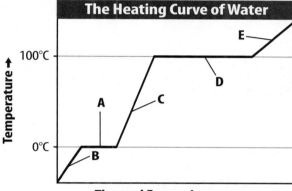

The Heating Curve of Water

4. Which point on the graph above represents water in a solid state with increasing kinetic energy? **LA.6.2.2.3**

   (F) Point A

   (G) Point B

   (H) Point C

   (I) Point E

Copyright © Glencoe/McGraw-Hill, a division of The McGraw-Hill Companies, Inc.

Benchmark Mini-Assessment   **Chapter 7 • Lesson 3**

mini **BAT**

Multiple Choice *Bubble the correct answer.*

A   B

1. The image above shows changes that occur when weight is added to a plunger on top of a cylinder of gas. How does the gas in cylinder B compare to the gas in cylinder A? **SC.8.P.8.1**

   Ⓐ Cylinder B has greater pressure and less volume than cylinder A.

   Ⓑ Cylinder B has greater volume and less pressure than cylinder A.

   Ⓒ Pressures are equal, but cylinder B has less volume than cylinder A.

   Ⓓ Volumes are equal, but cylinder B has less pressure than cylinder A.

2. According to the kinetic molecular theory, **SC.8.P.8.1**

   Ⓕ all particles are in constant motion.

   Ⓖ energy is lost when particles collide.

   Ⓗ gas particles do not collide with other particles or other objects.

   Ⓘ an increase in particle motion always increases gas volume.

3. If volume stays the same, the pressure of car tires would likely be highest **SC.8.P.8.3**

   Ⓐ on a cold day.

   Ⓑ on a warm day.

   Ⓒ on a rainy day.

   Ⓓ on a snowy day.

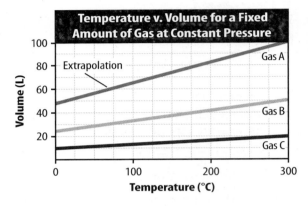

4. According to the graph above, **MA.6.S.6.2**

   Ⓕ Gas A has particles with the highest average kinetic energy.

   Ⓖ Gas B has particles with the highest average kinetic energy.

   Ⓗ Gas C has particles with the highest average kinetic energy.

   Ⓘ Gases A, B, and C have the same average kinetic energy.

Copyright © Glencoe/McGraw-Hill, a division of The McGraw-Hill Companies, Inc.

Notes

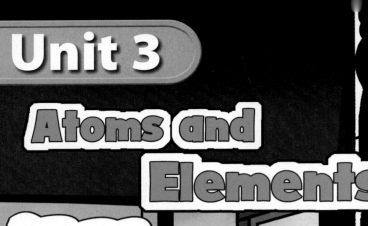

Unit 3
Atoms and Elements

Sent to her room, Molly Cool dreams of escaping.

If only she could change state and become a liquid, she could flow under the bedroom door and down the stairs...

...then flow to the fireplace where the heat would turn her into vapor and she could escape up the chimney.

I'm free!

Hello birds!

1000 B.C. — **1700** — **1800**

350 B.C.
Greek philosopher Aristotle defines an element as "one of those bodies into which other bodies can decompose, and that itself is not capable of being divided into another."

1704
Isaac Newton proposes that atoms attach to each other by some type of force.

1869
Dmitri Mendeleev publishes the first version of the periodic table.

1874
G. Johnstone Stoney proposes the existence of the electron, a subatomic particle that carries a negative electric charge, after experiments in electrochemistry.

1897
J.J. Thompson demonstrates the existence of the electron, proving Stoney's claim.

? Inquiry

Visit ConnectED for this unit's **STEM** activity.

1907
Physicists Hans Geiger and Ernest Marsden, under the direction of Ernest Rutherford, conduct the famous gold foil experiment. Rutherford concludes that the atom is mostly empty space and that most of the mass is concentrated in the atomic nucleus.

1918
Ernest Rutherford reports that the hydrogen nucleus has a positive charge, and he names it the proton.

1932
James Chadwick discovers the neutron, a subatomic particle with no electric charge and a mass slightly larger than a proton.

Patterns SC.7.N.1.5, SC.8.N.4.2

It's a bird! It's a plane! No, it's Venus! Besides the Sun, Venus is brighter than any other star or planet in the sky. It is often seen from Earth without the aid of a telescope, as shown in **Figure 1**.

Astronomers study the pattern of each planet's orbit. A **pattern** is a consistent plan or model used as a guide for understanding and predicting things. Studying the orbital patterns of planets allows scientists to predict the future position of each planet. By studying the pattern of Venus's orbit, astronomers can predict when Venus will travel between Earth and the Sun and be visible from Earth, as shown in **Figure 2**. Using patterns, scientists are able to predict the date when you will be able to see this event in the future.

Moon

Venus

▲ **Figure 1** Venus is often so bright in the morning sky that it has been nicknamed the morning star.

Active Reading

1. Analyze Explain how the pattern of Venus's orbit is used to determine its position in the sky.

_____

_____

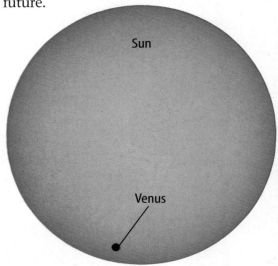

Figure 2 On June 8, 2004, observers around the world watched Venus pass in front of the Sun. This was the first time this event took place since 1882. ▶

Types of Patterns

Physical Patterns

A pattern that you can see and touch is a physical pattern. The crystalline structures of minerals are examples of physical patterns. When atoms form crystals, they produce structural, or physical, patterns. The crystal structure of the Star of India sapphire creates a pattern that reflects light in a stunning star shape.

Cyclic Patterns

An event that repeats many times again in a predictable order has a cyclic pattern. Since Earth's axis is tilted, the angle of the Sun's rays on your location on Earth changes as Earth orbits the Sun. This causes the seasons— winter, spring, summer, and fall— to occur in the same pattern every year.

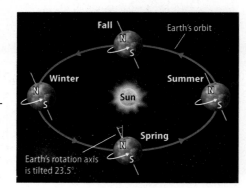

Patterns in Engineering

Engineers study patterns for many reasons, including to understand the physical properties of materials or to optimize the performance of their designs. Have you ever seen bricks with a pattern of holes through them? Clay bricks used in construction are fired, or baked, to make them stronger. Ceramic engineers understand that a regular pattern of holes in a brick assures that the brick is evenly fired and will not easily break.

Maybe you have seen a bridge constructed with a repeating pattern of large, steel triangles. Civil engineers, who design roads and bridges, know that the triangle is one of the strongest shapes in geometry. Engineers often use patterns of triangles in the structure of bridges to make them withstand heavy traffic and high winds.

Patterns in Physical Science

Scientists use patterns to explain past events or predict future events. At one time, only a few chemical elements were known. Chemists arranged the information they knew about these elements in a pattern according to the elements' properties. Scientists predicted the atomic numbers and the properties of elements that had yet to be discovered. These predictions made the discovery of new elements easier because scientists knew what properties to look for.

Look around. There are patterns everywhere—in art and nature, in the motion of the universe, in vehicles traveling on the roads, and in the processes of plant and animal growth. Analyzing patterns helps to understand the universe.

Patterns in Graphs

Scientists often graph their data to help identify patterns. For example, scientists might plot data from experiments on parachute nylon in graphs, such as the one below. Analyzing patterns on graphs then gives engineers information about how to design the strongest parachutes.

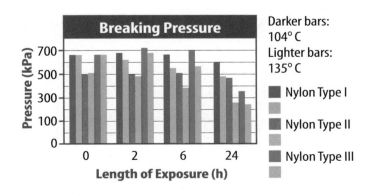

Breaking Pressure

Pressure (kPa)

Length of Exposure (h)

Darker bars: 104° C
Lighter bars: 135° C

Nylon Type I

Nylon Type II

Nylon Type III

Inquiry

LAB STATION

Try It!

SC.7.N.1.4, SC.6.N.1.5

MiniLab *How strong is your parachute?* at connectED.mcgraw-hill.com

Apply It! After you complete the lab, answer these questions.

1. **Describe** Think of a pattern you use in your daily life. Describe how this pattern is useful to you to predict future events.

2. **Analyze** List possible uses of graphs. These might include personal time in sporting events or use of electric power. Analyze how such graphic patterns reflect a change that can be useful information.

Name _____ Date _____

Seeing Inside an Atom

Five friends looked at a piece of aluminum foil. They wondered what they would see if they could see inside an atom of aluminum. This is what they said:

Jane: I think there would be a tiny, dense center surrounded by a lot of empty space where some particles are whizzing around.

Hal: I think there would be a large center made up mostly of empty space with particles whizzing around in it. It would be surrounded by a dense shell.

Katie: I think there would be all empty space with many tiny particles whizzing around through all that space.

Jeb: I think it would look like a tiny, dense ball with tiny particles tightly packed inside it.

Dara: I think it would look like a lot of tiny balls, all touching each other with no space in between them.

Circle the friend you agree with and explain why you agree.

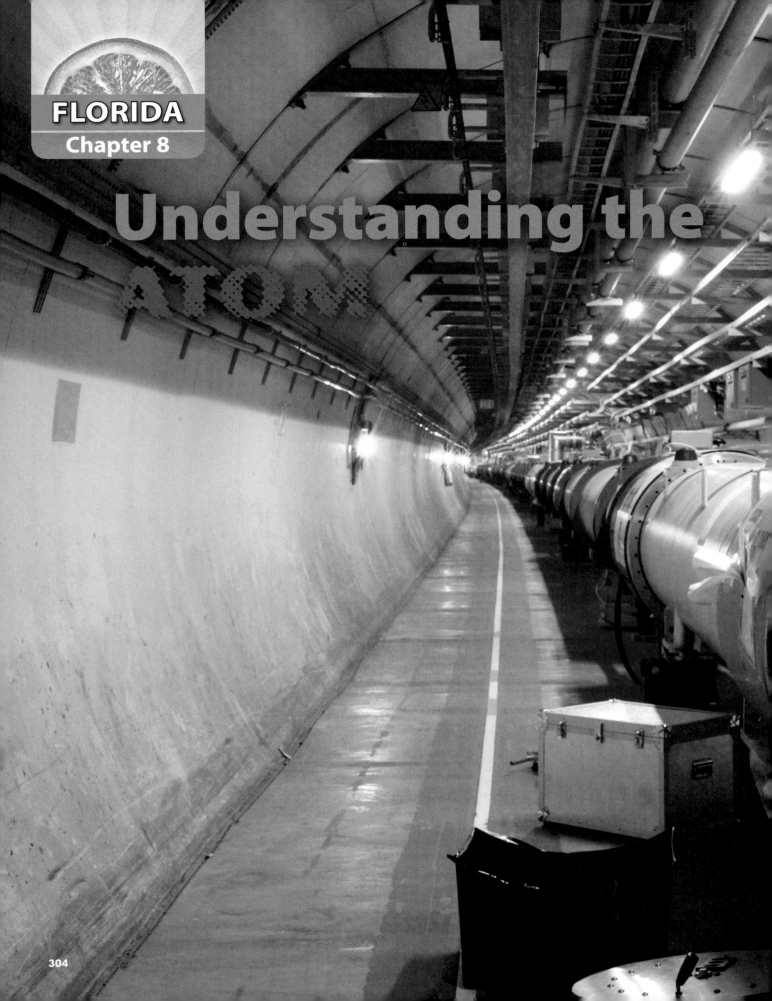

Understanding the
ATOM

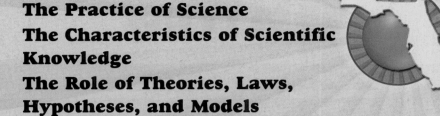

FLORIDA BIG IDEAS

1   The Practice of Science

2   The Characteristics of Scientific Knowledge

3   The Role of Theories, Laws, Hypotheses, and Models

8   Properties of Matter

11 Energy Transfer and Transformations

Think About It!

What are atoms, and what are they made of?

This huge machine is called the Large Hadron Collider (LHC). It's like a circular racetrack for particles and is about 27 km long. The LHC accelerates particle beams to high speeds and then smashes them into each other. The longer the tunnel, the faster the beams move and the harder they smash together. Scientists study the tiny particles produced in the crash.

1 How do you think scientists might have studied matter before colliders were invented?

_____

_____

2 What do you think are the smallest parts of matter?

_____

_____

3 What do you think atoms are, and what are they made of?

_____

_____

_____

Get Ready to Read

What do you think about the atom?

Before you read, decide if you agree or disagree with each of these statements. As you read this chapter, see if you change your mind about any of the statements.

	AGREE	DISAGREE
1 The earliest model of an atom contained only protons and electrons.	☐	☐
2 Air fills most of an atom.	☐	☐
3 In the present-day model of the atom, the nucleus of the atom is at the center of an electron cloud.	☐	☐
4 All atoms of the same element have the same number of protons.	☐	☐
5 Atoms of one element cannot be changed into atoms of another element.	☐	☐
6 Ions form when atoms lose or gain electrons.	☐	☐

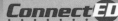 **There's More Online!**
Video • Audio • Review • ⓘLab Station • WebQuest • Assessment • Concepts in Motion • Multilingual eGlossary

305

Discovering Parts of an ATOM

ESSENTIAL QUESTIONS

 What is an atom?

How would you describe the size of an atom?

How has the atomic model changed over time?

Vocabulary

atom p. 309

electron p.311

nucleus p. 314

proton p. 314

neutron p. 315

electron cloud p. 316

 Florida NGSSS

SC.6.N.2.2 Explain that scientific knowledge is durable because it is open to change as new evidence or interpretations are encountered.

SC.6.N.2.3 Recognize that scientists who make contributions to scientific knowledge come from all kinds of backgrounds and possess varied talents, interests, and goals.

SC.7.N.1.5 Describe the methods used in the pursuit of a scientific explanation as seen in different fields of science such as biology, geology, and physics.

SC.7.N.1.6 Explain that empirical evidence is the cumulative body of observations of a natural phenomenon on which scientific explanations are based.

SC.7.N.1.7 Explain that scientific knowledge is the result of a great deal of debate and confirmation within the science community.

SC.7.N.2.1 Identify an instance from the history of science in which scientific knowledge has changed when new evidence or new interpretations are encountered.

SC.7.N.3.1 Recognize and explain the difference between theories and laws and give several examples of scientific theories and the evidence that supports them.

SC.8.P.8.7 Explore the scientific theory of atoms (also known as atomic theory) by recognizing that atoms are the smallest unit of an element and are composed of sub-atomic particles (electrons surrounding a nucleus containing protons and neutrons).

Also covers: SC.8.N.1.1, SC.8.N.1.6, LA.6.2.2.3, LA.6.4.2.2

SC.8.P.8.7

Inquiry **Launch Lab**
10 minutes

What's in there?

When you look at a sandy beach from far away, it looks like a solid surface. You can't see the individual grains of sand. What would you see if you zoomed in on one grain of sand?

Procedure

1. Read and complete a lab safety form.

2. Have your partner hold a **test tube** of a **substance,** filled to a height of 2–3 cm.

3. Observe the test tube from a distance of at least 2 m. Write a description of what you see below.

4. Pour about 1 cm of the substance onto a piece of **waxed pape**r. Record your observations.

5. Use a **toothpick** to separate out one particle of the substance. Suppose you could zoom in. What do you think you would see? Record your ideas.

Data and Observations

Think About This

1. Do you think one particle of the substance is made of smaller particles? Why or why not?

2. **Key Concept** Do you think you could use a microscope to see what the particles are made of? Why or why not?

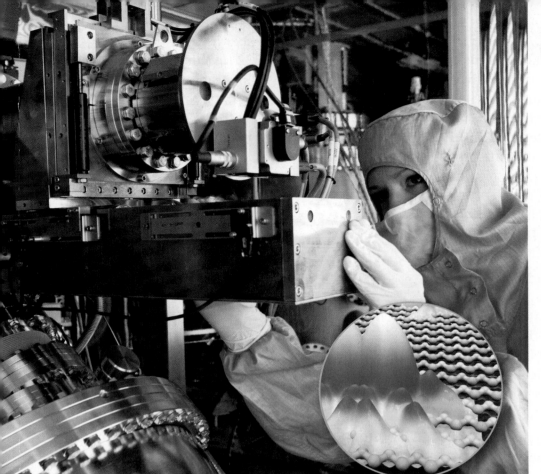

A Microscopic Mountain Range?

1. This photo shows a glimpse of the tiny particles that make up matter. A special microscope, invented in 1981, made this image. However, scientists knew these tiny particles existed long before they were able to see them. What are these tiny particles? How small do you think they are? How do you think scientists might have learned so much about them before being able to see them?

_____

_____

_____

_____

_____

_____

Early Ideas About Matter

Look at your hands. What are they made of? You might answer that your hands are made of things such as skin, bone, muscle, and blood. You might recall that each of these is made of even smaller structures called cells. Are cells made of even smaller parts? Imagine dividing something into smaller and smaller parts. What would you end up with?

Greek philosophers discussed and debated questions such as these more than 2,000 years ago. At the time, many thought that all matter is made of only four elements—fire, water, air, and earth, as shown in **Figure 1.** However, they weren't able to test their ideas because scientific tools and methods, such as experimentation, did not exist yet. The ideas proposed by the most influential philosophers usually were accepted over the ideas of less influential philosophers. One philosopher, Democritus (460–370 B.C.), challenged the popular idea of matter.

Figure 1 🔑 Most Greek philosophers believed that all matter is made of only four elements—fire, water, air, and earth.

Democritus

Democritus believed that matter is made of small, solid objects that cannot be divided, created, or destroyed. He called these objects *atomos,* from which the English word *atom* is derived. Democritus proposed that different types of matter are made from different types of atoms. For example, he said that smooth matter is made of smooth atoms. He also proposed that nothing is between these atoms except empty space. **Table 1** summarizes Democritus's ideas.

Although Democritus had no way to test his ideas, many of his ideas are similar to the way scientists describe the atom today. Because Democritus's ideas did not conform to the popular opinion and because they could not be tested scientifically, they were open for debate. One philosopher who challenged Democritus's ideas was Aristotle.

Active Reading

2. Infer According to Democritus, what might atoms of gold look like?

_____

_____

Aristotle

Aristotle (384–322 B.C.) did not believe that empty space exists. Instead, he favored the more popular idea—that all matter is made of fire, water, air, and earth. Because Aristotle was so influential, his ideas were accepted. Democritus's ideas about atoms were not studied again for more than 2,000 years.

Dalton's Atomic Model

In the late 1700s, English schoolteacher and scientist John Dalton (1766–1844) revisited the idea of atoms. Since Democritus's time, advancements had been made in technology and scientific methods. Dalton made careful observations and measurements of chemical reactions. He combined data from his own scientific research with data from the research of other scientists to propose the atomic theory. **Table 1** lists ways that Dalton's atomic theory supported some of the ideas of Democritus.

Table 1 Similarities Between Democritus's and Dalton's Ideas

Democritus
1. Atoms are small solid objects that cannot be divided, created, or destroyed.
2. Atoms are constantly moving in empty space.
3. Different types of matter are made of different types of atoms.
4. The properties of the atoms determine the properties of matter.

John Dalton
1. All matter is made of atoms that cannot be divided, created, or destroyed.
2. During a chemical reaction, atoms of one element cannot be converted into atoms of another element.
3. Atoms of one element are identical to each other but different from atoms of another element.
4. Atoms combine in specific ratios.

Figure 2 If you could keep dividing a piece of aluminum, you eventually would have the smallest possible piece of aluminum—an aluminum atom.

Active Reading **3. Restate** What are the theories in Dalton's atomic model?

All matter is made of atoms that...

cannot be

cannot be

are identical to

are different from

combine in

The Atom

Today, scientists agree that matter is made of atoms with empty space between and within them. What is an atom? Imagine dividing the piece of aluminum shown in **Figure 2** into smaller and smaller pieces. At first you would be able to cut the pieces with scissors. But eventually you would have a piece that is too small to see—much smaller than the smallest piece you could cut with scissors. This small piece is an aluminum atom. An aluminum atom cannot be divided into smaller aluminum pieces. *An* **atom** *is the smallest piece of an element that still represents that element.* The scientific theory of atoms, the smallest unit of an element, is called the atomic theory.

The Size of Atoms

Just how small is an atom? Atoms of different elements are different sizes, but all are very, very small. You cannot see atoms with just your eyes or even with most microscopes. Atoms are so small that about 7.5 trillion carbon atoms could fit into the period at the end of this sentence.

 4. NGSSS Check **Explain** How would you describe the size of an atom? SC.8.P.8.7

_____

_____

Seeing Atoms

Scientific experiments verified that matter is made of atoms long before scientists were able to see atoms. However, the 1981 invention of a high-powered microscope, called a scanning tunneling microscope (STM), enabled scientists to see individual atoms for the first time. **Figure 3** shows an STM image. An STM uses a tiny, metal tip to trace the surface of a piece of matter. The result is an image of atoms on the surface.

Even today, scientists still cannot see inside an atom. However, scientists have learned that atoms are not the smallest particles of matter. In fact, atoms are made of much smaller particles. What are these particles, and how did scientists discover them if they could not see them?

Figure 3 A scanning tunneling microscope created this image. The yellow sphere is a manganese atom on the surface of gallium arsenide.

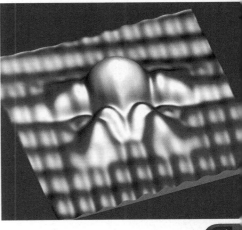

Thomson—Discovering Electrons

Not long after Dalton's findings, another English scientist, named J.J. Thomson (1856–1940), made some important discoveries. Thomson and other scientists of that time worked with cathode ray tubes. If you ever have seen a neon sign, an older computer monitor, or the color display on an ATM screen, you have seen a cathode ray tube. Thomson's cathode ray tube, shown in **Figure 4,** was a glass tube with pieces of metal, called electrodes, attached inside the tube. The electrodes were connected to wires, and the wires were connected to a battery. Thomson discovered that if most of the air was removed from the tube and electricity was passed through the wires, greenish-colored rays traveled from one electrode to the other end of the tube. What were these rays made of?

Negative Particles

Scientists called these rays cathode rays. Thomson wanted to know if these rays had an electric charge. To find out, he placed two plates on opposite sides of the tube. One plate was positively charged, and the other plate was negatively charged, as shown in **Figure 4.** Thomson discovered that these rays bent toward the positively charged plate and away from the negatively charged plate. Recall that opposite charges attract each other, and like charges repel each other. Thomson concluded that cathode rays are negatively charged.

Figure 4 As the cathode rays passed between the plates, they were bent toward the positive plate. Because opposite charges attract, the rays must be negatively charged.

Active Reading

5. Predict If the rays were positively charged, what would Thomson have observed as they passed between the plates?

_____

_____

Thomson's Cathode Ray Tube Experiment 🔑

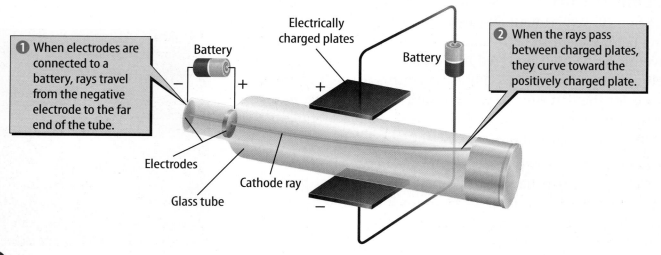

❶ When electrodes are connected to a battery, rays travel from the negative electrode to the far end of the tube.

Battery

Electrically charged plates

Battery

❷ When the rays pass between charged plates, they curve toward the positively charged plate.

Electrodes

Cathode ray

Glass tube

Parts of Atoms

Through more experiments, Thomson learned that these rays were made of particles that had mass. The mass of one of these particles was much smaller than the mass of the smallest atoms. This was surprising information to Thomson. Until then, scientists understood that the smallest particle of matter is an atom. But these rays were made of particles that were even smaller than atoms.

Where did these small, negatively charged particles come from? Thomson proposed that these particles came from the metal atoms in the electrode. Thomson discovered that identical rays were produced regardless of the kind of metal used to make the electrode. Putting these clues together, Thomson concluded that cathode rays were made of small, negatively charged particles. He called these particles electrons. *An* **electron** *is a particle with one negative charge (1−).* Because atoms are neutral, or not electrically charged, Thomson proposed that atoms also must contain a positive charge that balances the negatively charged electrons.

Thomson's Atomic Model

Thomson used this information to propose a new model of the atom. Instead of a solid, neutral sphere that was the same throughout, Thomson's model of the atom contained both positive and negative charges. He proposed that an atom was a sphere with a positive charge evenly spread throughout. Negatively charged electrons were mixed through the positive charge, similar to the way chocolate chips are mixed in cookie dough. **Figure 5** shows this model.

Active Reading 6. **Recall** How did Thomson's atomic model differ from Dalton's atomic model?

_____

_____

Active Reading

FOLDABLES LA.6.2.2.3

Use two sheets of paper to make a layered book. Label it as shown. Use it to organize your notes and diagrams on the parts of an atom.

Atom
Protons
Neutrons
Electrons

WORD ORIGIN

electron

from Greek *electron,* means "amber," the physical force so called because it first was generated by rubbing amber. Amber is a fossilized substance produced by trees.

Thomson's Atomic Model 🔑

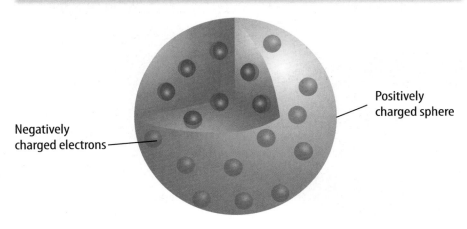

Negatively charged electrons

Positively charged sphere

Figure 5 Thomson's model of the atom contained a positively charged sphere with negatively charged electrons within it.

Rutherford—Discovering the Nucleus

The discovery of electrons stunned scientists. Ernest Rutherford (1871–1937) was a student of Thomson's who eventually had students of his own. Rutherford's students set up experiments to test Thomson's atomic model and to learn more about what atoms contain. They discovered another surprise.

Rutherford's Predicted Result

Imagine throwing a baseball into a pile of table-tennis balls. The baseball likely would knock the table-tennis balls out of the way and continue moving in a relatively straight line. This is similar to what Rutherford's students expected to see when they shot alpha particles into atoms. Alpha particles are dense and positively charged. Because they are so dense, only another dense particle could deflect the path of an alpha particle. According to Thomson's model, the positive charge of the atom was too spread out and not dense enough to change the path of an alpha particle. Electrons wouldn't affect the path of an alpha particle because electrons didn't have enough mass. The result that Rutherford's students expected is shown in **Figure 6.**

Active Reading

7. **Analyze** Explain why Rutherford's students did not think an atom could change the path of an alpha particle.

_____

_____

_____

_____

Figure 6 Rutherford predicted that the alpha particles would pass through the gold foil.

Rutherford's Predicted Result 🔑

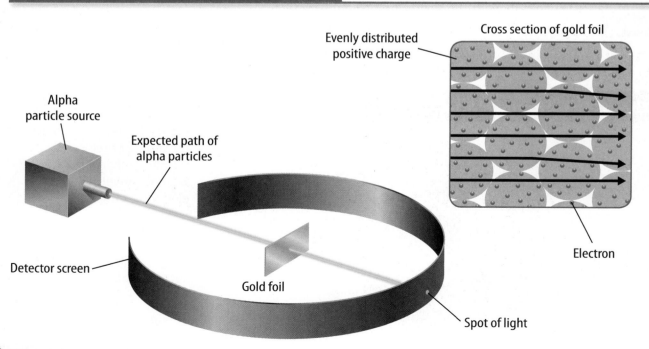

Alpha particle source

Expected path of alpha particles

Detector screen

Gold foil

Spot of light

Evenly distributed positive charge

Cross section of gold foil

Electron

The Gold Foil Experiment

Rutherford's students went to work. They placed a source of alpha particles near a very thin piece of gold foil. Recall that all matter is made of atoms. Therefore, the gold foil was made of gold atoms. A screen surrounded the gold foil. When an alpha particle struck the screen, it created a spot of light. Rutherford's students could determine the path of the alpha particles by observing the spots of light on the screen.

The Surprising Result

Figure 7 shows what the students observed. Most of the particles did indeed travel through the foil in a straight path. However, a few particles struck the foil and bounced off to the side. And one particle in 10,000 bounced straight back! Rutherford later described this surprising result, saying it was almost as incredible as if you had fired a 38-cm shell at a piece of tissue paper and it came back and hit you. The alpha particles must have struck something dense and positively charged inside the nucleus. Thomson's model had to be refined.

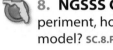 8. **NGSSS Check Predict** Given the results of the gold foil experiment, how do you think an actual atom differs from Thomson's model? SC.8.P.8.7

_____

_____

_____

_____

Figure 7 Some alpha particles traveled in a straight path, as expected. But some changed direction, and some bounced straight back.

Active Reading 9. **Identify** Label the parts of the gold atom below.

The Surprising Result

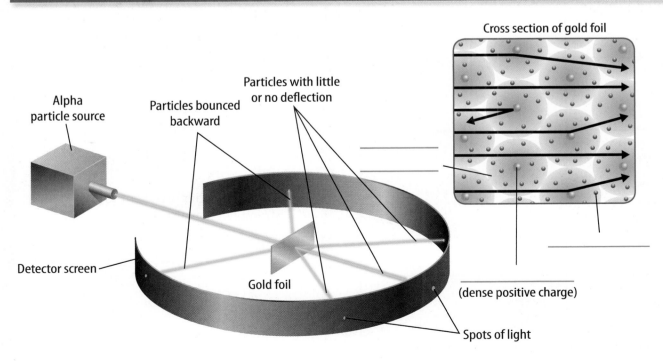

Alpha particle source

Particles bounced backward

Particles with little or no deflection

Detector screen

Gold foil

Cross section of gold foil

_____

_____ (dense positive charge)

Spots of light

Rutherford's Atomic Model

Because most alpha particles traveled through the foil in a straight path, Rutherford concluded that atoms are made mostly of empty space. The alpha particles that bounced backward must have hit a dense, positive mass. Rutherford concluded that *most of an atom's mass and positive charge is concentrated in a small area in the center of the atom called the* **nucleus. Figure 8** shows Rutherford's atomic model. Additional research showed that the positive charge in the nucleus was made of positively charged particles called protons. *A* **proton** *is an atomic particle that has one positive charge (1+).* Negatively charged electrons move in the empty space surrounding the nucleus.

Active Reading **10. Restate** <u>Underline</u> how Rutherford explained the observation that some of the alpha particles bounced directly backward.

Active Reading **11. Sequence** Show the path of discovery of the nucleus and the development of Rutherford's atomic model.

Event	Description
Experiment to test Thomson's atomic model	1.
Result of the gold foil experiment	2.
Conclusion	3.
Description of the dense, positive mass	4.
New model of the atom	5.
Results of further research	6.

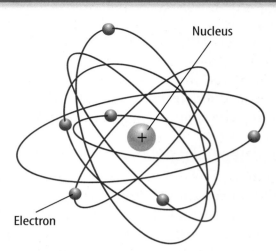

Rutherford's Atomic Model 🔑

Nucleus

Electron

Figure 8 Rutherford's model contains a small, dense, positive nucleus. Tiny, negatively charged electrons travel in empty space around the nucleus.

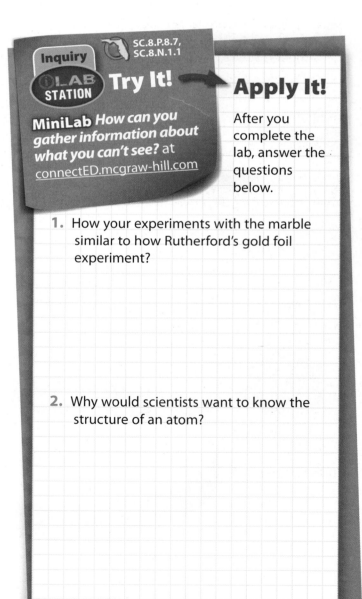

SC.8.P.8.7, SC.8.N.1.1

Inquiry

iLAB STATION **Try It!**

MiniLab *How can you gather information about what you can't see?* at connectED.mcgraw-hill.com

Apply It!

After you complete the lab, answer the questions below.

1. How your experiments with the marble similar to how Rutherford's gold foil experiment?

2. Why would scientists want to know the structure of an atom?

Discovering Neutrons

The modern model of the atom was beginning to take shape. Rutherford's colleague, James Chadwick (1891–1974), also researched atoms and discovered that in addition to protons, the nucleus contains neutrons. *A **neutron** is a neutral particle that exists in the nucleus of an atom.*

Bohr's Atomic Model

Rutherford's model explained much of his students' experimental evidence. However, there were several observations that the model could not explain. For example, scientists noticed that if certain elements were heated in a flame, they gave off specific colors of light. Each color of light had a specific amount of energy. Where did this light come from? Niels Bohr (1885–1962), another student of Rutherford's, proposed an answer. Bohr studied hydrogen atoms because they contain only one electron. He experimented with adding electric energy to hydrogen and studying the energy that was released. His experiments led to a revised atomic model.

Electrons in the Bohr Model

Bohr's model is shown in **Figure 9.** Bohr proposed that electrons move in circular orbits, or energy levels, around the nucleus. Electrons in an energy level have a specific amount of energy. Electrons closer to the nucleus have less energy than electrons farther away from the nucleus. When energy is added to an atom, electrons gain energy and move from a lower energy level to a higher energy level. When the electrons return to the lower energy level, they release a specific amount of energy as light. This is the light that is seen when elements are heated.

Limitations of the Bohr Model

Bohr reasoned that if his model were accurate for atoms with one electron, it would be accurate for atoms with more than one electron. However, this was not the case. More research showed that although electrons have specific amounts of energy, energy levels are not arranged in circular orbits. How do electrons move in an atom?

Active Reading

12. Contrast How did Bohr's atomic model differ from Rutherford's?

_____

_____

Figure 9 In Bohr's atomic model, electrons move in circular orbits around the atom. When an electron moves from a higher energy level to a lower energy level, energy is released— sometimes as light. Further research showed that electrons are not arranged in orbits.

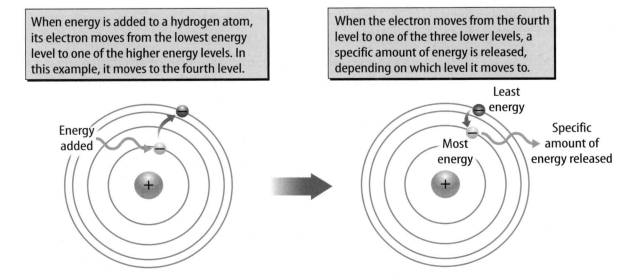

When energy is added to a hydrogen atom, its electron moves from the lowest energy level to one of the higher energy levels. In this example, it moves to the fourth level.

When the electron moves from the fourth level to one of the three lower levels, a specific amount of energy is released, depending on which level it moves to.

Energy added

Least energy

Most energy

Specific amount of energy released

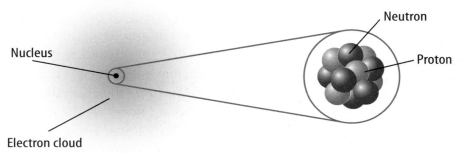

Nucleus

Neutron

Proton

Electron cloud

Figure 10 In this atom, electrons are more likely to be found closer to the nucleus than farther away.

Active Reading 13. **Hypothesize** Why do you think this model of the atom doesn't show the electrons?

_____

_____

_____

_____

15. **NGSSS Check Review** What are the parts that make up atoms?
SC.8.P.8.7

Particle	Smaller Parts
Electron	
Nucleus	
Proton	
Neutron	

The Modern Atomic Model

In the modern atomic model, electrons form an electron cloud. _An_ **electron cloud** _is an area around an atomic nucleus where an electron is most likely to be located._ Imagine taking a time-lapse photograph of bees around a hive. You might see a blurry cloud. The cloud might be denser near the hive than farther away because the bees spend more time near the hive.

In a similar way, electrons constantly move around the nucleus. It is impossible to know the speed and exact location of an electron at a given moment in time. Instead, scientists only can predict the likelihood that an electron is in a particular location. The electron cloud shown in **Figure 10** is mostly empty space but represents the likelihood of finding an electron in a given area. The darker areas represent areas where electrons are more likely to be.

Active Reading 14. **Describe** How has the model of the atom changed over time?

_____

_____

Quarks

You have read that atoms are made of smaller parts—protons, neutrons, and electrons. Are these particles made of even smaller parts? Scientists have discovered that electrons are not made of smaller parts. However, research has shown that protons and neutrons are made of smaller particles called quarks. Scientists theorize that there are six types of quarks. They have named these quarks up, down, charm, strange, top, and bottom. Protons are made of two up quarks and one down quark. Neutrons are made of two down quarks and one up quark. Just as the model of the atom has changed over time, the current model might also change with the invention of new technology that aids the discovery of new information.

If you were to divide an element into smaller and smaller pieces, the smallest piece would be an atom.

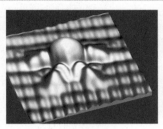

Atoms are so small that they can be seen only by using very powerful microscopes.

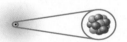

Scientists now know that atoms contain a dense, positive nucleus surrounded by an electron cloud.

Use Vocabulary

1 The smallest piece of the element gold is a gold _____.

2 **Write** a sentence that describes the nucleus of an atom.

_____

_____

3 **Define** *electron cloud* in your own words.

_____

_____

Understand Key Concepts 🗝️

4 What is an atom mostly made of? SC.8.P.8.7

  (A) air   (C) neutrons

  (B) empty space   (D) protons

5 Why have scientists only recently been able to see atoms? SC.8.P.8.7

  (A) Atoms are too small to see with ordinary microscopes.

  (B) Early experiments disproved the idea of atoms.

  (C) Scientists didn't know atoms existed.

  (D) Scientists were not looking for atoms.

6 **Explain** how Rutherford's students knew that Thomson's model of the atom needed to change.

_____

_____

_____

Interpret Graphics

7 **Contrast** Use the graphic organizer below to contrast the locations of electrons in Thomson's, Rutherford's, Bohr's, and the modern models of the atom. LA.6.2.2.3

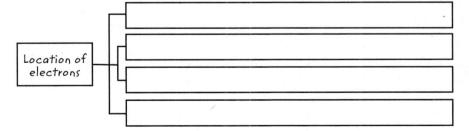

Critical Thinking

8 **Explain** what might have happened in Rutherford's experiment if he had used a thin sheet of copper instead of a thin sheet of gold.

_____

_____

_____

Subatomic Particles

Welcome to the Particle Zoo

QUARKS

BOSONS

LEPTONS

Much has changed since Democritus and Aristotle studied atoms.

When Democritus and Aristotle developed ideas about matter, they probably never imagined the kinds of research being performed today! From the discovery of electrons, protons, and neutrons to the exploration of quarks and other particles, the atomic model continues to change.

You've learned about quarks, which make up protons and neutrons. But quarks are not the only kind of particles! In fact, some scientists call the collection of particles that have been discovered the particle zoo, because different types of particles have unique characteristics, just like the different kinds of animals in a zoo.

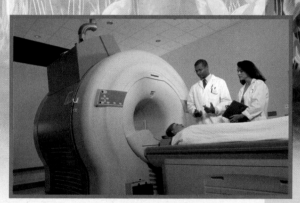

MRIs are just o ne way in which particle physics technology is applied.

In addition to quarks, scientists have discovered a group of particles called leptons, which includes the electron. Gluons and photons are examples of bosons—particles that carry forces. Some particles, such as the Higgs Boson, have been predicted to exist but have yet to be observed in experiments.

Identifying and understanding the particles that make up matter is important work. However, it might be difficult to understand why time and money are spent to learn more about tiny subatomic particles. How can this research possibly affect everyday life? Research on subatomic particles has changed society in many ways. For example, magnetic resonance imaging (MRI), a tool used to diagnose medical problems, uses technology that was developed to study subatomic particles. Cancer treatments using protons, neutrons, and X-rays are all based on particle physics technology. And, in the 1990s, the need for particle physicists to share information with one another led to the development of the World Wide Web!

It's Your Turn

RESEARCH AND REPORT Learn more about research on subatomic particles. Find out about one recent discovery. Make a poster to share what you learn with your classmates.

LA.6.4.2.2

Protons, Neutrons, and Electrons—
HOW ATOMS DIFFER

What happens during nuclear decay?

How does a neutral atom change when its number of protons, electrons, or neutrons changes?

Vocabulary

atomic number p. 321

isotope p. 322

mass number p. 322

average atomic mass p. 323

radioactive p. 324

nuclear decay p. 325

ion p. 326

Florida NGSSS

SC.6.N.2.1 Distinguish science from other activities involving thought.

SC.6.N.2.2 Explain that scientific knowledge is durable because it is open to change as new evidence or interpretations are encountered.

SC.6.N.2.3 Recognize that scientists who make contributions to scientific knowledge come from all kinds of backgrounds and possess varied talents, interests, and goals.

SC.7.N.1.3 Distinguish between an experiment (which must involve the identification and control of variables) and other forms of scientific investigation and explain that not all scientific knowledge is derived from experimentation.

SC.7.N.1.6 Explain that empirical evidence is the cumulative body of observations of a natural phenomenon on which scientific explanations are based.

SC.7.N.2.1 Identify an instance from the history of science in which scientific knowledge has changed when new evidence or new interpretations are encountered.

SC.7.P.11.2 Investigate and describe the transformation of energy from one form to another.

SC.8.P.8.7 Explore the scientific theory of atoms (also known as atomic theory) by recognizing that atoms are the smallest unit of an element and are composed of sub-atomic particles (electrons surrounding a nucleus containing protons and neutrons).

Also covers: SC.8.N.1.1, SC.8.N.1.6, LA.6.2.2.3, MA.6.A.3.6

 Launch Lab

SC.8.N.3.1

15 minutes

How many different things can you make?

Many buildings are made of just a few basic building materials, such as wood, nails, and glass. You can combine those materials in many different ways to make buildings of various shapes and sizes. How many things can you make from three materials?

Procedure 🥽

1. Read and complete a lab safety form.

2. Use **colored building blocks** to make as many different objects as you can with the following properties:

   • Each object must have a different number of red blocks.

   • Each object must have an equal number of red and blue blocks.

   • Each object must have at least as many yellow blocks as red blocks but can have no more than two extra yellow blocks.

3. As you complete each object, record the number of each color of block used to make it. For example, R = 1; B = 1; Y = 2.

4. When time is called, compare your objects with others in the class.

Think About This

1. How many different objects did you make? How many different objects did the class make?

2. 🔑 **Key Concept** In what ways does changing the number of building blocks change the properties of the objects?

Inquiry **Is this glass glowing?**

1. Under natural light, this glass vase is yellow. But when exposed to ultraviolet light, It glows green! What do you think causes the glass to glow green? Why would it react only in ultraviolet light?

_____

_____

_____

_____

_____

Active Reading

FOLDABLES® LA.6.2.2.3

Create a three-tab book and label it as shown. Use it to organize the three ways that atoms can differ.

Different Numbers of:
Protons | Neutrons | Electrons

The Parts of the Atom

If you could see inside any atom, you probably would see the same thing—empty space surrounding a very tiny nucleus. A look inside the nucleus would reveal positively charged protons and neutral neutrons. Negatively charged electrons would be whizzing by in the empty space around the nucleus.

Table 2 compares the properties of protons, neutrons, and electrons. Protons and neutrons have about the same mass. The mass of electrons is much smaller than the mass of protons or neutrons. That means most of the mass of an atom is found in the nucleus. In this lesson, you will learn that, while all atoms contain protons, neutrons, and electrons, the numbers of these particles are different for different types of atoms.

Table 2 Properties of Protons, Neutrons, and Electrons	Electron	Proton	Neutron
Symbol	e—	p	n
Charge	1—	1+	0
Location	electron cloud around the nucleus	nucleus	nucleus
Relative mass	1/1,840	1	1

Different Elements—Different Numbers of Protons

Look at the periodic table in the back of this book. Notice that there are more than 115 different elements. Recall that an element is a substance made from atoms that all have the same number of protons. For example, the element carbon is made from atoms that all have six protons. Likewise, all atoms that have six protons are carbon atoms. *The number of protons in an atom of an element is the element's* **atomic number.** The atomic number is the whole number listed with each element on the periodic table.

What makes an atom of one element different from an atom of another element? Atoms of different elements contain different numbers of protons. For example, oxygen atoms contain eight protons; nitrogen atoms contain seven protons. Different elements have different atomic numbers. **Figure 11** shows some common elements and their atomic numbers.

Neutral atoms of different elements also have different numbers of electrons. In a neutral atom, the number of electrons equals the number of protons. Therefore, the number of positive charges equals the number of negative charges.

Active Reading **3. Review** (Circle) the two numbers that can be used to identify an element.

Active Reading **2. Relate** Why is the atomic number so important?

determines

_____

Atomic Number

equals _____
and also equals

_____

in neutral atoms

Active Reading **4. Identify** Label the nitrogen atom and the oxygen atom based on the number of protons in the nucleus.

Different Elements 🔑

Figure 11 Atoms of different elements contain different numbers of protons.

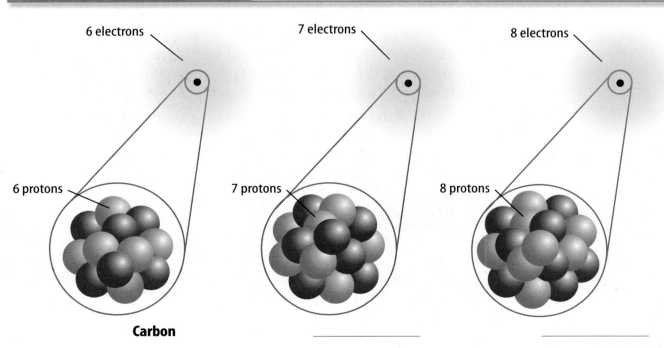

6 electrons

7 electrons

8 electrons

6 protons

7 protons

8 protons

Carbon

_____

_____

Use Percentages

You can calculate the average atomic mass of an element if you know the percentage of each isotope in the element. Lithium (Li) contains 7.5% Li-6 and 92.5% Li-7. What is the average atomic mass of Li?

1. Divide each percentage by 100 to change to decimal form.
$$\frac{7.5\%}{100} = 0.075$$
$$\frac{92.5\%}{100} = 0.925$$

2. Multiply the mass of each isotope by its decimal percentage.
$6 \times 0.075 = 0.45$
$7 \times 0.925 = 6.475$

3. Add the values together to get the average atomic mass.
$0.45 + 6.475 = 6.93$

Practice

5. Nitrogen (N) contains 99.63% N-14 and 0.37% N-15. What is the average atomic mass of nitrogen?

WORD ORIGIN

isotope
from Greek *isos,* means "equal", and *topos,* means "place"

Table 3 Naturally Occurring Isotopes of Carbon

Isotope	Carbon-12 Nucleus	Carbon-13 Nucleus	Carbon-14 Nucleus
Abundance	98.89%	<1.11%	<0.01%
Protons	6	6	6
Neutrons	+ 6	+ 7	+ 8
Mass Number	12	13	14

Neutrons and Isotopes

You have read that atoms of the same element have the same numbers of protons. However, atoms of the same element can have different numbers of neutrons. For example, carbon atoms all have six protons, but some carbon atoms have six neutrons, some have seven neutrons, and some have eight neutrons. These three different types of carbon atoms, shown in **Table 3,** are called isotopes. **Isotopes** *are atoms of the same element that have different numbers of neutrons.* Most elements have several isotopes.

Protons, Neutrons, and Mass Number

The **mass number** *of an atom is the sum of the number of protons and neutrons in an atom.* This is shown in the following equation.

Mass number = number of protons + number of neutrons

Any one of these three quantities can be determined if you know the value of the other two quantities. For example, to determine the mass number of an atom, you must know the number of neutrons and the number of protons in the atom.

The mass numbers of the isotopes of carbon are shown in **Table 3.** An isotope often is written with the element name followed by the mass number. Using this method, the isotopes of carbon are written carbon-12, carbon-13, and carbon-14.

Active Reading   6. **Differentiate** How do two different isotopes of the same element differ?

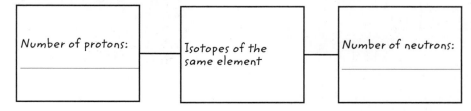

| Number of protons: | Isotopes of the same element | Number of neutrons: |

Average Atomic Mass

You might have noticed that the periodic table does not list mass numbers or the numbers of neutrons. This is because a given element can have several isotopes. However, you might notice that there is a decimal number listed with most elements, as shown in **Figure 12.** This decimal number is the average atomic mass of the element. The **average atomic mass** of an element is the average mass of the element's isotopes, weighted according to the abundance of each isotope.

Table 3 shows the three isotopes of carbon. The average atomic mass of carbon is 12.01. Why isn't the average atomic mass 13? After all, the average of the mass numbers 12, 13, and 14 is 13. The average atomic mass is weighted based on each isotope's abundance—how much of each isotope is present on Earth. Almost 99 percent of Earth's carbon is carbon-12. That is why the average atomic mass is close to 12.

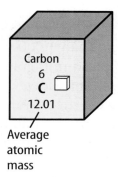

Figure 12 The element carbon has several isotopes. The decimal number 12.01 is the average atomic mass of these isotopes.

Active Reading

7. **Explain** What does the term *weighted average* mean?

_____

_____

_____

Inquiry SC.8.N.1.1, SC.8.P.8.7

LAB STATION **Try It!**

MiniLab *How many penny isotopes do you have?* at connectED.mcgraw-hill.com

Apply It! After you complete the lab, answer these questions.

1. **Describe** There are three isotopes of hydrogen: H-1, H-2, and H-3. However, the average atomic mass is 1.008, not 2. Explain why the average atomic mass of hydrogen is not 2. Which isotope of hydrogen is most abundant?

2. **Examine** Compare and contrast the nuclei of the three isotopes of hydrogen by discussing the number of protons and neutrons in each as listed in question 1 above.

Radioactivity

More than 1,000 years ago, people tried to change lead into gold by performing chemical reactions. However, none of their reactions were successful. Why not? Today, scientists know that a chemical reaction does not change the number of protons in an atom's nucleus. If the number of protons does not change, the element does not change. But in the late 1800s, scientists discovered that some elements change into other elements **spontaneously.** How does this happen?

An Accidental Discovery

In 1896, a scientist named Henri Becquerel (1852–1908) studied minerals containing the element uranium. When these minerals were exposed to sunlight, they gave off a type of energy that could pass through paper. If Becquerel covered a photographic plate with black paper, this energy would pass through the paper and expose the film. One day, Becquerel left the mineral next to a wrapped, unexposed plate in a drawer. Later, he opened the drawer, unwrapped the plate, and saw that the plate contained an image of the mineral, as shown in **Figure 13.** The mineral spontaneously emitted energy, even in the dark! Sunlight wasn't required. What was this energy?

Radioactivity

Becquerel shared his discovery with fellow scientists Pierre and Marie Curie. Marie Curie (1867–1934), shown in **Figure 14,** called *elements that spontaneously emit radiation* **radioactive.** Becquerel and the Curies discovered that the radiation released by uranium was made of energy and particles. This radiation came from the nuclei of the uranium atoms. When this happens, the number of protons in one atom of uranium changes. When uranium releases radiation, it changes to a different element.

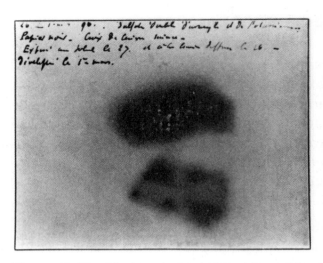

Figure 13 The black and white photo shows Henri Becquerel's photographic plate. The dark area on the plate was exposed to radiation given off by uranium in the mineral even though the mineral was not exposed to sunlight.

ACADEMIC VOCABULARY

spontaneous
(adjective) occurring without external force or cause

Figure 14 Marie Curie studied radioactivity and discovered two new radioactive elements—polonium and radium.

Types of Decay

Radioactive elements contain unstable nuclei. **Nuclear decay** *is a process that occurs when an unstable atomic nucleus changes into another more stable nucleus by emitting radiation.* Nuclear decay can produce three different types of radiation—alpha particles, beta particles, and gamma rays. **Figure 15** compares the three types of nuclear decay.

Alpha Decay An alpha particle is made of two protons and two neutrons. When an atom releases an alpha particle, its atomic number decreases by two. Uranium-238 decays to thorium-234 through the process of alpha decay.

Beta Decay When beta decay occurs, a neutron in an atom changes into a proton and a high-energy electron called a beta particle. The new proton becomes part of the nucleus, and the beta particle is released. In beta decay, the atomic number of an atom increases by one because it has gained a proton.

Gamma Decay Gamma rays do not contain particles, but they do contain a lot of energy. In fact, gamma rays can pass through thin sheets of lead. Because gamma rays do not contain particles, the release of gamma rays does not change one element into another element.

 9. NGSSS Check Restate <u>Underline</u> what happens during radioactive decay. SC.8.P.8.7

Uses of Radioactive Isotopes

The energy released by radioactive decay can be both harmful and beneficial to humans. Too much radiation can damage or destroy living cells, making them unable to function properly. Some organisms contain cells, such as cancer cells, that are harmful to the organism. Radiation therapy can be beneficial to humans by destroying these harmful cells.

Figure 15 🔑 Alpha and beta decay change one element into another element.

Active Reading **8. Illustrate** Show how the atomic number changes for each type of decay.

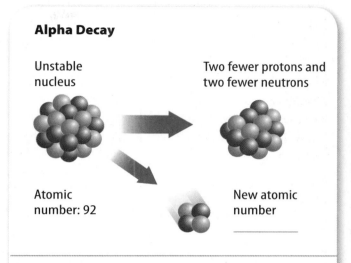

Alpha Decay

Unstable nucleus

Two fewer protons and two fewer neutrons

Atomic number: 92

New atomic number

_____

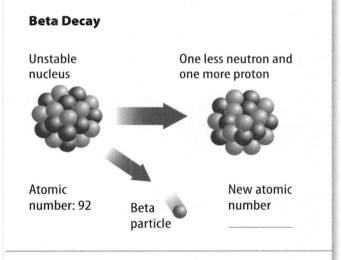

Beta Decay

Unstable nucleus

One less neutron and one more proton

Atomic number: 92

Beta particle

New atomic number

_____

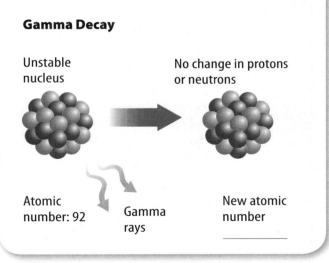

Gamma Decay

Unstable nucleus

No change in protons or neutrons

Atomic number: 92

Gamma rays

New atomic number

_____

10. NGSSS Check
Contrast How does a neutral atom change when the number of electrons changes? SC.8.P.8.7

Positive Ion
A neutral atom
Resulting charge:

Negative Ion
A neutral atom
Resulting charge:

Ions—Gaining or Losing Electrons

What happens to a neutral atom if it gains or loses electrons? Recall that a neutral atom has no overall charge. This is because it contains equal numbers of positively charged protons and negatively charged electrons. When electrons are added to or removed from an atom, that atom becomes an ion. *An **ion** is an atom that is no longer neutral because it has gained or lost electrons.* An ion can be positively or negatively charged depending on whether it has lost or gained electrons.

Positive Ions

When a neutral atom loses one or more electrons, it has more protons than electrons. As a result, it has a positive charge. An atom with a positive charge is called a positive ion. A positive ion is represented by the element's symbol followed by a superscript plus sign ($^+$). For example, **Figure 16** shows how sodium (Na) becomes a positive sodium ion (Na^+).

Negative Ions

When a neutral atom gains one or more electrons, it has more electrons than protons. As a result, the atom has a negative charge. An atom with a negative charge is called a negative ion. A negative ion is represented by the element's symbol followed by a superscript negative sign ($^-$). **Figure 16** shows how fluorine (F) becomes a fluorine ion (F^-).

Figure 16 An ion is formed when a neutral atom gains or loses an electron.

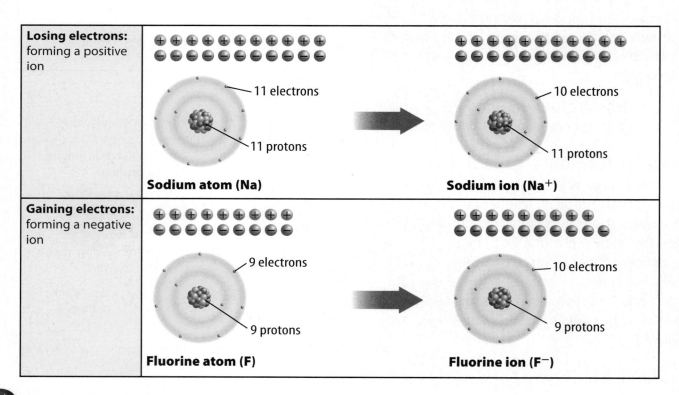

Losing electrons: forming a positive ion	11 electrons — 11 protons — **Sodium atom (Na)**	10 electrons — 11 protons — **Sodium ion (Na$^+$)**
Gaining electrons: forming a negative ion	9 electrons — 9 protons — **Fluorine atom (F)**	10 electrons — 9 protons — **Fluorine ion (F$^-$)**

Carbon **Nitrogen**

Different elements contain different numbers of protons.

Isotopes

Two isotopes of a given element contain different numbers of neutrons.

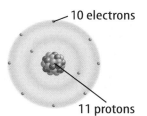

10 electrons

11 protons

Sodium ion (Na⁺)

When a neutral atom gains or loses an electron, it becomes an ion.

Use Vocabulary

1. The number of protons in an atom of an element is its

_____.

2. Nuclear decay occurs when an unstable atomic nucleus changes into another nucleus by emitting _____.

3. **Describe** how two isotopes of nitrogen differ from two nitrogen ions.

_____

Understand Key Concepts 🔑

4. An element's average atomic mass is calculated using the masses of its SC.8.P.8.7

  Ⓐ electrons. Ⓒ neutrons.

  Ⓑ isotopes. Ⓓ protons.

5. **Show** what happens to the electrons of a neutral calcium atom (Ca) when it is changed into a calcium ion (Ca^{2+}). SC.8.P.8.7

Interpret Graphics

6. **Contrast** Fill in this graphic organizer to contrast how different elements, isotopes, and ions are produced. LA.6.2.2.3

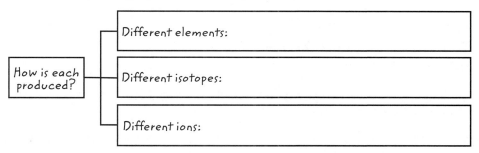

How is each produced?

Different elements:

Different isotopes:

Different ions:

Critical Thinking

7. **Consider** Find two neighboring elements on the periodic table whose positions would be reversed if they were arranged by atomic mass instead of atomic number.

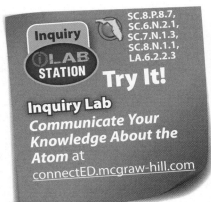

Inquiry SC.8.P.8.7, SC.6.N.2.1, SC.7.N.1.3, SC.8.N.1.1, LA.6.2.2.3

ⓘLAB STATION **Try It!**

Inquiry Lab

Communicate Your Knowledge About the Atom at connectED.mcgraw-hill.com

Math Skills MA.6.A.3.6

8. A sample of copper (Cu) contains 69.17% Cu-63. The remaining copper atoms are Cu-65. What is the average atomic mass of copper?

Chapter 8   Study Guide

 Think About It! An atom is the smallest unit of matter. An atom has a nucleus of protons and neutrons surrounded by an electron. The arrangement of the electrons determines an atom's chemical properties.

🔑 Key Concepts Summary

Vocabulary

LESSON 1 Discovering Parts of an Atom

- If you were to divide an element into smaller and smaller pieces, the smallest piece would be an **atom.**
- Atoms are so small that they can be seen only by powerful scanning microscopes.
- The first model of the atom was a solid sphere. Now, scientists know that an atom contains a dense positive nucleus surrounded by an **electron cloud.**

atom p. 309

electron p. 311

nucleus p. 314

proton p. 314

neutron p. 315

electron cloud p. 316

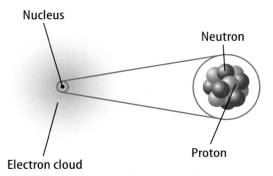

LESSON 2 Protons, Neutrons, and Electrons—How Atoms Differ

- **Nuclear decay** occurs when an unstable atomic nucleus changes into another more stable nucleus by emitting radiation.
- Different elements contain different numbers of protons. Two **isotopes** of the same element contain different numbers of neutrons. When a neutral atom gains or loses an electron, it becomes an **ion.**

atomic number p. 321

isotope p. 322

mass number p. 322

average atomic mass p. 323

radioactive p. 324

nuclear decay p. 325

ion p. 326

Nuclear Decay

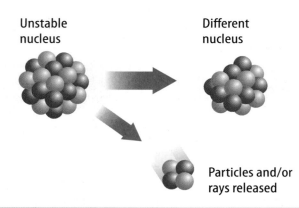

Active Reading

FOLDABLES® **Chapter Project**

Assemble your lesson Foldables as shown to make a Chapter Project. Use the project to review what you have learned in this chapter.

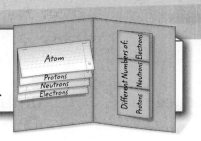

Use Vocabulary

1 A(n) _____ is a very small particle that is the basic unit of matter.

2 Electrons in an atom move throughout the

_____

surrounding the nucleus.

3 _____

is the weighted average mass of all of an element's isotopes.

4 All atoms of a given element have the same number of

_____.

5 When _____ occurs, one element is changed into another element.

6 Isotopes have the same _____ but different mass numbers.

Link Vocabulary and Key Concepts

Use vocabulary terms from the previous page to complete the concept map.

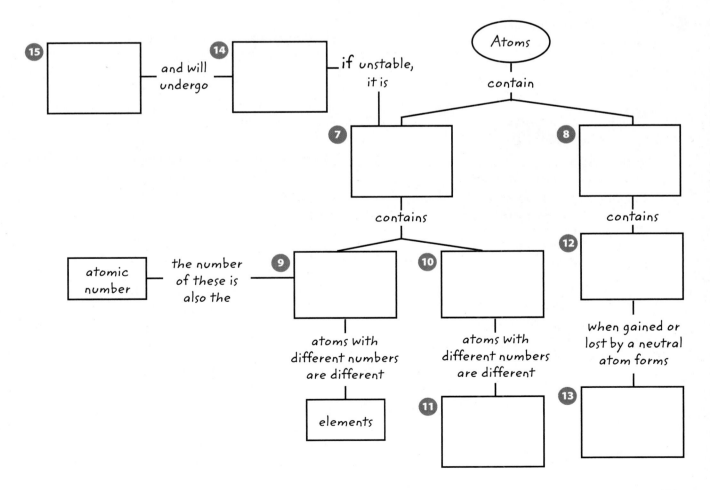

Fill in the correct answer choice.

🔑 Understand Key Concepts

1 Which part of an atom makes up most of its volume? SC.8.P.8.7

  Ⓐ its electron cloud
  Ⓑ its neutrons
  Ⓒ its nucleus
  Ⓓ its protons

2 What did Democritus believe an atom was? SC.8.P.8.7

  Ⓐ a solid, indivisible object
  Ⓑ a tiny particle with a nucleus
  Ⓒ a nucleus surrounded by an electron cloud
  Ⓓ a tiny nucleus with electrons surrounding it

3 If an ion contains 10 electrons, 12 protons, and 13 neutrons, what is the ion's charge? SC.8.P.8.7

  Ⓐ 2−
  Ⓑ 1−
  Ⓒ 2+
  Ⓓ 3+

4 J.J. Thomson's experimental setup is shown below. SC.8.P.8.7

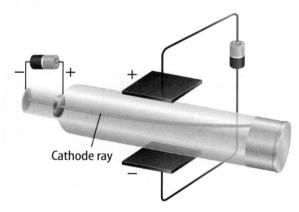

Cathode ray

What is happening to the cathode rays?
  Ⓐ They are attracted to the negative plate.
  Ⓑ They are attracted to the positive plate.
  Ⓒ They are stopped by the plates.
  Ⓓ They are unaffected by either plate.

Critical Thinking

5 **Consider** what would have happened in the gold foil experiment if Dalton's theory had been correct. SC.8.P.8.7

_____
_____
_____

6 **Contrast** How does Bohr's model of the atom differ from the present-day atomic model? SC.8.P.8.7

_____
_____
_____

7 **Describe** the electron cloud using your own analogy. SC.8.P.8.7

_____
_____
_____

8 **Summarize** how radioactive decay can produce new elements. SC.8.P.8.7

_____
_____
_____

9 **Hypothesize** What might happen if a negatively charged ion comes into contact with a positively charged ion? SC.8.P.8.7

_____
_____
_____

10 **Infer** Why isn't mass number listed with each element on the periodic table? SC.8.P.8.7

_____
_____
_____

11 **Explain** How is the average atomic mass calculated? SC.8.P.8.7

_____

_____

_____

12 **Infer** Oxygen has three stable isotopes.

Isotope	Average Atomic Mass
Oxygen-16	0.99757
Oxygen-17	0.00038
Oxygen-18	0.00205

What can you determine about the average atomic mass of oxygen without calculating it? SC.8.P.8.7

_____

Writing in Science

13 **Write** a newspaper article on a separate sheet of paper that describes how the changes in the atomic model provide an example of the scientific process in action. LA.6.2.2.3

Big Idea Review

14 **Describe** the current model of the atom. Explain the size of atoms. Also explain the charge, the location, and the size of protons, neutrons, and electrons. SC.8.P.8.7

_____

_____

_____

_____

_____

_____

15 **Summarize** The Large Hadron Collider, shown on page 304, is continuing the study of matter and energy. Use a set of four drawings to summarize how the model of the atom changed from Thomson, to Rutherford, to Bohr, to the modern model. SC.8.P.8.7

Math Skills   MA.6.A.3.6

Use Percentages

Use the information in the table to answer questions 21 and 22.

Magnesium (Mg) Isotope	Percent Found in Nature
Mg-24	78.9%
Mg-25	10.0%
Mg-26	

16 What is the percentage of Mg-26 found in nature?

17 What is the average atomic mass of magnesium?

Fill in the correct answer choice.

Multiple Choice

1 Which best describes an atom? SC.8.P.8.7

  (A) a particle with a single negative charge

  (B) a particle with a single positive charge

  (C) the smallest particle that still represents a compound

  (D) the smallest particle that still represents an element

Use the figure below to answer questions 2 and 3.

Structure X

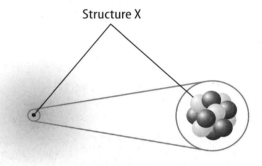

2 What is structure X? SC.8.P.8.7

  (F) an electron

  (G) a neutron

  (H) a nucleus

  (I) a proton

3 Which best describes structure X? SC.8.P.8.7

  (A) most of the atom's mass, neutral charge

  (B) most of the atom's mass, positive charge

  (C) very small part of the atom's mass, negative charge

  (D) very small part of the atom's mass, positive charge

4 Which is true about the size of an atom? SC.8.P.8.7

  (F) It can only be seen using a scanning tunneling microscope.

  (G) It is about the size of the period at the end of this sentence.

  (H) It is large enough to be seen using a magnifying lens.

  (I) It is too small to see with any type of microscope.

Use the figure below to answer question 5.

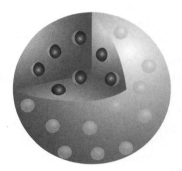

5 Whose model for the atom is shown? SC.8.P.8.7

  (A) Bohr's

  (B) Dalton's

  (C) Rutherford's

  (D) Thomson's

6 What structure did Rutherford discover? SC.8.P.8.7

  (F) the atom

  (G) the electron

  (H) the neutron

  (I) the nucleus

Use the table below to answer questions 7–9.

Particle	Number of Protons	Number of Neutrons	Number of Electrons
1	4	5	2
2	5	5	5
3	5	6	5
4	6	6	6

7 What is the atomic number of particle 3? SC.8.P.8.7

- (A) 3
- (B) 5
- (C) 6
- (D) 11

8 Which particles are isotopes of the same element? SC.8.P.8.7

- (F) 1 and 2
- (G) 2 and 3
- (H) 2 and 4
- (I) 3 and 4

9 Which particle is an ion? SC.8.P.8.7

- (A) 1
- (B) 2
- (C) 3
- (D) 4

10 Which reaction starts with a neutron and results in the formation of a proton and a high-energy electron? SC.8.P.8.7

- (F) alpha decay
- (G) beta decay
- (H) the formation of a positive ion
- (I) the formation of a negative ion

11 How many neutrons does iron-59 have? SC.8.P.8.7

- (A) 30
- (B) 33
- (C) 56
- (D) 59

12 Why were Rutherford's students surprised by the results of the gold foil experiment? SC.8.P.8.7

- (F) They didn't expect the alpha particles to bounce back from the foil.
- (G) They didn't expect the alpha particles to continue in a straight path.
- (H) They expected only a few alpha particles to bounce back from the foil.
- (I) They expected the alpha particles to be deflected by electrons.

13 Which determines the identity of an element? SC.8.P.8.7

- (A) its mass number
- (B) the charge of the atom
- (C) the number of its neutrons
- (D) the number of its protons

14 The figure below shows which of the following? SC.8.P.8.7

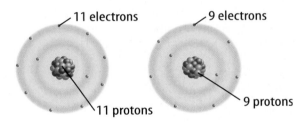

- (F) two different elements
- (G) two different ions
- (H) two different isotopes
- (I) two different protons

15 How is Bohr's atomic model different from Rutherford's model? SC.7.N.1.3

- (A) Bohr's model has a nucleus.
- (B) Bohr's model has electrons.
- (C) Electrons in Bohr's model are located farther from the nucleus.
- (D) Electrons in Bohr's model are located in circular energy levels.

NEED EXTRA HELP?

If You Missed Question...	1	2	3	4	5	6	7	8	9	10	11	12	13	14	15
Go to Lesson...	1	1	1	1	1	1	2	2	2	2	2	1	2	2	1

Benchmark Mini-Assessment   Chapter 8 • Lesson 1

Multiple Choice *Bubble the correct answer.*

Use the image below to answer questions 1 and 2.

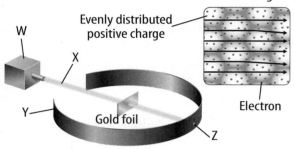

Cross section of gold foil

Evenly distributed positive charge

W

X

Y

Gold foil

Z

Electron

1. In the figure above, what does W represent? **SC.8.P.8.7**

- (A) alpha particle source
- (B) alpha particles path
- (C) detector screen
- (D) spot of light

2. Who performed the experiment shown in the figure above? **SC.8.P.8.7**

- (F) Bohr
- (G) Chadwick
- (H) Rutherford
- (I) Thomson

Use the image below to answer questions 3 and 4.

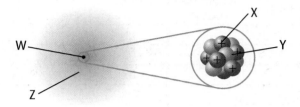

W

Z

X

Y

3. Which structure in the figure above has a negative charge? **SC.8.P.8.7**

- (A) W
- (B) X
- (C) Y
- (D) Z

4. Looking at the figure above, which statement is correct? **SC.8.P.8.7**

- (F) Structure W has a positive charge.
- (G) Structure X has a negative charge.
- (H) Structure Y has a negative charge.
- (I) Structure Z has a positive charge.

Copyright © Glencoe/McGraw-Hill, a division of The McGraw-Hill Companies, Inc.

Benchmark Mini-Assessment   **Chapter 8 • Lesson 2**

Multiple Choice *Bubble the correct answer.*

Properties of Protons, Neutrons, and Electrons			
	Electron	**Proton**	**Neutron**
	•		
Symbol	e−	p	n
Charge	1−	1+	0
Location	electron cloud around the nucleus	nucleus	nucleus
Relative mass	1/1,840	1	1

1. In the table above, which equation correctly describes the relative masses of electrons, protons, and neutrons? **SC.8.P.8.7**

(A) electron < proton = neutron

(B) electron = proton = neutron

(C) electron = proton < neutron

(D) electron < proton > neutron

2. What is the mass number of an atom? **SC.8.P.8.7**

(F) the number of protons minus the number of electrons

(G) the number of protons minus the number of neutrons

(H) the number of protons plus the number of electrons

(I) the number of protons plus the number of neutrons

Use the image below to answer questions 3 and 4.

Naturally Occurring Isotopes of Carbon			
Isotope	**A**	**B**	**C**
	⊙	⊙	⊙
Abundance	98.89%	<1.11%	<0.01%
Protons		6	6
Neutrons	6	7	
Mass Number	12		14

3. In the table above, which statement is true? **SC.8.P.8.7**

(A) Atom A has 6 protons.

(B) Atom A has 7 protons.

(C) Atom B has 7 protons.

(D) Atom C has 8 protons.

4. Which correctly identifies an isotope in the table above? **SC.8.P.8.7**

(F) A: carbon-6

(G) B: carbon-12

(H) B: carbon-13

(I) C: carbon-8

Copyright © Glencoe/McGraw-Hill, a division of The McGraw-Hill Companies, Inc.

Name _____ Date _____

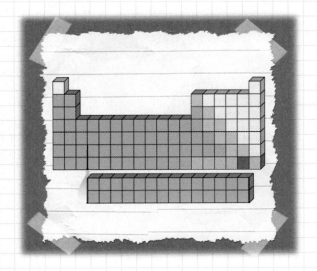

How is it arranged?

Five students were looking at a large poster of the periodic table. They noticed the elements were arranged in rows and columns. They also noticed the elements were arranged in order from 1–118. They had different ideas about the arrangement of the elements on the periodic table. This is what they said:

Damon: I think the elements are arranged by increasing mass.

Flo: I think the elements are arranged according to their properties.

Sienna: I think the elements are arranged by when they were discovered.

Kyle: I think the elements are arranged according to how common they are.

Glenda: I don't agree with any of you. I think there must be a different reason for the arrangement.

Which student do you agree with? Explain why you agree with that student.

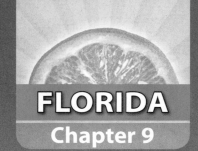

The Periodic TABLE

FLORIDA BIG IDEAS

1 The Practice of Science
2 The Characteristics of Scientific Knowledge
4 Science and Society
9 Properties of Matter

Think About It! **How is the periodic table used to classify and provide information about all known elements?**

Things are made out of specific materials for a reason. A weather balloon can rise high in the atmosphere and gather weather information. The plastic that forms this weather balloon and the helium gas that fills it were chosen after scientists researched and studied the properties of these materials.

1 What property of helium do you think makes the balloon rise through the air?

_____

_____

2 Why do you think the periodic table is a useful tool when determining properties of different materials?

_____

_____

_____

Get Ready to Read **What do you think about the periodic table?**

Before you read, decide if you agree or disagree with each of these statements. As you read this chapter, see if you change your mind about any of the statements.

	AGREE	DISAGREE
1 The elements on the periodic table are arranged in rows in the order they were discovered.	☐	☐
2 The properties of an element are related to the element's location on the periodic table.	☐	☐
3 Fewer than half of the elements are metals.	☐	☐
4 Metals are usually good conductors of electricity.	☐	☐
5 Most of the elements in living things are nonmetals.	☐	☐
6 Even though they look very different, oxygen and sulfur share some similar properties.	☐	☐

There's More Online!
Video • Audio • Review • ⓘLab Station • WebQuest • Assessment • Concepts in Motion • Multilingual eGlossary

Using the PERIODIC TABLE

How are elements arranged on the periodic table?

What can you learn about elements from the periodic table?

Vocabulary

periodic table p. 341

group p. 346

period p. 346

Florida NGSSS

LA.6.2.2.3 The student will organize information to show understanding (e.g., representing main ideas within text through charting, mapping, paraphrasing, summarizing, or comparing/contrasting);

MA.6.A.3.6 Construct and analyze tables, graphs, and equations to describe linear functions and other simple relations using both common language and algebraic notation.

SC.6.N.2.3 Recognize that scientists who make contributions to scientific knowledge come from all kinds of backgrounds and possess varied talents, interests, and goals.

SC.7.N.1.7 Explain that scientific knowledge is the result of a great deal of debate and confirmation within the science community.

SC.7.N.2.1 Identify an instance from the history of science in which scientific knowledge has changed when new evidence or new interpretations are encountered.

SC.8.N.1.1 Define a problem from the eighth grade curriculum using appropriate reference materials to support scientific understanding, plan and carry out scientific investigations of various types, such as systematic observations or experiments, identify variables, collect and organize data, interpret data in charts, tables, and graphics, analyze information, make predictions, and defend conclusions.

SC.8.P.8.6 Recognize that elements are grouped in the periodic table according to similarities of their properties.

SC.6.N.2.2 Explain that scientific knowledge is durable because it is open to change as new evidence or interpretations are encountered.

 Launch Lab

SC.8.N.1.6

15 minutes

How can objects be organized?

What would it be like to shop at a grocery store where all the products are mixed up on the shelves? Cereal might be next to the dish soap and bread might be next to the canned tomatoes. It would take a long time to find the groceries that you need. How does organizing objects help you find and use what you need?

Procedure

1. Read and complete a lab safety form.

2. Empty the **interlocking plastic bricks** from the **plastic bag** onto your desk and observe their properties. Think about ways you might group and sequence the bricks so that they are organized.

3. Organize the bricks according to your plan.

4. Compare your pattern of organization with those used by several other students.

Think About This

1. Describe the way you grouped your bricks. Why did you choose that way of grouping?

2. Describe how you sequenced your bricks.

3. **Key Concept** How does organizing things help you to use them more easily?

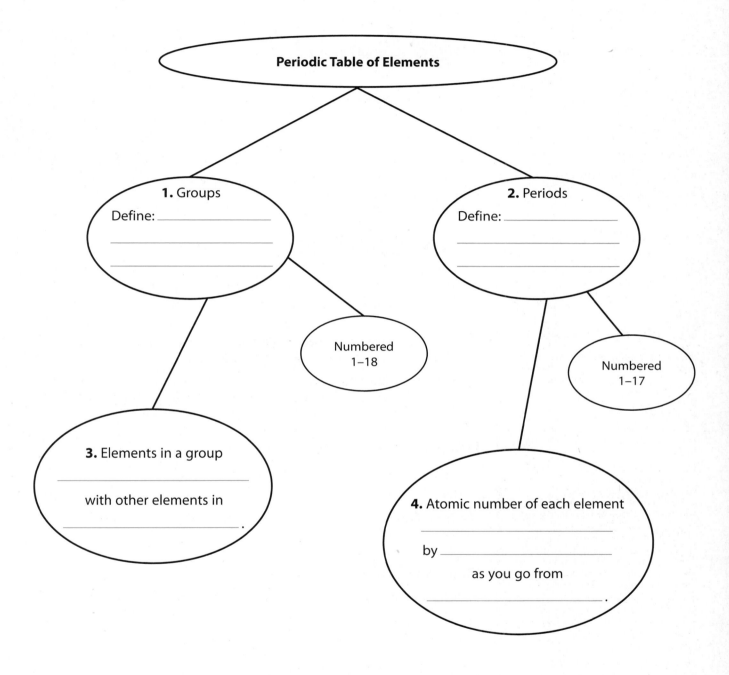

Organize Show how the periodic table is arranged by completing the concept map.

Periodic Table of Elements

1. Groups

Define: _____
_____
_____

2. Periods

Define: _____
_____
_____

Numbered
1–18

Numbered
1–17

3. Elements in a group
_____
with other elements in
_____.

4. Atomic number of each element
_____
by _____
as you go from
_____.

METALS

 What elements are metals?

 What are the properties of metals?

Vocabulary

metal p. 351

luster p. 351

ductility p. 352

malleability p. 352

alkali metal p. 353

alkaline earth metal p. 353

transition element p. 354

Florida NGSSS

LA.6.2.2.3 The student will organize information to show understanding (e.g., representing main ideas within text through charting, mapping, paraphrasing, summarizing, or comparing/contrasting);

SC.7.P.11.2 Investigate and describe the transformation of energy from one form to another.

SC.8.N.1.1 Define a problem from the eighth grade curriculum using appropriate reference materials to support scientific understanding, plan and carry out scientific investigations of various types, such as systematic observations or experiments, identify variables, collect and organize data, interpret data in charts, tables, and graphics, analyze information, make predictions, and defend conclusions.

SC.8.N.4.1 Explain that science is one of the processes that can be used to inform decision making at the community, state, national, and international levels.

SC.8.P.8.4 Classify and compare substances on the basis of characteristic physical properties that can be demonstrated or measured; for example, density, thermal or electrical conductivity, solubility, magnetic properties, melting and boiling points, and know that these properties are independent of the amount of the sample.

SC.8.P.8.6 Recognize that elements are grouped in the periodic table according to similarities of their properties.

 Inquiry Launch Lab

SC.8.N.1.1, SC.8.P.8.4

20 minutes

What properties make metals useful?

The properties of metals determine their uses. Copper conducts thermal energy, which makes it useful for cookware. Aluminum has low density, so it is used in aircraft bodies. What other properties make metals useful?

Procedure

1 Read and complete a lab safety form.

2 With your group, observe the **metal objects** in your **container.** For each object, discuss what properties allow the metal to be used in that way.

3 Observe the **photographs of gold and silver jewelry.** What properties make these two metals useful in jewelry?

4 Examine **other objects around the room** that you think are made of metal. Do they share the same properties as the objects in your container? Do they have other properties that make them useful?

Data and Observations

Think About This

1. What properties do all the metals share? What properties are different?

2. **Key Concept** List at least four properties of metals that determine their uses.

1. Lightning strikes the top of the Empire State Building approximately 100 times per year. Why do you think lightning hits the top of this building instead of the city streets or buildings below?

_____

_____

_____

_____

_____

_____

What is a metal?

What do stainless steel knives and forks, copper wire, aluminum foil, and gold jewelry have in common? They are all made from metals.

As you read in Lesson 1, most of the elements on the periodic table are metals. In fact, of all the known elements, more than three-quarters are metals. With the exception of hydrogen, all of the elements in groups 1–12 on the periodic table are metals. In addition, some of the elements in groups 13–15 are metals. To be a metal, an element must have certain properties.

 2. NGSSS Check Identify Highlight where metals are found on the periodic table. SC.8.P.8.6

Physical Properties of Metals

Recall that physical properties are characteristics used to describe or identify something without changing its makeup. All metals share certain physical properties.

A **metal** _is an element that is generally shiny. It is easily pulled into wires or hammered into thin sheets. A metal is a good conductor of electricity and thermal energy._ Gold exhibits the common properties of metals.

Luster and Conductivity People use gold for jewelry because of its beautiful color and metallic luster. **Luster** _describes the ability of a metal to reflect light._ Gold is also a good conductor of thermal energy and electricity. However, gold is too expensive to use in normal electrical wires or metal cookware. Copper is often used instead.

Figure 8 Gold has many uses based on its properties.

Gold

Active Reading **3. Connect** Match the properties of gold with the corresponding images below.

Ductility	Malleability	Luster	Conductivity	Unreactive

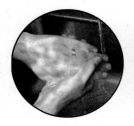    

WORD ORIGIN

ductility
from Latin *ductilis*, means "may be led or drawn"

REVIEW VOCABULARY

density
the mass per unit volume of a substance

Active Reading

FOLDABLES LA.6.2.2.3

Make a two-tab book. Label it as shown. Use it to record information about the properties of metals.

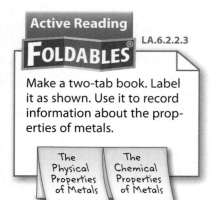

Ductility and Malleability Gold is the most ductile metal. **Ductility** (duk TIH luh tee) *is the ability to be pulled into thin wires.* A piece of gold with the mass of a paper clip can be pulled into a wire that is more than 3 km long.

Malleability (ma lee uh BIH luh tee) *is the ability of a substance to be hammered or rolled into sheets.* Gold is so malleable that it can be hammered into thin sheets. A pile of a million thin sheets would be only as high as a coffee mug.

Other Physical Properties of Metals In general the **density**, strength, boiling point, and melting point of a metal are greater than those of other elements. Except for mercury, all metals are solid at room temperature. Many uses of a metal are determined by the metal's physical properties, as shown in **Figure 8**.

4. NGSSS Check Find (Circle) some physical properties of metals. SC.8.P.8.4

Chemical Properties of Metals

Recall that a chemical property is the ability or inability of a substance to change into one or more new substances. The chemical properties of metals can differ greatly. However, metals in the same group usually have similar chemical properties. For example, gold and other elements in group 11 do not easily react with other substances.

Group 1: Alkali Metals

The elements in group 1 are called **alkali** *(AL kuh li)* **metals**. The alkali metals include lithium, sodium, potassium, rubidium, cesium, and francium.

Because they are in the same group, alkali metals have similar chemical properties. Alkali metals react quickly with other elements, such as oxygen. Therefore, in nature, they occur only in compounds. Pure alkali metals must be stored so that they do not come in contact with oxygen and water vapor in the air. **Figure 9** shows potassium and sodium reacting with water.

Alkali metals also have similar physical properties. Pure alkali metals have a silvery appearance. As shown in **Figure 9,** they are soft enough to cut with a knife. The alkali metals also have the lowest densities of all metals. A block of pure sodium metal could float on water because of its very low density.

Active Reading **5. Characterize** Assess information about alkali metals. (Circle) the correct choice in each set of parentheses.

Characteristics of Alkali Metals
- React (quickly, slowly) with other elements
- Found in nature (as elements, in compounds)
- Have a (dull, shiny) appearance
- (Soft, hard)
- Have the (highest, lowest) densities of all metals

Figure 9 Alkali metals react violently with water. They are also soft enough to be cut with a knife.

Potassium

Sodium

Lithium

Group 2: Alkaline Earth Metals

The elements in group 2 on the periodic table are called **alkaline** *(AL kuh lun)* **earth metals**. These metals are beryllium, magnesium, calcium, strontium, barium, and radium.

Alkaline earth metals also react quickly with other elements. However, they do not react as quickly as the alkali metals do. Like the alkali metals, pure alkaline earth metals do not occur naturally. Instead, they combine with other elements and form compounds. The physical properties of the alkaline earth metals are also similar to those of the alkali metals. Alkaline earth metals are soft and silvery. They also have low density, but they have greater density than alkali metals.

Active Reading **6. Predict** Which element reacts faster with oxygen—barium or potassium?

Figure 10 Transition elements are in blocks at the center of the periodic table. Many colorful materials contain small amounts of transition elements.

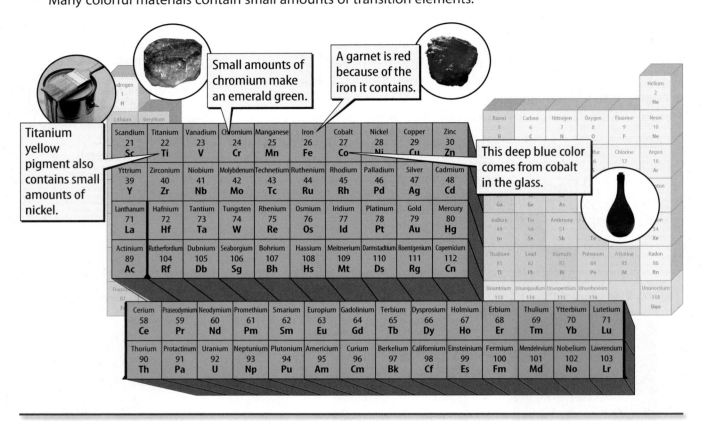

Groups 3–12: Transition Elements

The elements in groups 3–12 are called **transition elements**. The transition elements are in two blocks on the periodic table. The main block is in the center of the periodic table. The other block includes the two rows at the bottom of the periodic table, as shown in **Figure 10.**

Properties of Transition Elements

All transition elements are metals. They have higher melting points, greater strength, and higher densities than the alkali metals and the alkaline earth metals. Transition elements also react less quickly with oxygen. Some transition elements can exist in nature as free elements. An element is a free element when it occurs in pure form, not in a compound.

Uses of Transition Elements

Transition elements in the main block of the periodic table have many important uses. Because of their high densities, strength, and resistance to corrosion, transition elements such as iron make good building materials. Copper, silver, nickel, and gold are used to make coins. These metals are also used for jewelry, electrical wires, and many industrial applications.

Main-block transition elements can react with other elements and form many compounds. Many of these compounds are colorful. Artists use transition-element compounds in paints and pigments. The colors of many gems, such as garnets and emeralds, come from the presence of small amounts of transition elements, as illustrated in **Figure 10.**

Lanthanide and Actinide Series

Two rows of transition elements are at the bottom of the periodic table, as shown in **Figure 10.** These elements were removed from the main part of the table so that periods 6 and 7 were not longer than the other periods. If these elements were included in the main part of the table, the first row, called the lanthanide series, would stretch between lanthanum and halfnium. The second row, called the actinide series, would stretch between actinium and rutherfordium.

Some lanthanide and actinide series elements have valuable properties. For example, lanthanide series elements are used to make strong magnets. Plutonium, one of the actinide series elements, is used as a fuel in some nuclear reactors.

Patterns in Properties of Metals

Recall that the properties of elements follow repeating patterns across the periods of the periodic table. In general, elements increase in metallic properties such as luster, malleability, and electrical conductivity from right to left across a period, as shown in **Figure 11.** The elements on the far right of a period have no metallic properties at all. Potassium (K), the element on the far left in period 4, has the highest luster, is the most malleable, and conducts electricity better than all the elements in this period.

There are also patterns within groups. Metallic properties tend to increase as you move down a group, also shown in **Figure 11.** You could predict that the malleability of gold is greater than the malleability of either silver or copper because it is below these two elements in group 11.

Active Reading

7. Indicate Draw a line around the area on the periodic table in **Figure 11** where you would expect to find elements with few or no metallic properties.

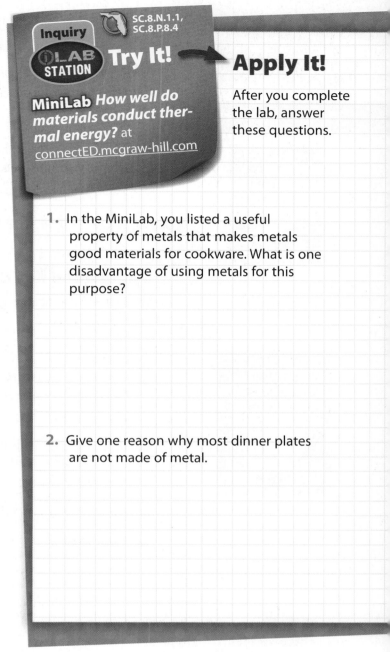

SC.8.N.1.1, SC.8.P.8.4

Inquiry
LAB STATION **Try It!**

MiniLab *How well do materials conduct thermal energy?* at connectED.mcgraw-hill.com

Apply It!

After you complete the lab, answer these questions.

1. In the MiniLab, you listed a useful property of metals that makes metals good materials for cookware. What is one disadvantage of using metals for this purpose?

2. Give one reason why most dinner plates are not made of metal.

Figure 11 Metallic properties of elements increase as you move to the left and down on the periodic table.

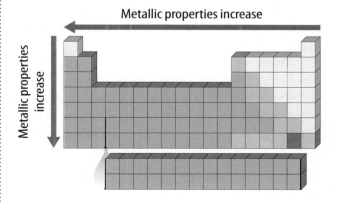

Metallic properties increase

Metallic properties increase

Properties of metals include conductivity, luster, malleability, and ductility.

Alkali metals and alkaline earth metals react easily with other elements. These metals make up groups 1 and 2 on the periodic table.

Transition elements make up groups 3–12 and the lanthanide and actinide series on the periodic table.

Use Vocabulary

1 **Use the term** *luster* in a sentence.

_____

_____

2 **Identify** the property that makes copper metal ideal for wiring.

_____

3 Elements that have the lowest densities of all the metals are called

_____.

Understand Key Concepts 🔑

4 **List** the physical properties that most metals have in common. SC.8.P.8.6

_____

_____

5 Which is a chemical property of transition elements?
- (A) brightly colored
- (C) denser than alkali metals
- (B) great ductility
- (D) reacts little with oxygen

6 **Organize** the following metals from least metallic to most metallic: barium, zinc, iron, and strontium. SC.8.P.8.6

_____

Interpret Graphics

7 **Examine** this section of the periodic table. What metal will have properties most similar to those of chromium (Cr)? Why? SC.8.P.8.6

_____

_____

_____

_____

_____

Vanadium 23 **V**	Chromium 24 **Cr**	Manganese 25 **Mn**
Niobium 41 **Nb**	Molybdenum 42 **Mo**	Technetium 43 **Tc**

Critical Thinking

8 **Investigate** your classroom and locate five examples of materials made from metal.

_____

_____

9 **Evaluate** the physical properties of potassium, magnesium, and copper. Select the best choice to use for a building project. Explain why this metal is the best building material to use. SC.8.P.8.4

_____

_____

_____

Fireworks

Metals add variety to color.

About 1,000 years ago, the Chinese discovered the chemical formula for gunpowder. Using this formula, they invented the first fireworks. One of the primary ingredients in gunpowder is saltpeter, or potassium nitrate. Find potassium on the periodic table. Notice that potassium is a metal. How does the chemical behavior of a metal contribute to a colorful fireworks show?

Purple: mix of strontium and copper compounds

Blue: copper compounds

Yellow: sodium compounds

Gold: iron burned with carbon

White-hot: barium-oxygen compounds or aluminum or magnesium burn

Metal compounds contribute to the variety of colors you see at a fireworks show. Recall that metals have special chemical and physical properties. Compounds that contain metals also have special properties. For example, each metal turns a characteristic color when burned. Lithium, an alkali metal, forms compounds that burn red. Copper compounds burn blue. Aluminum and magnesium burn white.

Orange: calcium compounds

Green: barium compounds

Red: strontium and lithium compounds

It's Your Turn

FORM AN OPINION Fireworks contain metal compounds. Are they bad for the environment or for your health? Research the effects of metals on human health and on the environment. Decide if fireworks are safe to use at holiday celebrations.

Nonmetals and METALLOIDS

ESSENTIAL QUESTIONS

 Where are nonmetals and metalloids on the periodic table?

 What are the properties of nonmetals and metalloids?

Vocabulary

nonmetal p. 359

halogen p. 361

noble gas p. 362

metalloid p. 363

semiconductor p. 363

 Florida NGSSS

LA.6.2.2.3 The student will organize information to show understanding (e.g., representing main ideas within text through charting, mapping, paraphrasing, summarizing, or comparing/contrasting);

SC.8.N.1.1 Define a problem from the eighth grade curriculum using appropriate reference materials to support scientific understanding, plan and carry out scientific investigations of various types, such as systematic observations or experiments, identify variables, collect and organize data, interpret data in charts, tables, and graphics, analyze information, make predictions, and defend conclusions.

SC.8.N.1.6 Understand that scientific investigations involve the collection of relevant empirical evidence, the use of logical reasoning, and the application of imagination in devising hypotheses, predictions, explanations and models to make sense of the collected evidence.

SC.8.P.8.4 Classify and compare substances on the basis of characteristic physical properties that can be demonstrated or measured; for example, density, thermal or electrical conductivity, solubility, magnetic properties, melting and boiling points, and know that these properties are independent of the amount of the sample.

SC.8.P.8.6 Recognize that elements are grouped in the periodic table according to similarities of their properties.

SC.8.P.8.8 Identify basic examples of and compare and classify the properties of compounds, including acids, bases, and salts.

Inquiry Launch Lab SC.8.P.8.4, SC.8.N.1.1

20 minutes

What are some properties of nonmetals?

You now know what the properties of metals are. What are the properties of nonmetals?

Procedure

1. Read and complete a lab safety form.

2. Examine pieces of **copper, carbon, aluminum,** and **sulfur**. Describe the appearance of these elements.

3. Use a **conductivity tester** to check how well these elements conduct electricity. Record your observations.

4. Wrap each element sample in a **paper towel**. Carefully hit the sample with a **hammer**. Unwrap the towel and observe the sample. Record your observations.

Data and Observations

Think About This

1. Locate these elements on the periodic table. From their locations, which elements are metals? Which elements are nonmetals?

2. **Key Concept** Using your results, compare the properties of metals and nonmetals.

3. **Key Concept** What property of a nonmetal makes it useful to insulate electrical wires?

Inquiry Why don't they melt?

1. What do you expect to happen to something when a flame is placed against it? As you can see, the material this flower sits on protects the flower from the flame. Do you think the material is made of metal? Why or why not?

_____

_____

_____

_____

_____

_____

The Elements of Life

Would it surprise you to learn that more than 96 percent of the mass of your body comes from just four elements? As shown in **Figure 12,** all four of these elements—oxygen, carbon, hydrogen, and nitrogen—are nonmetals. **Nonmetals** *are elements that have no metallic properties.*

Of the remaining elements in your body, the two most common elements also are nonmetals—phosphorus and sulfur. These six elements form the compounds in proteins, fats, nucleic acids, and other large molecules in your body and in all other living things.

 2. Recall List the six most common elements in the human body.

_____

_____

_____

_____

Figure 12 Like other living things, this woman's mass comes mostly from nonmetals.

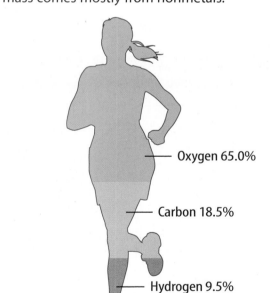

Oxygen 65.0%

Carbon 18.5%

Hydrogen 9.5%

Nitrogen 3.3%

Other elements 3.7%

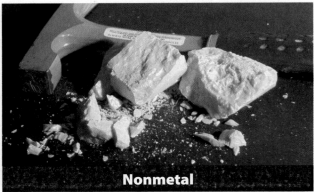

Metal

Nonmetal

Figure 13 Solid metals, such as copper, are malleable. Solid nonmetals, such as sulfur, are brittle.

Figure 14 Nonmetals have properties that are different from those of metals. Phosphorus and carbon are dull, brittle solids that do not conduct thermal energy or electricity.

How are nonmetals different from metals?

Recall that metals have luster. They are ductile, malleable, and good conductors of electricity and thermal energy. All metals except mercury are solids at room temperature.

The properties of nonmetals are different from those of metals. Many nonmetals are gases at room temperature. Those that are solid at room temperature have a dull surface, which means they have no luster. Because nonmetals are poor conductors of electricity and thermal energy, they are good insulators. For example, nose cones on space shuttles are insulated from the intense thermal energy of reentry by a material made from carbon, a nonmetal. **Figure 13** and **Figure 14** show several properties of nonmetals.

 3. NGSSS Check List Highlight the properties of nonmetals. SC.8.P.8.4

Properties of Nonmetals 🔑

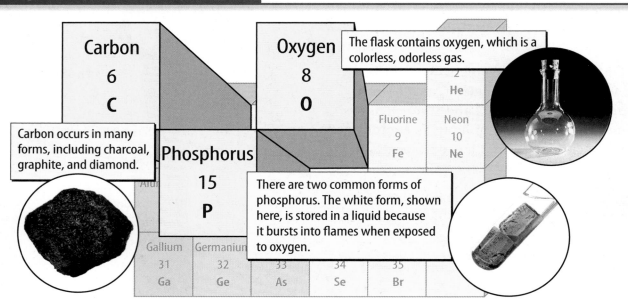

Carbon
6
C

Carbon occurs in many forms, including charcoal, graphite, and diamond.

Phosphorus
15
P

Oxygen
8
O

The flask contains oxygen, which is a colorless, odorless gas.

He

Fluorine
9
Fe

Neon
10
Ne

There are two common forms of phosphorus. The white form, shown here, is stored in a liquid because it bursts into flames when exposed to oxygen.

Gallium
31
Ga

Germanium
32
Ge

33
As

34
Se

35
Br

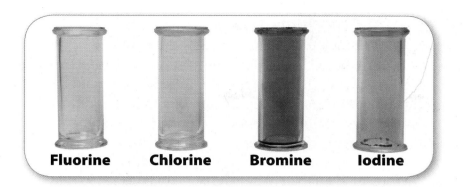

Fluorine   Chlorine   Bromine   Iodine

Nonmetals in Groups 14–16

Look back at the periodic table in **Figure 4**. Notice that groups 14–16 contain metals, nonmetals, and metalloids. The chemical properties of the elements in each group are similar. However, the physical properties of the elements can be quite different.

Carbon is the only nonmetal in group 14. It is a solid that has different forms. Carbon is in most of the compounds that make up living things. Nitrogen, a gas, and phosphorus, a solid, are the only nonmetals in group 15. These two elements form many different compounds with other elements, such as oxygen. Group 16 contains three nonmetals. Oxygen is a gas that is essential for many organisms. Sulfur and selenium are solids that have the physical properties of other solid nonmetals.

Group 17: The Halogens

An element in group 17 of the periodic table is called a **halogen** (HA luh jun). **Figure 15** shows the halogens fluorine, chlorine, bromine, and iodine. The term *halogen* refers to an element that can react with a metal and form a salt. For example, chlorine gas reacts with solid sodium and forms sodium chloride, or table salt. Calcium chloride is another salt often used on icy roads.

Halogens react readily with other elements and form compounds. They react so readily that halogens only can occur naturally in compounds. They do not exist as free elements. They even form compounds with other nonmetals, such as carbon. In general, the halogens are less reactive as you move down the group.

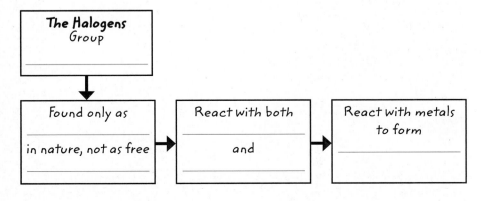

Figure 15 These glass containers each hold a halogen gas. Although they are different colors in their gaseous state, they react similarly with other elements.

Active Reading

FOLDABLES LA.6.2.2.3

Fold a sheet of paper to make a table with three columns and three rows. Label it as shown. Use it to organize information about nonmetals and metalloids.

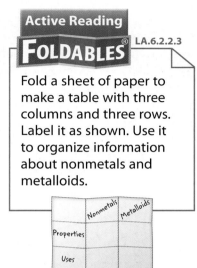

WORD ORIGIN

halogen
from Greek **hals**, means "salt"; and –**gen**, means "to produce"

Active Reading 4. **Explain** Will bromine react with sodium? Why or why not?

_____

_____

_____

Active Reading 5. **Describe** Using the organizer on the left, show the properties of halogens.

The Halogens
Group
_____

Found only as
_____
in nature, not as free
_____ → React with both
_____
and
_____ → React with metals
to form
_____

6. Organize
What properties do the noble gases have?

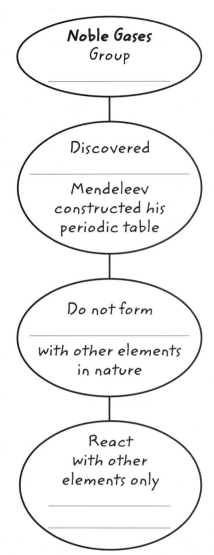

- **Noble Gases Group**
- **Discovered** — Mendeleev constructed his periodic table
- **Do not form** — with other elements in nature
- **React with other elements only**

Figure 16 More than 90 percent of all the atoms in the universe are hydrogen atoms. Hydrogen is the main fuel for the nuclear reactions that occur in stars.

Group 18: The Noble Gases

The elements in group 18 are known as the **noble gases**. The elements helium, neon, argon, krypton, xenon, and radon are the noble gases. Unlike the halogens, the only way elements in this group react with other elements is under special conditions in a laboratory. These elements were not yet discovered when Mendeleev constructed his periodic table because they do not form compounds naturally. Once they were discovered, they fit into a group at the far-right side of the table.

Hydrogen

Figure 16 shows the element key for hydrogen. Of all the elements, hydrogen has the smallest atomic mass. It is also the most common element in the universe.

Is hydrogen a metal or a nonmetal? Hydrogen is most often classified as a nonmetal because it has many properties like those of nonmetals. For example, like some nonmetals, hydrogen is a gas at room temperature. However, hydrogen also has some properties similar to those of the group 1 alkali metals. In its liquid form, hydrogen conducts electricity just like a metal does. In some chemical reactions, hydrogen reacts as if it were an alkali metal. However, under conditions on Earth, hydrogen usually behaves like a nonmetal.

7. Explain Why is hydrogen usually classified as a nonmetal?

_____

_____

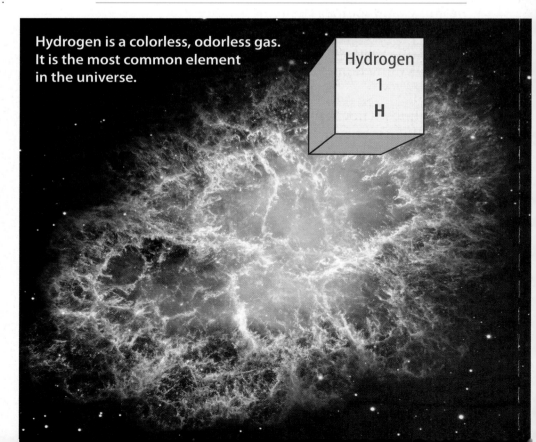

Hydrogen is a colorless, odorless gas. It is the most common element in the universe.

Hydrogen
1
H

Metalloids

Between the metals and the nonmetals on the periodic table are elements known as metalloids. *A* **metalloid** (MEH tul oyd) *is an element that has physical and chemical properties of both metals and nonmetals.* The elements boron, silicon, germanium, arsenic, antimony, tellurium, polonium, and astatine are metalloids. Silicon is the most abundant metalloid in the universe. Most sand is made of a compound containing silicon. Silicon is also used in many different products, some of which are shown in **Figure 17.**

 8. NGSSS Check **Identify** Where are metalloids found on the periodic table? SC.8.P.8.6

Semiconductors

Recall that metals are good conductors of thermal energy and electricity. Nonmetals are poor conductors of thermal energy and electricity but are good insulators. A property of metalloids is the ability to act as a semiconductor. *A* **semiconductor** *conducts electricity at high temperatures, but not at low temperatures.* At high temperatures, metalloids act like metals and conduct electricity. But at lower temperatures, metalloids act like nonmetals and stop electricity from flowing. This property is useful in electronic devices such as computers, televisions, and solar cells.

Active Reading

9. Classify Show how the characteristics of metalloids are like metals and like nonmetals.

Metalloids
Like Metals
Conduct electricity at
_____ temperatures
Like Nonmetals
Stop electricity from flowing at
_____ temperatures

WORD ORIGIN

semiconductor
from Latin *semi–*, means "half"; and *conducere,* means "to bring together"

Uses of Silicon **Figure 17** The properties of silicon make it useful for many different products.

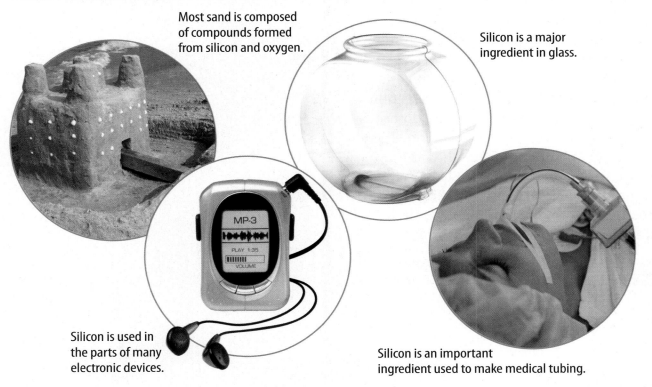

Most sand is composed of compounds formed from silicon and oxygen.

Silicon is a major ingredient in glass.

Silicon is used in the parts of many electronic devices.

Silicon is an important ingredient used to make medical tubing.

Properties and Uses of Metalloids

Pure silicon is used in making semiconductor devices for computers and other electronic products. Germanium is also used as a semiconductor. However, metalloids have other uses, as shown in **Figure 18.** Pure silicon and Germanium are used in semiconductors. Boron is used in water softeners and laundry products. Boron also glows bright green in fireworks. Silicon is one of the most abundant elements on Earth. Sand, clay, and many rocks and minerals are made of silicon compounds.

Figure 18 This microchip conducts electricity at high temperatures using a semiconductor.

Metals, Nonmetals, and Metalloids

You have read that all metallic elements have common characteristics, such as malleability, conductivity, and ductility. However, each metal has unique properties that make it different from other metals. The same is true for nonmetals and metalloids. How can knowing the properties of an element help you evaluate its uses?

Look again at the periodic table. An element's position on the periodic table tells you a lot about the element. By knowing that sulfur is a nonmetal, for example, you know that it breaks easily and does not conduct electricity. You would not choose sulfur to make a wire. You would not try to use oxygen as a semiconductor or sodium as a building material. You know that transition elements are strong, malleable, and do not react easily with oxygen or water. These metals make good building materials because they are strong, malleable, and less reactive than other elements. Understanding the properties of elements can help you decide which element to use in a given situation.

Active Reading

10. Explain Why would you not use an element on the right side of the periodic table as a building material?

_____

_____

_____

Inquiry

SC.8.N.1.1, SC.8.P.8.4

LAB STATION Try It!

MiniLab *Which insulates better?* at connectED.mcgraw-hill.com

Apply It!

After you complete the lab, answer the question below.

1. Insulation is placed inside the outside walls of most homes to hold a layer of air in the walls. Based on what you learned in the MiniLab, why is this beneficial?

Visual Summary

A nonmetal is an element that has no metallic properties. Solid nonmetals are dull, brittle, and do not conduct thermal energy or electricity.

Halogens and noble gases are nonmetals. These elements are found in group 17 and group 18 of the periodic table.

Metalloids have some metallic properties and some nonmetallic properties. The most important use of metalloids is as semiconductors.

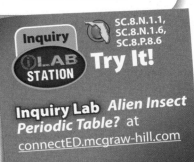

Inquiry SC.8.N.1.1, SC.8.N.1.6, SC.8.P.8.6

LAB STATION Try It!

Inquiry Lab *Alien Insect Periodic Table?* at connectED.mcgraw-hill.com

Use Vocabulary

1. **Distinguish** between a nonmetal and a metalloid.

_____

_____

2. An element in group 17 of the periodic table is called a(n) _____.

_____

3. An element in group 18 of the periodic table is called a(n) _____

Understand Key Concepts

4. The ability of a halogen to react with a metal to form a salt is an example of a _____ property.
   - (A) chemical
   - (B) noble gas
   - (C) periodic
   - (D) physical

5. **Classify** each of the following elements as a metal, a nonmetal, or a metalloid: boron, carbon, aluminum, and silicon. SC.8.P.8.6

_____

_____

6. **Infer** which group you would expect to contain element 117. Use the periodic table to help you answer this question. SC.8.P.8.6

_____

Interpret Graphics

7. **Sequence** nonmetals, metals, and metalloids in order from left to right across the periodic table by completing the graphic organizer below. LA.6.2.2.3

Critical Thinking

8. **Hypothesize** how your classroom would be different if there were no metalloids.

_____

_____

9. **Analyze** why hydrogen is sometimes classified as a metal. SC.8.P.8.6

_____

Chapter 9 Study Guide

Think About It! Elements are grouped and organized on the periodic table according to increasing atomic number and similarities of their properties.

Key Concepts Summary

Vocabulary

LESSON 1 Using the Periodic Table

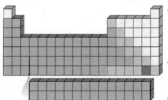

- Elements are organized on the **periodic table** by increasing atomic number and similar properties.
- Elements in the same **group,** or column, of the periodic table have similar properties.
- Elements' properties change across a **period,** which is a row of the periodic table.
- Each element key on the periodic table provides the name, symbol, atomic number, and atomic mass for an element.

periodic table p. 341

group p. 346

period p. 346

LESSON 2 Metals

- **Metals** are located on the left and middle side of the periodic table.
- Metals are elements that have **ductility, malleability, luster,** and conductivity.
- The **alkali metals** are in group 1 of the periodic table, and the **alkaline earth metals** are in group 2.
- **Transition elements** are metals in groups 3–12 of the periodic table, as well as the lanthanide and actinide series.

metal p. 351

luster p. 351

ductility p. 352

malleability p. 352

alkali metal p. 353

alkaline earth metal p. 353

transition element p. 354

LESSON 3 Nonmetals and Metalloids

- **Nonmetals** are on the right side of the periodic table, and **metalloids** are located between metals and nonmetals.
- Nonmetals are elements that have no metallic properties. Solid nonmetals are dull in appearance, brittle, and do not conduct electricity. Metalloids are elements that have properties of both metals and nonmetals.
- Some metalloids are **semiconductors.**
- Elements in group 17 are called **halogens,** and elements in group 18 are **noble gases.**

nonmetal p. 359

halogen p. 361

noble gas p. 362

metalloid p. 363

semiconductor p. 363

Active Reading

FOLDABLES **Chapter Project**

Assemble your lesson Foldables as shown to make a Chapter Project. Use the project to review what you have learned in this chapter.

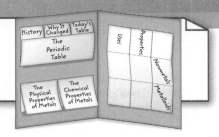

Use Vocabulary

1 The element magnesium (Mg) is in _____ 3 of the periodic table.

2 An element that is shiny, is easily pulled into wires or hammered into thin sheets, and is a good conductor of electricity and heat is a(n)

_____.

3 Copper is used to make wire because it has the property of _____.

4 An element that is sometimes a good conductor of electricity and sometimes a good insulator is a(n) _____.

5 An element that is a poor conductor of heat and electricity but is a good insulator is a(n) _____.

Link Vocabulary and Key Concepts

Use vocabulary terms from the previous page to complete the concept map.

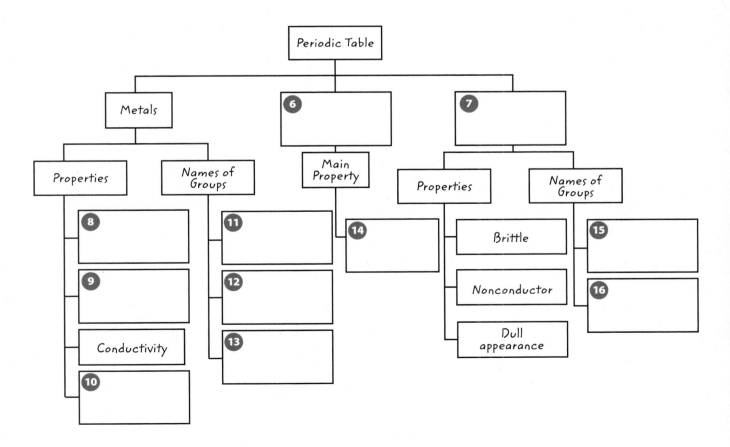

Chapter 9 Review

Fill in the correct answer choice.

🔑 Understand Key Concepts

1 What determines the order of elements on today's periodic table? SC.8.P.8.6

Ⓐ increasing atomic mass
Ⓑ decreasing atomic mass
Ⓒ increasing atomic number
Ⓓ decreasing atomic number

2 The element key for nitrogen is shown below.

From this key, determine the atomic mass of nitrogen. SC.8.P.8.6

Ⓐ 7
Ⓑ 7.01
Ⓒ 14.01
Ⓓ 21.01

3 Look at the periodic table in Lesson 1. Which of the following lists of elements forms a group on the periodic table? SC.8.P.8.6

Ⓐ Li, Be, B, C, N, O, F, and Ne
Ⓑ He, Ne, Ar, Kr, Xe, and Rn
Ⓒ B, Si, As, Te, and At
Ⓓ Sc, Ti, V, Cr, Mn, Fe, Co, Cu, Ni, and Zn

4 Which is NOT a property of metals? SC.8.P.8.4

Ⓐ brittleness
Ⓑ conductivity
Ⓒ ductility
Ⓓ luster

5 What are two properties that make a metal a good choice to use as wire in electronics? SC.8.P.8.4

Ⓐ conductivity, malleability
Ⓑ ductility, conductivity
Ⓒ luster, malleability
Ⓓ malleability, high density

Critical Thinking

6 **Recommend** an element to use to fill bottles that contain ancient paper. The element should be a gas at room temperature, should be denser than helium, and should not easily react with other elements. SC.8.P.8.4

_____

_____

7 **Apply** Why is mercury the only metal to have been used in thermometers? SC.8.P.8.4

_____

_____

_____

8 **Evaluate** the following types of metals as a choice to make a sun reflector: alkali metals, alkaline earth metals, or transition metals. The metal cannot react with water or oxygen and must be shiny and strong. SC.8.P.8.4

_____

_____

_____

9 The figure below shows a pattern of densities.

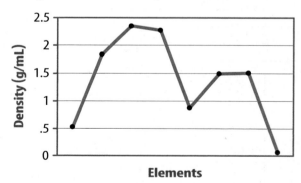

Elements

Infer whether you are looking at a graph of elements within a group or across a period. Explain your answer. SC.8.P.8.6

_____

_____

10 **Contrast** aluminum and nitrogen. Show why aluminum is a metal and nitrogen is not. SC.8.P.8.6

_____

_____

_____

_____

11 **Classify** A student sorted six elements. He placed iron, silver, and sodium in group A. He placed neon, oxygen, and nitrogen in group B. Name one other element that fits in group A and another element that belongs in group B. Explain your answer. SC.8.P.8.6

_____

_____

_____

Writing in Science

12 **Write** a plan on a separate sheet of paper that shows how a metal, a nonmetal, and a metalloid could be used when constructing a building. LA.6.2.2.3

Big Idea Review

13 Explain how atomic number and properties are used to determine where element 115 is placed on the periodic table. SC.8.P.8.6

_____

_____

_____

14 The photo on page 252 shows how the properties of materials determine their uses. How can the periodic table be used to help you find elements with properties similar to that of helium? SC.8.P.8.6

_____

_____

_____

Math Skills MA.6.A.3.6

Use Geometry

15 The table below shows the atomic radii of three elements in group 1 on the periodic table.

Element	Atomic radius
Li	152 pm
Na	186 pm
K	227 pm

a. What is the circumference of each atom?

b. Rubidium (Rb) is the next element in Group 1. What would you predict about the radius and circumference of a rubidium atom?

Fill in the correct answer choice.

Multiple Choice

1 Where are most nonmetals located on the periodic table? SC.8.P.8.6

(A) in the bottom row

(B) on the left side and in the middle

(C) on the right side

(D) in the top row

2 The illustration below shows elements in four groups in the periodic table.

1																	18
H	2															17	He
Li	Be														F	Ne
Na	Mg														Cl	Ar
K	Ca														Br	Kr
Rb	Sr														I	Xe
Cs	Ba														At	Rn

Which group of elements most readily combines with Group 17 elements? SC.8.P.8.6

(F) Group 1

(G) Group 2

(H) Group 17

(I) Group 18

3 Which element is most likely to react with potassium? SC.8.P.8.6

(A) bromine

(B) calcium

(C) nickel

(D) sodium

Use the table below about group 13 elements to answer question 4.

Element Symbol	Atomic Number	Density (g/cm³)	Atomic Mass
B	5	2.34	10.81
Al	13	2.70	26.98
Ga	31	5.90	69.72
In	49	7.30	114.82

4 How do density and atomic mass change as atomic number increases? SC.8.P.8.6

(F) Density and atomic mass decrease.

(G) Density and atomic mass increase.

(H) Density decreases and atomic mass increases.

(I) Density increases and atomic mass decreases.

5 Which elements have high densities, strength, and resistance to corrosion? SC.8.P.8.6

(A) alkali metals

(B) alkaline earth metals

(C) metalloids

(D) transition elements

6 Many elements that are essential for life, including nitrogen, oxygen, and carbon, are part of what classification? SC.8.P.8.6

(F) metalloids

(G) metals

(H) noble gases

(I) nonmetals

Use the figure below to answer questions 7 and 8.

17

7 The figure shows a group in the periodic table. What is the name of this group of elements? SC.8.P.8.6

Ⓐ halogens

Ⓑ metalloids

Ⓒ metals

Ⓓ noble gases

8 Which is a property of these elements? SC.8.P.8.6

Ⓕ They are conductors.

Ⓖ They are semiconductors.

Ⓗ They are nonreactive with other elements.

Ⓘ They react easily with other elements.

9 What is one similarity among elements in a group? SC.8.P.8.6

Ⓐ atomic mass

Ⓑ atomic weight

Ⓒ chemical properties

Ⓓ practical uses

Use the figure below to answer question 10.

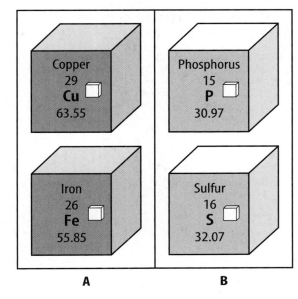

A   **B**

10 Groups A and B each contain two elements. Identify each group as metals, nonmetals, or metalloids. SC.8.P.8.6

Ⓕ Group A contains metals and Group B contains nonmetals.

Ⓖ Group A contains nonmetals and Group B contains metalloids.

Ⓗ Group A contains metalloids and Group B contains metals.

Ⓘ Group A contains metals and Group B contains metals.

NEED EXTRA HELP?

If You Missed Question...	1	2	3	4	5	6	7	8	9	10
Go to Lesson...	1	3	3	1	2	3	3	3	1	2,3

Multiple Choice *Bubble the correct answer.*

16

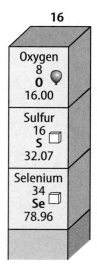

1. Using the image above, which statement is TRUE about oxygen, sulfur, and selenium? **SC.8.P.8.6**

(A) They belong to the same group.

(B) They belong to the same period.

(C) The mass number doubles between each element.

(D) The atomic number increases by eight between each element.

2. If the atomic number for oxygen is 8, how many protons are in the nucleus of each of its atoms? **SC.8.P.8.6**

(F) 2

(G) 4

(H) 8

(I) 16

3. Elements on the periodic table are listed according to **SC.8.P.8.6**

(A) atomic mass.

(B) atomic number.

(C) chemical symbol.

(D) state of matter.

Thorium	Protactinium	Uranium	Neptunium
90	91	92	93
Th	Pa	U	Np
232.04	231.04	238.03	(237)

4. Examine the section of the periodic table in the image above. Which element does NOT occur naturally on Earth? **SC.8.P.8.6**

(F) neptunium

(G) protactinium

(H) thorium

(I) uranium

Copyright © Glencoe/McGraw-Hill, a division of The McGraw-Hill Companies, Inc.

Multiple Choice *Bubble the correct answer.*

Use the image below to answer questions 1 and 2.

Potassium	Calcium	Scandium	Titanium
19	20	21	22
K	Ca	Sc	Ti
39.10	40.08	44.96	47.87

1. Identify which element is the MOST metallic. **SC.8.P.8.4**

(A) calcium

(B) potassium

(C) scandium

(D) titanium

2. Identify which element is the LEAST malleable. **SC.8.P.8.4**

(F) calcium

(G) potassium

(H) scandium

(I) titanium

3. On the periodic table, metallic properties increase from **SC.8.P.8.6**

(A) left to right and bottom to top.

(B) left to right and top to bottom.

(C) right to left and bottom to top.

(D) right to left and top to bottom.

4. Angelo wants to develop a new series of pots and pans for use in cooking. Which property of metals should he be MOST concerned about? **SC.8.P.8.4**

(F) conductivity

(G) ductility

(H) luster

(I) malleability

Copyright © Glencoe/McGraw-Hill, a division of The McGraw-Hill Companies, Inc.

Multiple Choice *Bubble the correct answer.*

Use the image below to answer questions 1 and 2.

					18
					Helium 2 **He** 4.00
13	**14**	**15**	**16**	**17**	
Boron 5 **B** 10.81	Carbon 6 **C** 12.01	Nitrogen 7 **N** 14.01	Oxygen 8 **O** 16.00	Fluorine 9 **F** 19.00	Neon 10 **Ne** 20.18
Aluminum 13 **Al** 26.98	Silicon 14 **Si** 28.09	Phosphorus 15 **P** 30.97	Sulfur 16 **S** 32.07	Chlorine 17 **Cl** 35.45	Argon 18 **Ar** 39.95
Gallium 31 **Ga** 69.72	Germanium 32 **Ge** 72.64	Arsenic 33 **As** 74.92	Selenium 34 **Se** 78.96	Bromine 35 **Br** 79.90	Krypton 36 **Kr** 83.80
Indium 49 **In** 114.82	Tin 50 **Sn** 118.71	Antimony 51 **Sb** 121.76	Tellurium 52 **Te** 127.60	Iodine 53 **I** 126.90	Xenon 54 **Xe** 131.29
Thallium 81 **Tl** 204.38	Lead 82 **Pb** 207.20	Bismuth 83 **Bi** 208.98	Polonium 84 **Po** (209)	Astatine 85 **At** (210)	Radon 86 **Rn** (222)
Ununtrium 113 **Uut** (284)	Ununquadium 114 **Uuq** (289)	Ununpentium 115 **Cn** (288)	Ununhexium 116 **Uuh** (293)		Ununoctium 118 **Uuo** (294)

1. Use the periodic table to identify to which group the noble gases belong. **SC.8.P.8.6**

   (A) 13

   (B) 14

   (C) 17

   (D) 18

2. Identify which element is a metalloid.
   SC.8.P.8.6

   (F) aluminum

   (G) arsenic

   (H) carbon

   (I) helium

3. Nonmetals often are used in insulating material because they do NOT **SC.8.P.8.4**

   (A) break apart easily.

   (B) conduct electricity and heat.

   (C) form compounds with other elements.

   (D) have luster.

4. How do halogens differ from noble gases?
   SC.8.P.8.4

   (F) Halogens are metalloids, whereas noble gases are nonmetals.

   (G) Halogens are semiconductors, whereas noble gases are conductors.

   (H) Halogens are solids at room temperature, whereas noble gases are gases.

   (I) Halogens form compounds, whereas noble gases generally do not.

Copyright © Glencoe/McGraw-Hill, a division of The McGraw-Hill Companies, Inc.

Notes

Name _____   Date _____

How do the atoms form bonds?

The picture above shows a spoon of sugar granules. The sugar granules are made up of countless molecules of sugar that contain the elements carbon, hydrogen, and oxygen. Chemical bonds hold atoms of these elements together and form sugar molecules. Which best describes how atoms form bonds?

A: When two atoms join, their nuclei form a bond.

B: An attractive force between atoms holds them together but they do not touch.

C: Each atom has a structure that enables it to join with one or more other atoms.

D: A sticky substance holds atoms together when they form a molecule.

Explain your thinking. Describe your ideas about how atoms form chemical bonds.

Elements and Chemical bonds

FLORIDA BIG IDEAS

1 The Practice of Science

3 The Role of Theories, Laws, Hypotheses, and Models

4 Science and Society

8 Properties of Matter

Think About It!

How do elements join together to form chemical compounds?

How many different words could you type using just the letters on a keyboard? The English alphabet has only 26 letters, but a dictionary lists hundreds of thousands of words using these letters! Similarly only about 115 different elements make all kinds of matter.

1 How do you think so few elements form so many different kinds of matter?

_____

_____

_____

2 Why do you think different types of matter have different properties?

_____

_____

_____

3 How do you think atoms are held together to produce different types of matter?

_____

_____

_____

Get Ready to Read

What do you think about chemical bonding?

Before you read, decide if you agree or disagree with each of these statements. As you read this chapter, see if you change your mind about any of the statements.

		AGREE	DISAGREE
1	Elements rarely exist in pure form. Instead, combinations of elements make up most of the matter around you.	☐	☐
2	Chemical bonds that form between atoms involve electrons.	☐	☐
3	The atoms in a water molecule are more chemically stable than they would be as individual atoms.	☐	☐
4	Many substances dissolve easily in water because opposite ends of a water molecule have opposite charges.	☐	☐
5	Losing electrons can make some atoms more chemically stable.	☐	☐
6	Metals are good electric conductors because they tend to hold onto their valence electrons very tightly.	☐	☐

There's More Online!
Video • Audio • Review • ⓘLab Station • WebQuest • Assessment • Concepts in Motion • Multilingual eGlossary

379

Lesson 1

Electrons and Energy
LEVELS

ESSENTIAL QUESTIONS

How is an electron's energy related to its distance from the nucleus?

Why do atoms gain, lose, or share electrons?

Vocabulary

chemical bond p. 382

valence electron p. 384

electron dot diagram p. 385

Florida NGSSS

SC.8.N.3.1 Select models useful in relating the results of their own investigations.

SC.8.P.8.5 Recognize that there are a finite number of elements and that their atoms combine in a multitude of ways to produce compounds that make up all of the living and nonliving things that we encounter.

SC.8.P.8.6 Recognize that elements are grouped in the periodic table according to similarities of their properties.

SC.8.P.8.7 Explore the scientific theory of atoms (also known as atomic theory) by recognizing that atoms are the smallest unit of an element and are composed of sub-atomic particles (electrons surrounding a nucleus containing protons and neutrons).

LA.6.2.2.3 The student will organize information to show understanding or relationships among facts, ideas, and events (e.g., representing key points within text through charting, mapping, paraphrasing, summarizing, or comparing/contrasting);

LA.6.4.2.2 The student will record information (e.g., observations, notes, lists, charts, legends) related to a topic, including visual aids to organize and record information, as appropriate, and attribute sources of information;

SC.8.P.8.6

inquiry Launch Lab

20 minutes

How is the periodic table organized?

How do you begin to put together a puzzle of a thousand pieces? You first sort similar pieces into groups. All edge pieces might go into one pile. All blue pieces might go into another pile. Similarly, scientists place the elements into groups based on their properties. The periodic table organizes information about all the elements.

Procedure

1. Obtain six **index cards** from your teacher. Using one card for each element name, write the names *beryllium, sodium, iron, zinc, aluminum,* and *oxygen* at the top of a card.

2. Using the periodic table, locate the element key for each element written on your cards.

3. For each element, find the following information and write it on the index card: symbol, atomic number, atomic mass, state of matter, and element type.

Think About This

1. What do the elements in the blue blocks have in common? In the green blocks? In the yellow blocks?

2. **Key Concept** Each element in a column on the periodic table has similar chemical properties and forms bonds in similar ways. Based on this, for the element you wrote on each card, name another element on the periodic table that has similar chemical properties.

inquiry Are pairs more stable?

1. Rowing can be hard work, especially if you are part of a racing team. The job is made easier because the rowers each pull on the water with a pair of oars. How do pairs make the boat more stable?

_____

_____

_____

_____

_____

_____

The Periodic Table

Imagine trying to find a book in a library if all the books were unorganized. Books are organized in a library to help you easily find the information you need. The periodic table is like a library of information about all chemical elements.

The periodic table has more than 100 blocks—one for each known element. Each block on the periodic table includes basic properties of each element such as the element's state of matter at room temperature and its atomic number. The atomic number is the number of protons in each atom of the element. Each block also lists an element's atomic mass, or the average mass of all the different isotopes of that element.

Periods and Groups

You can learn about some properties of an element from its position on the periodic table. Elements are organized in periods (rows) and groups (columns). The periodic table lists elements in order of atomic number. The atomic number increases from left to right as you move across a period. Elements in each group have similar chemical properties and react with other elements in similar ways. In this lesson, you will read more about how an element's position on the periodic table can be used to predict its properties.

Active Reading **2. Describe** As you read this page and the next, complete the sentences below about the periodic table.

The atomic number of an element is the

_____

_____

_____ .

On the periodic table, groups are arranged _____, and periods are arranged

_____ .

One difference between metals and nonmetals is

_____

_____

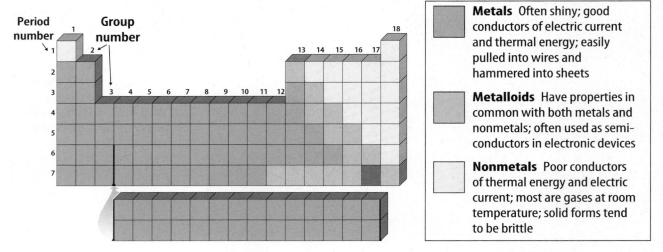

Metals Often shiny; good conductors of electric current and thermal energy; easily pulled into wires and hammered into sheets

Metalloids Have properties in common with both metals and nonmetals; often used as semi-conductors in electronic devices

Nonmetals Poor conductors of thermal energy and electric current; most are gases at room temperature; solid forms tend to be brittle

Figure 1 Elements on the periodic table are classified as metals, nonmetals, or metalloids.

REVIEW VOCABULARY

compound

matter that is made up of two or more different kinds of atoms joined together by chemical bonds

Figure 2 Protons and neutrons are in an atom's nucleus. Electrons move around the nucleus.

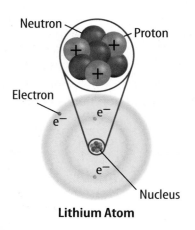

Neutron

Proton

Electron

e⁻

Nucleus

Lithium Atom

Metals, Nonmetals, and Metalloids

The three main regions of elements on the periodic table are shown in **Figure 1.** Except for hydrogen, elements on the left side of the table are metals. Nonmetals are on the right side of the table. Metalloids form the narrow stair-step region between metals and nonmetals.

 3. Label Place an *X* on all the metalloid blocks in the periodic table above.

Atoms Bond

In nature, pure elements are rare. Instead, atoms of different elements chemically combine and form **compounds**. Compounds make up most of the matter around you, including living and nonliving things. There are only about 115 elements, but these elements combine and form millions of compounds. Chemical bonds hold them together. *A* **chemical bond** *is a force that holds two or more atoms together.*

Electron Number and Arrangement

Recall that atoms contain protons, neutrons, and electrons, as shown in **Figure 2.** Each proton has a positive charge; each neutron has no charge; and each electron has a negative charge. The atomic number of an element is the number of protons in each atom of that element. In a neutral (uncharged) atom, the number of protons equals the number of electrons.

The exact position of electrons in an atom cannot be determined. This is because electrons are in constant motion around the nucleus. However, each electron is usually in a certain area of space around the nucleus. Some are in areas close to the nucleus, and some are in areas farther away.

Electrons and Energy Different electrons in an atom have different amounts of energy. An electron moves around the nucleus at a distance that corresponds to its amount of energy. Areas of space in which electrons move around the nucleus are called energy levels. Electrons closest to the nucleus have the least amount of energy. They are in the lowest energy level. Electrons farthest from the nucleus have the greatest amount of energy. They are in the highest energy level. The energy levels of an atom are shown in **Figure 3.** Notice that only two electrons can be in the lowest energy level. The second energy level can hold up to eight.

 4. NGSSS Check **Describe** How is an electron's energy related to the electron's position in an atom? **SC.8.P.8.7**

_____

_____

_____

Electrons and Bonding Imagine two magnets. The closer they are to each other, the stronger the attraction of their opposite ends. Negatively charged electrons have a similar attraction to the positively charged nucleus of an atom. The electrons in energy levels closest to the nucleus of the same atom have a strong attraction to that nucleus. However, electrons farther from that nucleus are weakly attracted to it. These outermost electrons can be attracted to the nucleus of other atoms. This attraction between the positive nucleus of one atom and the negative electrons of another is what causes a chemical bond.

Active Reading

FOLDABLES® LA.6.2.2.3

Make two quarter-sheet note cards from a sheet of paper. Use them to organize your notes on valence electrons and electron dot diagrams.

Valence Electrons

Electron Dot Diagrams

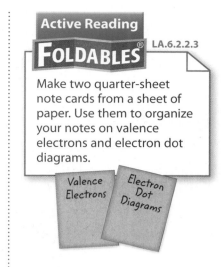

Active Reading **5. Identify** Correctly fill in the blanks in the diagram below with words from the following list.

- farthest
- positively
- closest
- negatively

Figure 3 Electrons are in certain energy levels within an atom.

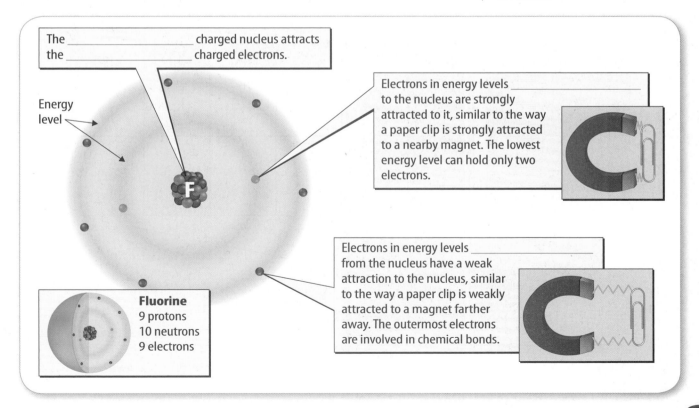

The _____ charged nucleus attracts the _____ charged electrons.

Energy level

Electrons in energy levels _____ to the nucleus are strongly attracted to it, similar to the way a paper clip is strongly attracted to a nearby magnet. The lowest energy level can hold only two electrons.

Electrons in energy levels _____ from the nucleus have a weak attraction to the nucleus, similar to the way a paper clip is weakly attracted to a magnet farther away. The outermost electrons are involved in chemical bonds.

Fluorine
9 protons
10 neutrons
9 electrons

Valence Electrons

You have read that electrons farthest from their nucleus are easily attracted to the nuclei of nearby atoms. These outermost electrons are the only electrons involved in chemical bonding. Even atoms that have only a few electrons, such as hydrogen or lithium, can form chemical bonds. This is because these electrons are still the outermost electrons and are exposed to the nuclei of other atoms. *A **valence electron** is an outermost electron of an atom that participates in chemical bonding.* Valence electrons have the most energy of all electrons in an atom.

The number of valence electrons in each atom of an element can help determine the type and the number of bonds it can form. How do you know how many valence electrons an atom has? The periodic table can tell you. Except for helium, elements in certain groups have the same number of valence electrons. **Figure 4** illustrates how to use the periodic table to determine the number of valence electrons in the atoms of groups 1, 2, and 13–18. Determining the number of valence electrons for elements in groups 3–12 is more complicated. You will learn about these groups in later chemistry courses.

WORD ORIGIN

valence
from Latin *valentia,* means "strength, capacity"

Figure 4 🔑 You can use the group numbers at the top of the columns to determine the number of valence electrons in atoms of groups 1, 2, and 13–18.

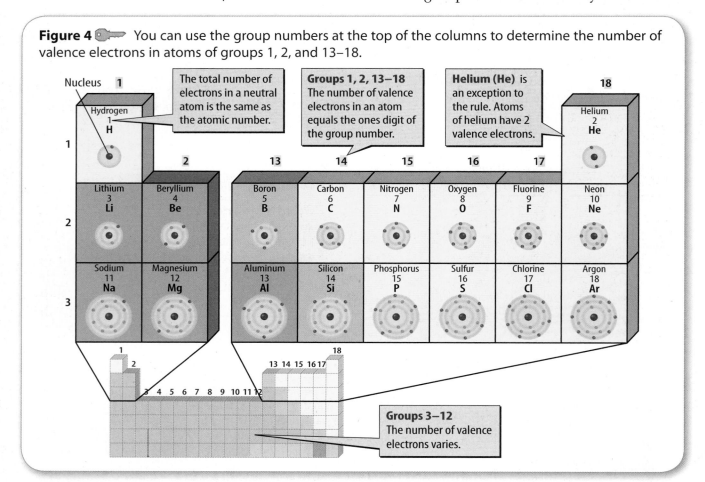

Active Reading

6. Decide How many valence electrons does an atom of phosphorus (P) have?

Writing and Using Electron Dot Diagrams

Steps for writing a dot diagram	Beryllium	Carbon	Nitrogen	Argon
1 Identify the element's group number on the periodic table.		14	15	
2 Identify the number of valence electrons. • This equals the ones digit of the group number.	2	4		
3 Draw the electron dot diagram. • Place one dot at a time on each side of the symbol (top, right, bottom, left). Repeat until all dots are used.	Be·		·N̈·	:Är:
4 Determine if the atom is chemically stable. • An atom is chemically stable if all dots on the electron dot diagram are paired.	Chemically Unstable	Chemically Unstable	Chemically Unstable	Chemically Stable
5 Determine how many bonds this atom can form. • Count the dots that are unpaired.	2		3	

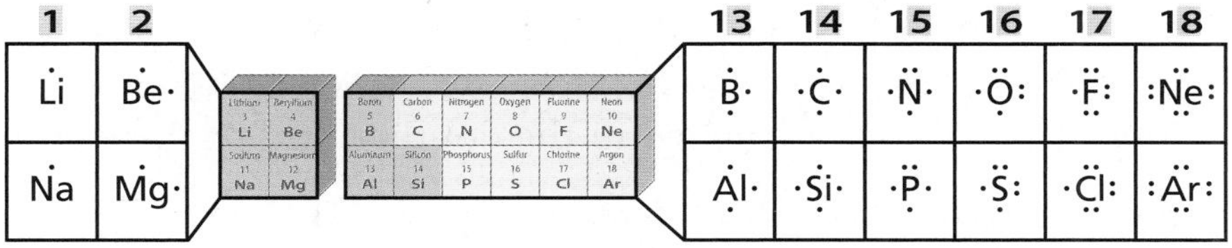

1	2			13	14	15	16	17	18
Li·	Be·			Ḃ·	·Ċ·	·N̈·	·Ö:	·F̈:	:N̈e:
Na·	Mg·			Äl·	·Si·	·P̈·	·S̈·	·C̈l:	:Är:

Figure 5 Electron dot diagrams show the number of valence electrons in an atom.

Electron Dot Diagrams

In 1916 an American chemist named Gilbert Lewis developed a method to show an element's valence electrons. He developed the **electron dot diagram**, *a model that represents valence electrons in an atom as dots around the element's chemical symbol.*

Electron dot diagrams can help you predict how an atom will bond with other atoms. Dots, representing valence electrons, are placed one-by-one on each side of an element's chemical symbol until all the dots are used. Some dots will be paired up; others will not. The number of unpaired dots is often the number of bonds an atom can form. The steps for writing dot diagrams are shown in **Figure 5**.

Recall that each element in a group has the same number of valence electrons. As a result, every element in a group has the same number of dots in its electron dot diagram.

Notice in **Figure 5** that an argon atom, Ar, has eight valence electrons, or four pairs of dots, in the diagram. There are no unpaired dots. Atoms with eight valence electrons do not easily react with other atoms. They are chemically stable. Atoms that have between one and seven valence electrons are reactive, or chemically unstable. These atoms easily bond with other atoms and form chemically stable compounds.

Atoms of hydrogen and helium have only one energy level. These atoms are chemically stable with two valence electrons.

Active Reading **7. Summarize** Complete the table above with the correct numbers and symbol.

Active Reading **8. Explain** Why are electron dot diagrams useful?

_____

_____

_____

_____

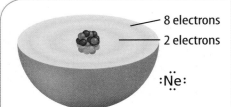
8 electrons
2 electrons
:N̈e:

Neon has 10 electrons: 2 inner electrons and 8 valence electrons. A neon atom is chemically stable because it has 8 valence electrons.

2 electrons
Ḧe

Helium has 2 electrons. Because an atom's lowest energy level can hold only 2 electrons, the 2 dots in the dot diagram are paired. Helium is stable.

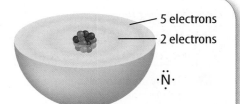
5 electrons
2 electrons
·N̈·

Nitrogen has 7 electrons: 2 inner electrons and 5 valence electrons. Its dot diagram has 3 unpaired dots. Nitrogen atoms become more stable by forming chemical bonds.

Figure 6 Atoms gain, lose, or share electrons to become chemically stable.

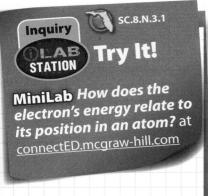

Inquiry
SC.8.N.3.1
iLAB STATION **Try It!**

MiniLab *How does the electron's energy relate to its position in an atom?* at connectED.mcgraw-hill.com

Apply It!

After you complete the lab, answer the question below.

1. Look at the periodic table. Notice that krypton (Kr) is in group 18. Why do atoms of krypton not easily bond with other atoms?

Active Reading **9. Compile** Complete the flow-chart about the behavior of atoms with unpaired valence electrons

Noble Gases

The elements in Group 18 are called noble gases. With the exception of helium, noble gases have eight valence electrons and are chemically stable. Chemically stable atoms do not easily react, or form bonds, with other atoms. The electron structures of two noble gases—neon and helium—are shown in **Figure 6.** Notice that all dots are paired in the dot diagrams of these atoms.

Stable and Unstable Atoms

Atoms with unpaired dots in their electron dot diagrams are reactive, or chemically unstable. For example, nitrogen, shown in **Figure 6,** has three unpaired dots in its electron dot diagram, and it is reactive. Nitrogen, like many other atoms, becomes more stable by forming chemical bonds with other atoms.

When an atom forms a bond, it gains, loses, or shares valence electrons with other atoms. By forming bonds, atoms become more chemically stable. Recall that atoms are most stable with eight valence electrons. Therefore, atoms with fewer than eight valence electrons form chemical bonds and become stable. In Lessons 2 and 3, you will read which atoms gain, lose, or share electrons when forming stable compounds.

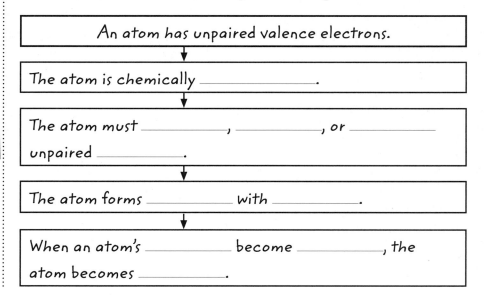

An atom has unpaired valence electrons.

↓

The atom is chemically _____.

↓

The atom must _____, _____, or unpaired _____.

↓

The atom forms _____ with _____.

↓

When an atom's _____ become _____, the atom becomes _____.

Lesson Review 1

Visual Summary

Electrons are less strongly attracted to a nucleus the farther they are from it, similar to the way a magnet attracts a paper clip.

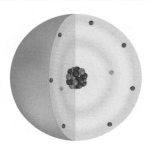

Electrons in atoms are in energy levels around the nucleus. Valence electrons are involved in chemical bonding.

All noble gases, except He, have four pairs of dots in their electron dot diagrams. Noble gases are chemically stable.

Use Vocabulary

1 **Use the term** *chemical bond* in a complete sentence.

_____

_____

2 The electrons of an atom that participate in chemical bonding are called _____.

Understand Key Concepts 🔑

3 **Identify** the number of valence electrons in each atom: calcium _____ carbon _____ and sulfur _____ . SC.8.P.8.7

4 Which part of the atom forms a chemical bond? SC.8.P.8.7

(A) electron   (C) nucleus

(B) neutron   (D) proton

5 **Draw** electron dot diagrams for oxygen, iodine, and beryllium. SC.8.P.8.6

Interpret Graphics

6 **Determine** the number of valence electrons in each diagram shown below.

Magnesium 12 **Mg** _____

Chlorine 17 **Cl** _____

7 **Organize Information** Fill in the graphic organizer below to describe one or more details for each concept: electron energy, valence electrons, stable atoms. LA.6.2.2.3

Concept	Description

Critical Thinking

8 **Compare** krypton and bromine in terms of chemical stability. SC.8.P.8.6

_____

_____

_____

New Green Airships

The Difference of One Valence Electron

Faster than ocean liners and safer than airplanes, airships used to be the best way to travel. The largest, the *Hindenburg,* was nearly the size of the *Titanic.* To this day, no larger aircraft has ever flown. So, what happened to the giant airship? The answer lies in a valence electron.

The builders of the *Hindenburg* filled it with a lighter-than-air gas, hydrogen, so that it would float. Their plan was to use helium, a noble gas. However, helium was scarce. They knew hydrogen was explosive, but it was easier to get. For nine years, hydrogen airships floated safely back and forth across the Atlantic. But in 1937, disaster struck. Just before it landed, the *Hindenburg* exploded in flames. The age of the airship was over.

Since the *Hindenburg,* airplanes have become the main type of air transportation. A big airplane uses hundreds of gallons of fuel to take off and fly. As a result, it releases large amounts of pollutants into the atmosphere. Some people are looking for other types of air transportation that will be less harmful to the environment. Airships may be the answer. An airship floats and needs very little fuel to take off and stay airborne. Airships also produce far less pollution than other aircraft.

Today, however, airships use helium—not hydrogen. With two valence electrons instead of one, as hydrogen has, helium is unreactive. Thanks to helium's chemical stability, someday you might be a passenger on a new, luxurious, but not explosive, version of the *Hindenburg.*

A new generation of big airships might soon be hauling freight and carrying passengers.

It's Your Turn

RESEARCH Precious documents deteriorate with age as their surfaces react with air. Parchment turns brown and crumbles. Find out how our founding documents have been saved from this fate by noble gases. LA.6.4.2.2

Compounds, Chemical Formulas, and Covalent BONDS

ESSENTIAL QUESTIONS

 How do elements differ from the compounds they form?

 What are some common properties of a covalent compound?

 Why is water a polar compound?

Vocabulary

covalent bond p. 391

molecule p. 392

polar molecule p. 393

chemical formula p. 394

 Florida NGSSS

SC.8.N.1.1 Define a problem from the eighth grade curriculum using appropriate reference materials to support scientific understanding, plan and carry out scientific investigations of various types, such as systematic observations or experiments, identify variables, collect and organize data, interpret data in charts, tables, and graphics, analyze information, make predictions, and defend conclusions.

SC.8.N.3.1 Select models useful in relating the results of their own investigations.

SC.8.P.8.5 Recognize that there are a finite number of elements and that their atoms combine in a multitude of ways to produce compounds that make up all of the living and nonliving things that we encounter.

SC.8.P.8.8 Identify basic examples of and compare and classify the properties of compounds, including acids, bases, and salts.

LA.6.2.2.3 The student will organize information to show understanding or relationships among facts, ideas, and events (e.g., representing key points within text through charting, mapping, paraphrasing, summarizing, or comparing/contrasting);

 Inquiry Launch Lab  SC.8.P.8.5

20 minutes

How is a compound different from its elements?

The sugar you use to sweeten foods at home is probably sucrose. Sucrose contains the elements carbon, hydrogen, and oxygen. How does table sugar differ from the elements that it contains?

Procedure

1 Read and complete a lab safety form.

2 Air is a mixture of several gases, including oxygen and hydrogen. Charcoal is a form of carbon. Write some properties of oxygen, hydrogen, and carbon.

3 Obtain from your teacher a piece of **charcoal** and a **beaker** with **table sugar** in it.

4 Observe the charcoal. Describe the way it looks and feels.

5 Observe the table sugar in the beaker. What does it look and feel like? Record your observations.

Data and Observations

Think About This

1. **Compare and contrast** the properties of charcoal, hydrogen, and oxygen.

2. **Key Concept** **Explain** How do you think the physical properties of carbon, hydrogen, and oxygen change when they combine to form sugar?

From Elements to Compounds

Have you ever baked cupcakes? First, combine flour, baking soda, and a pinch of salt. Then, add sugar, eggs, vanilla, milk, and butter. Each ingredient has unique physical and chemical properties. When you mix the ingredients together and bake them, a new product results—cupcakes. The cupcakes have properties that are different from the ingredients.

In some ways, compounds are like cupcakes. Recall that a compound is a substance made up of two or more different elements. Just as cupcakes are different from their ingredients, compounds are different from their elements. An element is made of one type of atom, but compounds are chemical combinations of different types of atoms. Compounds and the elements that make them up often have different properties.

Chemical **bonds** join atoms together. Recall that a chemical bond is a force that holds atoms together in a compound. In this lesson, you will learn that one way that atoms can form bonds is by sharing valence electrons. You will also learn how to write and read a chemical formula.

Inquiry **How do they combine?**

1. A jigsaw puzzle has pieces that connect in a certain way. The pieces fit together by sharing tabs with other pieces. All of the pieces combine and form a complete puzzle. Like pieces of a puzzle, atoms can join together and form a compound by sharing electrons. What do you think the world would be like if atoms could not join together?

_____

_____

_____

_____

SCIENCE USE V. COMMON USE

bond

Science Use a force that holds atoms together in a compound

Common Use a close personal relationship between two people

Active Reading **2. Recall** Read each statement in the table below. If it is true, write *T* in the center column. If it is false, write *F* in the center column. In the third column, replace the underlined words to make the statement true.

Statement	T or F	Correction
<u>Compounds</u> are chemical combinations of <u>elements</u>.		
Compounds <u>usually</u> have the same properties as the <u>bonds</u> they are made from.		
Atoms form bonds by sharing <u>physical</u> <u>properties</u>.		

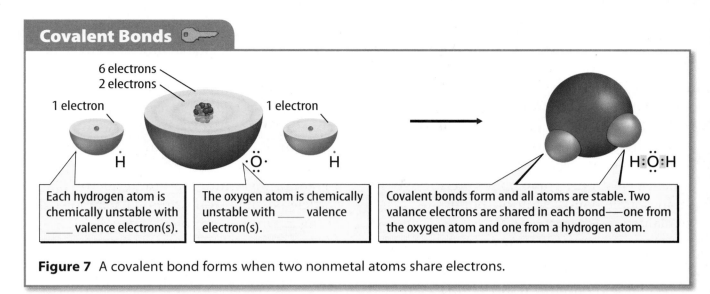

Covalent Bonds 🔑

6 electrons
2 electrons
1 electron
1 electron

H · · O · H

Each hydrogen atom is chemically unstable with ____ valence electron(s).

The oxygen atom is chemically unstable with ____ valence electron(s).

Covalent bonds form and all atoms are stable. Two valance electrons are shared in each bond—one from the oxygen atom and one from a hydrogen atom.

H:Ö:H

Figure 7 A covalent bond forms when two nonmetal atoms share electrons.

Active Reading 3. **Identify** Fill in the blanks in the two left boxes in **Figure 7** with number of valence electrons for each atom.

Covalent Bonds—Electron Sharing

As you read in Lesson 1, one way that atoms can become more chemically stable is by sharing valence electrons. When unstable, nonmetal atoms bond together, they bond by sharing valence electrons. *A* **covalent bond** *is a chemical bond formed when two atoms share one or more pairs of valence electrons.* The atoms then form a stable covalent compound.

A Noble Gas Electron Arrangement

Look at the reaction between hydrogen and oxygen in **Figure 7.** Before the reaction, each hydrogen atom has one valence electron. The oxygen atom has six valence electrons. Recall that most atoms are chemically stable with eight valence electrons—the same electron arrangement as a noble gas. An atom with less than eight valence electrons becomes stable by forming chemical bonds until it has eight valence electrons. Therefore, an oxygen atom forms two bonds to become stable. A hydrogen atom is stable with two valence electrons. It forms one bond to become stable.

Shared Electrons

If the oxygen atom and each hydrogen atom share their unpaired valence electrons, they can form two covalent bonds and become a stable covalent compound. Each covalent bond contains two valence electrons—one from the hydrogen atom and one from the oxygen atom. Since these electrons are shared, they count as valence electrons for both atoms in the bond. Each hydrogen atom now has two valence electrons. The oxygen atom now has eight valence electrons because it bonds to two hydrogen atoms. All three atoms have the electron arrangement of a noble gas, and the compound is stable.

Active Reading 4. **Find** How does an atom with less than eight valence electrons become stable? Highlight your answer in the text.

Active Reading

FOLDABLES LA.6.2.2.3

Make three quarter-sheet note cards from a sheet of paper to organize information about single, double, and triple covalent bonds.

Triple Covalent Bonds
Double Covalent Bonds
Single Covalent Bonds

Double and Triple Covalent Bonds

As shown in **Figure 8,** a single covalent bond exists when two atoms share one pair of valence electrons. A double covalent bond exists when two atoms share two pairs of valence electrons. Double bonds are stronger than single bonds. A triple covalent bond exists when two atoms share three pairs of valence electrons. Triple bonds are stronger than double bonds. Multiple bonds are explained in **Figure 8.**

Covalent Compounds

When two or more atoms share valence electrons, they form a stable covalent compound. The covalent compounds carbon dioxide, water, and sugar are very different, but they also share similar properties. Covalent compounds usually have low melting points and low boiling points. They are usually gases or liquids at room temperature, but they can also be solids. Covalent compounds are poor conductors of thermal energy and electric current.

Molecules

The chemically stable unit of a covalent compound is a molecule. *A* **molecule** *is a group of atoms held together by covalent bonding that acts as an independent unit.* Table sugar ($C_{12}H_{22}O_{11}$) is a covalent compound. One grain of sugar is made up of trillions of sugar molecules. Imagine breaking a grain of sugar into the tiniest microscopic particle possible. You would have a molecule of sugar. One sugar molecule contains 12 carbon atoms, 22 hydrogen atoms, and 11 oxygen atoms all covalently bonded together. The only way to further break down the molecule would be to chemically separate the carbon, hydrogen, and oxygen atoms. These atoms alone have very different properties from the compound sugar.

 5. NGSSS Check **Explain** What are some common properties of covalent compounds?
SC.8.P.8.5

_____

_____

_____

Figure 8 The more valence electrons that two atoms share, the stronger the covalent bond is between the atoms.

When two hydrogen atoms bond, they form a single covalent bond.	**One Single Covalent Bond** $H + H \longrightarrow H:H$	In a single covalent bond, 1 pair of electrons is shared between two atoms. Each H atom shares 1 valence electron with the other.
When one carbon atom bonds with two oxygen atoms, two double covalent bonds form.	**Two Double Covalent Bonds** $O: + C + O: \longrightarrow O::C::O$	In a double covalent bond, 2 pairs of electrons are shared between two atoms. One O atom and the C atom each share 2 valence electrons with the other.
When two nitrogen atoms bond, they form a triple covalent bond.	**One Triple Covalent Bond** $N + N \longrightarrow :N:::N:$	In a triple covalent bond, 3 pairs of electrons are shared between two atoms. Each N atom shares 3 valence electrons with the other.

Active Reading **6. Compare** Is the bond stronger between atoms in hydrogen gas (H_2) or nitrogen gas (N_2)? Why?

_____

_____

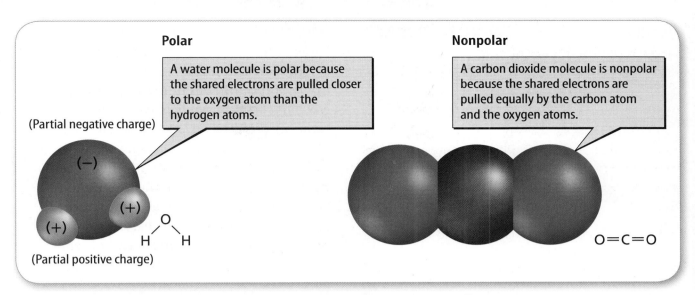

Figure 9 Atoms of a polar molecule share their valence electrons unequally. Atoms of a nonpolar molecule share their valence electrons equally.

Water and Other Polar Molecules

In a covalent bond, one atom can attract the shared electrons more strongly than the other atom can. Think about the valence electrons shared between oxygen and hydrogen atoms in a water molecule. The oxygen atom attracts the shared electrons more strongly than each hydrogen atom does. As a result, the shared electrons are pulled closer to the oxygen atom, as shown in **Figure 9.** Since electrons have a negative charge, the oxygen atom has a partial negative charge. The hydrogen atoms have a partial positive charge. *A molecule that has a partial positive end and a partial negative end because of unequal sharing of electrons is a* **polar molecule**.

The charges on a polar molecule affect its properties. Sugar, for example, dissolves easily in water because both sugar and water are polar. The negative end of a water molecule pulls on the positive end of a sugar molecule. Also, the positive end of a water molecule pulls on the negative end of a sugar molecule. This causes the sugar molecules to separate from one another and mix with the water molecules.

Nonpolar Molecules

A hydrogen molecule, H_2, is a nonpolar molecule. Because the two hydrogen atoms are identical, their attraction for the shared electrons is equal. The carbon dioxide molecule, CO_2, in **Figure 9** is also nonpolar. A nonpolar compound will not easily dissolve in a polar compound, but it will dissolve in other nonpolar compounds. Oil is an example of a nonpolar compound. It will not dissolve in water. Have you ever heard someone say "like dissolves like"? This means that polar compounds can dissolve in other polar compounds. Similarly, nonpolar compounds can dissolve in other nonpolar compounds.

WORD ORIGIN

polar
from Latin *polus*, means "pole"

Active Reading **7. Summarize** Fill in the flowchart below to describe the structure of polar molecules.

The _____ sharing of _____ between two or more _____

↓

results in one or more atoms having a _____ _____ charge, and one or more atoms having a _____ _____ charge.

↓

This type of _____ bond causes a molecule to be _____.

Inquiry

SC.8.N.1.1,
SC.8.P.8.5,
SC.8.N.3.1

LAB
STATION **Try It!**

**MiniLab How do com-
pounds form?** at
connectED.mcgraw-hill.com

Apply It!

After you
complete the lab,
answer these
questions.

1. CH_4 is methane. Methane is found
   naturally on Earth as a gas. This gas is
   often referred to as natural gas and is
   commonly used to heat homes. In the
   table below, draw the dot diagram
   and structural formula for this
   molecule.

Dot Diagram	Structural Formula

2. Study the two models that you drew in
   the above table. What does each dot
   and each line represent?

Chemical Formulas and Molecular Models

How do you know which elements make
up a compound? *A* **chemical formula** *is a
group of chemical symbols and numbers that
represent the elements and the number of atoms
of each element that make up a compound.* Just
as a recipe lists ingredients, a chemical
formula lists the elements in a compound.
For example, the chemical formula for
carbon dioxide shown in **Figure 10** is CO_2.
The formula uses chemical symbols that
show which elements are in the compound.
Notice that CO_2 is made up of carbon (C)
and oxygen (O). A subscript, or small
number after a chemical symbol, shows the
number of atoms of each element in the
compound. Carbon dioxide (CO_2) contains
two atoms of oxygen bonded to one atom of
carbon.

A chemical formula describes the types of
atoms in a compound or a molecule, but it
does not explain the shape or appearance of
the molecule. There are many ways to model
a molecule. Each one can show the molecule
in a different way. Common types of models
for CO_2 are shown in **Figure 10.**

Active
Reading **8. State** What information is given in a
chemical formula?

Figure 10 Chemical formulas and molecular models
provide information about molecules.

Chemical Formula

A carbon dioxide molecule is made up
of carbon (C) and oxygen (O) atoms.

A symbol without a subscript
indicates one atom. Each
molecule of carbon dioxide
has one carbon atom.

The subscript 2 indicates
two atoms of oxygen. Each
molecule of carbon dioxide
has two oxygen atoms.

Dot Diagram
• Shows atoms and valence
  electrons

$\ddot{O}::C::\ddot{O}$

Structural Formula
• Shows atoms and lines;
  each line represents one
  shared pair of electrons

$O=C=O$

Ball-and-Stick Model
• Balls represent atoms and
  sticks represent bonds;
  used to show bond angles

Space-Filling Model
• Spheres represent
  atoms; used to show
  three-dimensional
  arrangement of atoms

Visual Summary

$$CO_2$$

A chemical formula is one way to show the elements that make up a compound.

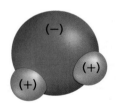

O=C=O

A covalent bond forms when atoms share valence electrons. The smallest particle of a covalent compound is a molecule.

Water is a polar molecule because the oxygen and hydrogen atoms unequally share electrons.

Inquiry

SC.8.P.8.5,
SC.8.P.8.8,
SC.8.N.1.1,
SC.8.N.3.1

LAB STATION Try It!

Skill Lab *How can you model compounds?* at connectED.mcgraw-hill.com

Use Vocabulary

1. **Define** *covalent bond* in your own words.

_____

_____

2. The group of symbols and numbers that shows the types and numbers of atoms that make up a compound is a _____.

Understand Key Concepts 🔑

3. **Contrast** Name at least one way water (H_2O) is different from the elements that make up water. SC.8.P.8.5

_____

_____

4. **Explain** why water is a polar molecule.

_____

_____

5. A sulfur dioxide molecule has one sulfur atom and two oxygen atoms. Which is its correct chemical formula? SC.8.P.8.5

(A) SO_2 (C) S_2O_2

(B) $(SO)_2$ (D) S_2O

Interpret Graphics

6. **Examine** the electron dot diagram for chlorine below.

In chlorine gas, two chlorine atoms join to form a Cl_2 molecule. How many pair(s) of valence electrons do the atoms share? SC.8.P.8.5

_____

7. **Compare and Contrast** Fill in the graphic organizer below to identify at least one way polar and nonpolar molecules are similar and one way they are different. LA.6.2.2.3

_____

Polar and Nonpolar Molecules	
Similarities	
Differences	

Critical Thinking

8. **Develop** an analogy to explain the unequal sharing of valence electrons in a water molecule.

_____

Complete the Lesson 1 and Lesson 2 graphic organizers now. As you read the following lesson come back to this page and complete the graphic organizers for Lesson 3.

Lesson 1

1. **Analyze** details about valence electrons.

farthest from _____	weakest attraction to _____

most energy	Valence Electrons	involved in _____ _____

same number for all elements in _____ (with the exception of _____)

Sequence the steps in constructing and interpreting an electron dot diagram.

2. Identify the element's _____.

3. Identify the number of _____,
   which is the same as the _____ of
   the _____.

4. Place _____ dot at a time on each
   _____ of the _____.
   Pair up the dots until all are used.

5. Identify an atom as _____ if all
   _____ are _____.

6. Count the _____ to determine
   how many _____ an unstable
   atom can form.

Lesson 2

7. **Summarize** the relationship between an electron's energy level and its location in an atom. (Circle) the word that makes each statement true.

The closer to the nucleus, the lower / higher an electron's energy level.	The farther from the nucleus, the lower / higher an electron's energy level.

8. **Describe** types of covalent bonds.

Covalent Bond	Description of Valence Electron Sharing	Comment on the Strength of the Bond
Single		weakest type of covalent bond
Double		
Triple		

9. **Identify** four common properties of covalent compounds.

_____

_____

_____

Lesson 3

10. **Describe** three properties of metallic compounds.

Properties of Metallic Compounds

11. **Contrast** three ways atoms can bond and become stable.

Process	Electron Pooling	Electron Transfer	Electron Sharing
Type of chemical bond	metallic	ionic	covalent
Description			

Ionic and Metallic
BONDS

ESSENTIAL QUESTIONS

 What is an ionic compound?

 How do metallic bonds differ from covalent and ionic bonds?

Vocabulary

ion p. 398

ionic bond p. 400

metallic bond p. 401

 Florida NGSSS

SC.6.N.1.4 Discuss, compare, and negotiate methods used, results obtained, and explanations among groups of students conducting the same investigation.

SC.6.N.1.5 Recognize that science involves creativity, not just in designing experiments, but also in creating explanations that fit evidence.

SC.8.N.1.6 Understand that scientific investigations involve the collection of relevant empirical evidence, the use of logical reasoning, and the application of imagination in devising hypotheses, predictions, explanations and models to make sense of the collected evidence.

SC.8.N.3.1 Select models useful in relating the results of their own investigations.

SC.8.N.4.1 Explain that science is one of the processes that can be used to inform decision making at the community, state, national, and international levels.

SC.8.P.8.5 Recognize that there are a finite number of elements and that their atoms combine in a multitude of ways to produce compounds that make up all of the living and nonliving things that we encounter.

SC.8.P.8.8 Identify basic examples of and compare and classify the properties of compounds, including acids, bases, and salts.

Also covers: SC.8.N.1.1, LA.6.2.2.3, MA.6.A.3.6

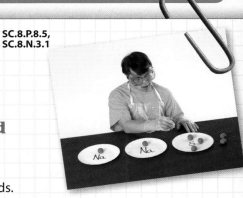

Inquiry **Launch Lab**

SC.8.P.8.5,
SC.8.N.3.1

15 minutes

How can atoms form compounds by gaining and losing electrons?

Metals on the periodic table often lose electrons when forming stable compounds. Nonmetals often gain electrons.

Procedure

1. Read and complete a lab safety form.

2. Make two model atoms of sodium and one model atom each of calcium, chlorine, and sulfur. To do this, write each element's chemical symbol with a **marker** on a **paper plate.** Surround the symbol with small balls of **clay** to represent valence electrons. Use one color of clay for the metals (groups 1 and 2 elements) and another color of clay for nonmetals (groups 16 and 17 elements).

3. To model sodium sulfide (Na_2S), place the two sodium atoms next to the sulfur atom. To form a stable compound, move each sodium atom's valence electron to the sulfur atom.

4. Form as many other compound models as you can by removing valence electrons from the groups 1 and 2 plates and placing them on the groups 16 and 17 plates.

Think About This

1. **Name** What other compounds were you able to form?

2. **Key Concept** **Differentiate** How do you think your models are different from covalent compounds?

1. This scene might look like snow along a shoreline, but it is actually thick deposits of salt on a lake. Over time, salt dissolved in river water flowed into this lake and built up as water evaporated. Salt is a compound that forms when elements form bonds by gaining or losing valence electrons, not sharing them. How do humans use this common substance?

Active Reading **2. Explain** Why do atoms that gain electrons become an ion with a negative charge?

Active Reading

FOLDABLES® LA.6.2.2.3

Make two quarter-sheet note cards as shown. Use the cards to summarize information about ionic and metallic compounds.

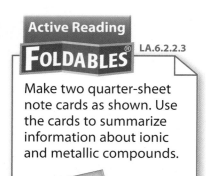

Understanding Ions

As you read in Lesson 2, the atoms of two or more nonmetals form compounds by sharing valence electrons. However, when a metal and a nonmetal bond, they do not share electrons. Instead, one or more valence electrons transfers from the metal atom to the nonmetal atom. After electrons transfer, the atoms bond and form a chemically stable compound. Transferring valence electrons results in atoms with the same number of valence electrons as a noble gas.

When an atom loses or gains a valence electron, it becomes an ion. *An **ion** is an atom that is no longer electrically neutral because it has lost or gained valence electrons.* Because electrons have a negative charge, losing or gaining an electron changes the overall charge of an atom. An atom that loses valence electrons becomes an ion with a positive charge. This is because the number of electrons is now less than the number of protons in the atom. An atom that gains valence electrons becomes an ion with a negative charge. This is because the number of protons is now less than the number of electrons.

Losing Valence Electrons

Look at the periodic table on the inside back cover of this book. What information about sodium (Na) can you infer from the periodic table? Sodium is a metal. Its atomic number is 11. This means each sodium atom has 11 protons and 11 electrons. Sodium is in group 1 on the periodic table. Therefore, sodium atoms have one valence electron, and they are chemically unstable.

Metal atoms, such as sodium, become more stable when they lose valence electrons and form a chemical bond with a nonmetal. If a sodium atom loses its one valence electron, it will have a total of ten electrons. Which element on the periodic table has atoms with ten electrons? Neon (Ne) atoms have a total of ten electrons. Eight of these are valence electrons. When a sodium atom loses one valence electron, the electrons in the next-lower energy level are now the new valence electrons. The sodium atom then has eight valence electrons, the same as the noble gas neon, and is chemically stable.

Gaining Valence Electrons

In Lesson 2, you read that nonmetal atoms can share valence electrons with other non-metal atoms. Nonmetal atoms can also gain valence electrons from metal atoms. Either way, they achieve the electron arrangement of a noble gas. Find the nonmetal chlorine (Cl) on the periodic table. Its atomic number is 17. Atoms of chlorine have seven valence electrons. If a chlorine atom gains one valence electron, it will have eight valence electrons. It will also have the same electron arrangement as the noble gas argon (Ar).

When a sodium atom loses a valence electron, it becomes a positively charged ion. This is shown by a plus (+) sign. When a chlorine atom gains a valence electron, it becomes a negatively charged ion. This is shown by a negative (−) sign. **Figure 11** illustrates this process.

Active Reading   **3. Choose** Are atoms of group 16 elements more likely to gain or lose valence electrons?

Losing and Gaining Electrons

Figure 11 Sodium atoms have a tendency to lose a valence electron. Chlorine atoms have a tendency to gain a valence electron.

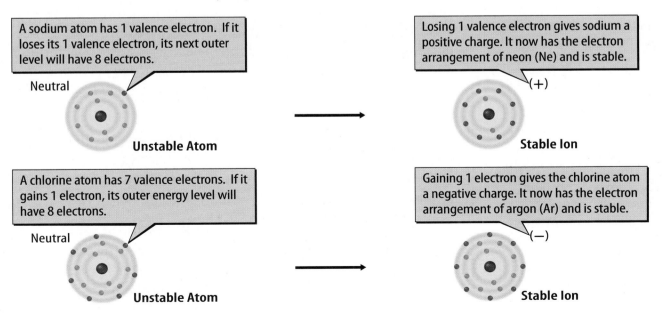

A sodium atom has 1 valence electron. If it loses its 1 valence electron, its next outer level will have 8 electrons.

Neutral

Unstable Atom

Losing 1 valence electron gives sodium a positive charge. It now has the electron arrangement of neon (Ne) and is stable.

(+)

Stable Ion

A chlorine atom has 7 valence electrons. If it gains 1 electron, its outer energy level will have 8 electrons.

Neutral

Unstable Atom

Gaining 1 electron gives the chlorine atom a negative charge. It now has the electron arrangement of argon (Ar) and is stable.

(−)

Stable Ion

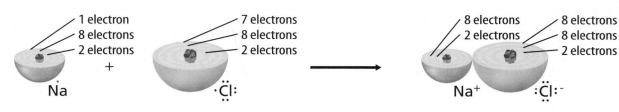

Figure 12 🔑 An ionic bond forms between Na and Cl when an electron transfers from Na to Cl.

A sodium atom loses one valence electron and becomes stable. A chlorine atom gains one valence electron and becomes stable.

The positively charged sodium ion and the negatively charged chlorine ion attract each other. Together they form a strong ionic bond.

Determining an Ion's Charge

Atoms are electrically neutral because they have the same number of protons and electrons. Once an atom gains or loses electrons, it becomes a charged ion. For example, the atomic number for nitrogen (N) is 7. Each N atom has 7 protons and 7 electrons and is electrically neutral. However, an N atom often gains 3 electrons when forming an ion. The N ion then has 10 electrons. To determine the charge, subtract the number of electrons in the ion from the number of protons.

7 protons − 10 electrons = −3 charge

A nitrogen ion has a −3 charge. This is written as N^{3-}.

Ionic Bonds—Electron Transferring

Recall that metal atoms typically lose valence electrons and nonmetal atoms typically gain valence electrons. When forming a chemical bond, the nonmetal atoms gain the electrons lost by the metal atoms. Take a look at **Figure 12.** In NaCl, or table salt, an atom of the metal sodium loses a valence electron. The electron is transferred to a nonmetal chlorine atom. The sodium atom becomes a positively charged ion. The chlorine atom becomes a negatively charged ion. These ions attract each other and form a stable ionic compound. *The attraction between positively and negatively charged ions in an ionic compound is an* **ionic bond**.

Active Reading **4. Identify** Underline Mark the sentences in the text that explain how metal atoms and nonmetal atoms chemically bond.

Ionic Compounds

Ionic compounds are usually solid and brittle at room temperature. They also have relatively high melting and boiling points. Many ionic compounds dissolve in water. Water that contains dissolved ionic compounds is a good conductor of electricity. This is because an electrical charge can pass from ion to ion in the solution.

Math Skills   MA.6.A.3.6

Use Percentage

A picometer (pm) is 1 trillion times smaller than a meter. When an atom becomes an ion, its radius increases or decreases. For example, a Na atom has a radius of **186 pm**. A Na⁺ ion has a radius of **102 pm**. By what percentage does the radius change?

Subtract the atom's radius from the ion's radius.

102 pm − 186 pm = _____

Divide the difference above by the atom's radius.

−84 pm ÷ 186 pm = _____

Multiply the answer above by 100 and add a % sign.

−0.45 × 100 = _____

A negative value is a decrease in size. A positive value is an increase.

Practice

5. The radius of an oxygen (O) atom is 73 pm. The radius of an oxygen ion (O^{2-}) is 140 pm. By what percentage does the radius change?

6. Analyze What happens to sodium and chlorine atoms in the formation of the compound sodium chloride?

	Na (sodium)	**Cl** (chlorine)
Type of element		
Atomic number		
Number of valence electrons		
Stable or unstable?		
Loses or gains electrons		
Type of ion		

Comparing Ionic and Covalent Compounds

Recall that in a covalent bond, two or more nonmetal atoms share electrons and form a unit, or molecule. Covalent compounds, such as water, are made up of many molecules. However, when nonmetal ions bond to metal ions in an ionic compound, there are no molecules. Instead, there is a large collection of oppositely charged ions. All of the ions attract each other and are held together by ionic bonds.

Metallic Bonds— Electron Pooling

Recall that metal atoms typically lose valence electrons when forming compounds. What happens when metal atoms bond to other metal atoms? Metal atoms form compounds with one another by combining, or pooling, their valence electrons. *A **metallic bond** is a bond formed when many metal atoms share their pooled valence electrons.*

The pooling of valence electrons in aluminum is shown in **Figure 13**. The aluminum atoms lose their valence electrons and become positive ions, indicated by the plus (+) signs. The negative (−) signs indicate the valence electrons, which move from ion to ion. Valence electrons in metals are not bonded to one atom. Instead, a "sea of electrons" surrounds the positive ions.

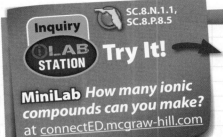

Inquiry **Try It!** **Apply It!**

MiniLab *How many ionic compounds can you make?* at connectED.mcgraw-hill.com

SC.8.N.1.1, SC.8.P.8.5

After you complete the lab, answer these questions.

1. Why are electric wires not made of ionic substances?

2. Explain why two sodium ions (Na^+) do not bond to form an ionic compound.

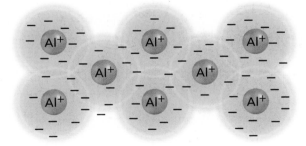

Figure 13 Valence electrons move among all the aluminum (Al) ions.

 7. NGSSS Check Explain How do metal atoms bond with one another? SC.8.P.8.5

_____

_____

6. Analyze What happens to sodium and chlorine atoms in the formation of the compound sodium chloride?

Lesson 3 • EXPLAIN **401**

Properties of Metallic Compounds

Metals are good conductors of thermal energy and electricity. Because the valence electrons can move from ion to ion, they can easily **conduct** an electric charge. When a metal is hammered into a sheet or drawn into a wire, it does not break. The metal ions can slide past one another in the electron sea and move to new positions. Metals are shiny because the valence electrons at the surface of a metal interact with light. **Table 1** compares the covalent, ionic, and metallic bonds that you studied in this chapter.

Active Reading

8. Paraphrase How does valence electron pooling explain why metals can be hammered into a sheet?

_____

_____

Active Reading

9. Choose Use the correct words to fill in the blanks to complete **Table 1.**

Table 1 Covalent, Ionic, and Metallic Bonds

Type of Bond	What is bonding?	Properties of Compounds
Water	nonmetal atoms; nonmetal atoms	• gas, liquid, or solid • low melting and boiling points • often not able to dissolve in water • poor conductors of thermal energy and electric current • dull appearance
Salt (Na⁺ Cl⁻)	_____ ions; _____ ions	• solid crystals • high melting and boiling points • dissolves in water • solids are poor conductors of thermal energy and electric current • ionic compounds in water solutions conduct electric current
Aluminum	_____ ions; _____ ions	• usually solid at room temperature • high melting and boiling points • do not dissolve in water • good conductors of thermal energy and electric current • shiny surface • can be hammered into sheets and pulled into wires

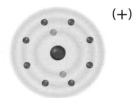

(+)

Metal atoms lose electrons, and nonmetal atoms gain electrons and form stable compounds. An atom that has gained or lost an electron is an ion.

(+)   (−)

Na⁺   :Cl:⁻

An ionic bond forms between positively and negatively charged ions.

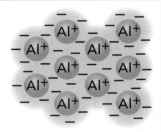

A metallic bond forms when many metal atoms share their pooled valence electrons.

Inquiry

SC.8.P.8.5, SC.8.N.1.1, SC.8.N.1.6, LA.6.2.2.3

LAB STATION Try It!

Inquiry Lab *Ions in Solution* at connectED.mcgraw-hill.com

Use Vocabulary

1 **Define** *ionic bond* in your own words.

_____

_____

2 An atom that changes so that it has an electric charge is a(n) _____ .

3 **Use the term** *metallic bond* in a sentence.

_____

_____

Understand Key Concepts 🔑

4 **Recall** What holds ionic compounds together? SC.8.P.8.5

_____

5 Which element would most likely bond with lithium and form an ionic compound? SC.8.P.8.5

(A) beryllium   (C) fluorine

(B) calcium   (D) sodium

6 **Contrast** Why are metals good conductors of electricity but covalent compounds are poor conductors? SC.8.P.8.6

_____

_____

_____

Interpret Graphics

7 **Organize** In each oval, list a common property of an ionic compound. LA.6.2.2.3

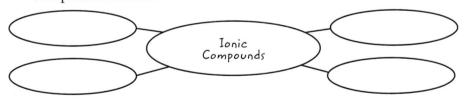

Ionic Compounds

Critical Thinking

8 **Evaluate** What type of bonding does a material most likely have if it has a high melting point, is solid at room temperature, and easily dissolves in water?

Math Skills   MA.6.A.3.6

9 The radius of the aluminum (Al) atom is 143 pm. The radius of the aluminum ion (Al³⁺) is 54 pm. By what percentage did the radius change as the ion formed?

Chapter 10 Study Guide

Think About It! There are a finite number of elements, and their atoms join together through sharing, transferring, or pooling their electrons to make chemical compounds.

Key Concepts Summary

LESSON 1 Electrons and Energy Levels

- Electrons with more energy are farther from the atom's nucleus and are in a higher energy level.
- Atoms with fewer than eight **valence electrons** gain, lose, or share valence electrons and form stable compounds. Atoms in stable compounds have the same electron arrangement as a noble gas.

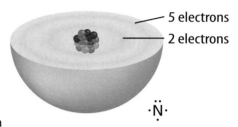

5 electrons

2 electrons

·N·̈

chemical bond p. 382
valence electron p. 384
electron dot diagram p. 385

LESSON 2 Compounds, Chemical Formulas, and Covalent Bonds

- A compound and the elements it is made from have different chemical and physical properties.
- A **covalent bond** forms when two nonmetal atoms share valence electrons. Common properties of covalent compounds include low melting points and low boiling points. They are usually gas or liquid at room temperature and poor conductors of electricity.
- Water is a polar compound because the oxygen atom pulls more strongly on the shared valence electrons than the hydrogen atoms do.

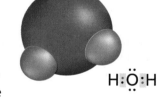

H:Ö:H

covalent bond p. 391
molecule p. 392
polar molecule p. 393
chemical formula p. 394

LESSON 3 Ionic and Metallic Bonds

- **Ionic bonds** form when valence electrons move from a metal atom to a nonmetal atom.
- An ionic compound is held together by ionic bonds, which are attractions between positively and negatively charged **ions**.
- A **metallic bond** forms when valence electrons are pooled among many metal atoms.

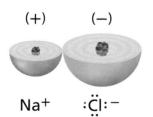

(+)   (−)

Na⁺   :C̈l:⁻

ion p. 398
ionic bond p. 400
metallic bond p. 401

FOLDABLES **Chapter Project**

Assemble your lesson Foldables as shown to make a Chapter Project. Use the project to review what you have learned in this chapter.

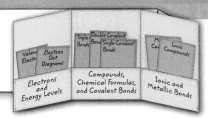

Use Vocabulary

1 The force that holds atoms together is called a(n) _____ .

2 You can predict the number of bonds an atom can form by drawing its _____ .

3 The nitrogen and hydrogen atoms that make up ammonia (NH_3) are held together by a(n) _____ because the atoms share valence electrons unequally.

4 Two hydrogen atoms and one oxygen atom together are a _____ of water.

5 A positively charged sodium ion and a negatively charged chlorine ion are joined by a(n) _____ to form the compound sodium chloride.

Link Vocabulary and Key Concepts

Use vocabulary terms from the previous page and other terms from the chapter to complete the concept map.

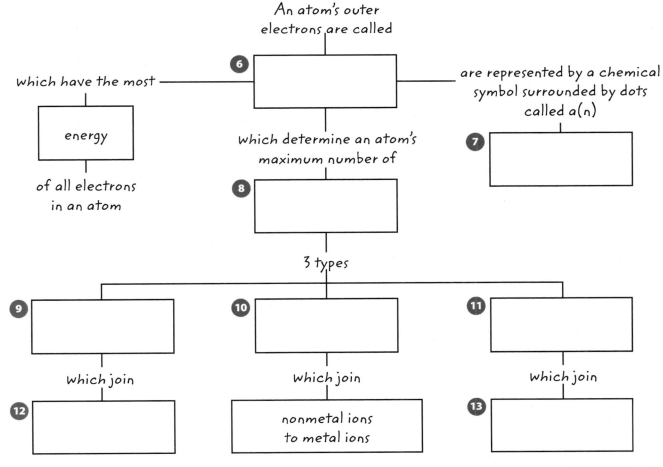

Fill in the correct answer choice.

🔑 Understand Key Concepts

1 Atoms lose, gain, or share electrons and become as chemically stable as SC.8.P.8.5
  Ⓐ an electron.
  Ⓑ an ion.
  Ⓒ a metal.
  Ⓓ a noble gas.

2 Which is the correct electron dot diagram for boron, one of the group 13 elements? SC.8.P.8.5

  Ⓐ Ḃ·

  Ⓑ ·B̤:

  Ⓒ :B̤:

  Ⓓ ·B̤·

3 If an electron transfers from one atom to another atom, what type of bond will most likely form? SC.8.P.8.5
  Ⓐ covalent
  Ⓑ ionic
  Ⓒ metallic
  Ⓓ polar

4 What change would make an atom represented by this diagram have the same electron arrangement as a noble gas? SC.8.P.8.5

  Ⓐ gaining two electrons
  Ⓑ gaining four electrons
  Ⓒ losing two electrons
  Ⓓ losing four electrons

5 What would make bromine, a group 17 element, more similar to a noble gas? SC.8.P.8.5
  Ⓐ gaining one electron
  Ⓑ gaining two electrons
  Ⓒ losing one electron
  Ⓓ losing two electrons

Critical Thinking

6 **Classify** Use the periodic table to classify the elements potassium (K), bromine (Br), and argon (Ar) according to how likely their atoms are to do the following: SC.8.P.8.6

  a. lose electrons to form positive ions
  _____

  b. gain electrons to form negative ions
  _____

  c. neither gain nor lose electrons
  _____

7 **Describe** the change that is shown in this illustration. How does this change affect the stability of the atom? SC.8.P.8.5

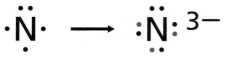

  _____
  _____
  _____

8 **Analyze** One of your classmates draws an electron dot diagram for a helium atom with two dots. He tells you that these dots mean each helium atom has two unpaired electrons and can gain, lose, or share electrons to have four pairs of valence electrons and become stable. What is wrong with your classmate's argument? SC.8.P.8.5

  _____
  _____
  _____
  _____

9 **Explain** why the hydrogen atoms in a hydrogen gas molecule (H_2) form nonpolar covalent bonds but the oxygen and hydrogen atoms in water molecules (H_2O) form polar covalent bonds. SC.8.P.8.5

  _____
  _____
  _____

10 **Contrast** Why is it possible for an oxygen atom to form a double covalent bond, but it is not possible for a chlorine atom to form a double covalent bond? SC.8.P.8.5

_____

_____

_____

_____

_____

_____

Writing in Science

11 **Compose** On a separate piece of paper, write a poem at least ten lines long that explains ionic bonding, covalent bonding, and metallic bonding. LA.6.2.2.3

Big Idea Review

12 Which types of atoms pool their valence electrons to form a "sea of electrons"? SC.8.P.8.5

_____

_____

13 Describe a way in which elements joining together to form chemical compounds is similar to the way the letters on a computer keyboard join together to form words. SC.8.P.8.5

_____

_____

_____

_____

_____

Math Skills MA.6.A.3.6

Element	Atomic Radius	Ionic Radius
Potassium (K)	227 pm	133 pm
Iodine (I)	133 pm	216 pm

14 What is the percent change when an iodine atom (I) becomes an ion (I-)?

15 What is the percent change when a potassium atom (K) becomes an ion (K^+)?

Fill in the correct answer choice.

Multiple Choice

1 Most atoms are chemically stable when they have how many valence electrons? SC.8.P.8.7

(A) 8

(B) 7

(C) 4

(D) 1

Use the diagram below to answer question 2.

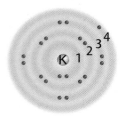

2 The diagram above shows a potassium atom. Which is the second-highest energy level? SC.8.P.8.7

(F) 1

(G) 2

(H) 3

(I) 4

3 On the periodic table, which groups of elements tend to form positive ions? SC.8.P.8.6

(A) Group 1 and Group 2

(B) Group 16 and Group 17

(C) Group 1 and Group 16

(D) Group 2 and Group 17

4 What is the electron diagram for the ionic compound sodium fluoride (NaF)? SC.8.P.8.6

(F) [Na]+ [F]−

(G) [Na]+ [:F:]−

(H) [Na:]+ [F:]−

(I) [Na:]+ [F:]−

Use the diagram below to answer question 5.

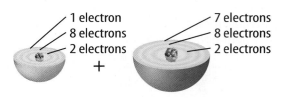

5 The atoms in the diagram above are forming a bond. Which represents that bond? SC.8.P.8.5

(A)

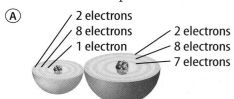

(B)

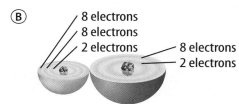

(C)

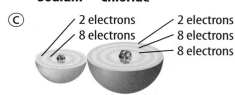

(D)

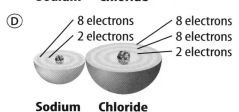

6 Covalent bonds typically form between the atoms of elements that share SC.8.P.8.5

(F) nuclei.

(G) oppositely charged ions.

(H) protons.

(I) valence electrons.

Use the diagram below to answer question 7.

Water Molecule

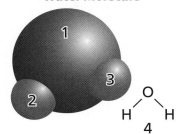

7 In the diagram above, which shows an atom with a partial negative charge? SC.8.P.8.7

Ⓐ 1

Ⓑ 2

Ⓒ 3

Ⓓ 4

8 Which compound is formed by the attraction between negatively and positively charged ions? SC.8.P.8.5

Ⓕ bipolar

Ⓖ covalent

Ⓗ ionic

Ⓘ nonpolar

9 The atoms of noble gases do NOT bond easily with other atoms because their valence electrons are SC.8.P.8.5

Ⓐ absent.

Ⓑ moving.

Ⓒ neutral.

Ⓓ stable.

Use the diagram below to answer question 10.

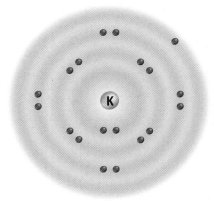

10 How many electrons does potassium, shown above, need to gain or lose to become stable? SC.8.P.8.6

Ⓕ gain 1

Ⓖ gain 2

Ⓗ lose 1

Ⓘ lose 2

11 What is the number of the group in which the elements have a stable outer energy level? SC.8.P.8.6

Ⓐ 1

Ⓑ 13

Ⓒ 16

Ⓓ 18

12 In what ways are the elements in a group similar? SC.8.P.8.6

Ⓕ atomic numbers

Ⓖ atomic masses

Ⓗ chemical properties

Ⓘ symbols

NEED EXTRA HELP?

If You Missed Question...	1	2	3	4	5	6	7	8	9	10	11	12
Go to Lesson...	2	1	3	1	3	2	2	3	1	1	1	1

Multiple Choice *Bubble the correct answer.*

Use the table below to answer questions 1 and 2.

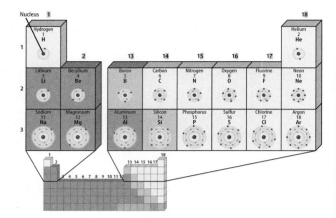

1. According to the table, how many valence electrons does phosphorus have? **SC.8.P.8.6**

(A) 3

(B) 4

(C) 5

(D) 6

2. Which element listed in the table is likely to be shiny and able to conduct electricity and heat? **SC.8.P.8.6**

(F) neon

(G) sulfur

(H) beryllium

(I) nitrogen

3. The element selenium is a nonmetal. It is an unstable atom that can form two bonds with other atoms. What group is selenium in? **SC.8.P.8.6**

(A) group 15

(B) group 16

(C) group 17

(D) group 18

4. The image above shows an electron dot diagram for silicon. How many total bonds can one atom of this metalloid make with other atoms? **SC.8.P.8.5**

(F) 1

(G) 2

(H) 3

(I) 4

Copyright © Glencoe/McGraw-Hill, a division of The McGraw-Hill Companies, Inc.

Benchmark Mini-Assessment   **Chapter 8 • Lesson 2**

Multiple Choice *Bubble the correct answer.*

1. Which electron dot diagram shows how hydrogen and oxygen are bonded together in the compound H_2O? **SC.8.P.8.5**

(A) :Ö:H:H

(B) H:Ö:H

(C) H: :O: :H

(D) H·Ö·H

2. Which has the strongest covalent bonds? **SC.8.P.8.5**

(F) CO_2

(G) H_2O

(H) H_2

(I) N_2

3. Which compound contains four atoms of hydrogen for every one atom of carbon? **SC.8.P.8.5**

(A) $2CH_4$

(B) $4CH$

(C) C_4H

(D) CH_4

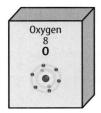

4. The diagram above shows the valence electrons in one atom of an element. If two atoms of this element joined together to create a molecule, they **SC.8.P.8.5**

(F) would form a double covalent bond.

(G) would form a quadruple covalent bond.

(H) would form a single covalent bond.

(I) would form a triple covalent bond.

Copyright © Glencoe/McGraw-Hill, a division of The McGraw-Hill Companies, Inc.

Benchmark Mini-Assessment   Chapter 8 • Lesson 3

Multiple Choice *Bubble the correct answer.*

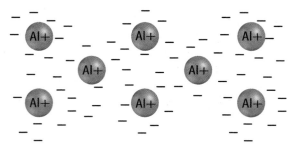

1. The atoms in the diagram above are **SC.8.P.8.5**

(A) losing valence electrons.

(B) pooling valence electrons.

(C) sharing pairs of valence electrons.

(D) transferring valence electrons.

2. How does an atom become an ion? **SC.8.P.8.5**

(F) An atom forms a compound with another atom.

(G) An atom is joined to another atom with a covalent bond.

(H) An atom loses or gains a valence electron.

(I) An atom shares a pair of valence electrons with another atom.

3. Which two elements are likely to form a compound that has ionic bonds? **SC.8.P.8.5**

(A) carbon and oxygen

(B) nitrogen and oxygen

(C) sodium and chlorine

(D) sulfur and hydrogen

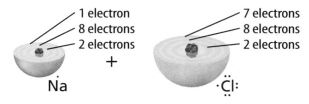

4. In the image above, if sodium loses one electron and chlorine gains one electron, how many valance electrons do sodium and chlorine now have? **SC.8.P.8.5**

(F) Sodium and chlorine each have 8 valence electrons.

(G) Sodium and chlorine each have 10 valence electrons.

(H) Sodium has 7 valence electrons, and chlorine has 9 valence electrons.

(I) Sodium has 11 valence electrons, and chlorine has 17 valence electrons.

Copyright © Glencoe/McGraw-Hill, a division of The McGraw-Hill Companies, Inc.

Name _____ Date _____

INTERACTIONS OF MATTER

WITH A GIANT LEMON HEADING FOR EARTH'S OCEANS...

...THE WORLD IS IN A PANIC.

THE PLANET'S TOP SCIENTISTS CALL AN EMERGENCY MEETING.

"WE NEED A BASE TO NEUTRALIZE THE ACID."

1000 B.C.
Chemistry is considered more of an art than a science. Chemical arts include the smelting of metals and the making of drugs, dyes, iron, and bronze.

1661
A clear distinction is made between chemistry and alchemy when *The Sceptical Chymist* is published by Robert Boyle. Modern chemistry begins to emerge.

1789
Antoine Lavoisier, the "father of modern chemistry," clearly outlines the law of conservation of mass.

1803
John Dalton publishes his atomic theory, which states that all matter is composed of atoms, which are small and indivisible and can join together to form chemical compounds. Dalton is considered the originator of modern atomic theory.

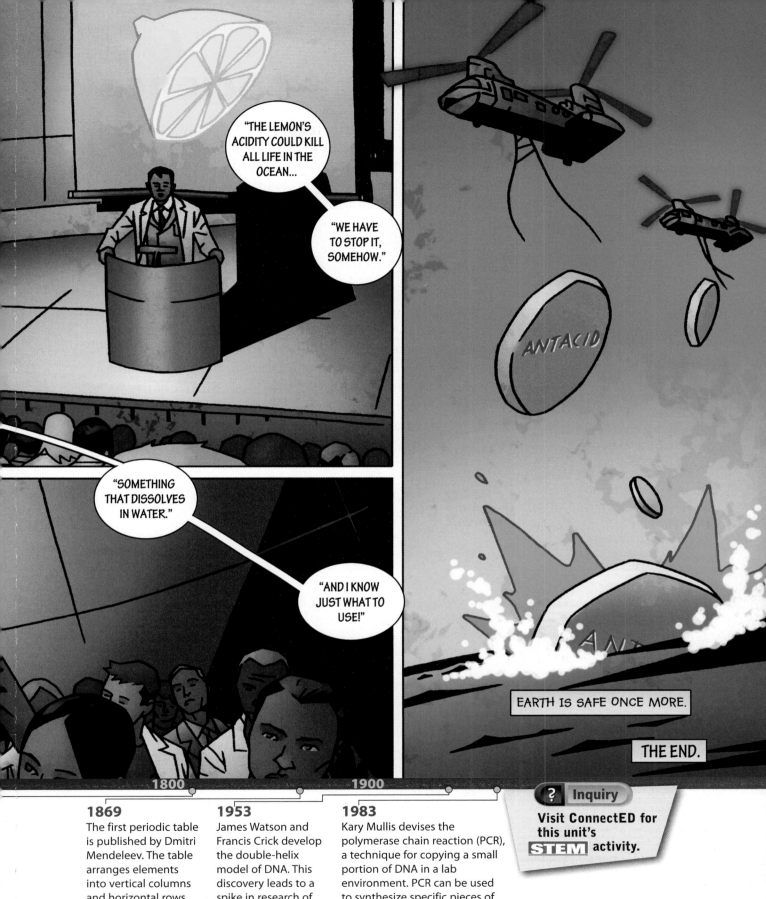

Health and Science

Have an upset stomach? Chew on some charcoal. Have a headache? Rub a little peppermint oil on your temples. As shown in **Figure 1,** people have used chemicals to fix physical ailments for thousands of years. Many cures were discovered by accident. People did not understand why the cures worked, only that they did work.

Asking Questions About Health

Over time, people asked questions about which cures worked and which cures did not work. They made observations, recorded their findings, and had discussions. This process was the start of the scientific investigation of health. **Health** is the overall condition of an organism or group of organisms at any given time. Early studies focused on treating the physical parts of the body. The study of how chemicals interact in organisms and affect health opened a whole new field of study known as biochemistry. The time line in **Figure 2** shows some of the discoveries people made that led to the development of medicines.

▲ **Figure 1** Thousands of years ago, people believed that evil spirits were responsible for illness. They often treated the physical symptoms with herbs or other natural materials.

Figure 2 The time line shows several significant discoveries and developments in the history of medicine.

Active Reading

1. Discuss How has the process of scientific investigation of health improved people's health?

_____

_____

4,200 years ago Clay tablets describe using sesame oil on wounds to treat infection.

More than 3,300 years later, scientists found that a chemical in mold broke down the cell membranes of bacteria, killing them. Similar discoveries led to the development of antibiotics.

1740s A doctor found that the disease called scurvy was caused by a lack of Vitamin C.

3,500 years ago An ancient papyrus described how Egyptians applied moldy bread to wounds to prevent infection.

Early explorers on long sea voyages often lost their teeth or developed deadly sores. Ships could not carry many fruits and vegetables, which contain Vitamin C, because they spoil quickly. Scientists suspect that many early explorers might have died because their diets did not include the proper vitamins.

Year 900 The first pharmacy opened in Persia, which is now Iraq.

2,500 years ago Hippocrates, the "Father of Medicine," is the first physician known to separate medical knowledge from myth and superstition.

Benefits and Risks of Medicines

Scientists might recognize that a person's body is missing a necessary chemical, but they cannot always fix the problem. For example, people used to get necessary vitamins and minerals by eating natural, whole foods. Today, food processing destroys many nutrients. Foods might last longer, but they might not provide nutrients the body needs.

Researchers still do not understand the role of many chemicals in the body. Taking a medicine to fix one problem sometimes causes others, called side effects. For example, antibiotics kill some disease-causing bacteria. However, widespread use of antibiotics has resulted in "super bugs"—bacteria that are resistant to treatment.

Histamines are chemicals that have many functions in the body, including regulating sleep and decreasing sensitivity to allergens. However, low levels of histamines have been linked to some serious illnesses. Many medicines have long-term effects on health. Before you take a medicine, recognize that you are adding a chemical to your body. Be as informed as possible about any side effects.

Active Reading

2. Recognize What are some possible advantages and disadvantages of taking medicines?

_____

_____

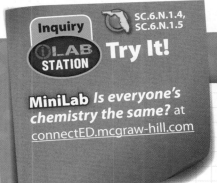

Inquiry SC.6.N.1.4, SC.6.N.1.5

iLAB STATION **Try It!**

MiniLab *Is everyone's chemistry the same?* at connectED.mcgraw-hill.com

Apply It! After you complete the lab, answer this question.

1. **Analyze** Discuss why doctors should not always prescribe the same medicine for every patient. Include information about personal biochemistry and pH, as well as prevention of side effects and medication-resistant "super bugs."

Scientists studying digestion in dogs noticed that ants were attracted to the urine of a dog whose pancreas had been removed. They determined the dog's urine contained sugar, which attracted ants. Eventually, scientists discovered that diabetes resulted from a lack of insulin, a chemical produced in the pancreas that regulates blood sugar. Today, some people with diabetes wear an insulin pump that monitors their blood sugar and delivers insulin to their bodies.

1770s The first vaccination is developed and administered.

1800s Nitrous oxide is first used as an anesthetic by dentists.

1920s Insulin is identified as the missing hormone in people with diabetes.

1920s Penicillin is discovered, but not developed for treatment of disease until the mid-1940s.

2000s First vaccine to target a cause of cancer

Name _____ Date _____

Chemical Reaction

When you mix two different substances together, sometimes a chemical reaction occurs that forms a new substance. Which best describes what happens during a chemical reaction when two different substances form a new substance?

    A. Some of the atoms are destroyed and replaced by new atoms.

    B. All of the atoms are destroyed and all new atoms form.

    C. None of the atoms are destroyed but some new atoms are added.

    D. None of the atoms are destroyed and no new atoms form.

Explain your thinking. Describe what you think happens to the atoms during a chemical reaction.

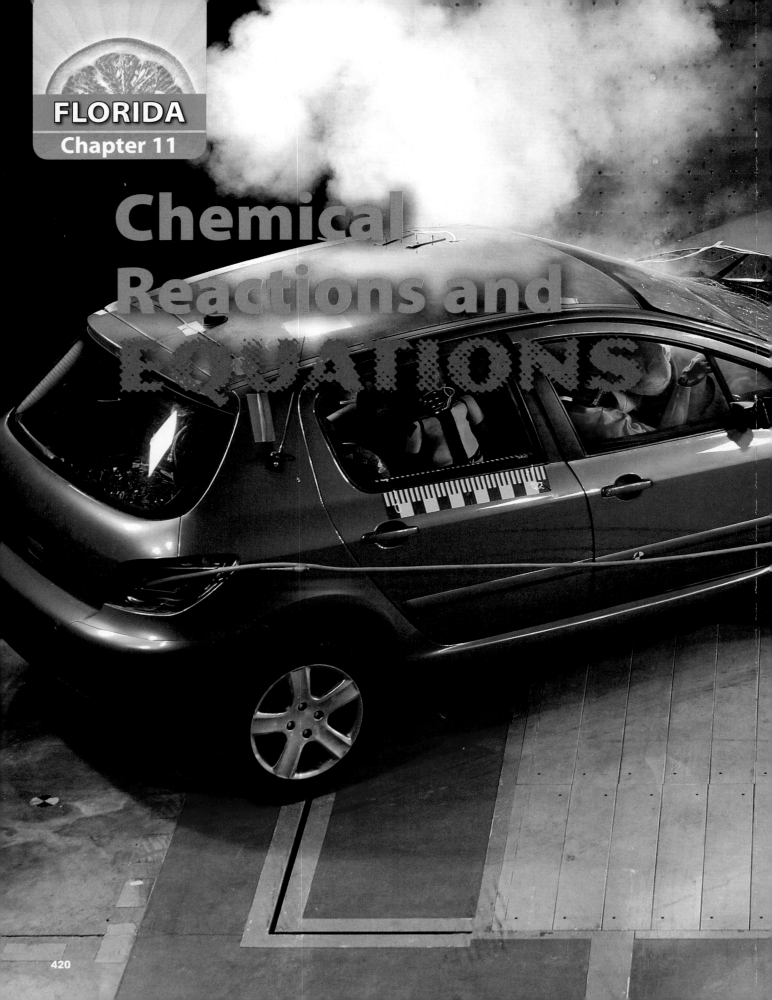

Chemical Reactions and EQUATIONS

FLORIDA BIG IDEAS

1 **The Practice of Science**
2 **The Characteristics of Scientific Knowledge**
3 **The Role of Theories, Laws, Hypotheses, and Models**
9 **Changes in Matter**
11 **Energy Transfer and Transformations**

| Think About It! | **What happens to atoms and energy during a chemical reaction?** |

An air bag deploys in less than the blink of an eye. How does the bag open so fast? At the moment of impact, a sensor triggers a reaction between two chemicals. This reaction quickly produces a large amount of nitrogen gas. This gas inflates the bag with a pop.

1 A chemical reaction can produce a gas. How is this different from a gas produced when a liquid boils?

_____

_____

_____

2 What do you think happens to atoms and energy during a chemical reaction?

_____

_____

_____

_____

| Get Ready to Read | **What do you think about chemical reactions?** |

Before you read, decide if you agree or disagree with each of these statements. As you read this chapter, see if you change your mind about any of the statements.

	AGREE	DISAGREE
1 If a substance bubbles, you know a chemical reaction is occurring.	☐	☐
2 During a chemical reaction, some atoms are destroyed and new atoms are made.	☐	☐
3 Reactions always start with two or more substances that react with each other.	☐	☐
4 Water can be broken down into simpler substances.	☐	☐
5 Reactions that release energy require energy to get started.	☐	☐
6 Energy can be created in a chemical reaction.	☐	☐

Connect ED **There's More Online!**
Video • Audio • Review • ⓘLab Station • WebQuest • Assessment • Concepts in Motion • Multilingual eGlossary

Understanding Chemical REACTIONS

ESSENTIAL QUESTIONS

 What are some signs that a chemical reaction might have occurred?

 What happens to atoms during a chemical reaction?

What happens to the total mass in a chemical reaction?

Vocabulary

chemical reaction p. 423

chemical equation p. 426

reactant p. 427

product p. 427

law of conservation of mass p. 428

coefficient p. 430

 Florida NGSSS

SC.7.N.1.6 Explain that empirical evidence is the cumulative body of observations of a natural phenomenon on which scientific explanations are based.

SC.8.N.1.1 Define a problem from the eighth grade curriculum using appropriate reference materials to support scientific understanding, plan and carry out scientific investigations of various types, such as systematic observations or experiments, identify variables, collect and organize data, interpret data in charts, tables, and graphics, analyze information, make predictions, and defend conclusions.

SC.8.N.3.1 Select models useful in relating the results of their own investigations.

SC.8.P.9.1 Explore the Law of Conservation of Mass by demonstrating and concluding that mass is conserved when substances undergo physical and chemical changes.

SC.8.P.9.2 Differentiate between physical changes and chemical changes.

LA.6.2.2.3 The student will organize information to show understanding or relationships among facts, ideas, and events (e.g., representing key points within text through charting, mapping, paraphrasing, summarizing, or comparing/contrasting);

 Inquiry **Launch Lab**

SC.8.P.9.1

15–20 minutes

Where did it come from?

Does a boiled egg have more mass than a raw egg? What happens when liquids change to a solid?

Procedure

1 Read and complete a lab safety form.

2 Use a **graduated cylinder** to add 25 mL of **solution A** to a **self-sealing plastic bag.** Place a **stoppered test tube** containing **solution B** into the bag. Be careful not to dislodge the stopper.

3 Seal the bag completely, and wipe off any moisture on the outside with a **paper towel.** Place the bag on the **balance.** Record the total mass.

4 Without opening the bag, remove the stopper from the test tube and allow the liquids to mix. Observe and record what happens.

5 Place the sealed bag and its contents back on the balance. Read and record the mass.

Data and Observations

Think About This

1. **Express** What did you observe when the liquids mixed? How would you account for this observation?

2. **Evaluate** Did the mass of the bag's contents change? If so, could the change have been due to the precision of the balance, or did the matter in the bag change its mass? Explain.

3. **Key Concept** **Assess** Do you think matter was gained or lost in the bag? How can you tell?

1. From Florida to Maine, east of the Mississippi River fireflies dot summer evening skies with flickering spots of light. But fireflies don't use batteries. Their light is produced using a two-step chemical process called bioluminescence (bi oh lew muh NE sents). In this process, chemicals in a firefly's body combine to make new chemicals and light. How do chemicals produce light? What are some other common chemical processes?

_____

_____

_____

_____

_____

_____

_____

_____

Changes in Matter

When you put liquid water in a freezer, it changes to solid water, or ice. When you pour brownie batter into a pan and bake it, the liquid batter changes to a solid, too. In both cases, a liquid changes to a solid. Are these changes the same?

Physical Changes

Recall that matter can undergo two types of changes—chemical or physical. A physical change does not produce new substances. The substances that exist before and after the change are the same, although they might have different physical properties. This is what happens when liquid water freezes. Its physical properties change from a liquid to a solid, but the water, H_2O, does not change into a different substance. Water molecules are always made up of two hydrogen atoms bonded to one oxygen atom regardless of whether they are solid, liquid, or gas.

Chemical Changes

Recall that during a chemical change, one or more substances change into new substances. The starting substances and the substances produced have different physical and chemical properties. For example, when brownie batter bakes, a chemical change occurs. Many of the substances in the baked brownies are different from the substances in the batter. As a result, baked brownies have physical and chemical properties that are different from those of brownie batter.

A chemical change also is called a chemical reaction. These terms mean the same thing. _A_ **chemical reaction** _is a process in which atoms of one or more substances rearrange to form one or more new substances._ In this lesson, you will read what happens to atoms during a reaction and how these changes can be described using equations.

Active Reading

2. Identify What types of properties change during a chemical reaction?

_____

_____

Signs of a Chemical Reaction

How can you tell if a chemical reaction has taken place? You have read that the substances before and after a reaction have different properties. You might think that you could look for changes in properties as a sign that a reaction occurred. In fact, changes in the physical properties of color, state of matter, and odor are all signs that a chemical reaction might have occurred. Another sign of a chemical reaction is a change in energy. If substances get warmer or cooler or if they give off light or sound, it is likely that a reaction has occurred. Some signs that a chemical reaction might have occurred are shown in **Figure 1**.

However, these signs are not proof of a chemical change. For example, bubbles appear when water boils. But, bubbles also appear when baking soda and vinegar react and form carbon dioxide gas. How can you be sure that a chemical reaction has taken place? The only way to know is to study the chemical properties of the substances before and after the change. If they have different chemical properties, then the substances have undergone a chemical reaction.

 3. NGSSS Check Recognize How can you know for sure that a chemical reaction has occurred? Highlight your answer in the text. SC.8.P.8.5

Figure 1 You can detect a chemical reaction by looking for changes in properties and changes in energy of the substances that reacted.

Change in Properties

Change in color

Bright copper changes to green when the copper reacts with certain gases in the air.

Formation of bubbles

Bubbles of carbon dioxide form when baking soda is added to vinegar.

Change in odor

When food burns or rots, a change in odor is a sign of chemical change.

Formation of a precipitate

A precipitate is a solid formed when two liquids react.

Change in Energy

Warming or cooling

Thermal energy is either given off or absorbed during a chemical change.

Release of light

A firefly gives off light as the result of a chemical change.

What happens in a chemical reaction?

During a chemical reaction, one or more substances react and form one or more new substances. How are these new substances formed?

Atoms Rearrange and Form New Substances

To understand what happens in a reaction, first review substances. Recall that there are two types of substances—elements and compounds. Substances have a fixed arrangement of atoms. For example, in a single drop of water, there are trillions of oxygen and hydrogen atoms. However, all of these atoms are arranged in the same way—two atoms of hydrogen are bonded to one atom of oxygen. If this arrangement changes, the substance is no longer water. Instead, a different substance forms with different physical and chemical properties. This is what happens during a chemical reaction. Atoms of elements or compounds rearrange and form different elements or compounds.

Bonds Break and Bonds Form

How does the rearrangement of atoms happen? Atoms rearrange when **chemical bonds** between atoms break. Recall that constantly moving particles make up all substances, including solids. As particles move, they collide with one another. If the particles collide with enough energy, the bonds between atoms can break. The atoms separate and rearrange, and new bonds can form. The reaction that forms hydrogen and oxygen from water is shown in **Figure 2**. Adding electric energy to water molecules can cause this reaction. The added energy causes bonds between the hydrogen atoms and the oxygen atoms to break. After the bonds between the atoms in water molecules break, new bonds can form between pairs of hydrogen atoms and between pairs of oxygen atoms.

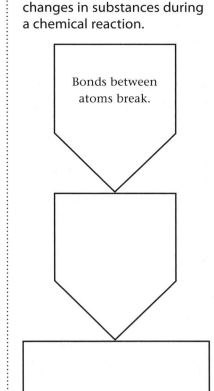

Active Reading 4. **Sequence** Fill in the graphic organizer below to list the changes in substances during a chemical reaction.

Bonds between atoms break.

REVIEW VOCABULARY

chemical bond
an attraction between atoms when electrons are shared, transferred, or pooled

Figure 2 🔑 Notice that no new atoms are created in a chemical reaction. The existing atoms rearrange and form new substances.

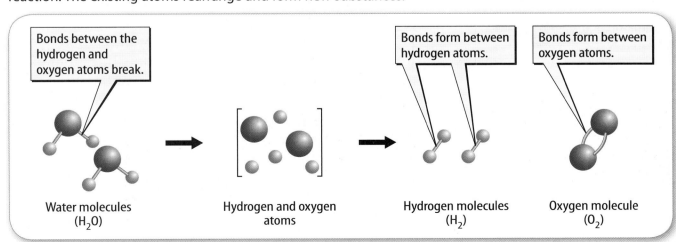

Bonds between the hydrogen and oxygen atoms break.

Bonds form between hydrogen atoms.

Bonds form between oxygen atoms.

Water molecules (H_2O)

Hydrogen and oxygen atoms

Hydrogen molecules (H_2)

Oxygen molecule (O_2)

Table 1 Symbols and Formulas of Some Elements and Compounds

Substance		Formula	# of atoms
Carbon		C	C: 1
Copper		Cu	Cu: ___
Cobalt		Co	Co: ___
Oxygen		O_2	O: 2
Hydrogen		H_2	H: ___
Chlorine		Cl_2	Cl: ___
Carbon dioxide		CO_2	C: 1 O: 2
Carbon monoxide		CO	C: ___ O: ___
Water		H_2O	H: ___ O: ___
Hydrogen peroxide		H_2O_2	H: 2 O: 2
Glucose		$C_6H_{12}O_6$	C: ___ H: ___ O: ___
Sodium chloride		NaCl	Na: 1 Cl: 1
Magnesium hydroxide		$Mg(OH)_2$	Mg: 1 O: 2 H: 2

Table 1 Symbols and subscripts describe the type and number of atoms in an element or a compound.

Active Reading **5. Write** In **Table 1**, fill in the correct number of atoms of each element.

Chemical Equations

Suppose your teacher asks you to produce a specific reaction in your science laboratory. How might your teacher describe the reaction to you? He or she might say something such as "react baking soda and vinegar to form sodium acetate, water, and carbon dioxide." It is more likely that your teacher will describe the reaction in the form of a chemical equation. *A* **chemical equation** *is a description of a reaction using element symbols and chemical formulas.* Element symbols represent elements. Chemical formulas represent compounds.

Element Symbols

Recall that symbols of elements are shown in the periodic table. For example, the symbol for carbon is C. The symbol for copper is Cu. Each element can exist as just one atom. However, some elements exist in nature as diatomic molecules—two atoms of the same element bonded together. A formula for one of these diatomic elements includes the element's symbol and the subscript *2*. A subscript describes the number of atoms of an element in a compound. Oxygen (O_2) and hydrogen (H_2) are examples of diatomic molecules. Some element symbols are shown above the blue line in **Table 1.**

Chemical Formulas

When atoms of two or more different elements bond, they form a compound. Recall that a chemical formula uses elements' symbols and subscripts to describe the number of atoms in a compound. If an element's symbol does not have a subscript, the compound contains only one atom of that element. For example, carbon dioxide (CO_2) is made up of one carbon atom and two oxygen atoms. Remember that two different formulas, no matter how similar, represent different substances. Some chemical formulas are shown below the blue line in **Table 1.**

Writing Chemical Equations

A chemical equation includes both the substances that react and the substances that are formed in a chemical reaction. *The starting substances in a chemical reaction are* **reactants**. *The substances produced by the chemical reaction are* **products**. **Figure 3** shows how a chemical equation is written. Chemical formulas are used to describe the reactants and the products. The reactants are written to the left of an arrow, and the products are written to the right of the arrow. Two or more reactants or products are separated by a plus sign. The general structure for an equation is:

reactant + reactant → product + product

When writing chemical equations, it is important to use correct chemical formulas for the reactants and the products. For example, suppose a certain chemical reaction produces carbon dioxide and water. The product carbon dioxide would be written as CO_2 and not as CO. CO is the formula for carbon monoxide, which is not the same compound as CO_2. Water would be written as H_2O and not as H_2O_2, the formula for hydrogen peroxide.

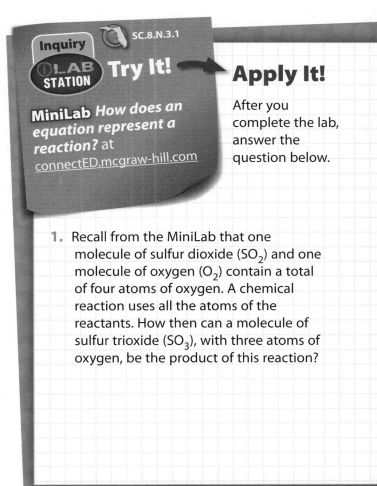

SC.8.N.3.1

Inquiry

LAB STATION Try It!

MiniLab *How does an equation represent a reaction?* at connectED.mcgraw-hill.com

Apply It!

After you complete the lab, answer the question below.

1. Recall from the MiniLab that one molecule of sulfur dioxide (SO_2) and one molecule of oxygen (O_2) contain a total of four atoms of oxygen. A chemical reaction uses all the atoms of the reactants. How then can a molecule of sulfur trioxide (SO_3), with three atoms of oxygen, be the product of this reaction?

Figure 3 An equation is read much like a sentence. This equation is read as "carbon plus oxygen produces carbon dioxide."

Active Reading 6. **Recognize** Fill in the blanks in **Figure 3** below.

Parts of an Equation

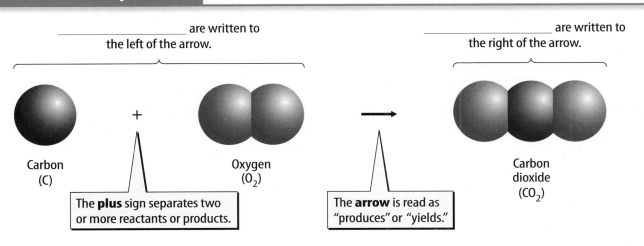

_____ are written to the left of the arrow.

_____ are written to the right of the arrow.

+

Carbon (C)

Oxygen (O_2)

The **plus** sign separates two or more reactants or products.

The **arrow** is read as "produces" or "yields."

Carbon dioxide (CO_2)

Conservation of Mass

A French chemist named Antoine Lavoisier (AN twan · luh VWAH see ay) (1743–1794) discovered something interesting about chemical reactions. In a series of experiments, Lavoisier measured the masses of substances before and after a chemical reaction inside a closed container. He found that the total mass of the reactants always equaled the total mass of the products. Lavoisier's results led to the law of conservation of mass. *The **law of conservation of mass** states that the total mass of the reactants before a chemical reaction is the same as the total mass of the products after the chemical reaction.*

Atoms are conserved.

The discovery of atoms provided an explanation for Lavoisier's observations. Mass is conserved in a reaction because atoms are conserved. Recall that during a chemical reaction, bonds break and new bonds form. However, atoms are not destroyed, and no new atoms form. All atoms at the start of a chemical reaction are present at the end of the reaction. **Figure 4** shows that mass is conserved in the reaction between baking soda and vinegar.

7. NGSSS Check Describe What happens to the total mass of the reactants in a chemical reaction? SC.8.P.9.1

_____

_____

Figure 4 As this reaction takes place, the mass on the balance remains the same, showing that mass is conserved.

Active Reading **8. Draw** For the products of the reaction shown in **Figure 4**, indicate the correct number of atoms of each element. Then use colored pencils to draw the correct number of atoms. Match the colors of the atoms of the product to the colors of the atoms of reactants.

Conservation of Mass 🔑

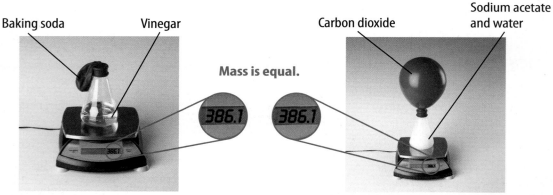

Baking soda is contained in a balloon. The balloon is attached to a flask that contains vinegar.

When the balloon is tipped up, the baking soda pours into the vinegar. The reaction forms a gas that is collected in the balloon.

Baking soda   Vinegar   Carbon dioxide   Sodium acetate and water

Mass is equal.

386.1   386.1

baking soda + vinegar → sodium acetate + water + carbon dioxide
$NaHCO_3$ $HC_2H_3O_2$ $NaC_2H_3O_2$ H_2O CO_2

1 Na: ○   4 H: ●●●●   ___ Na: ___ H: ___ C:
1 H: ●   2 C: ●●   ___ C: ___ O: ___ O:
1 C: ●   2 O: ●●   ___ H:
3 O: ●●●   ___ O:

Atoms are equal.

Is an equation balanced?

How does a chemical equation show that atoms are conserved? An equation is written so that the number of atoms of each element is the same, or balanced, on each side of the arrow. The equation showing the reaction between carbon and oxygen that produces carbon dioxide is shown below. Remember that oxygen is written as O_2 because it is a diatomic molecule. The formula for carbon dioxide is CO_2.

Active Reading
FOLDABLES® LA.6.2.2.3

Make a vertical four-tab book. Label it as shown. Use it to study the steps of balancing equations.

1. Write the unbalanced equation.
2. Count the atom.
3. Add coefficients.
4. Write the balanced equation.

Balancing Chemical Reactions

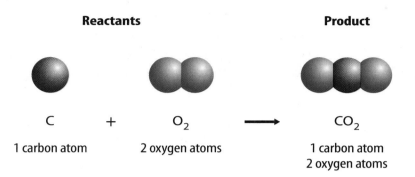

Reactants   **Product**

C   + O_2 ⟶ CO_2

1 carbon atom   2 oxygen atoms   1 carbon atom
                                      2 oxygen atoms

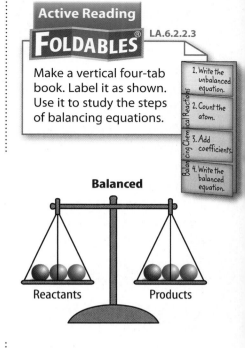

Balanced

Reactants   Products

Is there the same number of carbon atoms on each side of the arrow? Yes, there is one carbon atom on the left and one on the right. Carbon is balanced. Is oxygen balanced? There are two oxygen atoms on each side of the arrow. Oxygen also is balanced. The atoms of all elements are balanced. Therefore, the equation is balanced.

You might think a balanced equation happens automatically when you write the symbols and formulas for reactants and products. However, this usually is not the case. For example, the reaction between hydrogen (H_2) and oxygen (O_2) that forms water (H_2O) is shown below.

Active Reading **9. Evaluate** Why is this equation considered balanced?

_____

_____

_____

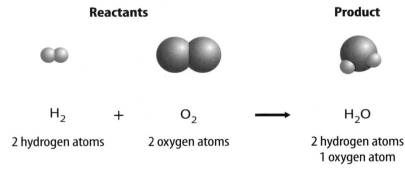

Reactants   **Product**

H_2 + O_2 ⟶ H_2O

2 hydrogen atoms   2 oxygen atoms   2 hydrogen atoms
                                          1 oxygen atom

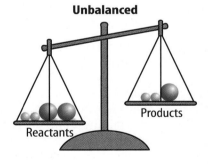

Unbalanced

Products

Reactants

Count the number of hydrogen atoms on each side of the arrow. There are two hydrogen atoms in the product and two in the reactants. They are balanced. Now count the number of oxygen atoms on each side of the arrow. Did you notice that there are two oxygen atoms in the reactants and only one in the product? Because they are not equal, this equation is not balanced. To accurately represent this reaction, the equation needs to be balanced.

Active Reading **10. Explain** Why is this equation considered unbalanced?

_____

_____

_____

Balancing Chemical Equations

When you balance a chemical equation, you count the atoms in the reactants and the products and then add coefficients to balance the number of atoms. *A* **coefficient** *is a number placed in front of an element symbol or chemical formula in an equation.* It is the number of units of that substance in the reaction. For example, in the formula $2H_2O$, the *2* in front of H_2O is a coefficient. This means that there are two molecules of water in the reaction. Only coefficients can be changed when balancing an equation. Changing subscripts changes the identities of the substances that are in the reaction.

If one molecule of water contains two hydrogen atoms and one oxygen atom, how many H and O atoms are in two molecules of water ($2H_2O$)? Multiply each by 2.

$$2 \times 2\,H \text{ atoms} = 4\,H \text{ atoms}$$
$$2 \times 1\,O \text{ atom} = 2\,O \text{ atoms}$$

When no coefficient is present, only one unit of that substance takes part in the reaction. **Table 2** shows the steps of balancing a chemical equation.

Active Reading

11. Write Fill in the correct numbers and chemical formulas to complete **Table 2**.

Table 2 Balancing a Chemical Equation 🔑

1 **Write the unbalanced equation.** Make sure that all chemical formulas are correct.	H_2 + O_2 → H_2O *reactants* *products*
2 **Count atoms of each element in the reactants and in the products.** **a.** Note which, if any, elements have a balanced number of atoms on each side of the equation. Which atoms are not balanced? **b.** If all of the atoms are balanced, the equation is balanced.	 H_2 + O_2 → H_2O *reactants* *products* $H = 2$ $H = \rule{1cm}{0.4pt}$ $O = 2$ $O = \rule{1cm}{0.4pt}$
3 **Add coefficients to balance the atoms.** **a.** Pick an element in the equation that is not balanced, such as oxygen. Write a coefficient in front of a reactant or a product that will balance the atoms of that element. **b.** Recount the atoms of each element in the reactants and the products. Note which atoms are not balanced. Some atoms that were balanced before might no longer be balanced. **c.** Repeat step 3 until the atoms of each element are balanced.	 H_2 + O_2 → $2H_2O$ *reactants* *products* $H = 2$ $H = \rule{1cm}{0.4pt}$ $O = 2$ $O = \rule{1cm}{0.4pt}$ $2H_2$ + O_2 → $2H_2O$ *reactants* *products* $H = \rule{1cm}{0.4pt}$ $H = \rule{1cm}{0.4pt}$ $O = \rule{1cm}{0.4pt}$ $O = \rule{1cm}{0.4pt}$
4 **Write the balanced chemical equation** including the coefficients.	$\rule{1cm}{0.4pt}$ + $\rule{1cm}{0.4pt}$ → $\rule{1cm}{0.4pt}$

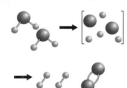

A chemical reaction is a process in which bonds break and atoms rearrange, forming new bonds.

$2H_2 + O_2 \rightarrow 2H_2O$

A chemical equation uses symbols to show reactants and products of a chemical reaction.

The mass and the number of each type of atom do not change during a chemical reaction. This is the law of conservation of mass.

Inquiry

SC.8.P.8.5, SC.8.P.9.2, SC.7.N.1.6, SC.8.N.1.1

iLAB STATION **Try It!**

Skill Lab *What can you learn from an experiment?* at connectED.mcgraw-hill.com

Use Vocabulary

1 **Define** *reactants* and *products*.

_____

_____

_____

Understand Key Concepts 🔑

2 Which is a definite sign of a chemical reaction?

(A) Chemical properties change. (C) A gas forms.

(B) Physical properties change. (D) A solid forms.

3 **Explain** why subscripts cannot change when balancing a chemical equation. SC.8.P.9.1

_____

_____

4 **Infer** Is the reaction below possible? Explain why or why not.

$$H_2O + NaOH \rightarrow NaCl + H_2$$

_____

_____

Interpret Graphics

5 **Interpret** Complete the table to determine if this equation is balanced:

$$CH_4 + 2O_2 \rightarrow CO_2 + 2H_2O$$

Is this reaction balanced? Explain. SC.8.P.9.1

_____

_____

Type of Atom	Number of Atoms in the Balanced Chemical Equation	
	Reactants	Products

Critical Thinking

6 Balance this chemical equation. Hint: Balance Al last and then use a multiple of 2 and 3.

$$Al + HCl \rightarrow AlCl_3 + H_2$$

1. Fill in the graphic organizer to distinguish the parts of a chemical equation.

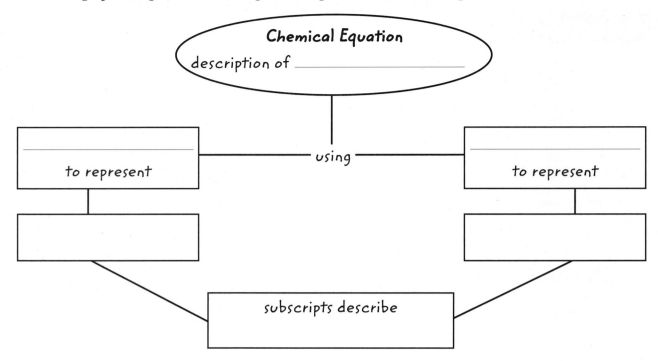

Complete the table below with information regarding the writing of chemical equations.

Define *reactant*.	**2.**
Define *product*.	**3.**
Write the general structure for a chemical equation.	**4.**
How is the arrow sign read?	**5.**
Write the equation for "carbon plus oxygen produces carbon dioxide."	**6.**

7. Restate the law of conservation of mass.

_____

_____

_____

Types of Chemical REACTIONS

How can you recognize the type of chemical reaction by the number or type of reactants and products?

What are the different types of chemical reactions?

Vocabulary

synthesis p. 435

decomposition p. 435

single replacement p. 436

double replacement p. 436

combustion p. 436

Florida NGSSS

SC.7.P.11.2 Investigate and describe the transformation of energy from one form to another.

LA.6.2.2.3 The student will organize information to show understanding or relationships among facts, ideas, and events (e.g., representing key points within text through charting, mapping, paraphrasing, summarizing, or comparing/contrasting);

LA.6.4.2.2 The student will record information (e.g., observations, notes, lists, charts, legends) related to a topic, including visual aids to organize and record information, as appropriate, and attribute sources of information;

 Launch Lab

15 minutes

What combines with what?

The reactants and the products in a chemical reaction can be elements, compounds, or both. In how many ways can these substances combine?

Procedure

1. Read and complete a lab safety form.

2. Divide a **sheet of paper** into four equal sections labeled *A, B, Y,* and *Z.* Place **red paper clips** in section A, **yellow clips** in section B, **blue clips** in section Y, and **green clips** in section Z.

3. Use another sheet of paper to copy the table shown to the right. Turn the paper so that a long edge is at the top. Print *REACTANTS → PRODUCTS* across the top, and then complete the table.

	REACTANTS → PRODUCTS
1	$AY \rightarrow A + Y$
2	$B + Z \rightarrow BZ$
3	$2A_2 + Y_2 \rightarrow 2A_2Y$
4	$A + BY \rightarrow B + AY$
5	$Z + BY \rightarrow Y + BZ$
6	$AY + BZ \rightarrow AZ + BY$

4. Using the paper clips, model the equations listed in the table. Hook the clips together to make diatomic elements or compounds. Place each clip model onto your paper over the matching written equation.

5. As you read Lesson 2, match the types of equations to your paper clip equations.

Think About This

1. **Select** the equation that represents hydrogen combining with oxygen and forming water. How did you determine which equation to choose?

2. **Key Concept** **Identify** How could you use the number and type of reactants to identify a type of chemical reaction?

1. When lead nitrate and potassium iodide—both clear liquids—combine, a yellow solid appears instantly. Where did the solid come from? Here's a hint—the name of the solid is lead iodide. Did you guess that parts of each reactant combined and formed it? What are chemical reactants? What chemical reactants were combined in **Figure 5** to cause the explosion?

_____

_____

_____

_____

_____

_____

_____

_____

Active Reading **2. Summarize** Fill in the graphic organizer below to describe the concept of patterns in chemical reactions.

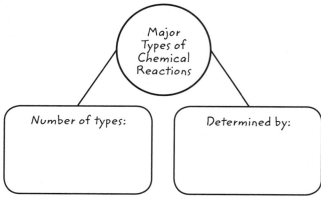

Patterns in Reactions

If you have ever used hydrogen peroxide, you might have noticed that it is stored in a dark bottle. This is because light causes hydrogen peroxide to change into other substances. Maybe you have seen a video of an explosion demolishing an old building, like in **Figure 5.** How is the reaction with hydrogen peroxide and light similar to a building demolition? In both, one reactant breaks down into two or more products.

The breakdown of one reactant into two or more products is one of four major types of chemical reactions. Each type of chemical reaction follows a unique pattern in the way atoms in reactants rearrange to form products. In this lesson, you will read how chemical reactions are classified by recognizing patterns in the way the atoms recombine.

Figure 5 When dynamite explodes, it chemically changes into several products and releases energy.

Types of Chemical Reactions

There are many different types of reactions. It would be impossible to memorize them all. However, most chemical reactions fit into four major categories. Understanding these categories of reactions can help you predict how compounds will react and what products will form.

Synthesis

A **synthesis** (SIHN thuh sus) *is a type of chemical reaction in which two or more substances combine and form one compound.* In the synthesis reaction shown in **Figure 6,** magnesium (Mg) reacts with oxygen (O_2) in the air and forms magnesium oxide (MgO). You can recognize a synthesis reaction because two or more reactants form only one product.

Decomposition

In a **decomposition** *reaction, one compound breaks down and forms two or more substances.* You can recognize a decomposition reaction because one reactant forms two or more products. For example, hydrogen peroxide (H_2O_2), shown in **Figure 6,** decomposes and forms water (H_2O) and oxygen gas (O_2). Notice that decomposition is the reverse of synthesis.

Figure 6 Synthesis and decomposition reactions are opposites of each other.

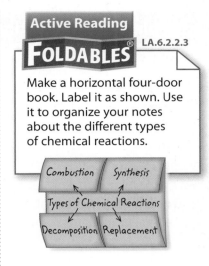

Active Reading

FOLDABLES ® LA.6.2.2.3

Make a horizontal four-door book. Label it as shown. Use it to organize your notes about the different types of chemical reactions.

Combustion | Synthesis

Types of Chemical Reactions

Decomposition | Replacement

Active Reading **3. Differentiate** How can you tell the difference between synthesis and decomposition reactions?

_____

_____

_____

_____

Synthesis and Decomposition Reactions 🔑

Synthesis Reactions

Examples:
$2Na + Cl_2 \rightarrow 2NaCl$
$2H_2 + O_2 \rightarrow 2H_2O$
$H_2O + SO_3 \rightarrow H_2SO_4$

$2Mg$ + O_2 → $2MgO$
magnesium   oxygen   magnesium oxide

Decomposition Reactions

Examples:
$CaCO_3 \rightarrow CaO + CO_2$
$2H_2O \rightarrow 2H_2 + O_2$
$2KClO_3 \rightarrow 2KCl + 3O_2$

$2H_2O_2$ → $2H_2O$ + O_2
hydrogen peroxide   water   oxygen

Single Replacement

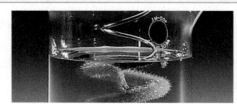

Examples:
$Fe + CuSO_4 \rightarrow FeSO_4 + Cu$
$Zn + 2HCl + ZnCl_2 + H_2$

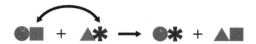

$$2AgNO_3 + Cu \rightarrow Cu(NO_3)_2 + 2Ag$$
silver nitrate   copper   copper nitrate   silver

Double Replacement

Examples:
$NaCl + AgNO_3 \rightarrow NaNO_3 + AgCl$
$HCl + FeS \rightarrow FeCl_2 + H_2S$

$$Pb(NO_3)_2 + 2KI \rightarrow 2KNO_3 + PbI_2$$
lead nitrate   potassium iodide   potassium nitrate   lead iodide

Figure 7 In each of these reactions, an atom or group of atoms replaces another atom or group of atoms.

Replacement

In a replacement reaction, an atom or group of atoms replaces part of a compound. There are two types of replacement reactions. *In a* **single-replacement** *reaction, one element replaces another element in a compound.* In this type of reaction, an element and a compound react and form a different element and a different compound. *In a* **double-replacement** *reaction, the negative ions in two compounds switch places, forming two new compounds.* In this type of reaction, two compounds react and form two new compounds. **Figure 7** describes these replacement reactions.

Combustion

Combustion *is a chemical reaction in which a substance combines with oxygen and releases energy.* This energy usually is released as thermal energy and light energy. For example, burning is a common combustion reaction. The burning of fossil fuels, such as the propane (C_3H_8) shown in **Figure 8,** produces the energy we use to cook food, power vehicles, and light cities.

Combustion Reactions

substance + O$_2$ → substance(s)

$$C_3H_8 + 5O_2 \rightarrow 3CO_2 + 4H_2O$$
propane   oxygen   carbon dioxide   water

Example:
$2C_4H_{10} + 13O_2 \rightarrow 8CO_2 + 10H_2O$

Figure 8 Combustion reactions always contain oxygen (O_2) as a reactant and often produce carbon dioxide (CO_2) and water (H_2O).

Active Reading **4. Select** What are the two types of energy typically released during a combustion reaction? (Circle) your answer in the text above.

Visual Summary

Chemical reactions are classified according to patterns seen in their reactants and products.

In a synthesis reaction, there are two or more reactants and one product. A decomposition reaction is the opposite of a synthesis reaction.

In replacement reactions, an element, or elements, in a compound is replaced with another element or elements.

Use Vocabulary

1 **Contrast** synthesis and decomposition reactions using a diagram.

2 A reaction in which parts of two substances switch places and make two new substances is a(n) _____.

Understand Key Concepts 🔑

3 **Classify** the following reaction: $2Na + Cl_2 \rightarrow 2NaCl$
(A) combustion   (C) single replacement
(B) decomposition   (D) synthesis

4 Write a balanced equation that produces Na and Cl_2 from NaCl. Classify this reaction.

5 **Classify** Which two types of reactions describe this reaction:

$$2SO_2 + O_2 \rightarrow 2SO_3$$

_____

_____

Interpret Graphics

6 **Complete** this table to identify four types of chemical reactions and the patterns shown by the reactants and the products. LA.6.2.2.3

Type of Reaction	Pattern of Reactants and Products
Synthesis	at least two reactants; one product

Critical Thinking

7 **Infer** The combustion of methane (CH_4) releases energy. Where do you think this energy comes from?

_____

_____

_____

_____

How Does a Light Stick Work?

What makes it glow?

A light stick consists of a plastic tube with a glass tube inside it. Hydrogen peroxide fills the glass tube.

A solution of phenyl oxalate ester and fluorescent dye surrounds the glass tube.

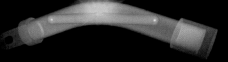

When you bend the outer plastic tube, the inner glass tube breaks, causing the hydrogen peroxide, ester, and dye to combine.

When the solutions combine, they react. Energy produced by the reaction causes the electrons in the dye to produce light.

Glowing necklaces, bracelets, or sticks—chances are you've worn or used them. Light sticks—also known as glow sticks—come in brilliant colors and provide light without electricity or batteries. Because they are lightweight, portable, and waterproof, they provide an ideal light source for campers, scuba divers, and other people participating in activities for which electricity is not readily available. Light sticks also are useful in emergency situations in which an electric current from battery-powered lights could ignite a fire.

Light sticks give off light because of a chemical reaction that happens inside the tube. During the reaction, energy is released as light. This is known as chemiluminescence (ke mee lew muh NE sunts).

It's Your Turn

RESEARCH AND REPORT Research bioluminescent organisms, such as fireflies and sea animals. How is the reaction that occurs in these organisms similar to or different from that in a glow stick? Work in small groups, and present your findings to the class. LA.6.4.2.2

Energy Changes and Chemical REACTIONS

ESSENTIAL QUESTIONS

 Why do chemical reactions always involve a change in energy?

 What is the difference between an endothermic reaction and an exothermic reaction?

 What factors can affect the rate of a chemical reaction?

Vocabulary

endothermic p. 441

exothermic p. 441

activation energy p. 442

catalyst p. 444

enzyme p. 444

inhibitor p. 444

 Florida NGSSS

SC.6.N.1.4 Discuss, compare, and negotiate methods used, results obtained, and explanations among groups of students conducting the same investigation.

SC.6.N.1.5 Recognize that science involves creativity, not just in designing experiments, but also in creating explanations that fit evidence.

SC.6.N.2.1 Distinguish science from other activities involving thought.

SC.7.N.1.4 Identify test variables (independent variables) and outcome variables (dependent variables) in an experiment.

SC.7.N.1.7 Explain that scientific knowledge is the result of a great deal of debate and confirmation within the science community.

SC.7.P.11.3 Cite evidence to explain that energy cannot be created nor destroyed, only changed from one form to another.

SC.8.P.9.3 Investigate and describe how temperature influences chemical changes.

Also covers: SC.7.N.1.3, SC.8.N.1.6, LA.6.2.2.3, MA.6.A.3.6

Inquiry Launch Lab
20 minutes

MA.6.S.6.2, SC.8.P.9.3

Where's the heat?

Does a chemical change always produce a temperature increase?

Procedure

1. Read and complete a lab safety form.
2. Copy the table on a sheet of paper.
3. Use a **graduated cylinder** to measure 25 mL of **citric acid solution** into a **foam cup**. Record the temperature with a **thermometer**.
4. Use a **spoon** to add a spoonful of **dry sodium bicarbonate** to the cup. Stir.
5. Record the temperature every 15 s until it stops changing. Record your observations during the reaction.
6. Add 25 mL of **sodium bicarbonate solution** to another cup. Record the temperature. Add a spoonful of **calcium chloride**. Repeat step 5.

Time	Temperature (°C)	
	Citric Acid Solution	Sodium Bicarbonate Solution
Starting temp.		
15 s		
30 s		
45 s		
1 min		
1 min, 15 s		
1 min, 30 s		
1 min, 45 s		
2 min		
2 min, 15 sec		

Think About This

1. **Explain** What evidence do you have that the changes in the two cups were chemical reactions?

2. **Describe** Account for the temperature changes in the two cups.

3. **Key Concept** **Infer** Based on your experiences, would a change in temperature convince you that a chemical change had taken place? Why or why not? What else could cause a temperature change?

Energy Changes

What is about 1,500 times heavier than a typical car and 300 times faster than a roller coaster? Do you need a hint? The energy it needs to move this fast comes from a chemical reaction that produces water. If you guessed a space shuttle, you are right!

It takes a large amount of energy to launch a space shuttle. The shuttle's main engines burn almost 2 million liters of liquid hydrogen and liquid oxygen. This chemical reaction produces water vapor and a large amount of energy. The energy produced heats the water vapor to high temperatures, causing it to expand rapidly. When the water expands, it pushes the shuttle into orbit. Where does all this energy come from?

Chemical Energy in Bonds

Recall that when a chemical reaction occurs, chemical bonds in the reactants break and new chemical bonds form. Chemical bonds contain a form of energy called chemical energy. Breaking a bond absorbs energy from the surroundings. The formation of a chemical bond releases energy to the surroundings. Some chemical reactions release more energy than they absorb. Some chemical reactions absorb more energy than they release. You can feel this energy change as a change in the temperature of the surroundings. Keep in mind that in all chemical reactions, energy is conserved.

Inquiry **Energy from Bonds?**

1. On April 12, 1981, the space shuttle *Columbia* became the first space shuttle to orbit Earth. What was the source of the energy that produced the deafening roar, the blinding light, and the power to lift the 2-million-kg spacecraft? The energy stored in chemical bonds and the reactions that released that energy carried *Columbia* on its mission. What is a chemical bond? Where can you find these chemical bonds that can release so much energy?

_____

_____

_____

_____

Active Reading **2. Explain** Fill in the blanks to describe the energy changes that occur during a chemical reaction.

When bonds between atoms break, energy

_____

When new bonds form, energy

_____

Active Reading **3. Identify** What type of energy is contained in a chemical bond?

_____

Endothermic Reactions—Energy Absorbed

Have you ever heard someone say that the sidewalk was hot enough to fry an egg? To fry, the egg must absorb energy. *Chemical reactions that absorb thermal energy are* **endothermic** *reactions*. For an endothermic reaction to continue, energy must be added constantly.

reactants + thermal energy → products

In an endothermic reaction, more energy is required to break the bonds of the reactants than is released when the products form. Therefore, the overall reaction absorbs energy. The reaction on the left in **Figure 9** is an endothermic reaction.

Exothermic Reactions—Energy Released

Most chemical reactions release energy as opposed to absorbing it. *An* **exothermic** *reaction is a chemical reaction that releases thermal energy.*

reactants → products + thermal energy

In an exothermic reaction, more energy is released when the products form than is required to break the bonds in the reactants. Therefore, the overall reaction releases energy. The reaction shown on the right in **Figure 9** is exothermic.

Figure 9 🔑 Whether a reaction is endothermic or exothermic depends on the amount of energy contained in the bonds of the reactants and in the bonds of the products.

Active Reading

FOLDABLES® LA.6.2.2.3

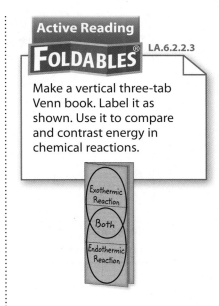

Make a vertical three-tab Venn book. Label it as shown. Use it to compare and contrast energy in chemical reactions.

WORD ORIGIN

exothermic
from Greek *exo–*, means "outside"; and *therm*, means "heat"

Active Reading **4. Summarize** What happens to energy during an endothermic and an exothermic reaction? Fill in the blanks in **Figure 9**.

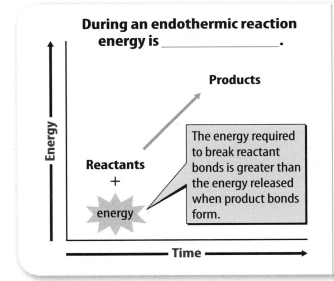

During an endothermic reaction energy is _____.

Energy

Products

Reactants
+
energy

The energy required to break reactant bonds is greater than the energy released when product bonds form.

Time

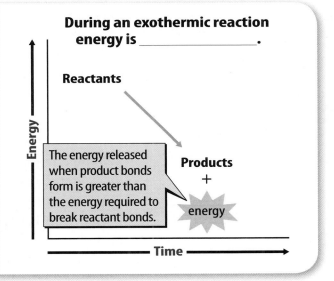

During an exothermic reaction energy is _____.

Energy

Reactants

Products
+
energy

The energy released when product bonds form is greater than the energy required to break reactant bonds.

Time

Active Reading **5. Explain** Why does one orange arrow point upward and the other point downward in these diagrams?

_____

_____

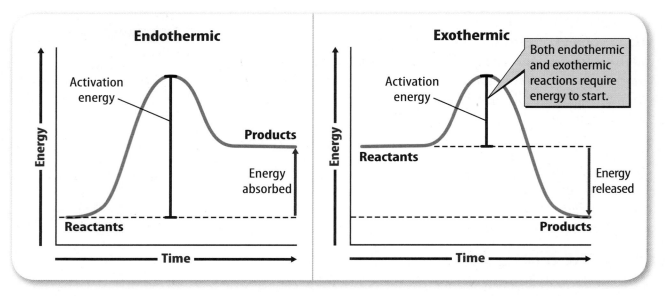

Endothermic

Activation energy

Energy

Products

Energy absorbed

Reactants

Time

Exothermic

Both endothermic and exothermic reactions require energy to start.

Activation energy

Energy

Reactants

Energy released

Products

Time

Figure 10 🔑 Both endothermic and exothermic reactions require activation energy to start the reaction.

Active Reading **6. Explain** How can a reaction absorb energy to start, but remain exothermic?

_____

_____

_____

_____

_____

Active Reading **7. Record** On the lines below, describe two factors that affect reaction rate.

_____

_____

_____

_____

_____

_____

Activation Energy

You might have noticed that some chemical reactions do not start by themselves. For example, a newspaper does not burn when it comes into contact with oxygen in air. However, if a flame touches the paper, it starts to burn.

All reactions require energy to start the breaking of bonds. This energy is called activation energy. **Activation energy** *is the minimum amount of energy needed to start a chemical reaction.* Different reactions have different activation energies. Some reactions, such as the rusting of iron, have low activation energy. The energy in the surroundings is enough to start these reactions. If a reaction has high activation energy, more energy is needed to start the reaction. For example, wood requires the thermal energy of a flame to start burning. Once the reaction starts, it releases enough energy to keep the reaction going. **Figure 10** shows the role activation energy plays in endothermic and exothermic reactions.

Reaction Rates

Some chemical reactions, such as the rusting of a bicycle wheel, happen slowly. Other chemical reactions, such as the explosion of fireworks, happen in less than a second. The rate of a reaction is the speed at which it occurs. What controls how fast a chemical reaction occurs? Recall that particles must collide before they can react. Chemical reactions occur faster if particles collide more often or move faster when they collide. Several factors affect how often particles collide and how fast particles move.

Surface Area

Surface area is the amount of exposed, outer area of a solid. Increased surface area increases reaction rate because more particles on the surface of a solid come into contact with the particles of another substance. For example, if you place a piece of chalk in vinegar, the chalk reacts slowly with the acid. This is because the acid contacts only the particles on the surface of the chalk. But, if you grind the chalk into powder, more chalk particles contact the acid, and the reaction occurs faster.

Temperature

Imagine a crowded hallway. If everyone in the hallway were running, they would probably collide with each other more often and with more energy than if everyone were walking. This is also true when particles move faster. At higher temperatures, the average speed of particles is greater. This speeds reactions in two ways. First, particles collide more often. Second, collisions with more energy are more likely to break chemical bonds.

Concentration and Pressure

Think of a crowded hallway again. Because the concentration of people is higher in the crowded hallway than in an empty hallway, people probably collide more often. Similarly, increasing the concentration of one or more reactants increases collisions between particles. More collisions result in a faster reaction rate. In gases, an increase in pressure pushes gas particles closer together. When particles are closer together, more collisions occur. Factors that affect reaction rate are shown in **Figure 11.**

Inquiry LAB STATION Try It!

MiniLab *Can you speed up a reaction?* at connectED.mcgraw-hill.com

Apply It! After you complete the lab, answer the question below.

1. Chemical reactions within some bacteria produce poisons that are harmful to humans. Explain why some foods are stored in a refrigerator or freezer.

Figure 11 🗝 Several factors can affect reaction rate.

Slower Reaction Rate

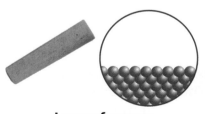

Less surface area

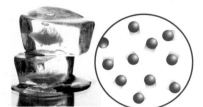

Lower temperature

Lower concentration

Faster Reaction Rate

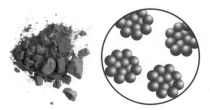

More surface area

Higher temperature

Higher concentration

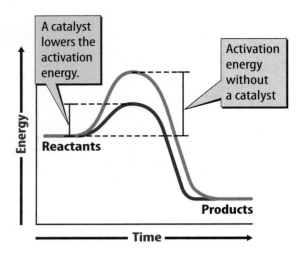

A catalyst lowers the activation energy.

Activation energy without a catalyst

Energy

Reactants

Products

Time

Figure 12 The blue line shows how a catalyst can increase the reaction rate.

Math Skills   MA.6.A.3.6

Use Geometry

The surface area (SA) of one side of a 1-cm cube is 1 cm × 1 cm, or 1 cm². The cube has 6 equal sides. Its total SA is 6 × 1 cm², or 6 cm². What is the total SA of the two solids when the cube is cut in half?

1
cm

1 cm

1 cm

1. The new surfaces made each have an area of 1 cm × 1 cm = 1 cm².

2. Multiply the area by the number of new surfaces. 2 ×1 cm² = 2 cm². The total SA is 8 cm².

Practice

8. Calculate the amount of SA gained when a 2-cm cube is cut in half.

Catalysts

A **catalyst** *is a substance that increases reaction rate by lowering the activation energy of a reaction.* One way catalysts speed reactions is by helping reactant particles contact each other more often. Look at **Figure 12.** Notice that the activation energy of the reaction is lower with a catalyst than it is without a catalyst. A catalyst isn't changed in a reaction, and it doesn't change the reactants or products. Also, a catalyst doesn't increase the amount of reactant used or the amount of product that is made. It only makes a given reaction happen faster. Therefore, catalysts are not considered reactants in a reaction.

You might be surprised to know that your body is filled with catalysts called enzymes. *An* **enzyme** *is a catalyst that speeds up chemical reactions in living cells.* For example, the enzyme protease (PROH tee ays) breaks the protein molecules in the food you eat into smaller molecules that can be absorbed by your intestine. Without enzymes, these reactions would occur too slowly for life to exist.

Inhibitors

Recall than an enzyme is a molecule that speeds reactions in organisms. However, some organisms, such as bacteria, are harmful to humans. Some medicines contain molecules that attach to enzymes in bacteria. This keeps the enzymes from working properly. If the enzymes in bacteria can't work, the bacteria die and can no longer infect a human. The active ingredients in these medicines are called inhibitors. *An* **inhibitor** *is a substance that slows, or even stops, a chemical reaction.* Inhibitors can slow or stop the reactions caused by enzymes.

Inhibitors are also important in the food industry. Preservatives in food are substances that inhibit, or slow down, food spoilage.

Visual Summary

Endothermic

Products

Reactants
+

energy

Chemical reactions that release energy are exothermic, and those that absorb energy are endothermic.

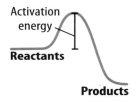

Activation energy must be added to a chemical reaction for it to proceed.

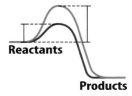

Catalysts, including enzymes, speed up chemical reactions. Inhibitors slow them down.

Inquiry

LAB STATION

SC.6.N.1.4,
SC.6.N.1.5,
SC.6.N.2.1,
SC.7.N.1.3,
SC.7.N.1.4,
SC.7.N.1.7,
SC.8.N.1.6,
SC.8.P.8.8

Try It!

Inquiry Lab *Design an Experiment to Test Advertising Claims* at connectED.mcgraw-hill.com

Use Vocabulary

1 The smallest amount of energy required by reacting particles for a chemical reaction to begin is the _____.

Understand Key Concepts

2 How does a catalyst increase reaction rate?
(A) by increasing the activation energy
(B) by increasing the amount of reactant
(C) by increasing the contact between particles
(D) by increasing the space between particles

3 **Explain** how increasing temperature affects reaction rate. SC.8.P.9.3

_____

_____

4 **Explain** When propane burns, heat and light are produced. Where does this energy come from?

_____

_____

Interpret Graphics

5 **List** Complete the graphic organizer to describe four ways to increase the rate of a reaction. LA.6.2.2.3

Increase reaction rate

Critical Thinking

6 **Infer** Explain why a catalyst does not increase the amount of product that can form.

_____

_____

_____

Math Skills   MA.6.A.3.6

7 An object measures 1 cm × 1 cm × 3 cm.

a. What is the surface area of the object?

b. What is the total surface area if you cut the object into three equal cubes?

Chapter 11 · Study Guide

 Think About It! Matter can undergo a variety of changes. When matter is changed chemically, a rearrangement of bonds between the atoms occurs. This results in new substances with new properties.

🔑 Key Concepts Summary

LESSON 1 Understanding Chemical Reactions

- There are several signs that a **chemical reaction** might have occurred, including a change in temperature, a release of light, a release of gas, a change in color or odor, and the formation of a solid from two liquids.
- In a chemical reaction, atoms of **reactants** rearrange and form **products.**
- The total mass of all the reactants is equal to the total mass of all the products in a reaction.

Reactants

1 Na: ●   4 H: ●●●●
1 H: ●   2 C: ●●
1 C: ●   2 O: ●●
3 O: ●●●

Atoms are equal.

Products

1 Na: ●   2 H: ●●   1 C: ●
2 C: ●●   1 O: ●   2 O: ●●
3 H: ●●●
2 O: ●●

LESSON 2 Types of Chemical Reactions

- Most chemical reactions fit into one of a few main categories—synthesis, decomposition, combustion, and single- or double-replacement.
- **Synthesis** reactions create one product. **Decomposition** reactions start with one reactant. **Single-** and **double-replacement** reactions involve replacing one element or group of atoms with another element or group of atoms. **Combustion** reactions involve a reaction between one reactant and oxygen, and they release thermal energy.

LESSON 3 Energy Changes and Chemical Reactions

- Chemical reactions always involve breaking bonds, which requires energy, and forming bonds, which releases energy.
- In an **endothermic** reaction, the reactants contain less energy than the products. In an **exothermic** reaction, the reactants contain more energy than the products.

Less surface area   **More surface area**

- The rate of a chemical reaction can be increased by increasing the surface area, the temperature, or the concentration of the reactants or by adding a **catalyst.**

FOLDABLES® Chapter Project

Assemble your lesson Foldables as shown to make a Chapter Project. Use the project to review what you have learned in this chapter.

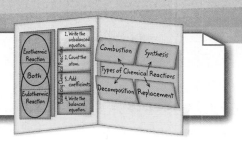

Use Vocabulary

1. When water forms from hydrogen and oxygen, water is the _____.

2. A(n) _____ uses symbols instead of words to describe a chemical reaction.

3. In a(n) _____ reaction, one element replaces another element in a compound.

4. When Na_2CO_3 is heated, it breaks down into CO_2 and Na_2O in a(n) _____ reaction.

5. The chemical reactions that keep your body warm are _____ reactions.

6. Even exothermic reactions require _____ to start.

Link Vocabulary and Key Concepts

Copy this concept map, and then use vocabulary terms from the previous page and other terms from the chapter to complete the concept map.

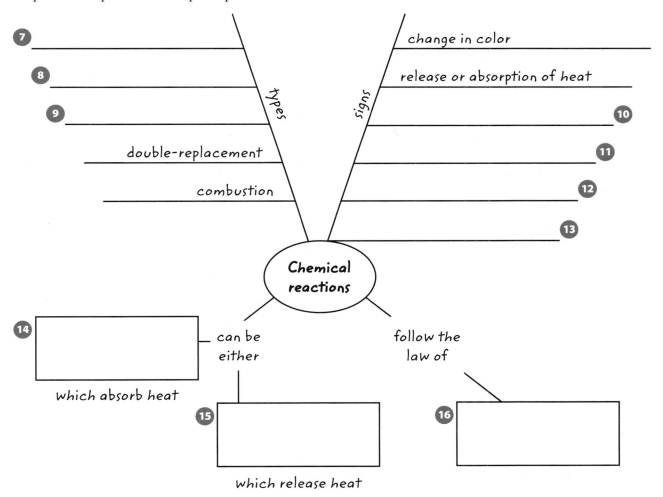

7. _____

8. _____

9. _____

double-replacement

combustion

change in color

release or absorption of heat

10. _____

11. _____

12. _____

13. _____

Chemical reactions

types

signs

14. _____
which absorb heat

can be either

15. _____
which release heat

follow the law of

16. _____

Fill in the correct answer choice.

🔑 Understand Key Concepts

1 How many carbon atoms react in this equation? SC.8.P.9.1

$$2C_4H_{10} + 13O_2 \rightarrow 8CO_2 + 10H_2O$$

- (A) 2
- (B) 4
- (C) 6
- (D) 8

2 The chemical equation below is unbalanced.

$$Zn + HCl \rightarrow ZnCl_2 + H_2$$

Which is the correct balanced chemical equation? SC.8.P.9.1

- (A) $Zn + H_2Cl_2 \rightarrow ZnCl_2 + H_2$
- (B) $Zn + HCl \rightarrow ZnCl + H$
- (C) $2Zn + 2HCl \rightarrow ZnCl_2 + H_2$
- (D) $Zn + 2HCl \rightarrow ZnCl_2 + H_2$

3 When iron combines with oxygen gas and forms rust, the total mass of the products SC.8.P.9.1

- (A) depends on the reaction conditions.
- (B) is less than the mass of the reactants.
- (C) is the same as the mass of the reactants.
- (D) is greater than the mass of the reactants.

4 Potassium nitrate forms potassium oxide, nitrogen, and oxygen in certain fireworks. SC.8.P.9.1

$$4KNO_3 \rightarrow 2K_2O + 2N_2 + 5O_2$$

This reaction is classified as a

- (A) combustion reaction.
- (B) decomposition reaction.
- (C) single-replacement reaction.
- (D) synthesis reaction.

5 Which type of reaction is the reverse of a decomposition reaction? SC.8.P.9.1

- (A) combustion
- (B) synthesis
- (C) double-replacement
- (D) single-replacement

Critical Thinking

6 **Predict** The diagram below shows two reactions—one with a catalyst (blue) and one without a catalyst (orange).

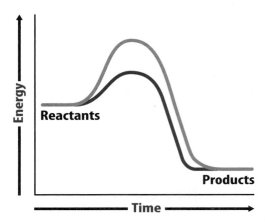

How would the blue line change if an inhibitor were used instead of a catalyst? MA.6.A.3.6

_____

_____

_____

7 **Analyze** A student observed a chemical reaction and collected the following data:

Observations before the reaction	A white powder was added to a clear liquid.
Observations during the reaction	The reactants bubbled rapidly in the open beaker.
Mass of reactants	4.2 g
Mass of products	4.0 g

The student concludes that mass was not conserved in the reaction. Explain why this is not a valid conclusion. What might explain the difference in mass? SC.8.P.9.1

_____

_____

_____

_____

_____

8 **Explain Observations** How did the discovery of atoms explain the observation that the mass of the products always equals the mass of the reactants in a reaction? SC.8.P.9.1

_____

_____

_____

_____

_____

_____

_____

Writing in Science

9 **Write** On a separate sheet of paper, write instructions that explain the steps in balancing a chemical equation. Use the following equation as an example. SC.8.P.9.1

$$MnO_2 + HCl \rightarrow MnCl_2 + H_2O + Cl_2$$

Big Idea Review

10 Explain how atoms and energy are conserved in a chemical reaction. SC.8.P.9.1

_____

_____

_____

_____

_____

11 When a car air bag inflates, sodium azide (NaN_3) decomposes and produces nitrogen gas (N_2) and another product. What element does the other product contain? How do you know? SC.8.P.9.1

_____

_____

_____

_____

_____

Math Skills   MA.6.A.3.6

12 What is the surface area of the cube shown below? What would the total surface area be if you cut the cube into 27 equal cubes?

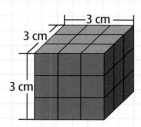

13 Suppose you have ten cubes that measure 2 cm on each side.
   a. What is the total surface area of the cubes?

   b. What would the surface area be if you glued the cubes together to make one object that is two cubes wide, one cube high, and five cubes long? Hint: Draw a picture of the final cube and label the length of each side.

Fill in the correct answer choice.

Multiple Choice

1 Which is NOT a chemical change? SC.8.P.8.5

Ⓐ copper turning green in air

Ⓑ baking a cake

Ⓒ drying clothes

Ⓓ exploding dynamite

Use the figure below to answer question 2.

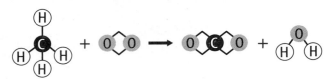

2 The figure above shows models of molecules in a chemical reaction. How would you describe this reaction? SC.8.P.8.5

Ⓕ a chemical change

Ⓖ a physical change

Ⓗ a chemical property

Ⓘ a physical property

3 The reaction between hydrogen chloride and sodium hydroxide is shown below. The name and the mass of each substance involved in the chemical reaction are shown. What mass of hydrogen chloride was used in this reaction? SC.8.P.9.1

? g	40.0 g	58.5 g	18.0 g
HCl +	NaOH →	NaCl +	H₂O
hydrogen chloride	sodium hydroxide	sodium chloride	water

Ⓐ 18.5 g

Ⓑ 36.5 g

Ⓒ 40.5 g

Ⓓ 116.5 g

4 Which occurs before new bonds can form during a chemical reaction? SC.8.P.9.1

Ⓕ The atoms in the original substances are destroyed.

Ⓖ The bonds between atoms in the original substances are broken.

Ⓗ The atoms in the original substances are no longer moving.

Ⓘ The bonds between atoms in the original substances get stronger.

5 Which can be used to speed up a reaction? SC.8.P.9.3

Ⓐ increase temperature

Ⓑ decrease concentration

Ⓒ add an inhibitor

Ⓓ decrease temperature

6 Preparing a meal involves both physical and chemical changes. Which involves a chemical change? SC.8.P.9.1

Ⓕ boiling water

Ⓖ making ice cubes

Ⓗ slicing a carrot

Ⓘ toasting bread

7 The law of conservation of mass states that in a chemical reaction, the mass of the reactants equals the mass of the products. If approximately 20 g of water reacts in the equation below, producing about 15 g of oxygen, what mass of hydrogen (H_2) is produced in this reaction? SC.8.P.9.1

$$2H_2O \rightarrow 2H_2 + O_2$$

Ⓐ 2 g

Ⓑ 5 g

Ⓒ 10 g

Ⓓ 15 g

8 Why does an oven's high temperature speed up chemical reactions? SC.8.P.9.3

(F) because heat lowers the activation energy

(G) because heat activates catalysts

(H) because heat makes molecules collide with each other

(I) because heat reduces the particle size of the reactants

9 Which is an example of a chemical change? SC.8.P.9.1

(A) boiling

(B) burning

(C) evaporation

(D) melting

10 Which statement best describes the law of conservation of mass? SC.8.P.9.1

(F) The mass of the products is always greater than the mass of the materials that react in a chemical change.

(G) The mass of the products is always less than the mass of the materials in a chemical change.

(H) A certain mass of material must be present for a reaction to occur.

(I) Matter is neither lost nor gained during a chemical change.

11 Antoine Lavoisier is credited with the discovery of the law of conservation of mass. The law states that in a chemical reaction, matter is not created or destroyed but preserved. Which equation correctly models this law? SC.8.P.9.1

(A) $H_2 + Cl_2 \rightarrow 2HCl$

(B) $H_2 + Cl_2 \rightarrow HCl$

(C) $H + Cl \rightarrow 2HCl$

(D) $2H_2 + Cl_2 \rightarrow 2HCl_2$

12 When a newspaper is left in direct sunlight for a few days, the paper begins to turn yellow? What is this change in color? SC.8.P.9.1

(F) physical property

(G) chemical property

(H) physical change

(I) chemical change

13 How does increasing the temperature affect a chemical reaction? SC.8.P.9.3

(A) The average speed of the particles speeds up causing more particle collisions.

(B) The average speed of the particles speeds up causing fewer particle collisions.

(C) The average speed of the particles slows down causing more particle collisions.

(D) The average speed of the particles slows down causing fewer particle collisions.

NEED EXTRA HELP?

If You Missed Question...	1	2	3	4	5	6	7	8	9	10	11	12	13
Go to Lesson...	1	1	1	1	2	1	1	3	1	1	1	1	2

Benchmark Mini-Assessment   Chapter 11 • Lesson 1

mini BAT

Multiple Choice *Bubble the correct answer.*

1. The diagram above is a model of the compound **SC.8.P.8.5**

(A) glucose, $C_6H_{12}O_6$.

(B) hydrogen peroxide, H_2O_2.

(C) magnesium hydroxide, $Mg(OH)_2$.

(D) sodium chloride, NaCl.

2. Jennifer and Cody are mixing substances during a lab experiment. They observe the four effects listed below. Which of these effects indicates that a chemical reaction has taken place between two substances? **SC.8.P.9.1**

(F) A clear, lemon-scented liquid continues to smell like lemon when it is mixed with water.

(G) A piece of shiny metal begins to bubble when placed in a clear liquid.

(H) A pink powdered substance turns water pink when they are mixed together.

(I) A white solid disappears when mixed with water.

3. In which equation is carbon dioxide a product? **SC.8.P.8.5**

(A) $CO_2 \rightarrow C + O_2$

(B) $CH_4 + 2O_2 \rightarrow CO_2 + 2H_2O$

(C) $6CO_2 + 6H_2O \rightarrow C_6H_{12}O_6 + 6O_2$

(D) $2H_2O \rightarrow 2H_2 + O_2$

Reactants   **Product**

___ H_2 + ___ O_2 $\longrightarrow$ ___ H_2O

4. What is the balanced equation that would represent the reaction above? **SC.8.P.9.1**

(F) $H_2 + O_2 = H_2O$

(G) $H_2 + 2O_2 = 2H_2O$

(H) $2H_2 + O_2 = 2H_2O$

(I) $4H_2 + 4O_2 = 4H_2O$

Copyright © Glencoe/McGraw-Hill, a division of The McGraw-Hill Companies, Inc.

Multiple Choice *Bubble the correct answer.*

1. Which reaction could be modeled by the illustration above? **SC.8.P.9.1**

(A) $CaCO_3 \rightarrow CaO + CO_2$

(B) $H_2O + SO_3 \rightarrow H_2SO_4$

(C) $NaCl + AgNO_3 \rightarrow NaNO_3 + AgCl$

(D) $Zn + 2HCl \rightarrow ZnCl_2 + H_2$

2. Which type of reaction forms water from hydrogen and oxygen atoms? **SC.8.P.8.5**

(F) decomposition reaction

(G) synthesis reaction

(H) double-replacement reaction

(I) single-replacement reaction

3. NaCl and $AgNO_3$ react in a double replacement reaction. Which is one of the products of this reaction? **SC.8.P.8.5**

(A) $AgClO_3$

(B) CO_2

(C) NaAg

(D) $NaNO_3$

$$2C_4H_{10} + 13O_2 \rightarrow 8CO_2 + 10H_2O$$

4. Which is true about the reaction shown above? **SC.8.P.8.4**

(F) The reaction is a single replacement.

(G) The reaction releases energy.

(H) The reaction has negative ions in two of the compounds that switch places.

(I) The reaction has two elements that combine to form a compound.

Copyright © Glencoe/McGraw-Hill, a division of The McGraw-Hill Companies, Inc.

Multiple Choice *Bubble the correct answer.*

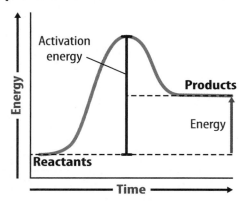

1. Which describes energy in the reaction represented by the graph above? **SC.6.N.1.5**

(A) Energy is required to start the reaction; energy is then absorbed.

(B) Energy is required to start the reaction; energy is then released.

(C) No energy is required to start the reaction; energy is then absorbed.

(D) No energy is required to start the reaction; energy is then released.

2. Which represents an endothermic reaction? **SC.7.N.1.6**

(F) burning wood

(G) exploding firecracker

(H) melting ice

(I) toasting bread

3. Which reaction has the lowest activation energy? **SC.8.P.9.1**

(A) combustion of oil

(B) explosion of fireworks

(C) rotting of bananas

(D) synthesis of water

4. What is one way to increase rate of a chemical reaction? **LA.6.2.2.3**

(F) decrease surface area

(G) increase surface area

(H) lower concentration

(I) lower temperature

Copyright © Glencoe/McGraw-Hill, a division of The McGraw-Hill Companies, Inc.

Notes

Notes

Is it a mixture?

All around you are things that are made of matter. Some of these things are made of a type of matter called a mixture. Put an X next to each thing that you think is a mixture.

_____ milk

_____ salt water

_____ salt (sodium chloride)

_____ paint

_____ blood

_____ bread

_____ soil

_____ iron

_____ shampoo

_____ sugar

_____ sun tan lotion

_____ catsup

_____ oxygen

_____ air

_____ ginger ale

_____ granite rock

_____ carbon dioxide

_____ copper

Explain your thinking. Describe your rule or reasoning for deciding whether something is a mixture.

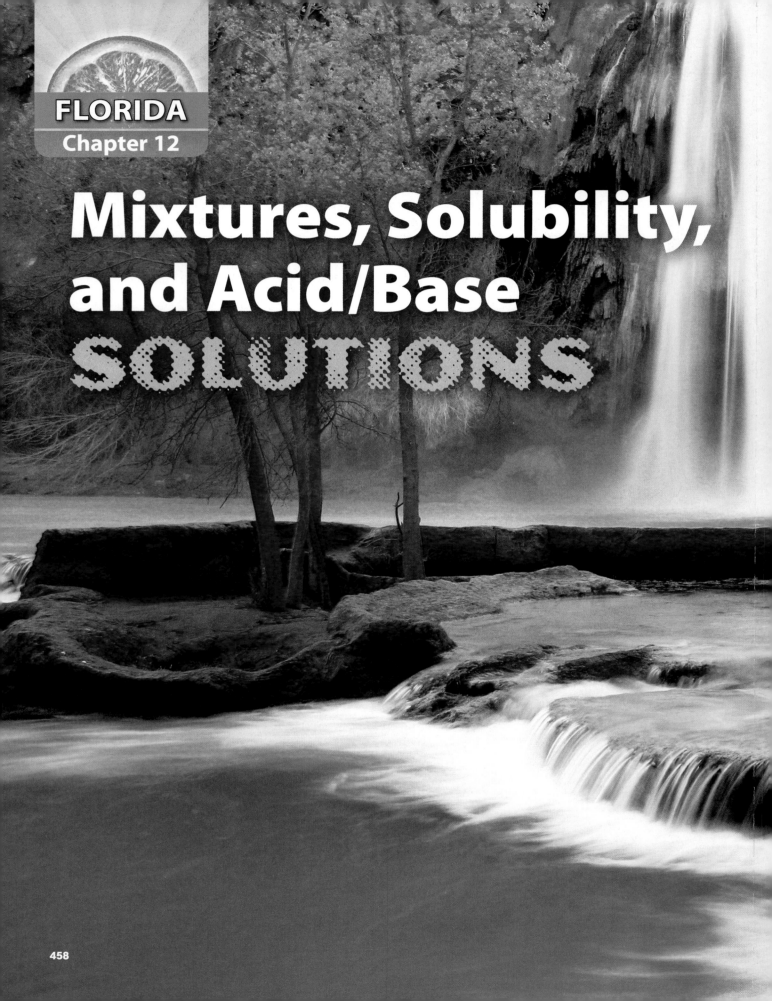

Mixtures, Solubility, and Acid/Base SOLUTIONS

FLORIDA BIG IDEAS

1 The Practice of Science

8 Properties of Matter

9 Changes in Matter

The Big Idea

Think About It!

What are solutions, and how are they described?

Havasu Falls is located in northern Arizona near the Grand Canyon. The creek that feeds these falls runs through a type of limestone called travertine. Small amounts of travertine are mixed evenly in the water, giving the water its unique blue-green color.

1 Can you think of other examples of something mixed evenly in water?

_____

_____

_____

2 What do you think of when you hear the word *solution*?

_____

_____

3 How would you describe a solution?

_____

_____

Get Ready to Read

What do you think about mixtures and solutions?

Before you read, decide if you agree or disagree with each of these statements. As you read this chapter, see if you change your mind about any of the statements.

	AGREE	DISAGREE
1 You can identify a mixture by looking at it without magnification.	☐	☐
2 A solution is another name for a homogeneous mixture.	☐	☐
3 Solutions can be solids, liquids, or gases.	☐	☐
4 A teaspoon of soup is less concentrated than a cup of the same soup.	☐	☐
5 Acids are found in many foods.	☐	☐
6 You can determine the exact pH of a solution by using pH paper.	☐	☐

  **There's More Online!**

Video • Audio • Review • ⓘLab Station • WebQuest • Assessment • Concepts in Motion • Multilingual eGlossary

Substances and MIXTURES

 How do substances and mixtures differ?

 How do solutions compare and contrast with heterogeneous mixtures?

 In what three ways do compounds differ from mixtures?

Vocabulary

substance p. 462

mixture p. 462

heterogeneous mixture p. 463

homogeneous mixture p. 463

solution p. 463

 Florida NGSSS

SC.8.N.1.1 Define a problem from the eighth grade curriculum using appropriate reference materials to support scientific understanding, plan and carry out scientific investigations of various types, such as systematic observations or experiments, identify variables, collect and organize data, interpret data in charts, tables, and graphics, analyze information, make predictions, and defend conclusions.

SC.8.N.1.6 Understand that scientific investigations involve the collection of relevant empirical evidence, the use of logical reasoning, and the application of imagination in devising hypotheses, predictions, explanations and models to make sense of the collected evidence.

SC.8.P.8.8 Identify basic examples of and compare and classify the properties of compounds, including acids, bases, and salts.

SC.8.P.8.9 Distinguish among mixtures (including solutions) and pure substances.

SC.8.P.9.2 Differentiate between physical changes and chemical changes.

LA.6.2.2.3 The student will organize information to show understanding or relationships among facts, ideas, and events (e.g., representing key points within text through charting, mapping, paraphrasing, summarizing, or comparing/contrasting);

LA.6.4.2.2 The student will record information (e.g., observations, notes, lists, charts, legends) related to a topic, including visual aids to organize and record information, as appropriate, and attribute sources of information;

 Inquiry Launch Lab

SC.8.N.1.6, SC.8.P.8.9

15 minutes

What makes black ink black?

Procedure

1. Read and complete a lab safety form.
2. Lay a **coffee filter** on your table.
3. Find the center of the coffee filter and mark it lightly with a **pencil.**
4. Use a **permanent marker** to draw a circle with a diameter of 5 cm around the center of the coffee filter. Do not fill this circle in.
5. Pour **rubbing alcohol** to a depth of 1 cm into a **beaker.**
6. Using the eraser end of your pencil, push the center of the coffee filter down into the beaker until the center, but not the ink, touches the liquid. Keep the ink above the surface of the liquid.
7. Observe the liquid in the bottom of the beaker and the circle on the coffee filter.
8. Dispose of experimental supplies as instructed by your teacher.

Think About This

1. What happened to the black ink circle on the coffee filter?

2. What was the purpose of the rubbing alcohol?

3. What do you think you would see if you used green ink instead?

4. **Key Concept** How do you think this shows that black ink is a mixture?

Figure 4 This organizational chart shows how different types of matter are classified. All matter can be classified as either a substance or a mixture.

7. **Label** Fill in the blanks below to complete the figure.

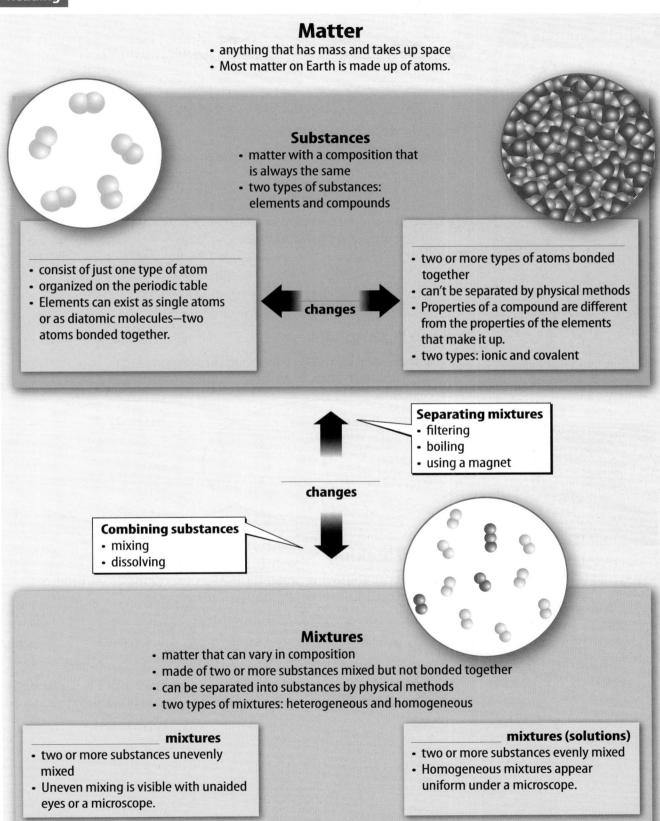

Matter
- anything that has mass and takes up space
- Most matter on Earth is made up of atoms.

Substances
- matter with a composition that is always the same
- two types of substances: elements and compounds

- consist of just one type of atom
- organized on the periodic table
- Elements can exist as single atoms or as diatomic molecules—two atoms bonded together.

changes

- two or more types of atoms bonded together
- can't be separated by physical methods
- Properties of a compound are different from the properties of the elements that make it up.
- two types: ionic and covalent

Separating mixtures
- filtering
- boiling
- using a magnet

changes

Combining substances
- mixing
- dissolving

Mixtures
- matter that can vary in composition
- made of two or more substances mixed but not bonded together
- can be separated into substances by physical methods
- two types of mixtures: heterogeneous and homogeneous

_____ **mixtures**
- two or more substances unevenly mixed
- Uneven mixing is visible with unaided eyes or a microscope.

_____ **mixtures (solutions)**
- two or more substances evenly mixed
- Homogeneous mixtures appear uniform under a microscope.

Visual Summary

Substance   Mixtures
Substances have a composition that does not change. The composition of mixtures can vary.

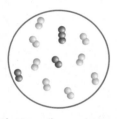

Solutions (homogeneous mixtures) are mixed at the atomic level.

Mixtures contain parts that are not bonded together. These parts can be separated using physical means.

Use Vocabulary

1 **Identify** What is another name for a homogeneous mixture?

_____

2 **Contrast** homogeneous and heterogeneous mixtures.

_____

_____

Understand Key Concepts 🔑

3 **Explain** why a compound is classified as a substance. SC.8.P.8.9

_____

4 **Describe** two tests that you can run to determine if something is a substance or a mixture. SC.8.P.8.9

_____

_____

_____

Interpret Graphics

5 **Compare and contrast** Use the graphic organizer to compare and contrast heterogeneous mixtures and solutions. LA.6.2.2.3

Heterogeneous mixtures   Solutions

Critical Thinking

6 **Explain** the following statement: All compounds are substances, but not all substances are compounds. SC.8.P.8.9

_____

_____

7 **Suppose** you have found an unknown substance in a laboratory. It has the formula H_2O_2 written on the bottle. Is it water? How do you know?

_____

_____

_____

Sports Drinks and Your Body

The Importance of Electrolytes

When you exercise, you sweat. When you're thirsty, do you drink water or a sports drink? Chances are you have seen ads that claim sports drinks are better for you than water because they contain electrolytes. What are electrolytes, and how are they used in your body?

What are electrolytes?

An electrolyte is a charged particle, scientifically known as an ion. Electrolytes can conduct electric charges. Pure water cannot conduct electric charges, but water containing electrolytes can.

Electrolytes in Your Body

Why does your body need electrolytes? Solutions of water and electrolytes surround all of the cells in your body. Electrolytes enable these solutions to carry nerve impulses from one cell to another. Your body's voluntary movements, such as walking, and involuntary movements, such as your heart beating, are caused by nerve impulses. Without electrolytes these nerve impulses cannot move normally.

Replenishing Fluids

Sweat is a solution of water and electrolytes, including sodium. Sports drinks can replace water and electrolytes. Some foods, such as bananas and oranges, also contain electrolytes. However many sports drinks contain ingredients your body doesn't need, such as caffeine and sugar. Unless you are sweating for an extended period of time, you don't need to replace electrolytes, but replacing water is always essential.

Origin of the Sports Drink

Did you know that the first sports drink was created at the University of Florida in 1965? The assistant coach of the football team wondered why players lost so much weight and rarely felt the need to urinate, even though they were eating and drinking. Scientists began researching the effects of heat on the human body. Soon they understood that all the sweat was taking the players' energy, strength, and endurance. The researchers speculated that the electrolytes—primarily sodium and potassium—the players were losing through their sweat were upsetting the body's delicate chemical balance. After careful research and testing, the now famous sports drink was created and had great success. Over the next five years, only one player was taken to the hospital—and he didn't drink any of the sports drink!

Heat

Sweat

It's Your Turn

RESEARCH Study three different sports drinks. Create a bar graph that compares ingredients such as water, sugar, electrolytes, and caffeine in each brand. Draw a conclusion about which type of drink is best for your body. LA.6.4.2.2

Properties of SOLUTIONS

ESSENTIAL QUESTIONS

 Why do some substances dissolve in water and others do not?

 How do concentration and solubility differ?

 How can the solubility of a solute be changed?

Vocabulary

solvent p. 469

solute p. 469

polar molecule p. 470

concentration p. 472

solubility p. 474

saturated solution p. 474

unsaturated solution p. 474

 Florida NGSSS

SC.7.N.1.4 Identify test variables (independent variables) and outcome variables (dependent variables) in an experiment.

SC.8.N.1.2 Design and conduct a study using repeated trials and replication.

SC.8.N.1.6 Understand that scientific investigations involve the collection of relevant empirical evidence, the use of logical reasoning, and the application of imagination in devising hypotheses, predictions, explanations and models to make sense of the collected evidence.

SC.8.P.8.4 Classify and compare substances on the basis of characteristic physical properties that can be demonstrated or measured; for example, density, thermal or electrical conductivity, solubility, magnetic properties, melting and boiling points, and know that these properties are independent of the amount of the sample.

SC.8.P.8.8 Identify basic examples of and compare and classify the properties of compounds, including acids, bases, and salts.

SC.8.P.8.9 Distinguish among mixtures (including solutions) and pure substances.

Also covers: SC.8.N.1.1, LA.6.2.2.3, MA.6.A.3.6

 Launch Lab

SC.8.N.1.6, SC.8.P.8.9

15 minutes

How are they different?

If you have ever looked at a bottle of Italian salad dressing, you know that some substances do not easily form solutions. The oil and vinegar do not mix, and the spices sink to the bottom. How can you describe the difference quantitatively?

Procedure

1. Read and complete a lab safety form.
2. Label one **beaker** A and another **beaker** B.
3. Measure 100 mL of water and pour it into beaker A.
4. Measure 100 mL of water and pour it into beaker B.
5. Add 10 g of **baking soda** to beaker A, and stir with a **plastic spoon** for 2 min or until all the baking soda dissolves, whichever happens first.
6. Add 25 g of **sugar** to beaker B and stir with a plastic spoon for 2 min or until all of the sugar dissolves, whichever happens first.
7. Observe the mixtures in each beaker.

Think About This

1. Predict what would happen if you were to use 200 mL of water instead of 100 mL.

2. Do you think more baking soda might dissolve if you stirred the solution longer?

3. **Key Concept** Why do you think one substance dissolved more easily in water than the other substance? What factors do you think contribute to this difference?

Inquiry Large Fingers?

1. These stalactites and stalagmites, found at the Florida Caverns State Park in Marianna, Florida, were formed over thousands of years. What do you think causes these structures to form? What substances might be involved?

_____

_____

_____

_____

_____

_____

_____

Parts of Solutions

You've read that a solution is a homogeneous mixture. Recall that in a solution, substances are evenly mixed on the atomic level. How does this mixing occur? Dissolving is the process of mixing one substance into another to form a solution. Scientists use two terms to refer to the substances that make up a solution. Generally, the **solvent** _is the substance that exists in the greatest quantity in a solution. All other substances in a solution are_ **solutes**. Recall that air is a solution of 78 percent nitrogen, 21 percent oxygen, and 1 percent other substances. Which substance is the solvent? In air, nitrogen exists in the greatest quantity. Therefore, it is the solvent. The oxygen and other substances are solutes. In this lesson, you will read the terms _solute_ and _solute_ often. Refer back to this page if you forget what these terms mean.

Active Reading **2. Contrast** How do a solute and a solvent differ?

Solute   Solvent

Active Reading

FOLDABLES LA.6.2.2.3

Make a four-tab shutterfold. Label it as shown. Collect information about which solvents dissolve which solutes.

Polar solvents dissolve:	Nonpolar solvents dissolve:

Like Dissolves Like

Polar solvents do not dissolve:	Nonpolar solvents do not dissolve:

Table 1 Types of Solutions

State of Solution	Solvent Is:	Solute Can Be:
Solid	solid	**gas or solid (called alloys)** This saxophone is a solid solution of solid copper and solid zinc.
Liquid	liquid	**solid, liquid, and/or gas** Soda is a liquid solution of liquid water, gaseous carbon dioxide, and solid sugar and other flavorings.
Gas	gas	**gas** This lighted sign contains a gaseous mixture of gaseous argon and gaseous mercury.

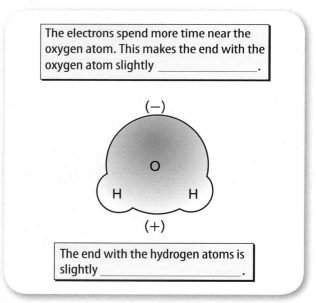

The electrons spend more time near the oxygen atom. This makes the end with the oxygen atom slightly _____.

(−)

O

H   H

(+)

The end with the hydrogen atoms is slightly _____.

Figure 5 🔑 Water is a polar molecule.

Active Reading
3. Recall Label the figure above indicating the charges of the atom.

Types of Solutions

When you think of a solution, you might think of a liquid. However, solutions can exist in all three states of matter—solid, liquid, or gas. The state of the solvent, because it exists in the greatest quantity, determines the state of the solution. **Table 1** contrasts solid, liquid, and gaseous solutions.

Water as a Solvent

Did you know that over 75 percent of your brain and almost 90 percent of your lungs are made of water? Water is one of the few substances on Earth that exists naturally in all three states—solid, liquid, and gas. However, much of this water is not pure water. In nature, water almost always exists as a solution; it contains dissolved solutes. All of Florida's 11,761 miles of water area contain solutes such as sodium chloride ($NaCl$). Why does nearly all water on Earth contain dissolved solutes? The answer has to do with the structure of the water molecule.

The Polarity of Water

A water molecule, such as the one shown in **Figure 5,** is a covalent compound. Recall that atoms are held together with covalent bonds when sharing electrons. In a water molecule, one oxygen atom shares electrons with two hydrogen atoms. However, these electrons are not shared equally. The electrons in the oxygen-hydrogen bonds more often are closer to the oxygen atom than they are to the hydrogen atoms. This unequal sharing of electrons gives the end with the oxygen atom a slightly negative charge. And it gives the end with the hydrogen atoms a slightly positive charge. Because of the unequal sharing of electrons, a water molecule is said to be polar. *A **polar molecule** is a molecule with a slightly negative end and a slightly positive end.* Nonpolar molecules have an even distribution of charge. Solutes and solvents can be polar or nonpolar.

Like Dissolves Like

Water is often called the universal solvent because it dissolves many different substances. But water can't dissolve everything. Why does water dissolve some substances but not others? Water is a polar solvent. Polar solvents dissolve polar solutes easily. Nonpolar solvents dissolve nonpolar solutes easily. This is summarized by the phrase "like dissolves like." Because water is a polar solvent, it dissolves most polar and ionic solutes.

Active Reading **4. Diagram** Why do some substances dissolve in water and some do not?

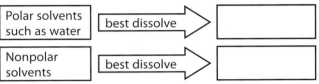

| Polar solvents such as water | best dissolve → | |
| Nonpolar solvents | best dissolve → | |

Polar Solvents and Polar Molecules

Because water molecules are polar, water dissolves groups of other polar molecules. **Figure 6** shows what rubbing alcohol, a substance used as a disinfectant, looks like when it is in a solution with water. Molecules of rubbing alcohol also are polar. Therefore, when rubbing alcohol and water mix, the positive ends of the water molecules are attracted to the negative ends of the alcohol molecules. Similarly, the negative ends of the water molecules are attracted to the positive ends of the alcohol. In this way, alcohol molecules dissolve in the solvent.

Polar Solvents and Ionic Compounds

Many ionic compounds are also soluble in water. Recall that ionic compounds are composed of alternating positive and negative ions. Sodium chloride (NaCl) is an ionic compound composed of sodium ions (Na^+) and chloride ions (Cl^-). When sodium chloride dissolves, these ions are pulled apart by the water molecules. This is shown in **Figure 7**. The negative ends of the water molecules attract the positive sodium ions. The positive ends of the water molecules attract the negative chloride ions.

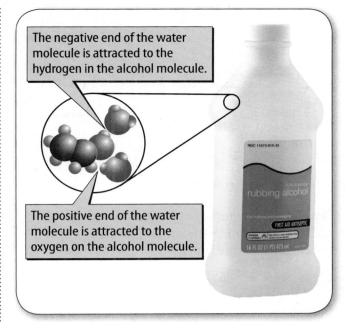

The negative end of the water molecule is attracted to the hydrogen in the alcohol molecule.

The positive end of the water molecule is attracted to the oxygen on the alcohol molecule.

Figure 6 When a polar solute, such as rubbing alcohol, dissolves in a polar solvent, such as water, the poles of the solvent are attracted to the oppositely charged poles of the solute.

The negative ends of the water molecules are attracted to the positive ion.

The positive ends of the water molecules are attracted to the negative ion.

Figure 7 When ionic solutes dissolve, the positive poles of the solvent are attracted to the negative ions. The negative poles of the solvent are attracted to the positive ions.

More solute   Less solute

Equal amounts of water

Concentrated   Dilute

Figure 8 The volumes of both drinks are the same, but the glass on the left contains more solute than the solution on the right.

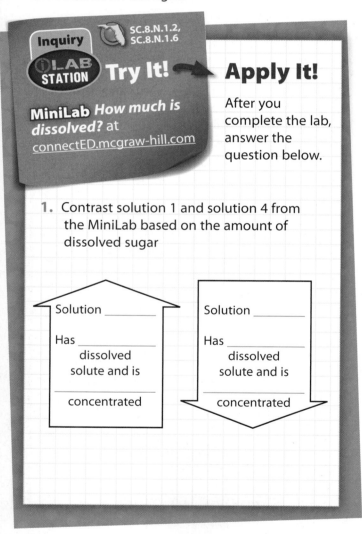

SC.8.N.1.2, SC.8.N.1.6

Inquiry

LAB STATION **Try It!**

MiniLab *How much is dissolved?* at connectED.mcgraw-hill.com

Apply It!

After you complete the lab, answer the question below.

1. Contrast solution 1 and solution 4 from the MiniLab based on the amount of dissolved sugar

Solution _____

Has _____ dissolved solute and is _____ concentrated

Solution _____

Has _____ dissolved solute and is _____ concentrated

Active Reading
5. Infer Why is the term *dilute* not a precise way to describe concentration?

_____

_____

_____

_____

Concentration—How much is dissolved?

Have you ever tasted a spoonful of soup and wished it had more salt in it? In a way, your taste buds were measuring the amount, or concentration, of salt in the soup. **Concentration** *is the amount of a particular solute in a given amount of solution.* In the soup, salt is a solute. Saltier soup has a higher concentration of salt. Soup with less salt has a lower concentration of salt. Look at the glasses of fruit drink in **Figure 8.** Which drink has a higher concentration of solute? The darker blue drink has a higher concentration of solute.

Concentrated and Dilute Solutions

One way to describe the saltier soup is to say that it is more concentrated. The less-salty soup is more dilute. The terms *concentrated* and *dilute* are one way to describe how much solute is dissolved in a solution. However, these terms don't state the exact amount of solute dissolved. What one person thinks is concentrated might be what another person thinks is dilute. Soup that tastes too salty to you might be perfect for someone else. How can concentration be described more precisely?

Describing Concentration Using Quantity

A more precise way to describe concentration is to state the quantity of solute in a given quantity of solution. When a solution is made of a solid dissolved in a liquid, such as salt in water, concentration is the mass of solute in a given volume of solution. Mass usually is stated in grams, and volume usually is stated in liters. For example, concentration can be stated as grams of solute per 1 L of solution. However, concentration can be stated using any units of mass or volume.

Calculating Concentration—Mass per Volume

One way that concentration can be calculated is by the following equation:

$$\text{Concentration } (C) = \frac{\text{mass of solute } (m)}{\text{volume of solution } (V)}$$

To calculate concentration, you must know both the mass of solute and the volume of solution that contains this mass. Then divide the mass of solute by the volume of solution.

Concentration—Percent by Volume

Not all solutions are made of a solid dissolved in a liquid. If a solution contains only liquids or gases, its concentration is stated as the volume of solute in a given volume of solution. In this case, the units of volume must be the same—usually mL or L. Because the units match, the concentration can be stated as a percentage. Percent by volume is calculated by dividing the volume of solute by the total volume of solution and then multiplying the quotient by 100. For example, if a container of orange drink contains 3 mL of acetic acid in a 1,000-mL container, the concentration is 0.3 percent.

Math Skills   MA.6.A.3.6

Solve for Concentration

Suppose you want to calculate the concentration of salt in a **0.4 L** can of soup. The back of the can says it contains **1.6 g** of salt. What is its concentration in g/L? In other words, how much salt would be contained in 1 L of soup?

1. This is what you know: mass: **1.6 g**
                                 volume: **0.4 L**

2. This is what you need to find: concentration: C

3. Use this formula: $C = \dfrac{m}{V}$

4. Substitute: $C = \dfrac{1.6\text{ g}}{0.4\text{ L}} = 4 \text{ g/L}$
the values for m and v
into the formula and divide.

Answer: The concentration is 4 g/L. As you might expect, 0.4 L of soup contains less salt (1.6 g) than 1 L of soup (4 g). However, the concentration of both amounts of soup is the same—4 g/L.

6. What is the concentration of 5 g of sugar in 0.2 L of solution?

For more Concentration Calculation practice, check out the Math Skill Review on page 477.

Active Reading
7. Contrast How do concentration and solubility differ?

Concentration

Solubility

Active Reading
8. Explain Using arrows, show what happens to the concentration of a solution as the amount of solvent or solute changes.

↑ Solvent		Concentration
Solvent		Concentration
Solute	↓	Concentration
Solute		Concentration

9. Visual Check
Extrapolate How many grams of KNO₃ will dissolve in 100 g of water at 10°C?

Solubility—How much can dissolve?

Have you ever put too much sugar into a glass of iced tea? What happens? Not all of the sugar dissolves. You stir and stir, but there is still sugar at the bottom of the glass. That is because there is a limit to how much solute (sugar) can be dissolved in a solvent (water). **Solubility** (sahl yuh BIH luh tee) *is the maximum amount of solute that can dissolve in a given amount of solvent at a given temperature and pressure.* If a substance has a high solubility, more of it can dissolve in a given solvent.

Saturated and Unsaturated Solutions

If you add water to a dry sponge, the sponge absorbs the water. If you keep adding water, the sponge becomes saturated. It can't hold any more water. This is analogous (uh NA luh gus), or similar, to what happens when you stir too much sugar into iced tea. Some sugar dissolves, but the excess sugar does not dissolve. The solution is saturated. *A* **saturated solution** *is a solution that contains the maximum amount of solute the solution can hold at a given temperature and pressure.* An **unsaturated solution** *is a solution that can still dissolve more solute at a given temperature and pressure.*

Factors that Affect How Much Can Dissolve

Can you change the amount of a particular solute that can dissolve in a solvent? Yes. Recall the definition of solubility—the maximum amount of solute that can dissolve in a given amount of solvent at a given temperature and pressure. Changing either temperature or pressure changes how much solute can dissolve in a solvent.

Figure 9 Some solids are more soluble in warmer liquids than cooler ones. Other solids are less soluble in warmer liquids than cooler ones.

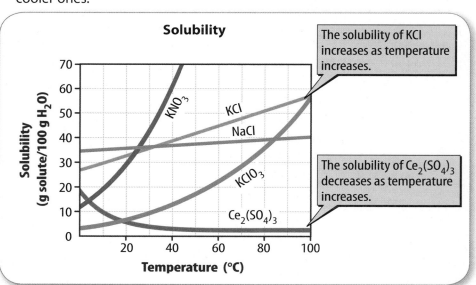

Effect of Temperature Have you noticed that more sugar dissolves in hot tea than in iced tea? The solubility of sugar in water increases with temperature. This is true for many solid solutes, as shown in **Figure 9.** Notice that some solutes become less soluble when temperature is increased.

How does temperature affect the solubility of a gas in a liquid? Recall that soda, or soft drinks, contains carbon dioxide, a gaseous solute, dissolved in liquid water. The bubbles you see in soda are made of undissolved carbon dioxide. Have you ever noticed that more carbon dioxide bubbles out when you open a warm can of soda than when you open a cold can? This is because the solubility of a gas in a liquid decreases when the temperature of the solution increases.

Effect of Pressure What keeps carbon dioxide dissolved in an unopened can of soda? In a can, the carbon dioxide in the space above the liquid soda is under pressure. This causes the gas to move to an area of lower pressure—the solvent. The gas moves into the solvent, and a solution is formed. When the can is opened, as shown in **Figure 10,** this pressure is released, and the carbon dioxide gas leaves the solution. Pressure does not affect the solubility of a solid solute in a liquid.

 10. Describe How can the solubility of a solute be changed?

_____

_____

How Fast a Solute Dissolves

Temperature and pressure can affect how much solute dissolves. If solute and solvent particles come into contact more often, the solute dissolves faster. **Figure 11** shows three ways to increase how often solute particles contact solvent particles. Each of these methods will make a solute dissolve faster. However, it is important to note that stirring the solution or crushing the solute will not make more solute dissolve.

Figure 10 When the pressure of a gas is increased, it becomes more soluble in a liquid. When the can is opened, this pressure is lowered and the gas leaves the solution.

Stirring the solution

Crushing the solute

Increasing the temperature

Figure 11 Several factors can affect how quickly a solute will dissolve in a solution. However, dissolving more quickly won't necessarily make more solute dissolve.

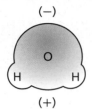

(−)

O

H   H

(+)

Polar molecule

Substances dissolve in other substances that have similar polarity. In other words, like dissolves like.

Concentration is the amount of substance that is dissolved. Solubility is the maximum amount that can dissolve.

Both temperature and pressure affect the solubility of solutes in solutions.

Use Vocabulary

1 **Define** *polar molecule* in your own words.

_____

_____

Understand Key Concepts 🔑

2 **Explain** how you could use the solubility of a substance to make a saturated solution.

_____

_____

3 **Predict** whether an ionic compound will dissolve in a nonpolar solvent.

_____

_____

Interpret Graphics

4 **Organize** Use the graphic organizer below to organize three factors that increase the speed a solute dissolves in a liquid. LA.6.2.2.3

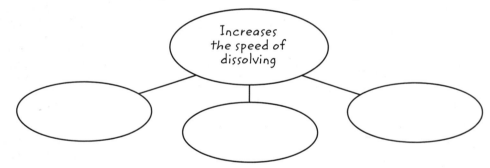

Increases the speed of dissolving

Critical Thinking

5 **Explain** A student wants to increase the maximum amount of sugar that can dissolve in water. She crushes the sugar and then stirs it into the water. Does this work? Why or why not?

_____

_____

Inquiry 🌀 SC.8.P.8.4, SC.8.P.8.8, SC.7.N.1.4, SC.8.N.1.1, SC.8.N.1.6

ⓘLAB STATION **Try It!**

Skill Lab *How does a solute affect the conductivity of a solution?* at connectED.mcgraw-hill.com

Math Skills   MA.6.A.3.6

6 Use ratios to explain how a tablespoon of soup and a cup of the same soup have the same concentration.

Math Skills MA.6.A.3.6

Calculate Concentration Practice

1 How many grams of salt are in 5 L of a solution with a concentration of 3 g/L?

2 Suppose you add water to a 6 g of sugar to make a solution with a concentration of 3 g/L. What is the total volume of the solution?

3 What is the concentration of KCl if 10 g are dissolved in enough water to make 12 L?

4 How many grams of KMnO4 would you get if you evaporated the water from 85.75 mL of 1.27 g/L solution?

5 Your teacher created a 2 g/L salt water solution (solution A) and a 2.5 g/L salt water solution (solution B), both in 500 mL of water. Using the concentration equation, which solution is more concentrated? How do you know?

Math Skills MA.6.A.3.6

Difference in pH Practice

After learning about pH in Lesson 3, come back and practice calculating pH strength.

6 Suppose you have two solutions that have a pH of 1 and a pH of 4. How much more acidic is the solution with a pH of 1 than the solution with a pH of 4?

7 What is the difference in basicity between drain cleaner (pH 13) and ammonia (pH 11.9)?

8 Which is more acidic—milk of magnesia or lemon juice? What is the difference in acidity between the two? (milk of magnesia = pH 10.5, lemon juice = pH 2.5)

Acid and Base
SOLUTIONS

Vocabulary

acid p. 480

hydronium ion p. 480

base p. 480

pH p. 482

indicator p. 484

 Florida NGSSS

SC.6.N.1.4 Discuss, compare, and negotiate methods used, results obtained, and explanations among groups of students conducting the same investigation.

SC.8.N.1.6 Understand that scientific investigations involve the collection of relevant empirical evidence, the use of logical reasoning, and the application of imagination in devising hypotheses, predictions, explanations and models to make sense of the collected evidence.

SC.8.P.8.8 Identify basic examples of and compare and classify the properties of compounds, including acids, bases, and salts.

SC.8.P.8.9 Distinguish among mixtures (including solutions) and pure substances.

MA.6.A.3.6 Construct and analyze tables, graphs, and equations to describe linear functions and other simple relations using both common language and algebraic notation.

MA.6.S.6.2 Select and analyze the measures of central tendency or variability to represent, describe, analyze, and/or summarize a data set for the purposes of answering questions appropriately.

Also covers: SC.8.N.1.1, LA.6.2.2.3

inquiry Launch Lab SC.8.N.1.6, SC.8.P.8.8

15 minutes

What color is it?

Did you know that all rain is naturally acidic? As raindrops fall through the air, they pick up molecules of carbon dioxide. An acid called carbonic acid is formed when the water molecules react with the carbon dioxide molecules. An indicator is a substance that can be used to tell if a solution is acidic, basic, or neutral.

Procedure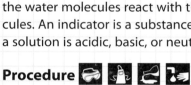

1. Read and complete a lab safety form.

2. Half fill a **beaker** with the **colored solution**.

3. Place one end of a **straw** into the solution.
   ⚠ **Caution:** *Do not suck liquid through the straw.*

4. Blow through the straw, making bubbles in the solution. Continue blowing, and count how many times you have to blow bubbles until you observe a change.

Think About This

1. Describe what change you saw take place.

2. What do you think made this change occur?

3. How do you think the results would have been different if you had held your breath for several seconds before blowing through the straw?

4. 🔑 **Key Concept** Using the terms *acidic* and *basic,* explain what you have learned about the colored solution being used.

Inquiry What's eating her?

1. When this statue was first carved, it didn't have any of these odd-shaped marks on its surface. What do you think caused these marks? What would happen to the statue if this process were to continue?

_____

_____

_____

_____

Active Reading **2. Predict** What do you already know about acids and bases? Place an _A_ before properties of acids and a _B_ before properties of bases. Note that some properties apply to both.

_____ Provide sour taste in food

_____ Found in saliva

_____ Can damage skin and eyes

_____ Slippery

_____ Provide bitter taste in food

_____ Found in milk

_____ Helps plants grow

What are acids and bases?

Would someone ever drink an acid? At first thought, you might answer no. After all, when people think of acids, they often think of acids such as those found in batteries or in acid rain. However, acids are found in other items, including milk, vinegar, fruits, and green leafy vegetables. Some examples of acids that you might eat are shown in **Figure 12**. Along with the word _acid_, you might have heard the word _base_. Like acids, you can also find bases in your home. Detergent, antacids, and baking soda are examples of items that contain bases. But acids and bases are found in more than just household goods. As you will learn in this lesson, they are necessities for our daily life.

Figure 12 You might be surprised to learn that acids are common in the foods you eat.

Active Reading **3. Differentiate** What happens when acids and bases dissolve in water?

> **Acids**
>
> **Bases**

4. Visual Check

Infer How is dissolving an acid, shown above on the right similar to dissolving ammonia, shown below?

_____

_____

_____

Acids in Water

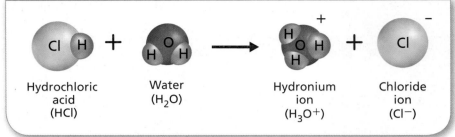

Figure 13 🔑 Acids, such as hydrochloric acid, produce hydronium ions when they dissolve in water.

Acids

Have you ever tasted the sourness of a lemon or a grapefruit? This sour taste is due to the acid in the fruit. *An* **acid** *is a substance that produces a hydronium ion (H_3O^+) when dissolved in water.* Nearly all acid molecules contain one or more hydrogen atoms (H). When an acid mixes with water, this hydrogen atom separates from the acid. It quickly combines with a water molecule, resulting in a hydronium ion. This process is shown in **Figure 13.** *A* **hydronium ion**, H_3O^+, *is a positively charged ion formed when an acid dissolves in water.*

Bases

A **base** *is a substance that produces hydroxide ions (OH^-) when dissolved in water.* When a hydroxide compound such as sodium hydroxide (NaOH) mixes with water, hydroxide ions separate from the base and form hydroxide ions (OH^-) in water. Some bases, such as ammonia (NH_3), do not contain hydroxide ions. These bases produce hydroxide ions by taking hydrogen atoms away from water, leaving hydroxide ions (OH^-). This process is shown in **Figure 14.** Some properties and uses of acids and bases are shown in **Table 2.**

Figure 14 🔑 Bases, such as sodium hydroxide and ammonia, produce hydroxide ions when they dissolve in water.

Bases in Water

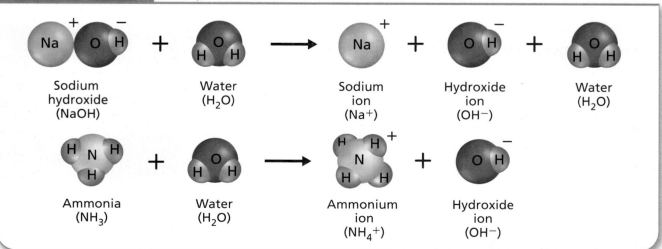

Table 2 Properties and Uses of Acids and Bases

	Acids	Bases
Ions produced	Acids produce H_3O^+ in water.	Bases produce OH^- ions in water.
Examples	• hydrochloric acid, HCl • acetic acid, CH_3COOH • citric acid, $H_3C_6H_5O_7$ • lactic acid, $C_3H_6O_3$	• sodium hydroxide, NaOH • ammonia, NH_3 • sodium carbonate, Na_2CO_3 • calcium hydroxide, $Ca(OH)_2$
Some properties	• Acids provide the sour taste in food (never taste acids in the laboratory). • Most can damage skin and eyes. • Acids react with some metals to produce hydrogen gas. • H_3O^+ ions can conduct electricity in water. • Acids react with bases to form neutral solutions.	• Bases provide the bitter taste in food (never taste bases in the laboratory). • Most can damage skin and eyes. • Bases are slippery when mixed with water. • OH^- ions can conduct electricity in water. • Bases react with acids to form neutral solutions.
Some uses	• Acids are responsible for for natural and artificial flavoring in foods, such as fruits. • Lactic acid is found in milk. • Acid in your stomach breaks down food. • Blueberries, strawberries, and many vegetable crops grow better in acidic soil. • Acids are used to make products such as fertilizers, detergents, and plastics.	• Bases are found in natural and artificial flavorings in food, such as cocoa beans. • Antacids neutralize stomach acid, alleviating heartburn. • Bases are found in cleaners such as shampoo, dish detergent, and window cleaner. • Many flowers grow better in basic soil. • Bases are used to make products such as rayon and paper.

What is pH?

Have you ever seen someone test the water in a swimming pool? It is likely that the person was testing the pH of the water. Swimming pool water should have a pH around 7.4. If the pH of the water is higher or lower than 7.4, the water might become cloudy, burn swimmers' eyes, or contain too many bacteria. What does a pH of 7.4 mean?

Hydronium Ions

*The **pH** is an inverse measure of the concentration of hydronium ions (H_3O^+) in a solution.* What does *inverse* mean? It means that as one thing increases, another thing decreases. In this case, as the concentration of hydronium ions increases, pH decreases. A solution with a lower pH is more acidic. As the concentration of hydronium ions decreases, the pH increases. A solution with a higher pH is more basic. This relationship is shown in **Figure 15.**

Balance of Hydronium and Hydroxide Ions

All acid and base solutions contain both hydronium and hydroxide ions. Acids have a greater concentration of hydronium ions (H_3O^+) than hydroxide ions (OH^-). Bases have a greater concentration of hydroxide ions than hydronium ions. Brackets around a chemical formula mean *concentration*.

Acids	$[H_3O^+]$ > $[OH^-]$
Neutral	$[H_3O^+]$ = $[OH^-]$
Bases	$[H_3O^+]$ < $[OH^-]$

Active Reading **5. Model** How does the concentration of hydronium ions affect pH?

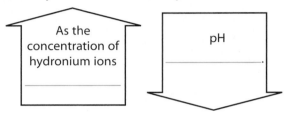

As the concentration of hydronium ions _____

pH _____

Salt Formation

A salt is a compound formed when the negative ions from an acid combine with the positive ions from a base. In the reaction between HCl and NaOH, a salt, sodium chloride (NaCl), is formed.

pH Scale

Figure 15 Notice that as hydronium concentration increases, the pH decreases.

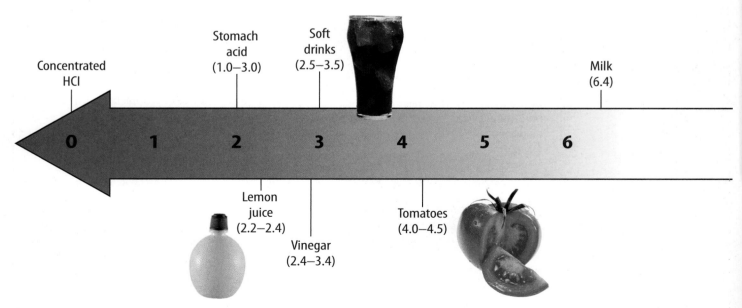

Concentrated HCl

Stomach acid (1.0–3.0)

Soft drinks (2.5–3.5)

Milk (6.4)

0   1   2   3   4   5   6

Lemon juice (2.2–2.4)

Vinegar (2.4–3.4)

Tomatoes (4.0–4.5)

The pH Scale

The pH scale is used to indicate how acidic or basic a solution is. Notice in **Figure 15** that the pH scale contains values that range from below 0 to above 14. Acids have a pH below 7. Bases have a pH above 7. Solutions that are neutral have a pH of 7—they are neither acidic nor basic.

You might be wondering what the numbers on the pH scale mean. How is the concentration of hydronium ions different in a solution with a pH of 1 from the concentration in a solution with a pH of 2? A change in one pH unit represents a tenfold change in the acidity or basicity of a solution. For example, if one solution has a pH of 1 and a second solution has a pH of 2, then the first solution is not twice as acidic as the second solution; it is ten times more acidic.

The difference in acidity or basicity between two solutions is represented by 10^n, where n is the difference between the two pH values. For example, how much more acidic is a solution with a pH of 1 than a solution with pH of 3? First, calculate the difference, n, between the two pH values: $n = 3 - 1 = 2$. Then use the formula, 10^n, to calculate the difference in acidity: $10^2 = 100$. A solution with a pH of 1 is 100 times more acidic than a solution with a pH of 3.

To practice differences in pH, go to page 477.

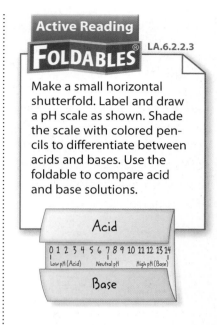

Active Reading

FOLDABLES LA.6.2.2.3

Make a small horizontal shutterfold. Label and draw a pH scale as shown. Shade the scale with colored pencils to differentiate between acids and bases. Use the foldable to compare acid and base solutions.

Acid

0 1 2 3 4 5 6 7 8 9 10 11 12 13 14
Low pH (Acid)   Neutral pH   High pH (Base)

Base

✓ **6. Visual Check**
Review Is a tomato more or less acidic than detergent?

_____

What is the difference in acidity?

_____

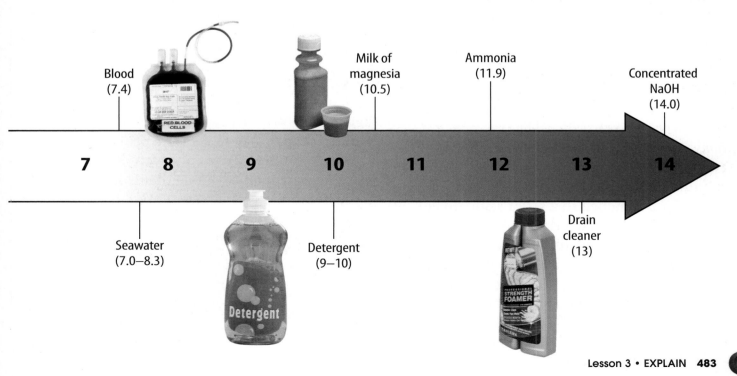

Blood (7.4)

Milk of magnesia (10.5)

Ammonia (11.9)

Concentrated NaOH (14.0)

7   8   9   10   11   12   13   14

Seawater (7.0–8.3)

Detergent (9–10)

Drain cleaner (13)

Apply It! After you complete the lab, answer these questions below.

1. Choose two acids that you tested in the MiniLab and calculate how much more acidic one is from the other.

2.   Why would the pH of soils be important to Florida farmers who grow orange crops?

How is pH measured?

How is the pH of a solution, such as swimming pool water, measured? Water test kits contain chemicals that change color when an acid or a base is added to them. These chemicals are called indicators.

pH Indicators

Indicators can be used to measure the approximate pH of a solution. *An* **indicator** *is a compound that changes color at different pH values when it reacts with acidic or basic solutions.* The pH of a solution is measured by adding a drop or two of the indicator to the solution. When the solution changes color, this color is matched to a set of standard colors that correspond to certain pH values. There are many different indicators—each indicator changes color over a specific range of pH values. For example, bromthymol blue is an indicator that changes from yellow to green to blue between pH 6 and pH 7.6.

pH Testing Strips

pH also can be measured using pH testing strips. The strips contain an indicator that changes to a variety of colors over a range of pH values. To use pH strips, dip the strip into the solution. Then match the resulting color to the list of standard colors that represent specific pH values.

pH Meters

Although pH strips are quick and easy, they provide only an approximate pH value. A more accurate way to measure pH is to use a pH meter. A pH meter is an electronic instrument with an electrode that is sensitive to the hydronium ion concentration in solution.

Active Reading 7. **List** What are two methods that can be used to measure the pH of a solution?

_____

_____

_____

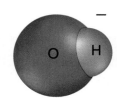

Acids contain hydrogen ions that are released and form hydronium ions in water. Bases are substances that form hydroxide ions when dissolved in water.

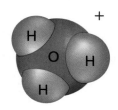

Hydronium ion concentration changes inversely with pH. This means that as hydronium ion concentration increases, the pH decreases.

pH can be measured using indicators or digital pH meters.

Inquiry

⊕ LAB STATION

SC.8.P.8.8,
SC.6.N.1.4,
SC.8.N.1.1,
SC.8.N.1.6,
MA.6.A.3.6,
MA.6.S.6.2

Try It!

Inquiry Lab *Can the pH of a solution be changed?* at connectED.mcgraw-hill.com

Use Vocabulary

1 A measure of the concentration of hydronium ions (H_3O^+) in a solution is _____ .

2 A(n) _____ is used to determine the approximate pH of a solution.

Understand Key Concepts 🔑

3 **Describe** What happens to a hydrogen atom in an acid when the acid is dissolved in water?

_____

_____

4 **Explain** How does pH vary with hydronium ion and hydroxide ion concentrations in water?

_____

_____

5 **Show** Does an acidic solution contain hydroxide ions? Explain your answer with a diagram.

Interpret Graphics

6 **Contrast** Use the graphic organizer to describe and contrast three ways to measure pH. In the organizer, describe which methods are most and least accurate. LA.6.2.2.3

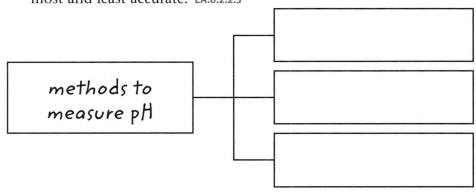

Critical Thinking

7 **Describe** the concentration of hydronium ions and hydroxide ions when a base is added slowly to a white vinegar solution. The pH of white vinegar is 3.1. SC.8.P.8.8

_____

_____

_____

Chapter 12 Study Guide

Think About It! Mixtures and substances are the two main classifications of matter. A solution is a type of mixture. Solutions can be described by the concentration and type of solute they contain.

 Key Concepts Summary

Vocabulary

LESSON 1 Substances and Mixtures

- **Substances** have a fixed composition. The composition of **mixtures** can vary.
- **Solutions** and **heterogeneous mixtures** are both types of mixtures. Solutions are mixed at the atomic level.
- Mixtures contain parts that are not bonded together. These parts can be separated using physical means, and their properties can be seen in the solution.

substance p. 462

mixture p. 462

heterogeneous mixture p. 463

homogeneous mixture p. 463

solution p. 463

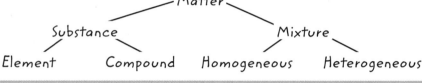

LESSON 2 Properties of Solutions

- Substances dissolve other substances that have a similar polarity. In other words, like dissolves like.
- **Concentration** is the amount of a **solute** that is dissolved. **Solubility** is the maximum amount of a solute that can dissolve.
- Both temperature and pressure affect the solubility of solutes in solutions.

solvent p. 469

solute p. 469

polar molecule p. 470

concentration p. 472

solubility p. 474

saturated solution p. 474

unsaturated solution p. 474

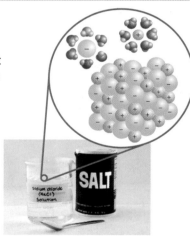

LESSON 3 Acid and Base Solutions

- **Acids** contain hydrogen ions that are released and form **hydronium ions** in water. **Bases** are substances that form hydroxide ions when dissolved in water.
- Hydronium ion concentration changes inversely with **pH.** This means that as hydronium ion concentration increases, the pH decreases.
- pH can be measured using **indicators** or digital pH meters.

acid p. 480

hydronium ion p. 480

base p. 480

pH p. 482

indicator p. 484

acidic   basic

0 1 2 3 4 5 6 7 8 9 10 11 12 13 14

Active Reading

FOLDABLES® Chapter Project

Assemble your lesson Foldables as shown to make a Chapter Project. Use the project to review what you have learned in this chapter.

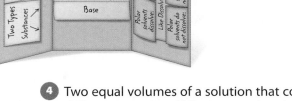

Use Vocabulary

1 The parts of a _____ can be seen with unaided eyes or with a microscope.

2 It is impossible to tell the difference between a solution and a _____ just by looking at them.

3 Water dissolves other _____ easily.

4 Two equal volumes of a solution that contain different amounts of the same solute have a different _____.

5 As _____ concentration decreases, pH increases.

6 A(n) _____ can be added to milk to neutralize it.

Link Vocabulary and Key Concepts

Use vocabulary terms from the previous page to complete the concept map.

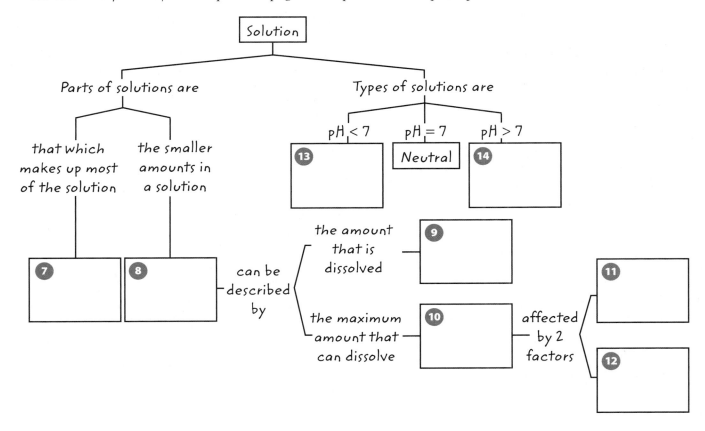

Fill in the correct answer choice.

🔑 Understand Key Concepts

1 Which is a solution? SC.8.P.8.9

  Ⓐ copper
  Ⓑ vinegar
  Ⓒ pure water
  Ⓓ a raisin cookie

2 The graph below shows the solubility of sodium chloride (NaCl) in water. MA.6.A.3.6

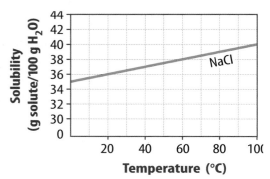

What mass of sodium chloride is needed to form a saturated solution at 80°C?

  Ⓐ 30 g
  Ⓑ 40 g
  Ⓒ 50 g
  Ⓓ 60 g

3 What would you add to a solution with a pH of 1.5 to obtain a solution with a pH of 7? SC.8.P.8.8

  Ⓐ milk (pH 6.4)
  Ⓑ vinegar (pH 3.0)
  Ⓒ lye (pH 13.0)
  Ⓓ coffee (pH 5.0)

4 Which can change the solubility of a solid in a liquid? SC.8.P.8.4

  Ⓐ crushing the solute
  Ⓑ stirring the solute
  Ⓒ increasing the pressure of the solution
  Ⓓ increasing the temperature of the solution

Critical Thinking

5 Infer How can you tell which component in a solution is the solvent? SC.8.P.8.4

_____

_____

_____

6 Predict The graph below shows the solubility of potassium chloride (KCl) in water.

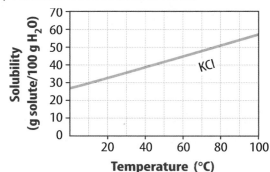

Imagine you have made a solution that contains 50 g of potassium chloride (KCl) in 100 g of solution. Predict what you would observe as you gradually increased the temperature from 0°C to 100°C. MA.6.A.3.6

_____

_____

_____

7 Organize The pH of three solutions is shown below. SC.8.P.8.8

    Milk (pH 6.7)
    Coffee (pH 5)
    Ammonia (pH 11.6)

Place these solutions in order of

  Ⓐ most acidic to least acidic

  _____

  Ⓑ most basic to least basic

  _____

  Ⓒ highest OH⁻ concentration to lowest OH⁻ concentration

8 **Explain** The pH of a solution is inversely related to the concentration of hydronium ions in solution. Explain what this means. MA.6.A.3.6

_____

_____

_____

9 **Design** a method to determine the solubility of an unknown substance at 50°C. SC.8.P.8.4

_____

_____

_____

_____

Writing in Science

10 **Compose** A haiku is a poem containing three lines of five, seven, and five syllables, respectively. Write a haiku on a separate sheet of paper describing what happens when an acid is dissolved in water. LA.6.2.2.3

Big Idea Review

11 What are solutions? List at least three ways a solution can be described. SC.8.P.8.9

_____

_____

_____

_____

12 How do solutions differ from other types of matter? SC.8.P.8.9

_____

_____

_____

Math Skills   MA.6.A.3.6

Calculate Concentration

13 Calculate the concentration of sugar in g/L in a solution that contains 40 g of sugar in 100 mL of solution. There are 1,000 mL in 1 L.

14 There are many ways to make a solution of a given concentration. What are two ways you could make a sugar solution with a concentration of 100 g/L?

15 A salt solution has a concentration of 200 g/L. How many grams of salt are contained in 500 mL of this solution? How many grams of salt would be contained in 2 L of this solution?

Fill in the correct answer choice.

Multiple Choice

Use the figures below to answer question 1.

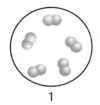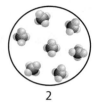

1   2

1 Which statement describes the two figures? SC.8.P.8.9

  Ⓐ Both 1 and 2 are mixtures.

  Ⓑ Both 1 and 2 are substances.

  Ⓒ 1 is a mixture and 2 is a substance.

  Ⓓ 1 is a substance and 2 is a mixture.

2 Which statement is an accurate comparison of solutions and homogeneous mixtures? SC.8.P.8.9

  Ⓕ They are the same.

  Ⓖ They are opposites.

  Ⓗ Solutions are more evenly mixed than homogeneous mixtures.

  Ⓘ Homogeneous mixtures are more evenly mixed than solutions.

3 A worker uses a magnet to remove bits of iron from a powdered sample. Which describes the sample before the worker used the magnet to remove the iron? SC.8.P.8.9

  Ⓐ The sample is a compound because the iron was removed using a physical method.

  Ⓑ The sample is a compound because the iron was removed using a chemical change.

  Ⓒ The sample is a mixture because the iron was removed using a chemical change.

  Ⓓ The sample is a mixture because the iron was removed using a physical method.

4 A beaker contains a mixture of sand and small pebbles. What kind of mixture is this? SC.8.P.8.9

  Ⓕ compound

  Ⓖ heterogeneous

  Ⓗ homogeneous

  Ⓘ solution

5 What ions must be present in the greatest amount in a solution with a pH of 8.5? SC.8.P.8.8

  Ⓐ hydrogen ions

  Ⓑ oxygen ions

  Ⓒ hydronium ions

  Ⓓ hydroxide ions

6 According to the pH scale, which pH measurement is basic? SC.8.P.8.8

pH Scale

0 1 2 3 4 5 6 7 8 9 10 11 12 13 14

  Ⓕ 7.0

  Ⓖ 9.5

  Ⓗ 5.5

  Ⓘ 1.2

7 Which is a property of an acidic solution? SC.8.P.8.8

  Ⓐ It tastes sour.

  Ⓑ It feels slippery.

  Ⓒ It is used in many cleaning products.

  Ⓓ It tastes bitter.

8 The neutralization reaction between sodium hydroxide (NaOH) and hydrogen chloride (HCl) forms a salt. What is the correct chemical formula? SC.8.P.8.8

   (F) NaCl

   (G) ClNa

   (H) Na_2Cl

   (I) Cl_2Na

Use the table below to answer question 9.

Sample solution	Change in blue litmus	Change in red litmus
1	turns red	no change
2	no change	turns blue
3	turns red	no change
4	no change	no change

9 A scientist collects the data above using litmus paper. Blue litmus paper is a type of pH indicator that turns red when placed in an acidic solution. Red litmus paper is an indicator that turns blue when placed in a basic solution. Neutral solutions cause no change in either color of litmus paper. Which sample solution must be a base? SC.8.P.8.8

   (A) solution 1

   (B) solution 2

   (C) solution 3

   (D) solution 4

10 Which statement is an accurate description for bases? SC.8.P.8.8

   (F) They decrease the concentration of hydroxide ions when dissolved in water.

   (G) They increase the concentration of hydroxide ions when dissolved in water.

   (H) They increase the concentration of hydronium ions when dissolved in water.

   (I) They have no effect on the concentration of hydroxide ions when dissolved in water.

11 Which product has a bitter taste, is slippery when mixed with water, and can damage skin and eyes? SC.8.P.8.8

   (A) sodium hydroxide

   (B) citric acid

   (C) lactic acid

   (D) acetic acid

12 How does a solution with a pH of 2 compare to a solution with a pH of 1? SC.8.P.8.8

   (F) The pH 2 solution is two times more acidic than that with a pH of 1.

   (G) The pH 1 solution is ten times more acidic than that with a pH of 2.

   (H) The pH 1 solution is two times more acidic than that with a pH of 2.

   (I) The pH 2 solution is ten times more acidic than that with a pH of 1.

NEED EXTRA HELP?

If You Missed Question...	1	2	3	4	5	6	7	8	9	10	11	12
Go to Lesson...	1	1	1	1	3	3	3	3	3	3	3	3

Benchmark Mini-Assessment   **Chapter 12 • Lesson 1**

Multiple Choice *Bubble the correct answer.*

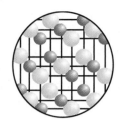

1. The image above represents a(n) **SC.8.P.8.9**

   (A) compound.

   (B) element.

   (C) heterogeneous mixture.

   (D) homogeneous mixture.

2. Which is another name for a homogeneous mixture? **SC.8.P.8.9**

   (F) compound

   (G) element

   (H) solution

   (I) substance

3. Which is a homogeneous mixture? **SC.8.P.8.9**

   (A) blood

   (B) granite

   (C) lemonade

   (D) pizza

**Blood
Samples**

4. Which is true about the substances shown in the figure above? **SC.8.P.8.9**

   (F) The different substances in each sample are bonded together.

   (G) The different substances in each sample are evenly mixed at the atomic level.

   (H) The different substances in each sample are not evenly mixed.

   (I) The different substances in each sample cannot be separated from each other.

Copyright © Glencoe/McGraw-Hill, a division of The McGraw-Hill Companies, Inc.

Multiple Choice *Bubble the correct answer.*

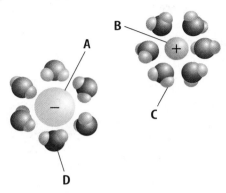

1. The image above shows sodium and chloride ions in solution. Which label in the diagram represents the chloride ion? **SC.8.P.8.9**

    (A) Label A

    (B) Label B

    (C) Label C

    (D) Label D

2. Which does NOT describe a water molecule? **SC.8.P.8.9**

    (F) The molecule has a bent shape.

    (G) The molecule is an ionic compound.

    (H) The molecule dissolves most polar solutes.

    (I) The oxygen end of the molecule has a slightly negative charge.

3. Rosalinda is thirsty and wants to make a glass of lemonade as quickly as possible. She needs to add sugar to lemon juice and water to make the lemonade. Which type of sugar should she use? **SC.8.P.8.4**

    (A) granulated sugar with fine grains

    (B) raw sugar with coarse grains

    (C) sugar from rock candy

    (D) sugar formed into cubes

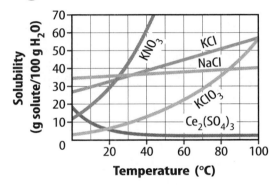

4. The solubility of which substance in the figure above increases MOST quickly as temperature increases? **SC.8.P.8.4**

    (F) KCl

    (G) $KClO_3$

    (H) KNO_3

    (I) NaCl

Copyright © Glencoe/McGraw-Hill, a division of The McGraw-Hill Companies, Inc.

Multiple Choice *Bubble the correct answer.*

Use the images below to answer questions 1–4.

Reactants and Products in Two Reactions

Reaction A

Reaction B

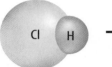

1. One of the products of Reaction A is
SC.8.P.8.8

  (A) an ammonium ion.

  (B) an oxygen ion.

  (C) a hydronium ion.

  (D) a sodium ion.

2. Which product could contain a substance formed by Reaction B? **SC.8.P.8.8**

  (F) antacid

  (G) milk

  (H) baking soda

  (I) dark chocolate

3. Reaction B could produce a substance with a pH of **SC.8.P.8.8**

  (A) 4.

  (B) 7.

  (C) 10.

  (D) 14.

4. Which is one of the reactants in Reaction A? **SC.8.P.8.8**

  (F) copper

  (G) silver

  (H) sodium

  (I) water

Copyright © Glencoe/McGraw-Hill, a division of The McGraw-Hill Companies, Inc.

Notes

Name _____ Date _____

Atom of Life

When you walk through an ecosystem, there are several different atoms that make up the living things in the ecosystem. Of all these different atoms, which atom do you think forms the basis of the molecules that make up living things?

A. Hydrogen

B. Oxygen

C. Carbon

D. Nitrogen

E. Sulfur

F. Phosphorous

G. Water

Explain your thinking. Why do you think that atom is considered the basis of the molecules that make up living things?

Carbon
CHEMISTRY

FLORIDA BIG IDEAS

1 The Practice of Science

3 The Role of Theories, Laws, Hypotheses, and Models

4 Science and Society

8 Properties of Matter

Think About It!

What is carbon's role in the chemistry of living things?

When you walk through a forest, you probably do not think about carbon atoms. However, all the living things surrounding you contain carbon atoms.

① Where do you think carbon atoms can be found?

_____

_____

_____

② What other elements do you think are in living things?

_____

_____

_____

③ What do you think is carbon's role in the chemistry of living things?

_____

_____

_____

Get Ready to Read

What do you think about carbon's role in our lives?

Before you read, decide if you agree or disagree with each of these statements. As you read this chapter, see if you change your mind about any of the statements.

		AGREE	DISAGREE
①	Charcoal and diamonds are made of carbon atoms.	☐	☐
②	Carbon atoms often bond with hydrogen atoms in chemical compounds.	☐	☐
③	Rubbing alcohol, cheese, and the venom of stinging ants contain carbon compounds.	☐	☐
④	Carbon atoms cannot bond to other carbon atoms.	☐	☐
⑤	Foods that you eat provide chemical elements that are needed to make carbon compounds in your body.	☐	☐
⑥	Carbohydrates should be totally eliminated from a healthful diet.	☐	☐

 **Connect ED** **There's More Online!**
Video • Audio • Review • ⓘLab Station • WebQuest • Assessment • Concepts in Motion • Multilingual eGlossary

499

Elemental Carbon and Simple Organic COMPOUNDS

Vocabulary
organic compound p. 502

hydrocarbon p. 504

isomer p. 504

saturated hydrocarbon p. 505

unsaturated hydrocarbon p. 505

 Florida NGSSS

SC.8.N.1.1 Define a problem from the eighth grade curriculum using appropriate reference materials to support scientific understanding, plan and carry out scientific investigations of various types, such as systematic observations or experiments, identify variables, collect and organize data, interpret data in charts, tables, and graphics, analyze information, make predictions, and defend conclusions.

SC.8.N.3.1 Select models useful in relating the results of their own investigations.

SC.8.N.3.2 Explain why theories may be modified but are rarely discarded.

SC.8.P.8.5 Recognize that there are a finite number of elements and that their atoms combine in a multitude of ways to produce compounds that make up all of the living and nonliving things that we encounter.

LA.6.2.2.3 The student will organize information to show understanding or relationships among facts, ideas, and events (e.g., representing key points within text through charting, mapping, paraphrasing, summarizing, or comparing/contrasting);

LA.6.4.2.2 The student will record information (e.g., observations, notes, lists, charts, legends) related to a topic, including visual aids to organize and record information, as appropriate, and attribute sources of information;

 Launch Lab

SC.8.N.1.1, SC.8.N.3.1, SC.8.P.8.5

15 minutes

Why is carbon a unique element?

A carbon atom is unique because it can easily form four bonds with other atoms, including other carbon atoms. Because of this property, carbon forms many different compounds.

Procedure 🖐️ 🧤 ⚠️ *Do not eat the gumdrops.*

1. Read and complete a lab safety form.

2. Use **gumdrops** and **toothpicks** to make as many different carbon molecules as you can. Keep these rules in mind.
   - Each gumdrop represents one carbon atom.
   - Each toothpick represents one chemical bond.
   - Each molecule must contain four carbon atoms (gumdrops).
   - Each carbon atom must have four chemical bonds (toothpicks).
   - One carbon atom can share up to three bonds with another carbon atom.

3. Make a sketch of each molecule you make below.

Data and Observations

Think About This

1. How many different molecules were you able to build?

2. If you had five gumdrops, would you be able to build more molecules? Explain your answer.

3. 🔑 **Key Concept** How do you think carbon bonds with other carbon atoms?

Inquiry **A Diamond?**

1. The yellow rock probably does not look like a diamond that you have seen. It lacks sparkle because it is rough and uncut. It formed deep within Earth under intense pressure and heating. Believe it or not, diamonds are pure carbon. What do you think makes a diamond so hard?

_____

_____

_____

_____

Elements in Living Things

What do you have in common with the fish and sea anemones in **Figure 1?** You might be surprised to learn that you, a fish, and a sea anemone have several things in common. Each of you is made of cells that contain carbon, hydrogen, oxygen, nitrogen, and a few other elements. In fact, the mass of all living organisms contains about 18 percent carbon compounds.

Except for water and some salts, most things you put in or on your body—food, clothing, cosmetics, and medicines— consist of compounds that contain carbon. This chapter explores various types of carbon compounds that make up living things.

Active Reading 3. **Identify** Describe the major elements in living things by filling out the graphic organizer.

> Elements in Living Things

Element that makes up about 18 percent of the mass of all living things:

_____

Other elements found in large amounts in living things:

1. _____

2. _____

3. _____

Active Reading 2. **Identify** As you read, highlight the main idea in each paragraph. Then use this to review later.

Figure 1 All living things are made of similar carbon-containing compounds.

SCIENCE USE v. COMMON USE

organic

Science Use a chemical compound that contains carbon and usually contains at least one carbon-hydrogen bond

Common Use pertaining to or grown with natural fertilizers and pesticides; often used to refer to foods, such as fruits, vegetables, and meats

REVIEW VOCABULARY

covalent bond

a chemical bond formed by sharing one or more pairs of electrons between atoms

Figure 2 Carbon often bonds with four hydrogen atoms and forms a stable compound.

 Active Reading

4. Identify For each carbon-hydrogen bond below, how many electrons come from a hydrogen atom?

Organic Compounds

Scientists once thought that all carbon compounds came from living or once-living organisms, and they called these compounds organic. Scientists now know that carbon is also in many nonliving things. Today, scientists define an **organic compound** as a *chemical compound that contains carbon atoms usually bonded to at least one hydrogen atom.* Organic compounds can also contain other elements such as oxygen, phosphorus, or sulfur. However, compounds such as carbon dioxide (CO_2) and carbon monoxide (CO) are not organic because they do not have a carbon-hydrogen bond.

Understanding Carbon

A carbon atom is unique because it can easily combine with other atoms and form millions of compounds. Find carbon on the periodic table in the back of this book. Carbon has an atomic number of 6. Therefore, a neutral carbon atom has six protons and six electrons. Four of these electrons are valence electrons, or are in the outermost energy level. Recall that many atoms are chemically stable when they have eight valence electrons. Carbon atoms become more chemically stable through **covalent bonding**, as shown in **Figure 2.** In a covalent bond, carbon atoms have eight valence electrons, like a stable, unreactive noble gas.

The Carbon Group

Look again at the periodic table. Notice that silicon and germanium are in the same group as carbon. They each have four valence electrons. Silicon and germanium atoms also become stable by forming four covalent bonds. However, it takes more energy for them to do this. The more energy it takes, the less likely it is that bonding will occur.

 5. NGSSS Check Explain Why is carbon unique? SC.8.P.8.5

_____

_____

_____

Bonding with Carbon 🗝

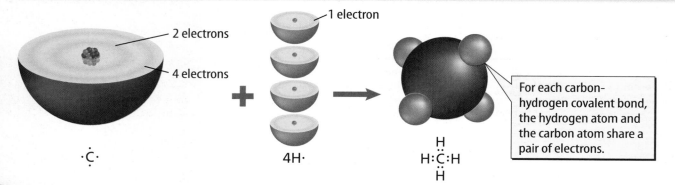

2 electrons

4 electrons

·Ċ·

1 electron

4H·

H
H:C:H
H

For each carbon-hydrogen covalent bond, the hydrogen atom and the carbon atom share a pair of electrons.

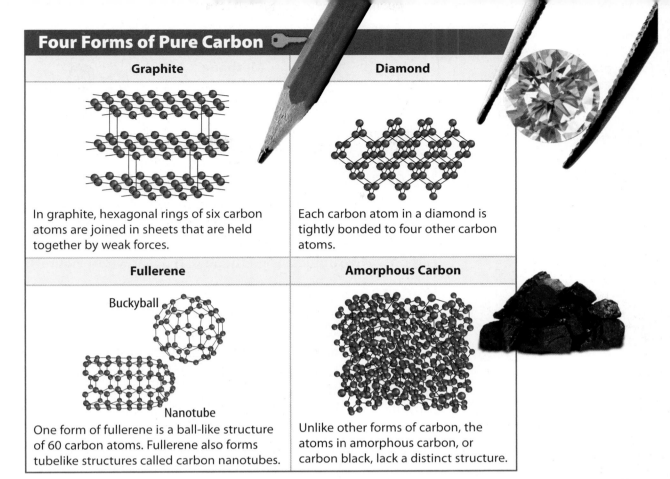

Four Forms of Pure Carbon 🔑

Graphite	Diamond
In graphite, hexagonal rings of six carbon atoms are joined in sheets that are held together by weak forces.	Each carbon atom in a diamond is tightly bonded to four other carbon atoms.
Fullerene	**Amorphous Carbon**
Buckyball / Nanotube — One form of fullerene is a ball-like structure of 60 carbon atoms. Fullerene also forms tubelike structures called carbon nanotubes.	Unlike other forms of carbon, the atoms in amorphous carbon, or carbon black, lack a distinct structure.

The Forms of Pure Carbon

When carbon atoms bond together, they form one of several different arrangements, such as those shown in **Figure 3**. Forms of carbon are described below.

- One form of carbon is graphite (GRA fite). In graphite, carbon atoms form thin sheets that can slide over one another or bend. Graphite is used as a lubricant and in making golf clubs, tennis rackets, pencil lead, and other items.

- Diamonds, another form of carbon, are used in jewelry, on the tips of drill bits, and on the edges of some saw blades. The carbon atoms bond to one another in a rigid and orderly structure, making diamonds extremely strong. This makes diamonds one of the hardest materials known.

- Carbon atoms in fullerene (FOOL uh reen) form various cage-like structures. Fullerene was discovered late in the twentieth century, and uses for fullerene are still being explored. However, future fullerene uses might include the development of faster, smaller electronic components.

- The atoms in **amorphous** (uh MOR fus) carbon lack an orderly arrangement. Amorphous carbon is found in soot, coal, and charcoal.

Figure 3 Four common forms of pure carbon are graphite, diamond, fullerene, and amorphous.

🛠 **6. Visual Check**
Explain Why is graphite used in pencil lead but diamonds are not?

_____

_____

_____

WORD ORIGIN

amorphous
from *a*– and Greek *morphe*, means "without form"

Figure 4 Methane consists of one carbon atom bonded to four hydrogen atoms.

Methane CH$_4$

Structural formula   Ball-and-stick model

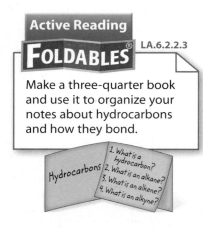

Active Reading

FOLDABLES® LA.6.2.2.3

Make a three-quarter book and use it to organize your notes about hydrocarbons and how they bond.

Hydrocarbons
1. What is a hydrocarbon?
2. What is an alkane?
3. What is an alkene?
4. What is an alkyne?

Table 1 Butane and isobutane have four carbons and ten hydrogens, but their structures and properties are different.

Active Reading **7. Identify** Highlight the three ways a hydrocarbon can be modeled.

Hydrocarbons

Many organic compounds contain only carbon and hydrogen atoms. *A compound that contains only carbon and hydrogen atoms is called a* **hydrocarbon.** There are many different hydrocarbons. The simplest is methane (CH_4), shown in **Figure 4.**

Hydrocarbon Chains

When carbon atoms form hydrocarbons, the carbon atoms can link together in different ways. They can form straight chains, branched chains, or rings. **Table 1** shows examples of each type of arrangement. Look closely at the molecular formula for each compound. Notice that butane and isobutane have the same molecular formula. They have the same ratio of carbon atoms to hydrogen atoms. *Compounds that have the same molecular formula but different structural arrangements are called* **isomers** (I suh murz). Each isomer is a different molecule with its own unique name and properties.

 8. NGSSS Check **Explain** In what ways can carbon form hydrocarbons? SC.8.P.8.5

_____

Table 1 Hydrocarbon Arrangements		
Butane	**Isobutane**	**Cyclobutane**
Molecular formula: C$_4$H$_{10}$	**Molecular formula: C$_4$H$_{10}$**	**Molecular formula: C$_4$H$_8$**
Structural formula	Structural formula	Structural formula
Ball-and-stick model	Ball-and-stick model	Ball-and-stick model

Carbon-to-Carbon Bonding

When a carbon atom bonds to another carbon atom, the two atoms can share two, four, or six electrons, as shown in **Figure 5.** In all cases, the carbon atoms in each molecule have eight valence electrons and are stable. When two carbon atoms share two electrons, it is called a single bond. A hydrocarbon that contains only single bonds is called an alkane. When two carbon atoms share four electrons, it is called a double bond. A hydrocarbon that contains at least one double bond is called an alkene. If two carbon atoms share six electrons, it is called a triple bond. A hydrocarbon that contains at least one triple bond is called an alkyne.

 9. NGSSS Check <u>Underline</u> the types of bonds carbon atoms form with other carbon atoms. SC.8.P.8.5

Saturated Hydrocarbons

Hydrocarbons are often classified by the type of bonds the carbon atoms share. *A hydrocarbon that contains only single bonds is called a* **saturated hydrocarbon.** It is called saturated because no more hydrogen atoms can be added to the molecule. Look at the top image in **Figure 5.** Notice that three of the valence electrons in each carbon atom bond with hydrogen atoms. The carbon atoms are saturated with hydrogen atoms.

Unsaturated Hydrocarbons

A hydrocarbon that contains one or more double or triple bonds is called an **unsaturated hydrocarbon.** Look at the double bond and triple bond examples in **Figure 5.** If the double and triple bonds are broken, additional hydrogen atoms could bond to the carbon atoms. Therefore, molecules containing double and triple bonds are not saturated with hydrogen atoms.

 10. NGSSS Check **Explain** What is the difference between saturated and unsaturated compounds? SC.8.P.8.5

_____

_____

_____

_____

 11. NGSSS Check **Relate** Is an alkene a saturated or unsaturated hydrocarbon? SC.8.P.8.5

_____

_____

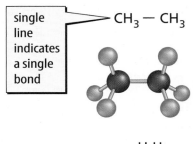

Alkane
- single bond
- two electrons shared between two carbon atoms

single line indicates a single bond → $CH_3 - CH_3$

Alkene
- double bond
- four electrons shared between two carbon atoms

double line indicates a double bond → $CH_2 = CH_2$

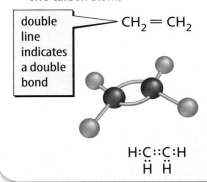

Alkyne
- triple bond
- six electrons shared between two carbon atoms

triple line indicates a triple bond → $CH \equiv CH$

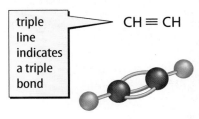

Figure 5 Two carbon atoms can form a single, double, or triple bond.

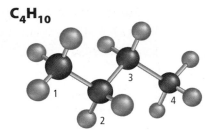

C₄H₁₀

Figure 6 This hydrocarbon has four carbon atoms in its chain.

Table 2 The number of carbon atoms in the longest continuous chain determines the root word for the name of the hydrocarbon.

Active Reading **12. Identify** (Circle) the number of carbon atoms in a molecule with the root word *hept–*.

Table 2 Root Words

Carbon Atoms	Name	Carbon Atoms	Name
1	meth–	6	hex–
2	eth–	7	hept–
3	prop–	8	oct–
4	but–	9	non–
5	pent–	10	dec–

Naming Hydrocarbons

What type of shape is a stop sign? It is an octagon. An octagon is a figure with eight sides. Its name comes from the root *oct–*, which means "eight." Most geometric shapes have names that refer to the number of sides they have, such as a triangle. Similarly, hydrocarbons have names that indicate how many carbon atoms are in each molecule.

Carbon Chains

When naming a hydrocarbon, the first thing you need to do is find the longest carbon chain and count the number of carbon atoms in it. Look at **Figure 6.** Find the carbon chain and count the carbon atoms. In this molecule, there are four carbon atoms. The number of carbon atoms gives you the root word of the name. Now, look at **Table 2.** This table shows the root word for any hydrocarbon that has one through ten carbon atoms. What is the root name for the molecule in **Figure 6?** The root name is *but–* (BYEWT). What would be the root name if the carbon chain had eight carbon atoms like a stop sign? The root name would be *oct–*.

Inquiry SC.8.P.8.5, SC.8.N.1.1, SC.8.N.3.1

LAB STATION **Try It!**

MiniLab *How do carbon atoms bond with carbon and hydrogen atoms?* at connectED.mcgraw-hill.com

Apply It! After you complete the lab, answer these questions.

1. **Formulate** How can you change your model of propene to propane?

2. **Draw** In the space below, draw the structural formula of octane.

Determine the Suffix

Now that you know how to find the root word of a hydrocarbon, you must also find the suffix, or end, of the name. Recall that carbon atoms bond to other carbon atoms by single bonds, double bonds, or triple bonds. **Table 3** shows which suffix to use when the type of bonds in the molecule is determined. Look at **Figure 6** again. The molecule has all single bonds and should have the suffix –*ane*. Put the root and the suffix together and you get *butane*.

Active Reading 13. **Identify** What is the name of a hydrocarbon with six carbon atoms that contain only single bonds?

Determine the Prefix

Sometimes hydrocarbons have a prefix, and sometimes they do not. Recall that hydrocarbons form chains, branched chains, and rings. If a hydrocarbon contains a ring structure, the prefix *cyclo*– is added before the root name. Hydrocarbons sometimes have other prefixes and numbers added before their name. You might read about this naming system in more advanced chemistry courses. For this lesson, only hydrocarbons in the form of a ring will get a prefix. **Table 4** summarizes the steps used to name a hydrocarbon.

Table 3 The type of bonds in the hydrocarbon chain determines the suffix in the name.

Table 3 Bond Type and Hydrocarbon Suffix

Bond Type	Suffix
All single bonds $-C-C-$	–ane
At least one double bond $-C=C-$	–ene
At least one triple bond $-C\equiv C-$	–yne

Table 4 Naming Hydrocarbons

Steps for Naming Hydrocarbons	Example A	Example B
1 Examine the compound.	$H-C\equiv C-C-C-H$ (with H atoms)	(ring structure with 6 carbons)
2 Count the number of carbon atoms in the longest continuous chain.	There are 4 carbon atoms in the longest chain.	There are 6 carbon atoms in the longest chain.
3 Determine the root name of the hydrocarbon using **Table 2.**	The root name is *but–*.	The root name is *hex–*.
4 Determine the type of bonds in the hydrocarbon, and then use **Table 3** to find the suffix.	There is a triple bond, so add the suffix –*yne*.	There are only single bonds, so add the suffix –*ane*.
5 Put the root and suffix together to name the hydrocarbon.	Combining the root and suffix gives the name *butyne*.	Combining the root and suffix gives the name *hexane*.
6 If the hydrocarbon is a ring, add *cyclo*– to the beginning of the name.	No prefix is needed because the structure is not a ring. The name of the hydrocarbon is butyne.	The structure is a ring so the prefix *cyclo*– is added to the name. The name of the hydrocarbon is cyclohexane.

Methane

Chemical compounds that contain carbon and usually at least one hydrogen atom are called organic compounds.

Carbon atoms form several different substances including graphite, diamonds, fullerene, and amorphous carbon.

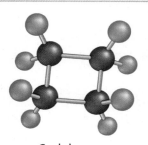

Cyclobutane

A hydrocarbon can be in the form of a straight chain, branched chain, or a ring structure.

Use Vocabulary

1 Define *saturated hydrocarbon* and *unsaturated hydrocarbon*. SC.8.P.8.5

_____

_____

_____

2 **Use the term** *isomer* in a complete sentence.

_____

_____

Understand Key Concepts 🔑

3 **Summarize** how carbon bonds with other carbon atoms. SC.8.P.8.5

_____

_____

4 **Examine** each molecular formula. Which is propane? SC.8.P.8.5

   (A) CH_3 (C) C_3H_8

   (B) C_2H_6 (D) C_4H_{10}

5 **Explain** why carbon is unique compared to other elements. SC.8.P.8.5

_____

_____

Interpret Graphics

6 **Explain** Fill in the table below to explain how the four forms of carbon differ. LA.6.2.2.3

Graphite	
Diamond	
Fullerene	
Amorphous carbon	

Critical Thinking

SC.8.P.8.5

7 **Explain** why so many different compounds are made from carbon.

_____

_____

8 **Draw** the molecular structure for pentane (C_5H_{12}). SC.8.P.8.5

Carbon

Will it replace the silicon in your computer?

You have a computer inside your head—your brain. Your brain has 100 billion tiny switches that allow you to process information every day. Those switches are your brain cells. Like your brain, computers have billions of tiny "brain cells" called silicon transistors. These tiny electronic components have been changing society since they were first developed about 50 years ago.

Transistors are devices that can strengthen an electronic signal. Transistors can rapidly turn computer circuits off and on. They are efficient and produce very little heat, which makes them useful in cell phones, radios, and computers. Silicon transistors first were used in computers in 1955. At the time, computers were about the size of three or four adults. Now, of course, computers are much smaller.

Scientists developed the first miniature silicon transistor in 1965. They attached several of the tiny transistors, along with other electronic components, to a piece of plastic. This was the invention of the circuit board. The circuit board allowed designers to fit many more transistors into a computer. This also allowed computers to be made much smaller.

The number of transistors that can be placed on a single circuit board has doubled every two years since 1965. Computers have become smaller, faster, and more powerful. But silicon transistors cannot be made much smaller. How might scientists create smaller transistors? The answer is to use carbon instead of silicon. Carbon nanotubes are cylindrical structures made of pure carbon. They conduct both heat and electricity and are many times faster and more efficient than silicon transistors. And nanotubes are tiny—100,000 of them side by side would be about as thick as a human hair. As if that weren't enough, the tubes are ten times stronger than steel. Once again, society will change as electronics you can't even imagine today become everyday items of the future.

The first silicon transistor, shown here on top of a postage stamp, was small for its time but huge by today's standards.

TRIDAC was the first fully transistorized computer.

By the 1980s, computers had become small enough to fit on a desktop.

Carbon nanotube

It's Your Turn

RESEARCH AND REPORT What do scientists think is the next step in using carbon nanotube transistors? If nanotube transistors lead to even smaller, faster, and more powerful devices, how might those devices impact your everyday life?

LA.6.4.2.2

Other Organic COMPOUNDS

SC.8.P.8.5

ESSENTIAL QUESTIONS

 What are the four common functional groups of organic compounds?

 What are polymers?

Vocabulary

substituted hydrocarbon p. 511

functional group p. 512

hydroxyl group p. 512

halide group p. 513

carboxyl group p. 513

amino group p. 514

polymer p. 515

monomer p. 515

polymerization p. 515

 Florida NGSSS

SC.7.N.1.6 Explain that empirical evidence is the cumulative body of observations of a natural phenomenon on which scientific explanations are based.

SC.8.N.1.1 Define a problem from the eighth grade curriculum using appropriate reference materials to support scientific understanding, plan and carry out scientific investigations of various types, such as systematic observations or experiments, identify variables, collect and organize data, interpret data in charts, tables, and graphics, analyze information, make predictions, and defend conclusions.

SC.8.P.8.5 Recognize that there are a finite number of elements and that their atoms combine in a multitude of ways to produce compounds that make up all of the living and nonliving things that we encounter.

LA.6.2.2.3 The student will organize information to show understanding or relationships among facts, ideas, and events (e.g., representing key points within text through charting, mapping, paraphrasing, summarizing, or comparing/contrasting);

MA.6.A.3.6 Construct and analyze tables, graphs, and equations to describe linear functions and other simple relations using both common language and algebraic notation.

 Launch Lab SC.8.P.8.5

20 minutes

How do functional groups affect compounds?

In some hydrocarbons, a hydrogen atom is removed and another atom or group of atoms takes its place. Rubbing alcohol and glycerin are two examples.

Propane

Rubbing alcohol

Glycerin

Procedure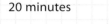

1. Read and complete a lab safety form.

2. Use a **plastic spoon** to measure two spoons of **rubbing alcohol** and pour the liquid into a **clear plastic cup.** Observe the properties of the alcohol. Use the wafting method to check the odor. Record your observations below.

3. Repeat step 2 with **glycerin** using a clean spoon and cup. Add **distilled water** to both cups until they are one-third full. Stir gently using the same spoon in each cup that you used before.

4. Twist three **chenille stems** to make bubble wands like the ones shown. Dip a clean bubble wand into each cup. Check to see if a film forms within the circle for each mixture. Record your observations below.

Data and Observations

Think About This

1. 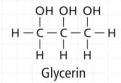 **Key Concept** Propane is a colorless gas. What changes occur when a hydrogen atom in propane is replaced by an oxygen atom and a hydrogen atom and rubbing alcohol forms?

1. This interesting caterpillar is not feared because of its bite. It shoots an organic compound called formic acid at its enemies. While you might not have caterpillars shooting formic acid at you, you do have other organic compounds all around you. Why might the caterpillar benefit from being able to spit acid?

_____

_____

_____

_____

Substituted Hydrocarbons

Think for a moment about any sports team. Teams often substitute players in and out of a game. In a similar way, other atoms can be substituted for a hydrogen atom in a hydrocarbon. *A **substituted hydrocarbon** is an organic compound in which a carbon atom is bonded to an atom, or group of atoms, other than hydrogen.* Just as a team might function differently when new players are substituted into the game, organic compounds function differently when hydrogen atoms are substituted with other atoms.

You might be familiar with the substituted hydrocarbon ethanol. It is often mixed with gasoline and used as a fuel for cars. Ethanol is also in food flavorings, such as vanilla extract. The chemical formula for ethanol is CH_3CH_2OH. In this compound, one hydrogen atom of ethane (CH_3CH_3) has been replaced with –OH. This is just one type of substituted hydrocarbon that you will read about in this lesson.

Active Reading **2. Dissect** Analyze the formation of a substituted hydrocarbon by completing the cause-and-effect diagram. SC.8.P.8.5

Cause		Effect
A carbon atom bonds to an atom or a group of atoms other than _____.	→	A _____ hydrocarbon is formed.

Active Reading
FOLDABLES LA.6.2.2.3

Make a four-door book and label it as shown. Use it to organize your notes about the four main types of functional groups.

Hydroxyl Group	Halide Group
Carboxyl Group	Amino Group

Active Reading **3. Summarize** As you read, summarize the key information under each heading in one or two sentences.

Use Ratios

A ratio expresses the relationship between two or more things. In the formula for methane, CH_4, the ratio of carbon atoms to hydrogen atoms is 1:4, read as "1 to 4."

Ethanol is written CH_3CH_2OH. One molecule contains **2** carbon atoms, **6** hydrogen atoms, and **1** oxygen atom. The ratio is:

C:H:O = **2:6:1**

Practice

4. What is the ratio of carbon to hydrogen to oxygen atoms in table sugar, $C_{12}H_{22}O_{11}$?

Functional Groups

A hydrogen atom, in an organic compound, can be substituted with other atoms. This causes the substituted hydrocarbon to have new properties. *A **functional group** is an atom or group of atoms that determine the function and properties of the compound.* The substituted hydrocarbon is renamed to indicate which functional group has been substituted. There are many functional groups, each with specific characteristics. Four functional groups are discussed in this lesson.

Hydroxyl Group

Have you ever used rubbing alcohol to clean a cut or a scrape? Rubbing alcohol is the common name for the compound 2-propanol. 2-Propanol is a substituted hydrocarbon of propane and contains the hydroxyl (hi DRAHK sul) functional group, as shown in **Figure 7**. *The **hydroxyl group** contains two atoms—oxygen and hydrogen. Its formula is –OH.* Organic compounds that contain the hydroxyl group are called alcohols. Alcohols are polar compounds and can dissolve in water. They have high melting and boiling points and are commonly used as disinfectants, fuel, and solvents. **Figure 7** also shows how the substituted hydrocarbon ethanol differs from ethane.

Figure 7 🔑 Substituting a H atom for a functional group in a hydrocarbon changes its properties.

 5. NGSSS Check Differentiate (Circle) the functional group that makes 2-propanol different from propane? SC.8.P.8.5

Hydrocarbon		Alcohol	
Ethane	• **Melting point:** —181.7°C • **Boiling point:** —88.6°C • **Appearance:** colorless, odorless gas • **Uses:** automotive fuel; refrigerant in extremely low-temperature systems	**Ethanol**	• **Melting point:** —117.3°C • **Boiling point:** 78.5°C • **Appearance:** colorless liquid with mild odor • **Uses:** solvent in perfumes and paints; automotive fuel; fluid in low-temperature thermometers
Propane	• **Melting point:** —189.7°C • **Boiling point:** —42.1°C • **Appearance:** colorless, odorless gas • **Uses:** fuel for cooking, hot-air balloons, and some automobiles; raw material for other products	**2-Propanol**	• **Melting point:** —89.5°C • **Boiling point:** 82.4°C • **Appearance:** colorless liquid with a strong odor • **Uses:** solvent in cleaning fluid; preservative for biological specimens; fuel additive to keep gasoline from freezing

Halide Group

Group 17 elements—the halogens—can also be substitutions in hydrocarbons. This functional group is called the halide group. *The* **halide group** *contains group 17 halogens—fluorine, chlorine, bromine, and iodine.* Bromomethane (broh moh MEH thayn) is an example of a substance with halide substitution. In bromomethane, a bromine atom replaces one of the hydrogen atoms in methane. This process is similar to the one shown in **Figure 7.** Notice that the prefix *bromo–* is before the hydrocarbon name to form the name bromomethane as illustrated in **Figure 8.**

As shown in **Figure 8,** bromomethane is a pesticide. It can be used to kill pests in the soil before strawberries are planted. Its use is strictly regulated because of its environmental hazards. Some countries have banned or are phasing out its use.

Carboxyl Group

If you had orange juice for breakfast or a salad for lunch, you ate compounds containing one or more carboxyl (kar BAHK sul) functional groups. *A* **carboxyl group** *consists of a carbon atom with a single bond to a hydroxyl group and a double bond to an oxygen atom. Its formula is -COOH.* When a carboxyl group replaces a hydrogen atom in a hydrocarbon, the result is a carboxylic acid.

Citric acid in citrus fruits, such as oranges, lemons, and limes, is a carboxylic acid. Dairy products such as buttermilk and yogurt also contain a carboxylic acid called lactic acid. Two simple carboxylic acids are methanoic acid and ethanoic acid, as shown in **Figure 9.**

 Active Reading

6. Calculate How many electrons are shared between each carbon atom and oxygen atom in a carboxyl group?

_____

_____

_____

Figure 8  Bromomethane, sometimes called methyl bromide, contains one carbon atom, three hydrogen atoms, and one bromine atom. It is used for rodent control.

Bromomethane, CH_3Br

Figure 9 Methanoic and ethanoic acid are simple carboxylic acids.

Methanoic acid, HCOOH

Methanoic acid (HCOOH)
Methanoic acid is in the toxin of stinging ants. It is also known as formic acid.

Ethanoic acid, CH_3COOH

Ethanoic acid (CH_3COOH)
Ethanoic acid is in vinegar, which is used in many food items including salad dressings and pickles. Ethanoic acid is also known as acetic acid.

Figure 10 🔑 Methylamine is found in cheese.

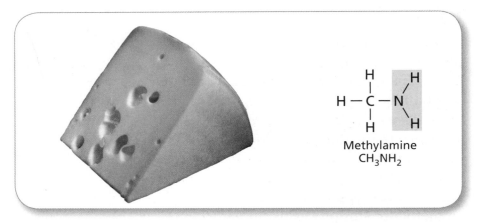

Methylamine
CH_3NH_2

Amino Group

Have you ever smelled cheese with a strong odor? The strong odor is due to the presence of an amino (uh MEE noh) group. *The **amino group** consists of a nitrogen atom covalently bonded to two hydrogen atoms, and its formula is* $-NH_2$. The suffix *–amine* is added to the end of each root name to indicate that the amino group is in the compound. The amine, methylamine (meh thuh luh MEEN), forms when an amino group is substituted for a hydrogen in methane. **Figure 10** illustrates the structure and chemical formula of methylamine.

Notice that *–yl* follows the root name *meth*. If a hydrocarbon, such as methane, loses a hydrogen atom, its name changes to methyl. If ethane loses a hydrogen atom, its name becomes ethyl.

7. NGSSS Check Summarize What are four common functional groups of organic compounds? SC.8.P.8.5

_____

_____

_____

Shapes of Molecules

Molecules come in different shapes and sizes. Scientists often make three-dimensional models of molecules to study their shapes. Knowing a molecule's shape helps scientists understand how it interacts with other molecules, how strong the bonds are between atoms, and what type of bonds are in the molecule. The molecular shapes in **Table 5** show you how some molecules might look in three dimensions.

❶ **Tetrahedral**—Methane is an example of a tetrahedral molecule. The atoms in a tetrahedral molecule form a pyramid.

❷ **Planar**—Ethene is an example of a planar molecule. The atoms in a planar molecule are all on the same plane.

❸ **Linear**—Ethyne is an example of a linear molecule. The atoms in a linear molecule form a straight line.

Table 5 Molecules are not flat. They are three-dimensional.

 **8. Visual Check
Label** (Circle) the name of the compound that has a planar shape.

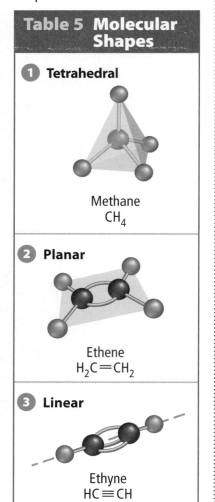

Table 5 Molecular Shapes
❶ **Tetrahedral**
Methane CH_4
❷ **Planar**
Ethene $H_2C{=}CH_2$
❸ **Linear**
Ethyne $HC{\equiv}CH$

Polymers

What do a bottle of water, a toy car, a marker, and a video game have in common? They all contain some amount of plastic. The word *plastic* is a common term that refers to a type of substance called a polymer (PAH luh mur). *A* **polymer** *is a molecule made up of many of the same small organic molecules covalently bonded together, forming a long chain. A* **monomer** *(MAH nuh mur) is one of the small organic molecules that makes up the long chain of a polymer.* Some polymers occur naturally, but many are made in laboratories. Polymers occurring in nature are called natural polymers. Polymers made in laboratories are called synthetic polymers.

Many synthetic polymers are made from simple hydrocarbons by a process called polymerization (pah luh muh ruh ZAY shun). **Polymerization** *is the chemical process in which small organic molecules, or monomers, bond together to form a chain.* Polyethylene is a polymer used to make shampoo bottles, grocery bags, and toys. It is made by the polymerization of ethene—also known as ethylene. As shown in **Figure 11,** first the double bonds are broken in the ethene molecules. Then the carbon atoms bond and form long chains.

Active Reading 9. **Describe** <u>Underline</u> the process of polymerization.

Inquiry **①LAB STATION** **Try It!**

SC.8.P.8.5

MiniLab *How can you make a polymer?* at connectED.mcgraw-hill.com

Apply It!

After you complete the lab, answer the questions below.

1. **Imagine** If 1,000 paper clips attached to each other represent a polymer, what part represents the monomer?

2. **Analyze** Scientists predict that someday there will be no more oil. What effect will this have on the plastics industry and the production of synthetic polymers?

Formation of a Polymer 🔑

Figure 11 Ethene, also called ethylene, molecules form polyethylene during polymerization.

Active Reading 10. **Pick** Highlight the name of the monomer shown in the diagram.

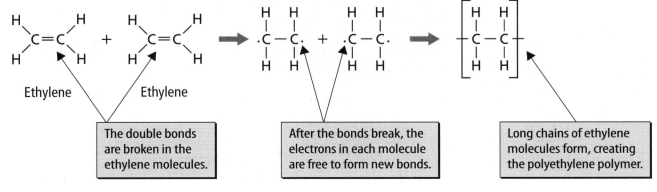

Ethylene   Ethylene

| The double bonds are broken in the ethylene molecules. | After the bonds break, the electrons in each molecule are free to form new bonds. | Long chains of ethylene molecules form, creating the polyethylene polymer. |

Synthetic Polymers

Polyethylene and many other synthetic polymers are made from petroleum. Petroleum is a thick, oily, flammable mixture of solid, liquid, and gaseous hydrocarbons. Petroleum, an example of a fossil fuel, occurs naturally beneath Earth's surface. It formed from the remains of ancient, microscopic marine organisms.

Examples of polymers and some of their applications are shown in **Table 6.** This is only a small sample of the many polymers and polymer applications that are used today.

11. **Explain** Why are the polymers in **Table 6** classified as synthetic polymers?

_____

_____

Table 6 Many common objects are made of synthetic polymers.

12. **Visual Check** **Apply** What does the tetra– in polytetrafluoroethylene refer to?

_____

_____

Table 6 Sample Polymers and Applications

Polymer	Examples
Polyethylene (PE)	Bales of hay are rolled in polyethylene to protect them from rain. The hay is used to feed farm animals such as cows, horses, and sheep.
Polyvinyl chloride (PVC)	Pipes made of polyvinyl chloride, or PVC, are used for plumbing. Some rainwear, home siding, and garden hoses are also made of polyvinyl chloride.
Polytetrafluoroethylene (PTFE)	Polytetrafluoroethylene (pah lee teh truh flor oh ETH uh leen) is used for nonstick coating on cookware.
Polypropylene (PP)	Polymer bank notes are made of polypropylene. These bank notes last longer than traditional paper notes. Also, many ropes are made of polypropylene.

Visual Summary

Ethanol

When a functional group replaces a hydrogen atom in a hydrocarbon, it forms a substituted hydrocarbon.

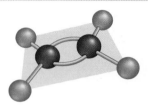

Ethene
$H_2C = CH_2$
Planar

Hydrocarbon molecules form many different three-dimensional shapes such as planar or tetrahedral.

Polymers, made of small, linked, organic molecules, are used to make common items such as money.

Inquiry
SC.8.P.8.5,
SC.7.N.1.6,
SC.8.N.1.1

LAB STATION Try It!

Skill Lab *How do you test for Vitamin C?* at
connectED.mcgraw-hill.com

Use Vocabulary

1. **Define** *polymerization* in your own words.

_____

_____

Understand Key Concepts

2. **List** the atoms that make up each of the four common functional groups. SC.8.P.8.5

_____

_____

_____

3. **Select** the correct formula for an example of the halide group.
   - (A) $-Cl$
   - (B) $-COOH$
   - (C) $-H$
   - (D) $-OH$

4. **Describe** What are polymers?

_____

_____

Interpret Graphics

5. **Organize Information** List the functional groups mentioned in this lesson and their formulas. LA.6.2.2.3

Functional Group	Formula

Critical Thinking

6. **Compare and contrast** the structures of butane and butanol. SC.8.P.8.5

_____

_____

Math Skills   MA.6.A.3.6

7. What is the ratio of carbon to hydrogen to oxygen atoms in ethanoic acid, CH_3COOH?

Summarize Fill in the table below to compare the four functional groups.

Functional Group	Example	Substitution	Naming Convention
Amino Group	methylamine	1.	2.
Carboxyl Group	ethanoic acid	3.	4.
Halide Group	bromomethane	5.	6.
Hydroxyl Group	propanol	7.	8.

Create Draw the structural formulas for your examples.

Example	Structural Formula
methylamine	9.
ethanoic acid	10.
bromomethane	11.
propanol	12.

Compounds of LIFE

ESSENTIAL QUESTIONS

 What are biological molecules?

 What are some groups of carbon compounds found in living organisms?

Vocabulary

biological molecule p. 520

protein p. 521

amino acid p. 521

carbohydrate p. 522

nucleic acid p. 523

lipid p. 524

 Florida NGSSS

SC.7.N.1.6 Explain that empirical evidence is the cumulative body of observations of a natural phenomenon on which scientific explanations are based.

SC.8.N.1.1 Define a problem from the eighth grade curriculum using appropriate reference materials to support scientific understanding, plan and carry out scientific investigations of various types, such as systematic observations or experiments, identify variables, collect and organize data, interpret data in charts, tables, and graphics, analyze information, make predictions, and defend conclusions.

SC.8.N.1.5 Analyze the methods used to develop a scientific explanation as seen in different fields of science.

SC.8.N.4.1 Explain that science is one of the processes that can be used to inform decision making at the community, state, national, and international levels.

SC.8.P.8.5 Recognize that there are a finite number of elements and that their atoms combine in a multitude of ways to produce compounds that make up all of the living and nonliving things that we encounter.

LA.6.2.2.3 The student will organize information to show understanding or relationships among facts, ideas, and events (e.g., representing key points within text through charting, mapping, paraphrasing, summarizing, or comparing/contrasting);

Inquiry Launch Lab   SC.8.P.8.5

15 minutes

What does a carbon compound look like when its structure changes?

Like all proteins, the proteins in milk have a three-dimensional structure. You can observe the result of a structural change of one protein when acid is added to milk.

Procedure

1. Read and complete a lab safety form.

2. Add **skim milk** to a **clear plastic cup** until it is one-third filled. Observe the milk with a **magnifying lens.** Record your observations below.

3. Use a **plastic spoon** to add two spoonfuls of **white vinegar (acetic acid)** to the skim milk. After 1 min, observe the mixture with the magnifying lens. Record your observations.

4. Place a **coffee filter** in a **funnel.** Hold the funnel over another **clear plastic cup** and pour the milk-vinegar mixture into the filter-lined funnel. Observe the contents of the cup and the funnel. Record your observations.

Data and Observations

Think About This

1. Compare and contrast the milk before and after the vinegar was added.

2. **Key Concept** What is one group of carbon compounds found in living organisms?

Inquiry **Healthful Diet?**

1. The man and the dog get their energy from the foods they eat. Foods contain compounds that the body uses to function. What carbon compounds are known as the compounds of life?

_____

_____

_____

_____

Active Reading 2. **Define** What are biological molecules?

_____

_____

Biological Molecules

Many of the objects you use, such as CDs, DVDs, sandwich bags, plastic bowls, combs, and hairbrushes, are made of synthetic polymers. The bodies of the man and the dog shown above also contain polymers. Recall that polymers in nature are called natural polymers. Cells, tissues, and organs in your body and in the bodies of other living things contain natural polymers.

Individual cells of a living thing contain polymers that carry genetic information and pass this information to new cells. The chemical energy stored in your muscles is a polymer too. All of these natural polymers are called biological molecules. *A* **biological molecule** *is a large organic molecule in any living organism.* Biological molecules help determine the structure and function of many different body parts. They also provide the energy needed to run, to pedal a bicycle, and to do many other activities that you do. The chemical elements that make up these molecules come from the variety of foods you eat and the air you breathe.

Active Reading 3. **Analyze** Complete the boxes about biological molecules.

Active Reading

FOLDABLES® LA.6.2.2.3

Fold a sheet of paper into a four-column chart. Label it as shown. Use it to record information about the four biological molecules and their functions.

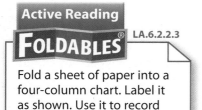

Biological	Molecules		
Proteins	Carbohydrates	Nucleic Acids	Lipids

Biological Molecules

Definition:	Responsible for:	Where the body gets them:

Figure 12 Amino acids are the monomers that form proteins.

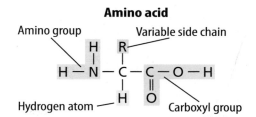

Amino acid

Amino group — Variable side chain

$$H - N - C - C - O - H$$

Hydrogen atom — Carboxyl group

Proteins

What do spiderwebs, plant leaves and roots, and the feathers of a peacock have in common? They contain natural polymers called proteins. Much of your body is made of proteins too, including your hair, muscles, blood, organs, immune system, and fingernails. A **protein** (PROH teen) *is a biological polymer made of amino acid monomers. An* **amino acid** *(uh MEE noh • A sud) is a carbon compound that contains the two functional groups—amino and carboxyl.*

Amino Acid Chains

Amino acids link together and form long chains. **Figure 12** shows the basic chemical structure of an amino acid. The R represents a side chain of molecules that can differ. There are 20 different side chains. Therefore, 20 different amino acids can link and form proteins, as shown in **Figure 13.**

Proteins and the Human Body

Proteins are important to the body. Of the 20 different amino acids, the human body can make 11 of them. The other nine must be included in the foods you eat. These nine amino acids are often referred to as essential amino acids. They are in a variety of foods including fish, dairy products, beans, and meat.

 5. NGSSS Check **Describe** What does the R stand for on the molecule? SC.8.P.8.5

Figure 13 Proteins are polymers that contain hundreds of amino acids linked together in a chain.

Active Reading **4. State** What functional groups do amino acids have?

_____

_____

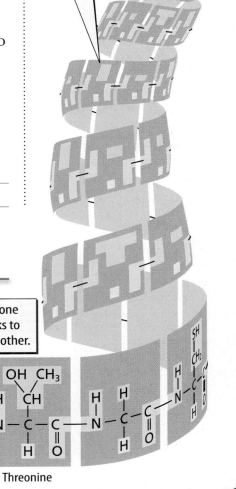

Some proteins form helical, or spiral, shapes.

Protein Formation 🔑

Individual amino acids link and release a water molecule.

The carboxyl group of one amino acid always links to the amino group of another.

$$H - N - C - C - OH$$

Glycine

H_2O

$$H - N - C - C - N - C - C - N - C - C - N - C - C$$

Alanine   Cysteine   Threonine

Figure 14 🔑 Glucose and fructose are simple sugars in fruits. Sucrose, also known as table sugar, forms from the chemical reaction between glucose and fructose.

CH₂OH + CH₂OH → CH₂OH CH₂OH + H₂O

Glucose   Fructose   Sucrose   Water

Carbohydrates

When you eat pasta or sugary snacks, you probably do not think about the carbohydrates in your food. *A* **carbohydrate** (kar boh HI drayt) *is a group of organic molecules that includes sugars, starches, and cellulose.* They are natural polymers that contain carbon, hydrogen, and oxygen atoms. Carbohydrates are a source of energy in cells.

Sugars

Simple sugars, such as glucose and fructose, usually contain five or six carbon atoms. The carbon atoms can be arranged in a ring, as shown in **Figure 14,** or in a straight chain. Your cells can easily break apart simple sugars, which provide quick energy. Glucose is in foods such as fruits and honey. It is also in your blood.

Glucose and fructose combine and form a sugar called sucrose, also shown in **Figure 14.** Sucrose is used to sweeten many foods.

Starch and Cellulose

When simple sugar molecules form chains, they form polymers called complex carbohydrates. Starch and cellulose are complex carbohydrates made of glucose monomers. The chemical bonds in starches take longer to break apart than simple sugars. They provide energy over a longer period. Human digestive systems cannot break the bonds in cellulose, but the digestive systems of animals, such as cows and horses, can.

Active Reading 6. **Locate** Underline the elements that make up carbohydrates.

Inquiry 🔬 SC.7.N.1.6

🔬LAB STATION **Try It!**

MiniLab *How many carbohydrates do you consume?* at connectED.mcgraw-hill.com

Apply It! After you complete the lab, answer these questions.

Foods high in carbohydrates are sources of energy. The chart below shows some foods and their carbohydrate count. Look at the differences in how much energy they might provide, given their carbohydrate count.

Carbohydrate Counts for Common Foods					
Main Dish		**Side Dish**		**Drink**	
chicken and noodles	39 g	apple	21 g	orange juice	27 g
lasagna	50 g	blueberry muffin	27 g	apple juice	29 g
hot oatmeal	25 g	cooked carrots	8 g	whole milk	12 g
plain bagel	38 g	corn on the cob	14 g	sports drink	24 g
macaroni and cheese	29 g	brown rice	22 g	lemonade	28 g

1. **Determine** Calculate a high-energy meal with the most carbohydrates. Include one choice from each category.

Nucleotide 🔑

O
‖
HO—P—O—CH₂
|
HO

Phosphate group

Sugar

Nitrogen-containing base

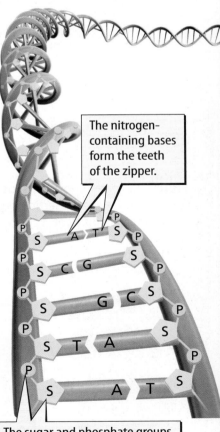

Figure 15 Nucleotide monomers make up nucleic acid polymers.

Active Reading **7. Determine** In the nitrogen-containing base, how many covalent bonds does each nitrogen atom form?

Nucleic Acids

A *biological polymer that stores and* **transmits** *genetic information is a* **nucleic acid** (new KLEE ihk • A sud). Genetic information includes instructions for cells on how to make proteins, produce new cells, and transfer genetic information. It is genetic information that determines how you look and how your body functions.

The monomer in a nucleic acid polymer is called a nucleotide, as shown in **Figure 15.** Each nucleotide monomer contains a phosphate group, a sugar, and a nitrogen-containing base. All nucleotides contain the same phosphate group. However, the sugar and nitrogen base can vary in nucleic acids. The elements needed by your body to make nucleic acids come from the foods you eat.

DNA

Two common nucleic acids, DNA and RNA, control cellular function and heredity. DNA is deoxyribonucleic acid (dee AHK sih rib oh noo klay ihk • A sud). DNA is a spiral-shaped molecule that resembles a twisted zipper, as shown in **Figure 16.** Each DNA monomer contains the five-carbon sugar deoxyribose. Deoxyribose and phosphate groups form the outside of the zipper. Pairs of nitrogen-containing bases—adenine (A) and thymine (T) or cytosine (C) and guanine (G)—form the teeth of the zipper.

RNA

RNA is ribonucleic acid. It contains the five-carbon sugar ribose. RNA is usually single stranded, not double stranded like DNA. It contains the nitrogen bases adenine, cytosine, guanine, and uracil. DNA provides information to make RNA, and then RNA makes the proteins that a cell needs to function.

 8. NGSSS Check **Review** What are two groups of carbon compounds found in living organisms? SC.8.P.8.5

Figure 16 🔑 DNA is a spiral-shaped molecule that is often called a double helix.

The nitrogen-containing bases form the teeth of the zipper.

The sugar and phosphate groups form the backbone of the zipper.

Oleic acid

$$\underset{HO}{\overset{O}{\|}}CCH_2CH_2CH_2CH_2CH_2CH_2CH_2CH = CHCH_2CH_2CH_2CH_2CH_2CH_2CH_2CH_3$$

Stearic acid

$$\underset{HO}{\overset{O}{\|}}CCH_2CH_2CH_2CH_2CH_2CH_2CH_2CH_2CH_2CH_2CH_2CH_2CH_2CH_2CH_2CH_2CH_3$$

Figure 17 🔑 Oleic acid is an unsaturated lipid found in olive oil. Stearic acid is a saturated lipid found in bacon.

🖊 **9. Visual Check**
Find Underline the lipid that is unsaturated.

WORD ORIGIN

lipid
from Greek *lipos,* means "fat, grease"

Figure 18 🔑 Phospholipids form a two-layer cell membrane that controls what enters and leaves cells, such as nutrients, waste, and water.

Lipids

Examples of lipids are shown in **Figure 17.** Lipids are biological molecules, but they are not polymers. *A* **lipid** *is a type of biological molecule that includes fats, oils, hormones, waxes, and components of cellular membranes.* Lipids have two major functions in living organisms. They store energy, and they make up cellular membranes.

Saturated and Unsaturated Lipids

There are two main groups of lipids—saturated and unsaturated. Saturated lipids contain only single bonds. Just like saturated hydrocarbons, they contain carbon bonded to the maximum number of hydrogen atoms possible. Unsaturated lipids contain at least one double bond. If an unsaturated lipid has one double bond, it is called monounsaturated. If it has more than one double bond, it is called polyunsaturated.

Lipids in Organisms

Lipids, such as fats and oils, store energy for organisms. Another function of lipids is to control what enters and leaves individual cells. These lipids are called phospholipids because they contain a phosphate functional group in their structure. Phospholipids form the cell membrane around individual cells, as shown in **Figure 18.**

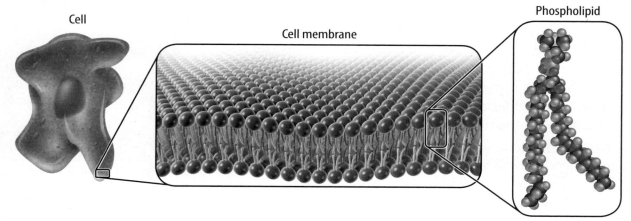

Cell

Cell membrane

Phospholipid

Visual Summary

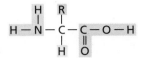

Amino acid

A protein is a long chain of amino acids.

The carbohydrate group contains sugars, such as glucose, starches, and cellulose.

Oleic acid

Lipids can be saturated or unsaturated.

Use Vocabulary

1 **Write** a complete sentence correctly using the term *biological molecule.* SC.8.P.8.5

_____

_____

2 **Define** *lipid* in your own words. SC.8.P.8.5

_____

Understand Key Concepts 🔑

3 **Explain** What are the four types of biological molecules? SC.8.P.8.5

_____

_____

4 **Describe** the function of two carbon compounds in living organisms. SC.8.P.8.5

_____

_____

5 Which monomer makes up nucleic acids? SC.8.P.8.5
- (A) amino acids
- (C) glucose
- (B) carbohydrates
- (D) nucleotide

Interpret Graphics

6 **Organize Information** Fill in the table below with details about each biological molecule from this lesson. SC.8.P.8.5

Molecule	Details

Critical Thinking

7 **Explain** why a meal containing boiled chicken is more healthful than a meal containing fried chicken.

_____

_____

8 **Evaluate** Many weight loss diets stress eliminating carbohydrates and fats from the diet. Explain why eliminating these foods completely from the diet over long periods of time might not be a good idea.

_____

_____

_____

Inquiry SC.8.P.8.5, SC.8.N.1.1, SC.8.N.1.3

iLAB STATION **Try It!**

Inquiry Lab *Testing for Carbon Compounds* at connectED.mcgraw-hill.com

Chapter 13 Study Guide

Think About It! Carbon is an element necessary for life. It has the chemical property of being able to bond easily with other atoms to form the compounds that make up many living and nonliving things.

 ## Key Concepts Summary

Vocabulary

LESSON 1 Elemental Carbon and Simple Organic Compounds

- Carbon atoms can form four covalent bonds using less energy than any other element in group 14. Carbon atoms can also form four different arrangements when bonded to other carbon atoms.

- When carbon bonds with only hydrogen atoms, it forms a **hydrocarbon.**

- A hydrocarbon that contains only single bonds is a **saturated hydrocarbon.** If a hydrocarbon contains at least one double or triple bond, it is an **unsaturated hydrocarbon.**

organic compound p. 502

hydrocarbon p. 504

isomer p. 504

saturated hydrocarbon p. 505

unsaturated hydrocarbon p. 505

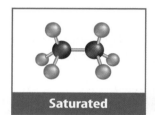

Saturated

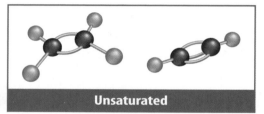

Unsaturated

LESSON 2 Other Organic Compounds

- When a hydrogen atom is replaced by a **functional group,** it forms a **substituted hydrocarbon.**

- Alcohols, halides, carboxylic acids, and amines form when specific functional groups are added to a hydrocarbon.

- **Monomers** link to form long chains called **polymers.**

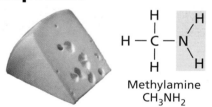

Methylamine
CH_3NH_2

substituted hydrocarbon p. 511

functional group p. 512

hydroxyl group p. 512

halide group p. 513

carboxyl group p. 513

amino group p. 514

polymer p. 515

monomer p. 515

polymerization p. 515

LESSON 3 Compounds of Life

- A **biological molecule** is a large organic molecule found in any living organism.

- **Proteins, carbohydrates, nucleic acids,** and **lipids** are groups of carbon compounds found in living organisms.

biological molecule p. 520

protein p. 521

amino acid p. 521

carbohydrate p. 522

nucleic acid p. 523

lipid p. 524

Phospholipid

Active Reading

FOLDABLES® **Chapter Project**

Assemble your lesson Foldables as shown to make a Chapter Project. Use the project to review what you have learned in this chapter.

Carbon Chemistry

Use Vocabulary

1 A chemical compound that contains carbon and at least one carbon-hydrogen bond is a(n)

_____ .

2 A(n) _____ contains the maximum number of hydrogen atoms possible.

3 Compounds that have the same chemical formula but different structural arrangements are called

_____ .

4 A molecule made up of many small organic molecules covalently bonded together forming a long chain is called a(n)

_____ .

5 A biological molecule made of amino acids is called a(n)

_____ .

6 RNA and DNA are two different types of

_____ .

7 A group of polymers that includes sugars, starches, and cellulose is called a(n)

_____ .

Link Vocabulary and Key Concepts

Use vocabulary terms from the previous page to complete the concept map.

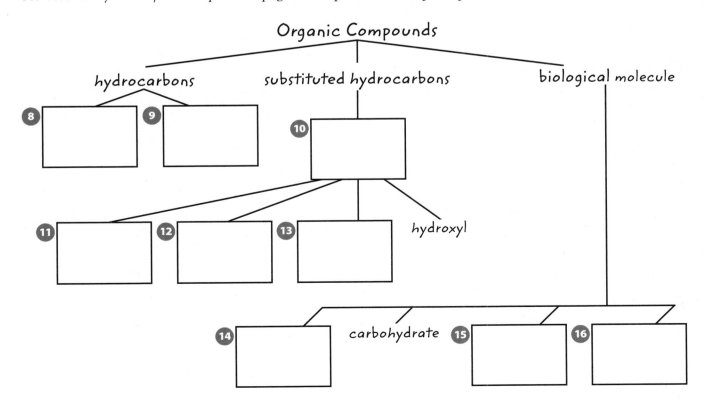

Fill in the correct answer choice.

🔑 Understand Key Concepts

1 What is the maximum number of covalent bonds that carbon can form? SC.8.P.8.5
- Ⓐ 1
- Ⓑ 2
- Ⓒ 3
- Ⓓ 4

2 Which is the correct name for the following carbon compound pictured below? SC.8.P.8.5

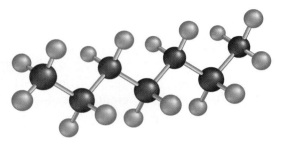

- Ⓐ heptane
- Ⓑ hexane
- Ⓒ pentane
- Ⓓ pentene

3 Which is NOT a functional group? SC.8.P.8.5
- Ⓐ carboxyl
- Ⓑ halide
- Ⓒ hydroxyl
- Ⓓ protein

4 Which functional group contains nitrogen in its structural formula? SC.8.P.8.5
- Ⓐ amino
- Ⓑ carboxyl
- Ⓒ halide
- Ⓓ hydroxyl

5 For which purpose does the body primarily use carbohydrates? SC.8.P.8.5
- Ⓐ muscle formation
- Ⓑ transmit genetic information
- Ⓒ to store energy
- Ⓓ cell membrane formation

6 Which item below does not contain lipids? SC.8.P.8.5
- Ⓐ beeswax
- Ⓑ cellular membranes
- Ⓒ cooking oil
- Ⓓ table sugar

Critical Thinking

7 **Draw** three isomers for pentane. SC.8.P.8.5

8 **Differentiate** between saturated and unsaturated hydrocarbons. SC.8.P.8.5

9 **Compare and contrast** the properties of diamond and graphite. SC.8.P.8.5

10 **Create** a poster on a separate sheet of paper that illustrates why hydrocarbons are present in so many different compounds. SC.8.P.8.5

11 **Evaluate** which of these formulas contains at least one double bond: C_7H_{16}, C_4H_8, C_3H_8. Explain your reasoning. SC.8.P.8.5

12 **Compare** substituted hydrocarbons with regular hydrocarbons. SC.8.P.8.5

13 **Describe** the process used to make most synthetic polymers. SC.8.P.8.5

14 **Describe** similarities and differences between proteins and carbohydrates. SC.8.P.8.5

_____

_____

_____

15 **Examine** the monomer below. Which biological molecules does it form, and what is the function of the biological molecule? SC.8.P.8.5

_____

_____

_____

$$HO-\overset{\overset{\displaystyle O}{\|}}{\underset{\underset{\displaystyle HO}{|}}{P}}-O-CH_2$$

16 **Evaluate** Some diet plans suggest eating only one type of food over a long period of time. Is this a wise practice for maintaining a healthy body? Explain why or why not. SC.8.P.8.5

_____

_____

_____

17 **Explain** the function of phospholipids in cells. SC.8.P.8.5

_____

_____

_____

Writing in Science

18 **Design** a brochure on a separate sheet of paper that explains information about each biological molecule discussed in this chapter. LA.6.2.2.3

Big Idea Review

19 What role does carbon have in the chemistry of living things? Explain a carbon atom's electron structure, how it bonds with other atoms, and how it forms substituted hydrocarbons. SC.8.P.8.5

_____

_____

_____

20 The photo below shows a collection of many living things. Describe as many carbon compounds as possible in the living things in the photo. SC.8.P.8.5

_____

_____

Math Skills   MA.6.A.3.6

Use Ratios

21 Many plastics, known as PVCs, are polymers of vinyl chloride, CH_2CHCl. What is the ratio of carbon to hydrogen to chlorine atoms in vinyl chloride?

22 Isopropy alcohol, or rubbing alcohol, has the formula $CH_3CH_2CH_2OH$. What is the ratio of carbon to hydrogen to oxygen atoms in this compound?

Fill in the correct answer choice.

Multiple Choice

Use the diagram below to answer question 1.

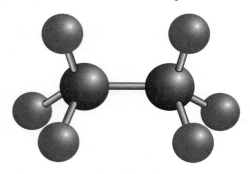

1 Which does the above diagram illustrate? SC.8.P.8.5

Ⓐ an alkane

Ⓑ an alkyne

Ⓒ a polymer

Ⓓ an unsaturated hydrocarbon

2 Which describes the structure of amorphous carbon? SC.8.P.8.5

Ⓕ cage-like

Ⓖ cross-bonded

Ⓗ formless

Ⓘ pyramidal

3 Which group includes a sugar, a phosphate group, and a nitrogen-containing base? SC.8.P.8.5

Ⓐ hydrocarbons

Ⓑ monomers

Ⓒ nucleotides

Ⓓ polymers

4 Which is NOT true about essential amino acids? SC.8.P.8.5

Ⓕ Foods supply them to the body.

Ⓖ Eleven are known to exist.

Ⓗ They can link to form proteins.

Ⓘ They form long chains.

Use the charts below to answer question 5.

Carbon Atoms	Name	Carbon Atoms	Name
1	meth–	6	hex–
2	eth–	7	hept–
3	prop–	8	oct–
4	but–	9	non–
5	pent–	10	dec–

Bond Type	Suffix
All single bonds —C—C—	–ane
At least one double bond —C=C—	–ene
At least one triple bond —C≡C—	–yne

5 Based on the charts, which hydrocarbon has four carbon atoms and only single bonds? SC.8.P.8.5

Ⓐ butane

Ⓑ butene

Ⓒ methane

Ⓓ methyne

6 Which property makes a carbon atom unique? SC.8.P.8.5

Ⓕ It has protons, neutrons, and electrons.

Ⓖ It has the ability to bond easily.

Ⓗ It is a nonmetal.

Ⓘ It is solid at room temperature.

7 Which is characteristic of biological molecules? SC.8.P.8.5

Ⓐ They are very small molecules.

Ⓑ They are found in living things.

Ⓒ They are classified as inorganic.

Ⓓ They form synthetic polymers.

Use the diagram below to answer question 8.

8 How many bonds link the two carbon atoms in the diagram? SC.8.P.8.5

  Ⓕ one

  Ⓖ two

  Ⓗ three

  Ⓘ four

9 Which is true about the formation of polymers? SC.8.P.8.5

  Ⓐ Laboratories produce most natural polymers.

  Ⓑ Most polymers result from splitting large molecules.

  Ⓒ Many synthetic polymers are products of petroleum.

  Ⓓ Typical polymers result from physical changes in compounds.

10 What is true about an unsaturated hydrocarbon? SC.8.P.8.5

  Ⓕ It contains only single bonds.

  Ⓖ It gives off molecules of water.

  Ⓗ It has double or triple bonds.

  Ⓘ Its name ends with the suffix *–ane*.

11 Which functional group does this molecule contain? SC.8.P.8.5

$$\begin{array}{cccccc} & & H & & & \\ & & | & & & \\ H & - & C & - & C & - O - H \\ & & | & & \| & \\ & & H & & O & \end{array}$$

  Ⓐ amino

  Ⓑ carboxyl

  Ⓒ halide

  Ⓓ hydroxyl

12 Which biological molecule is single-stranded and contains the instructions for making the proteins that cells need to function? SC.8.P.8.5

  Ⓕ DNA

  Ⓖ RNA

  Ⓗ lipid

  Ⓘ carbohydrate

13 Which is true about the carbon compound cyclooctane? SC.8.P.8.5

  Ⓐ It is a straight-chain hydrocarbon with eight carbon atoms and only single bonds.

  Ⓑ It is a branched-chain hydrocarbon with eight carbon atoms and at least one double bond.

  Ⓒ It is a ringed hydrocarbon with eight carbon atoms and at least one double bond.

  Ⓓ It is a ringed hydrocarbon with eight carbon atoms and only single bonds.

14 Which sugar is in DNA? SC.8.P.8.5

  Ⓕ deoxyribose

  Ⓖ fructose

  Ⓗ glucose

  Ⓘ ribose

NEED EXTRA HELP?

If You Missed Question...	1	2	3	4	5	6	7	8	9	10	11	12	13	14
Go to Lesson...	1	1	3	3	1	1	3	1	2	1	2	3	3	3

Multiple Choice *Bubble the correct answer.*

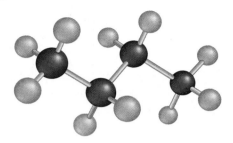

1. What is the name of the hydrocarbon shown in the image above? **SC.8.P.8.5**

   (A) alkane

   (B) butane

   (C) methane

   (D) propane

2. Cassandra is trying to name a hydrocarbon that is shown in her lab book. The hydrocarbon has eight carbon atoms in its longest chain. It has one triple bond. It does not have a ring shape. The hydrocarbon is called **SC.8.P.8.5**

   (F) butane.

   (G) ethane.

   (H) hexane.

   (I) octyne.

3. Which is a saturated hydrocarbon? **SC.8.P.8.5**

   (A) cyclohexyne

   (B) heptene

   (C) nonene

   (D) octane

4. Which type of carbon has the strongest bonds? **SC.8.P.8.5**

   (F)

   (G)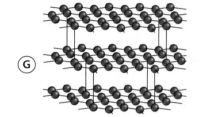

   (H)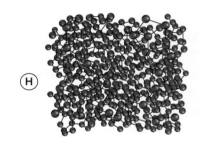

   (I)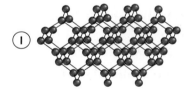

Copyright © Glencoe/McGraw-Hill, a division of The McGraw-Hill Companies, Inc.

Benchmark Mini-Assessment   Chapter 13 • Lesson 2

mini
BAT

Multiple Choice *Bubble the correct answer.*

1. Which molecule has a planar shape? **SC.7.N.1.6**

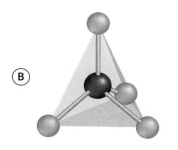

(A)

(B)

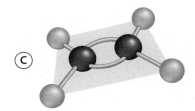

(C)

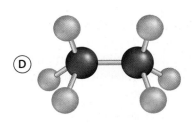

(D)

2. Which functional group replaces a hydrogen atom when ethane becomes ethanol? **SC.8.P.8.5**

(F) –CH

(G) –Cl

(H) –COOH

(I) –OH

3. Chlorohexane is a substance that contains **SC.8.P.8.5**

(A) an amino group.

(B) a carboxyl group.

(C) a halide group.

(D) a hydroxyl group.

4. Which polymer is formed from hydrocarbons that each contain three carbon atoms in their longest chain? **SC.8.N.1.6**

(F)

$$\left[\begin{array}{cc} F & F \\ | & | \\ C & C \\ | & | \\ F & F \end{array}\right]$$

Polytetrafluoroethylene (PTFE)

(G)

$$\left[\begin{array}{cc} H & H \\ | & | \\ C & C \\ | & | \\ H & H \end{array}\right]$$

Polyethylene (PE)

(H)

$$\left[\begin{array}{c} CH_2 - CH \\ | \\ CH_3 \end{array}\right]$$

Polypropylene (PP)

(I)

$$\left[\begin{array}{cc} H & H \\ | & | \\ C & C \\ | & | \\ H & Cl \end{array}\right]$$

Polyvinyl chloride (PVC)

Copyright © Glencoe/McGraw-Hill, a division of The McGraw-Hill Companies, Inc.

Benchmark Mini-Assessment Chapter 7 • Lesson 3

mini BAT

Multiple Choice *Bubble the correct answer.*

1. The biological molecule shown above belongs to which group? **SC.8.N.1.6**

(A) carbohydrates

(B) lipids

(C) nucleic acids

(D) proteins

2. A lipid that contains four double bonds is called **SC.8.P.8.5**

(F) monounsaturated.

(G) phospholipid.

(H) polyunsaturated.

(I) unsaturated.

3. Which functional groups are found in the monomer cysteine? **SC.8.P.8.5**

(A) amino and halide groups

(B) carboxyl and amino groups

(C) halide and hydroxyl groups

(D) hydroxyl and carboxyl groups

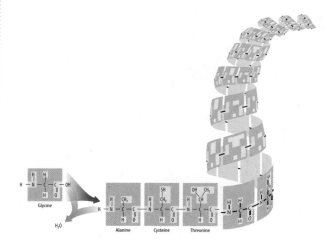

4. Terrell orders a turkey sub with mayonnaise, lettuce, and tomato. The biological molecule shown above can be found in this sub. In which part of the sub is this molecule most likely to be found? **SC.8.N.1.6**

(F) in the bread (carbohydrate)

(G) in the mayonnaise (fat)

(H) in the tomato (vegetable)

(I) in the turkey (protein)

Copyright © Glencoe/McGraw-Hill, a division of The McGraw-Hill Companies, Inc.

Notes

Unit 5

WAVES, ELECTRICITY, & MAGNETISM

1600

1660
Robert Hooke publishes the wave theory of light, comparing light's movement to that of waves in water.

1705
Francis Hauksbee experiments with a clock in a vacuum and proves that sound cannot travel without air.

1700

1820
Danish physicist Hans Christian Ørsted publishes his discovery that an electric current passing through a wire produces a magnetic field.

1878
Thomas Edison develops a system to provide electricity to homes and businesses using locally generated and distributed direct current (DC) electricity.

1800

1882
Thomas Edison develops and builds the first electricity-generating plant in New York City, which provides 110 V of direct current to 59 customers in lower Manhattan.

1883
The first standardized incandescent electric lighting system using overhead wires begins service in Roselle, New Jersey.

1890s
Physicist Nikola Tesla introduces alternating current (AC) by inventing the alternating current generator, allowing electricity to be transmitted at higher voltages over longer distances.

1947
Chuck Yeager becomes the first pilot to travel faster than the speed of sound

? Inquiry
Visit ConnectED for this unit's **STEM** activity.

Graphs SC.7.N.1.6, SC.8.N.4.1

Have you ever felt a shock from static electricity? That electric energy is similar to the energy in lightning, such as in **Figure 1**, only smaller. Scientists investigate what causes lightning and where it will occur. They use graphs to learn about lightning in different places and at different times. A **graph** is a type of chart that shows relationships between variables. Graphs organize and summarize data in a visual way. Three of the most common graphs are circle line, bar graphs, and circle graphs.

Types of Graphs

Line Graphs

A line graph is used when you want to analyze how a change in one variable affects another variable. This line graph shows how the average number of lightning flashes changes over time in Illinois. Time is plotted on the *x*-axis. The average number of lightning flashes is plotted on the *y*-axis. Each data point indicates the average number of flashes recorded during that hour. A line connects the data points to help determine any trends.

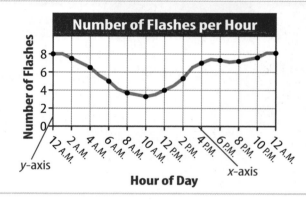

Bar Graphs

When you want to compare amounts in different categories, you use a bar graph. The horizontal axis often contains categories instead of numbers. This bar graph shows the average number of lightning flashes that occur in different states. On average, about 9.8 lightning flashes strike each square kilometer of land in Florida every year. Florida has more lightning flashes per square kilometer than all other states shown.

Circle Graphs

To show how the parts of something relate to the whole, use a circle graph. This circle graph shows the average percentage of lightning flashes each U.S. region receives in a year.

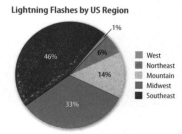

Active Reading **1. Complete** Fill in the blanks to complete the paragraph below.

The graph shows that the _____ region receives about 14 percent of all lightning flashes that strike each year. You can also determine that the _____ receives the most lightning in a given year.

Active Reading **2. Infer** Determine in which region of the U.S. Florida is located. How does the Florida information from the bar graph above relate to the circle graph data?

_____

_____

Line Graphs and Trends

Suppose you are planning a picnic in an area that experiences quite a bit of lightning. When would be the safest time to go? First, you gather data about the average number of lightning flashes per hour. Next, you plot the data on a line graph and analyze trends. Trends are patterns in data that help you find relationships among the data and make predictions.

Active Reading 3. **Determine** Highlight the definition of *trend* as used with a graph.

Follow the orange line on the line graph from 12 A.M. to 10 A.M. in **Figure 2**. Notice that the line slopes downward, indicated by the green arrow. A downward slope means that as measurements on the *x*-axis increase, measurements on the *y*-axis decrease. So, as time passes from 12 A.M. to 10 A.M., the number of lightning flashes decreases.

Follow the orange line on the graph from 12 P.M. to 5 P.M. Notice that the line slopes upward, indicated by the blue arrow. An upward slope means that as the measurements on the *x*-axis increase, the measurements on the *y*-axis also increase. So, as time passes from 12 P.M. to 5 P.M., the number of lightning flashes increases.

Active Reading 4. **Predict** Analyze the line graph to determine when you would have the least risk of lightning during your picnic.

Figure 2 The slope of a line in a line graph shows the relationship between the variable on the *x*-axis and the variable on the *y*-axis. ▼

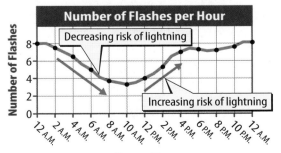

▲ **Figure 1** Scientists study lightning to get a better understanding of what causes it and to predict when it will occur.

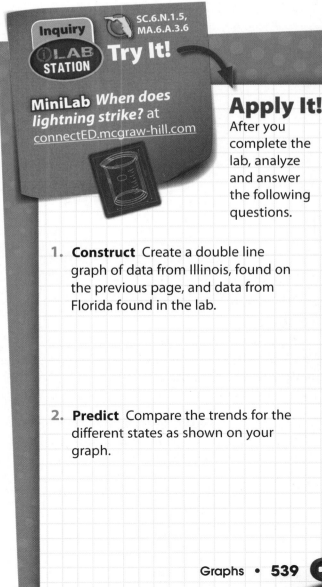

Inquiry SC.6.N.1.5, MA.6.A.3.6

①LAB STATION **Try It!**

MiniLab *When does lightning strike?* at connectED.mcgraw-hill.com

Apply It!
After you complete the lab, analyze and answer the following questions.

1. **Construct** Create a double line graph of data from Illinois, found on the previous page, and data from Florida found in the lab.

2. **Predict** Compare the trends for the different states as shown on your graph.

Name _____   Date _____

Getting the Ball Back

Two brothers are playing soccer on the beach. One brother kicks the ball really hard and the ball lands in the water, about 50 meters from the beach. They wonder if the ball will float back to the beach. This is what they said:

Todd: Waves carry objects as they travel through water. If we wait, the waves will move the ball back onto the beach

Brian: Waves don't carry things as they travel through water. I think we need to swim out and get the ball.

Who do you agree with?

Explain why you agree using ideas about waves.

WAVES

FLORIDA BIG IDEAS

1 **The Practice of Science**
3 **The Role of Theories, Laws, Hypotheses, and Models**
10 **Forms of Energy**

Think About It!

How do waves travel through matter?

South of Cocoa Beach is "Surf City." That's where you will find the best surfers and some of Florida's best waves to ride. Waves are actually energy moving through matter. Think about the amount of energy this wave must be carrying.

1 What do you think caused this giant wave?

_____

_____

_____

2 Do you think this is the only large wave in the area?

_____

_____

_____

_____

3 How do you think this wave moves through water?

_____

_____

_____

_____

Get Ready to Read

What do you think about energy and wave motion?

Before you read, decide if you agree or disagree with each of these statements. As you read this chapter, see if you change your mind about any of the statements.

		AGREE	DISAGREE
1	Waves carry matter as they travel from one place to another.	☐	☐
2	Sound waves can travel where there is no matter.	☐	☐
3	Waves that carry more energy cause particles in a material to move a greater distance.	☐	☐
4	Sound waves travel fastest in gases.	☐	☐
5	When light waves strike a mirror, they change direction.	☐	☐
6	Light waves travel at the same speed in all materials.	☐	☐

 **There's More Online!**
Video • Audio • Review • ⓘLab Station • WebQuest • Assessment • Concepts in Motion • Multilingual eGlossary

543

What are WAVES?

ESSENTIAL QUESTIONS

 What is a wave?

 How do different types of waves make particles of matter move?

 Can waves travel through empty space?

Vocabulary

wave p. 545

mechanical wave p. 547

medium p. 547

transverse wave p. 547

crest p. 547

trough p. 547

longitudinal wave p. 548

compression p. 548

rarefaction p. 548

electromagnetic wave p. 551

 Florida NGSSS

SC.7.N.1.3 Distinguish between an experiment (which must involve the identification and control of variables) and other forms of scientific investigation and explain that not all scientific knowledge is derived from experimentation.

SC.7.P.10.1 Illustrate that the sun's energy arrives as radiation with a wide range of wavelengths, including infrared, visible, and ultraviolet, and that white light is made up of a spectrum of many different colors.

SC.8.N.1.5 Analyze the methods used to develop a scientific explanation as seen in different fields of science.

LA.6.2.2.3 The student will organize information to show understanding or relationships among facts, ideas, and events (e.g., representing key points within text through charting, mapping, paraphrasing, summarizing, or comparing/contrasting);

Inquiry Launch Lab   SC.7.N.1.3

20 minutes

How can you make waves?

Oceans, lakes, and ponds aren't the only places you can find waves. Can you create waves in a cup of water?

Procedure

1. Read and complete a lab safety form.
2. Add **water** to a **clear plastic cup** until it is about two-thirds full. Place the cup on a **paper towel.**
3. Explore ways of producing water waves by touching the cup. Do not move the cup.
4. Explore ways of producing water waves without touching the cup. Do not move the cup.

Think About This

1. **Describe** How did the water's surface change when you produced water waves in the cup?

2. **Key Concept** **Explain** What did the different ways of producing water waves have in common?

Assemble your lesson Foldables as shown to make a Chapter Project. Use the project to review what you have learned in this chapter.

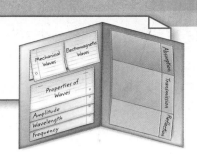

Use Vocabulary

1 A material though which a wave travels is a(n) _____ .

2 A(n) _____ is a region where matter is more closely spaced in a longitudinal wave.

3 The Sun gives off energy that travels through space in the form of _____ .

4 The product of _____ and wavelength is the speed of the wave _____ .

5 _____ is a property of waves that is measured in hertz.

6 The highest point on a transverse wave is a(n) _____ .

7 _____ is when two waves pass through each other and keep going.

Link Vocabulary and Key Concepts

Use vocabulary terms from the previous page to complete the concept map.

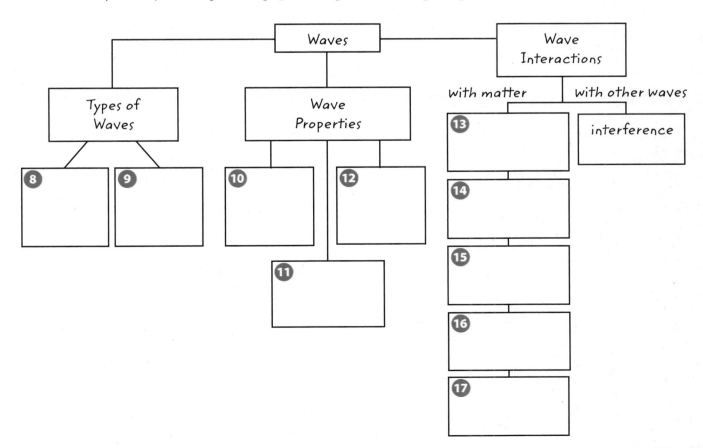

Chapter 14 Review

Fill in the correct answer choice.

🔑 Understand Key Concepts

1 What is transferred by a radio wave? SC.7.P.10.1
- Ⓐ air
- Ⓑ energy
- Ⓒ matter
- Ⓓ space

2 In a longitudinal wave, where are the particles most spread out? SC.7.P.10.3
- Ⓐ compression
- Ⓑ crest
- Ⓒ rarefaction
- Ⓓ trough

3 Which would produce mechanical waves?
- Ⓐ burning a candle   SC.7.P.10.3
- Ⓑ hitting a wall with a hammer
- Ⓒ turning on a flashlight
- Ⓓ tying a rope to a doorknob

4 Which is an electromagnetic wave? SC.7.P.10.1
- Ⓐ a flag waving in the wind
- Ⓑ a vibrating guitar string
- Ⓒ the changes in the air that result from blowing a horn
- Ⓓ the waves that heat a cup of water in a microwave oven

5 Identify the crest of the wave in the illustration below. SC.7.P.10.3

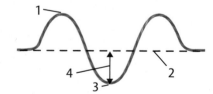

- Ⓐ 1
- Ⓑ 2
- Ⓒ 3
- Ⓓ 4

6 If the energy carried by a wave increases, which other wave property also increases. SC.7.P.10.3
- Ⓐ amplitude
- Ⓑ medium
- Ⓒ wavelength
- Ⓓ wave speed

Critical Thinking

7 **Assess** A student sets up a line of dominoes so that each is standing vertically next to another. He then pushes the first one and each falls down in succession. How does this demonstration represent a wave? How is it different? SC.7.N.1.3

_____

_____

_____

_____

8 **Infer** In the figure below, suppose wave 1 and wave 2 have the same amplitude. Describe the wave that forms when destructive interference occurs. LA.6.2.2.3

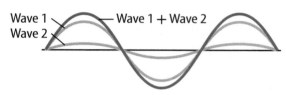

_____

_____

9 **Compare** A category 5 hurricane has more energy than a category 3 hurricane. Which hurricane will create water waves with greater amplitude? Why? LA.6.2.2.3

_____

_____

_____

10 **Infer** At a baseball game when you are far from the batter, you might see the batter hit the ball before you hear the sound of the bat hitting the ball. Explain why this happens. SC.7.P.10.3

_____

_____

11 **Evaluate** Geologists measure the amplitude of seismic waves using the Richter scale. If an earthquake of 7.3 has a greater amplitude than an earthquake of 4.4, which one carries more energy? Explain your answer. SC.7.N.1.6

_____

_____

_____

12 Recommend Some medicines lose their potency when exposed to ultraviolet light. Recommend the type of container in which these medicines should be stored. SC.7.P.10.2

_____

_____

13 Explain why the noise level rises in a room full of many talking people. LA.6.2.2.3

_____

_____

_____

Writing in Science

14 Write a short essay, on a separate sheet of paper, explaining how an earthquake below the ocean floor can affect the seas near the earthquake area. LA.6.2.2.3

Big Idea Review

15 What are waves and how do they travel? Describe the movement of particles from their resting positions for transverse and longitudinal waves. SC.7.P.10.3

_____

_____

_____

16 The chapter opener photo shows waves in the ocean. Describe the waves using vocabulary terms from the chapter. LA.6.2.2.3

_____

_____

_____

Math Skills   MA.6.A.3.6

Use Numbers

17 A hummingbird can flap its wings 200 times per second. If the hummingbird produces waves that travel at 340 m/s by flapping its wings, what is the wavelength of these waves?

18 A student did an experiment in which she collected the data shown in the table. What can you conclude about the wave speed and rope diameter in this experiment? What can you conclude about frequency and wavelength?

Wave Speed and Diameter			
Trial	**Rope Diameter**	**Frequency**	**Wavelength**
1	2.0 cm	2.0 Hz	8.0 m
2	2.0 cm	8.0 Hz	2.0 m
3	4.0 cm	2.0 Hz	10.0 m
4	4.0 cm	4.0 Hz	5.0 m

Florida NGSSS

Benchmark Practice

Record your answers on the answer sheet provided by your teacher or on a sheet of paper.

Multiple Choice

1 Through which medium would sound waves move most slowly? SC.7.P.10.3

   (A) air

   (B) aluminum

   (C) glass

   (D) water

Use the illustration below to answer question 2.

2 Which process enables the boy to see over the wall? SC.7.P.10.2

   (F) diffraction

   (G) interference

   (H) reflection

   (I) refraction

3 Which is an electromagnetic wave? SC.7.P.10.1

   (A) light

   (B) seismic

   (C) sound

   (D) water

4 Which statement best defines radiant energy? SC.7.P.10.1

   (F) Radiant energy is the energy that must travel through certain types of media.

   (G) Radiant energy is the energy carried by an electromagnetic wave.

   (H) Radiant energy is the energy carried only by magnetic waves.

   (I) Radiant energy travels at short wavelengths.

Use the table below to answer question 5.

Speed of Light Through Different Mediums			
Medium	air	water	glass
Speed of light (m/s × 10^8)	3.0	2.2	1.5

5 According to the information in the table above, how does the speed of light in air compare to the speed of light through glass? SC.7.P.10.3

   (A) Light travels slower through air than through glass.

   (B) Light travels three times faster through air than through glass.

   (C) Light travels two times faster through air than through glass.

   (D) Light travels at the same speed through air and through glass.

6 How does solar energy travel to Earth? SC.7.P.10.1

   (F) through conduction

   (G) through convection

   (H) through emission

   (I) through radiation

7 Sound waves travel fastest through what type of material? SC.7.P.10.3

   (A) gases

   (B) liquids

   (C) solids

   (D) empty space

8 Which is an example of refraction? SC.7.P.10.2

Ⓕ a flashlight beam hitting a mirror

Ⓖ a shout crossing a crowded room

Ⓗ a sunbeam striking a window

Ⓘ a water wave bending around a rock

Use the table below to answer question 9.

Boulder Temperature (°C)	Speed of Sound (m/s)
−5	562
0	564
10	566
15	568
20	?

9 A student is studying how sound travels through a solid medium by timing the movement of sound through a boulder when the boulder is at different temperatures. Using the information above, how fast will sound move through the boulder when it reaches 20°C? SC.7.P.10.3

Ⓐ 569 m/s

Ⓑ 570 m/s

Ⓒ 571 m/s

Ⓓ 572 m/s

10 Given that sound waves are formed by vibrating molecules, which statement is true? SC.7.P.10.3

Ⓕ Sound travels fastest through space.

Ⓖ Sound travels faster through water than through air.

Ⓗ Sound travels at the same rate through all mediums.

Ⓘ Sound travels faster through air than through water.

Use the graph below to answer question 11.

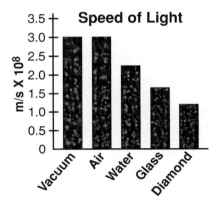

11 According to the graph, through which medium is the speed of light nearly half as fast as it is in a vacuum? SC.7.P.10.3

Ⓐ air

Ⓑ diamond

Ⓒ glass

Ⓓ water

NEED EXTRA HELP?

If You Missed Question...	1	2	3	4	5	6	7	8	9	10	11
Go to Lesson...	2	3	1	1	2	1	2	3	3	2	2

Multiple Choice *Bubble the correct answer.*

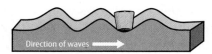

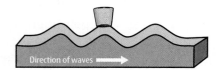

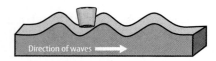

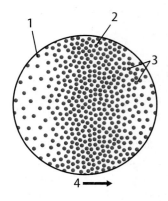

1. Identify the type of wave shown in the images above. **SC.7.P.10.1**

   (A) asymmetric wave

   (B) elliptical wave

   (C) longitudinal wave

   (D) transverse wave

2. Which type of electromagnetic waves do most organisms emit? **SC.7.P.10.1**

   (F) infrared waves

   (G) light waves

   (H) radio waves

   (I) ultraviolet waves

3. At which point in the diagram above are the particles of air closest together? **SC.7.P.10.1**

   (A) Point 1

   (B) Point 2

   (C) Point 3

   (D) Point 4

Copyright © Glencoe/McGraw-Hill, a division of The McGraw-Hill Companies, Inc.

Benchmark Mini-Assessment — Chapter 14 • Lesson 2

FLORIDA NGSSS

mini BAT

Multiple Choice *Bubble the correct answer.*

Use the images below to answer questions 1–3.

1

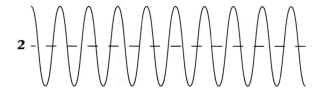

2

3

4

1. In which image is the wave amplitude the greatest? **SC.7.P.10.3**

(A) 1

(B) 2

(C) 3

(D) 4

2. Imagine that wave 1 is traveling through air. What happens to the wave as it moves into a brick wall? **SC.7.P.10.3**

(F) It slows down.

(G) It speeds up.

(H) It changes amplitude.

(I) It changes frequency.

3. In which image is the energy the greatest? **SC.7.P.10.3**

(A) 1

(B) 2

(C) 3

(D) 4

4. In which medium will a sound wave move more quickly? **SC.7.P.10.3**

(F) air

(G) rock

(H) ethanol

(I) vacuum

Copyright © Glencoe/McGraw-Hill, a division of The McGraw-Hill Companies, Inc.

Multiple Choice *Bubble the correct answer.*

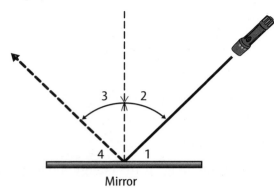

Mirror

1. The diagram above shows the angle of incidence from a light source. What will be the angle of reflection? **SC.7.P.10.1**

(A) Point 1

(B) Point 2

(C) Point 3

(D) Point 4

2. Transmission occurs when waves pass through **SC.7.P.10.2**

(F) aluminum foil.

(G) a mirror.

(H) plastic wrap.

(I) a steel beam.

3. Each diagram below shows two waves overlapping. Which diagram shows an example of a standing wave? **SC.7.P.10.2**

(A)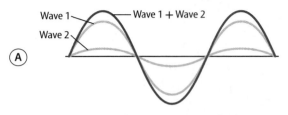

(B)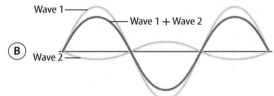

(C)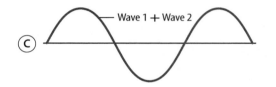

(D)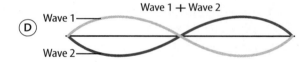

Copyright © Glencoe/McGraw-Hill, a division of The McGraw-Hill Companies, Inc.

Notes

Name _____ Date _____

Good Vibrations

Three friends argued about what causes sound. This is what they thought:

Ben: I don't think vibrations are necessary in order to produce sound.

Anna: I think vibrations are always necessary in order to produce sound.

Leah: I think vibrations are sometimes necessary in order to produce sound.

Circle the friend you agree with and explain why you agree.

SOUND

FLORIDA BIG IDEAS

1 The Practice of Science
10 Forms of Energy

How can you produce, describe, and use sound?

Recording a song involves more than just a band playing music. Often, each instrument records alone. Then, a singer records the vocals. Next, an audio engineer blends together the vocals and the sounds from the instruments. Some sounds may need to be adjusted to sound louder, softer, higher, or lower.

1 What other properties of sound do you think the soundboard knobs control?

_____

_____

2 Why do you think musicians wear headphones?

_____

_____

3 How can you produce, describe, and use sound?

_____

_____

_____

What do you think about sound?

Before you read, decide if you agree or disagree with each of these statements. As you read this chapter, see if you change your mind about any of the statements.

	AGREE	DISAGREE
1 Sound waves travel fastest through empty space.	☐	☐
2 Hearing loss is mainly caused by brief, loud sounds, such as a firecracker.	☐	☐
3 Sound waves from different sources can affect one another when they meet.	☐	☐
4 You can tell whether an ambulance is moving toward or away from you by listening to changes in the sound of the siren.	☐	☐
5 Animals use sounds that humans cannot hear to locate objects around them.	☐	☐
6 The method that ships use to locate underwater objects can be used to locate organs or tumors inside the human body.	☐	☐

 **Connect ED**

There's More Online!
Video • Audio • Review • ①Lab Station • WebQuest • Assessment • Concepts in Motion • Multilingual eGlossary

583

Producing and Detecting SOUND

ESSENTIAL QUESTIONS

How is sound produced?

How does sound move from one place to another?

Why does sound travel at different speeds through various materials?

What are the functions of the different parts of the human ear?

Vocabulary

sound wave p. 585
longitudinal wave p. 585
vibration p. 585
medium p. 586
compression p. 586
rarefaction p. 586

Florida NGSSS

SC.7.N.1.6 Explain that empirical evidence is the cumulative body of observations of a natural phenomenon on which scientific explanations are based.

SC.7.P.10.3 Recognize that light waves, sound waves, and other waves move at different speeds in different materials.

LA.6.2.2.3 The student will organize information to show understanding or relationships among facts, ideas, and events (e.g., representing key points within text through charting, mapping, paraphrasing, summarizing, or comparing/contrasting);

LA.6.4.2.2 The student will record information (e.g., observations, notes, lists, charts, legends) related to a topic, including visual aids to organize and record information, as appropriate, and attribute sources of information;

 Launch Lab

10 minutes

What causes sound?

Sound travels through air as vibrations. When those vibrations reach the ear, sound is heard.

Procedure

1. Read and complete a lab safety form.

2. Stretch **waxed paper** over the top of a **beaker.** Wrap a **rubber band** around the top to hold it tight.

3. Strike the center of the waxed paper gently with the eraser end of a **pencil.** Then strike it harder. How did the sound change? Write your observations.

4. Sprinkle a few grains of **rice** onto the waxed paper. Strike the paper gently and then harder. Observe how the rice moves each time. Record your observations.

Data and Observations

Think About This

1. How was the change in the rice's motion related to the change in sound?

2. **Key Concept** Based on your results, what do you think causes sound?

inquiry What is he listening to?

1. Think about the different ways people use sounds. Like the boy in the picture, many people use headphones to listen to their favorite songs. Others enjoy attending concerts or playing musical instruments. The sound of a car horn or a person shouting can alert people to danger. How do you think different types of sounds are produced? How are your ears able to detect the sounds?

_____

_____

_____

What is sound?

Everywhere you look, it seems that people have something on their ears! Some are talking on cell phones or listening to music. Some, such as people who work around airplanes, wear ear protection to prevent damage to their hearing. All of these devices have something to do with sound. The sounds you hear are produced by **sound waves**—*longitudinal waves that can only travel through matter.* A **longitudinal wave** *is a wave that makes the particles in the material that carries the wave move back and forth along the direction the wave travels.*

Sources of Sound

Every sound, from the buzzing of a bee to a loud siren, is the result of a vibration. *A* **vibration** *is a rapid, back-and-forth motion that can occur in solids, liquids, or gases.* The energy carried by a sound wave is caused by vibration. For example, as you pull on a guitar string, you transfer energy to the string. When you let go, the string snaps back and vibrates. As the string vibrates, it collides with nearby air particles. The string transfers energy to these particles. The air particles collide with other air particles and pass on energy, as shown in **Figure 1.**

As the string vibrates, its back-and-forth motion causes a disturbance in the air that carries energy outward from the source of the sound. The disturbance is a sound wave.

2. NGSSS Check Explain How is sound produced? SC.7.P.10.3

_____

_____

Figure 1  As the guitar string vibrates, it transfers energy to nearby air particles.

Active Reading **3. Draw** How would the picture of the particles be different if the string vibrated faster?

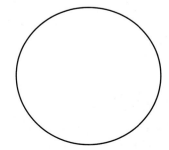

Figure 2 🔑 Sound waves move away from a source as compressions and rarefactions.

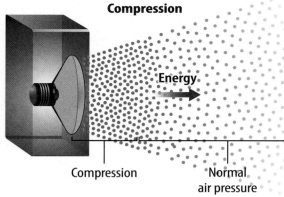

Compression

Energy

Compression | Normal air pressure

When the speaker cone moves out, it forces particles in the air closer together. This produces a high-pressure area, or compression.

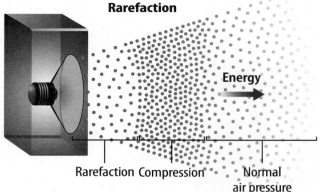

Rarefaction

Energy

Rarefaction Compression | Normal air pressure

When the speaker cone moves back, it leaves behind an area with fewer particles. This is a low-pressure area called a rarefaction.

Active Reading

4. Describe What are the parts of a sound wave? Include the term *medium* in each of your descriptions.

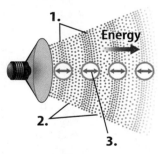

1.

Energy

2.

3.

Wave Part & Description
1. Wave Part: Description
2. Wave Part: Description
3. Wave Part: Description

How Sound Waves Travel

Vibrating objects cause sound waves that occur only in matter. For this reason, sound waves must travel through a solid, a liquid, or a gas. *A material in which a wave travels is a* **medium.** You usually hear sound through the medium of air, but sound waves also can travel through other **media,** such as water, wood, and metal. Sound waves cannot travel through empty space because there is no medium to carry the energy.

From a Sound Source to Your Ear If you touch a speaker, like the one in **Figure 2,** you can feel it vibrate as it produces sound waves. Air particles fill a room. Each time the speaker cone moves forward, it pushes air particles ahead of it in the room. This push forces the particles closer together, increasing air pressure in that area. *A region of a longitudinal wave where the particles in the medium are closest together is a* **compression.** With each vibration, the speaker cone moves forward and then back. This motion leaves behind a low-pressure region with fewer air particles, as shown on the right in **Figure 2.** *A* **rarefaction** (rayr uh FAK shun) *is a region of a longitudinal wave where the particles are farthest apart.*

Energy in Sound Waves Suppose you are in the ticket line at the movies. Someone at the back of the line bumps into the next person in line. That person stumbles and bumps the next person before returning to his or her place in line. The energy of the bump continues down the line as each person bumps the next one. In the same way, particles of a medium vibrate back and forth as a sound wave carries energy away from a source. This process is shown in **Figure 3.**

Speed of Sound

You are swimming in a pool when someone taps on the side nearby. Would you hear the sound if your head were under water? Yes, and you would probably hear the sound better than if your head were above water. Sound waves travel faster in water than in air. **Table 1** compares the speed of sound in different media.

Material Two factors that affect the speed of sound waves are the density and the stiffness of the material, or medium. Density is how closely the particles of a medium are packed. Gas particles are far apart and do not collide as often as do the particles of a liquid or a solid. Therefore, sound energy transfers more slowly in a gas than in a liquid or a solid. Notice in **Table 1** that sound travels fastest in solids. In a stiff or rigid solid where particles are packed very close together, the particles collide and transfer energy quickly.

Sound waves also travel faster in seawater than in freshwater. Seawater contains dissolved salts and has a higher density than freshwater. The fin whale in **Figure 4** emits sounds heard by other whales hundreds of kilometers away.

Figure 4 The low-pulse sounds of a fin whale travel more than four times faster through water than they would through air.

Temperature The temperature of a medium also affects the speed of sound. As the temperature of a gas increases, the particles move faster and collide more often. This increase in collisions transfers more energy in less time. Notice that sound waves travel faster in air at 20°C than in air at 0°C.

In liquids and solids, temperature has the opposite effect. Why do sound waves travel faster in water at 0°C than in water at 20°C? As water cools, the molecules move closer together, so they collide more often. Sound waves travel even faster when the water freezes into ice because it is rigid.

Table 1	The Speed of Sound
Material	**Speed (m/s)**
Air (0°C)	331
Air (20°C)	343
Water (20°C)	1,481
Water (0°C)	1,500
Seawater (25°C)	1,533
Ice (0°C)	3,500
Iron	5,130
Glass	5,640

5. NGSSS Check
Predict Based on the data in the table, do you think sound will move faster in cool pond water or in warm bath water?
SC.7.P.10.3

Active Reading

FOLDABLES LA.6.2.2.3

Make a vertical three-tab book and label it as shown. Use it to organize your notes about sound.

How Sound Is Produced

How Sound Travels

How Sound Is Detected

6. NGSSS Check
Analyze Why does sound travel at different speeds through various materials? SC.7.P.10.3

Figure 5 The large ears of the fennec fox can detect prey moving underground.

Detecting Sound

How do you think having large ears helps the fennec fox in **Figure 5?** Sound waves fill the air, and the large outer ear helps funnel sound waves to the inner ear, where sound is detected. Ears also tell you the direction a sound comes from. With its large ears, the fennec fox is better able to hear predators approach from greater distances.

The Human Ear

Have you ever cupped your hand around the back of your ear so you could hear better? Why does that work? The human ear has three main parts. The outer ear collects sound waves. By cupping your hand around your ear, you extend the outer ear and therefore collect more waves. The middle ear amplifies sound. The inner ear sends signals about sound to the brain. As shown in **Figure 6,** each part of the ear has a special shape with different parts that help it perform its function.

Active Reading **7. Label** <u>Underline</u> on the figure below how the eardrum helps you hear.

Figure 6 Each part of the ear has an important job that helps you hear.

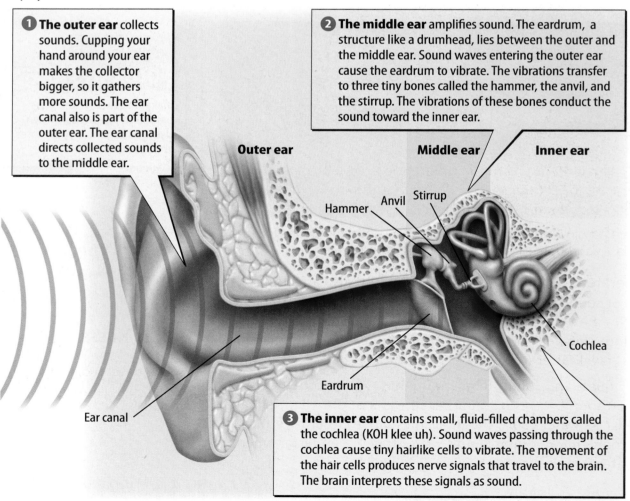

❶ **The outer ear** collects sounds. Cupping your hand around your ear makes the collector bigger, so it gathers more sounds. The ear canal also is part of the outer ear. The ear canal directs collected sounds to the middle ear.

❷ **The middle ear** amplifies sound. The eardrum, a structure like a drumhead, lies between the outer and the middle ear. Sound waves entering the outer ear cause the eardrum to vibrate. The vibrations transfer to three tiny bones called the hammer, the anvil, and the stirrup. The vibrations of these bones conduct the sound toward the inner ear.

Outer ear   **Middle ear**   **Inner ear**

Hammer   Anvil   Stirrup

Ear canal

Eardrum

Cochlea

❸ **The inner ear** contains small, fluid-filled chambers called the cochlea (KOH klee uh). Sound waves passing through the cochlea cause tiny hairlike cells to vibrate. The movement of the hair cells produces nerve signals that travel to the brain. The brain interprets these signals as sound.

Apply It! After you complete the lab, answer these questions.

1. **Explain** How can you tell where a sound is produced?

2. **Describe** How do stereo headphones create the illusion of musical instruments and vocalists being in different places around you?

Hearing Loss

The harder you pound on a drum, the farther the drumhead travels as it vibrates. What would a drum sound like if the drumhead had a big tear in it? Like a drumhead, the eardrum can be damaged. The eardrum vibrates as pressure changes in the ear. The louder the sound, the farther the eardrum moves in and out as it vibrates.

A very loud sound can make the eardrum vibrate so hard that it tears. Damage to the eardrum can cause hearing loss. Also, the tear can allow bacteria into the ear, causing infection. The tear may heal, but thick, uneven scar tissue can make the eardrum less sensitive to sounds.

Listening to loud music over a long period of time also can damage the ears. Look again at **Figure 6** and locate the cochlea. Infection or loud sounds can damage tiny hair cells in the cochlea. Cells that are damaged or die do not grow back, so hearing becomes less sensitive. Many people who work around loud machines, construction, or traffic wear ear protection to prevent damage, as shown in **Figure 7**. Wearing a headset while listening to loud music, however, traps the pressure changes in the ear; this can lead to permanent hearing loss.

Figure 7 This woman is protecting two of her most important senses—sight and hearing.

Active Reading 8. **Name** What are some things that might happen to your ear to cause hearing loss?

_____

_____

_____

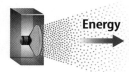

Sound waves carry energy away from a sound source.

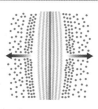

Sound waves move out from a source as a series of compressions and rarefactions.

Ears detect vibrations and interpret them as different sounds.

Use Vocabulary

1 **Define** *longitudinal wave* in your own words.

_____

_____

2 **Identify** the region in a sound wave where the particles are farthest apart.

_____

3 The energy carried by a sound wave comes from a(n)

_____ in a medium.

Understand Key Concepts

4 Through which medium would sound travel fastest? SC.7.P.10.3

  (A) air   (C) cold water

  (B) iron   (D) warm water

5 **Summarize** how you produce sound when you tap a pencil against your desk.

_____

_____

6 **Compare and contrast** the functions of the three main parts of the human ear.

_____

_____

7 **Describe** the motion of an air particle as a sound wave passes through it.

_____

_____

Interpret Graphics

8 **Sequence** Fill in the graphic organizer below, identifying the path of a sound wave as it enters the ear. LA.6.2.2.3

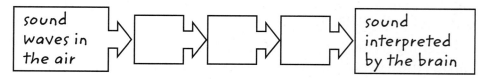

sound waves in the air → ☐ → ☐ → ☐ → sound interpreted by the brain

Critical Thinking

9 **Evaluate** A spaceship in a science fiction movie explodes. People in a nearby spaceship hear a loud sound. Is this realistic? Explain.

SC.7.P.10.3

_____

_____

Cochlear Implants

Helping Damaged Ears Hear Again

Is there any way to hear again after hair cells in your cochlea are destroyed? Not long ago, the answer was no. But over the last 20 to 30 years, scientists developed a way to bypass the damaged cells. It is called a cochlear implant—a device that uses electrical signals to stimulate the nerves that go from the ear to the brain. How does it work?

First a surgeon implants the interior part of the device under the scalp and into the inner ear. Then the exterior part of the device is put to work. With hearing restored, you are once again connected to the sounds in the world around you!

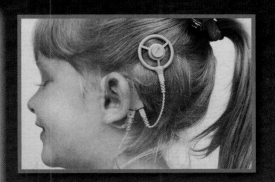

1 A microphone receives sound waves from the environment. A speech processor changes sounds into electrical signals.

2 The electrical signals are then sent across the scalp from the transmitter to a receiver.

3 The receiver sends the signals through a wire to electrodes implanted in the cochlea.

4 Nerves in the inner ear pick up the signals and send them to the brain. The brain interprets the signals as sound.

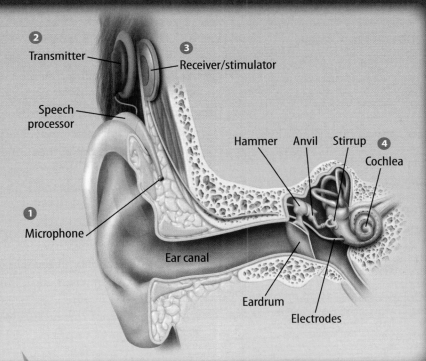

2 Transmitter
3 Receiver/stimulator
Speech processor
Hammer   Anvil   Stirrup   **4**
Cochlea
1
Microphone
Ear canal
Eardrum
Electrodes

It's Your Turn

MAKE A POSTER How does a hearing aid work? How is it different from a cochlear implant? Research these questions and create a poster from what you discover.

LA.6.4.2.2

Properties of SOUND WAVES

Vocabulary

amplitude p. 593

intensity p. 594

wavelength p. 595

frequency p. 595

pitch p. 595

Doppler effect p. 596

interference p. 596

resonance p. 598

Florida NGSSS

SC.7.N.1.3 Distinguish between an experiment (which must involve the identification and control of variables) and other forms of scientific investigation and explain that not all scientific knowledge is derived from experimentation.

SC.7.N.1.4 Identify test variables (independent variables) and outcome variables (dependent variables) in an experiment.

SC.7.N.1.6 Explain that empirical evidence is the cumulative body of observations of a natural phenomenon on which scientific explanations are based.

LA.6.2.2.3 The student will organize information to show understanding or relationships among facts, ideas, and events (e.g., representing key points within text through charting, mapping, paraphrasing, summarizing, or comparing/contrasting);

MA.6.S.6.2 Select and analyze the measures of central tendency or variability to represent, describe, analyze, and/or summarize a data set for the purposes of answering questions appropriately.

Inquiry Launch Lab

SC.7.N.1.6

15 minutes

How can sound blow out a candle?

Procedure

1. Read and complete a lab safety form.

2. Cut off the neck of a **balloon** with **scissors.** Stretch the remaining part of the balloon over the wide end of a **small funnel.**

3. Set a ball of **modeling clay** on a table. Insert a small **candle** in the clay, and use a **safety match** to light it.

4. Hold the funnel with the narrow end pointing toward the lit candle, about 2 cm away. ⚠ *Do not let the funnel touch the flame.*

5. Sharply strike the rim of the funnel several times with a **ruler** so that it makes a loud sound. What happens to the candle flame?

6. Strike the funnel several more times. Each time, vary the amount of energy you use.

Think About This

1. What was different about the flame when you made a soft sound compared to when you made a loud sound? Why do you think this happened?

2. **Key Concept** How did the amount of energy you used to strike the funnel affect the sound? How did it affect the flame?

 **What do they sound like?**

1. High in the mountains of Switzerland, you might hear the clear, mellow tones of an alphorn. At one time, herders used the horns to call or soothe cows. How do you think the shape of the horn affects the sound? Why do you think the sound is able to travel so far?

_____

_____

_____

Energy of Sound Waves

Shhhhh! How do you change your voice from a yell to a whisper? The energy a sound wave carries depends on the amount of energy that caused the original vibration. To speak softly, just use less energy!

Amplitude

How do the sound waves produced when you yell and when you whisper differ? The more energy you put into your voice, the farther the air particles move as they vibrate back and forth. *For a longitudinal wave,* **amplitude** *is the maximum distance the particles in a medium move from their rest positions as the wave passes through the medium.* As the energy in a sound wave increases, its amplitude increases. Sound waves with small and large amplitudes are shown in **Figure 8.**

Active Reading

2. Indicate Using arrows show how the amplitude of sound is related to energy.

↑	Energy		Amplitude
	Energy		Amplitude

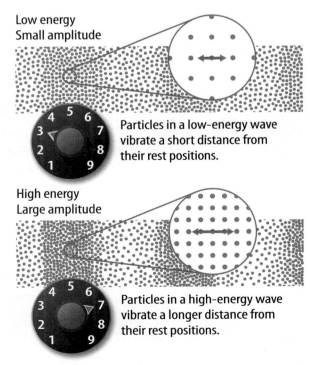

Low energy
Small amplitude

Particles in a low-energy wave vibrate a short distance from their rest positions.

High energy
Large amplitude

Particles in a high-energy wave vibrate a longer distance from their rest positions.

Figure 8 ⚿ Particle spacing differs in high-energy and low-energy waves.

Amplitude, Intensity, and Loudness

Imagine blowing into a trumpet, like the one in **Figure 9.** Sound waves leave the horn with a certain amount of energy and, therefore, a certain amplitude. Recall that amplitude is the distance the particles of air vibrate back and forth. Loudness is how you perceive the energy of a sound wave. So, a wave with a greater amplitude will produce a louder sound.

Why do sounds get quieter as you get farther from the source? Think of what happens as a sound wave travels away from the horn. As the particles of air in front of the horn vibrate back and forth, they collide with, and transfer energy to, surrounding particles of air. As the energy spreads out among more and more particles, the intensity of the wave decreases. **Intensity** *is the amount of sound energy that passes through a square meter of space in one second.*

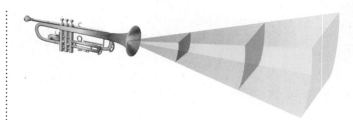

Figure 9 As sound waves spread out, the amount of energy in each area of space decreases.

As a sound wave travels farther from the horn, there is a larger area of particles sharing the same amount of energy that left the horn. Therefore, the farther you are from the horn, the less energy there is passing through one square meter. This results in less intensity of the wave. As intensity decreases, amplitude decreases. Therefore, loudness decreases.

The Decibel Scale

The unit decibel (dB) describes the intensity and, in turn, the loudness of sound. Decibel levels of common sounds are shown in **Figure 10.** Each increase of 10 dB indicates the sound is about twice as loud and has about 10 times the energy. For example, the decibel level of city traffic is about 85 dB, and the level of a rock concert is about 105 dB. This means a concert, which is 20 dB higher, has about 10×10, or 100 times, more energy than traffic. Recall that a loud sound can make the eardrum vibrate so hard that it tears. As sounds get louder, the amount of time you can listen without hearing loss gets shorter.

| Active Reading | 3. Contrast What is the difference in intensity between these two sounds? |

Normal Conversation	Hair Dryer	
Intensity: 60 dB	Intensity: _____	
	about _____ times energy	

Decibel Scale

Figure 10 The decibel scale rates the loudness of some common sounds.

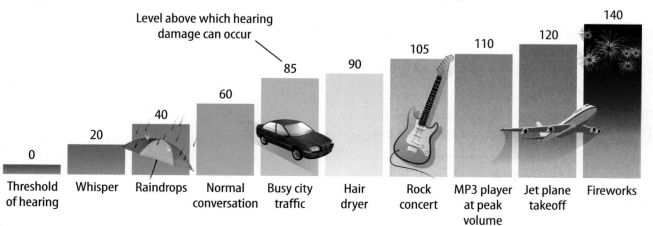

Level above which hearing damage can occur

Threshold of hearing	Whisper	Raindrops	Normal conversation	Busy city traffic	Hair dryer	Rock concert	MP3 player at peak volume	Jet plane takeoff	Fireworks
0	20	40	60	85	90	105	110	120	140

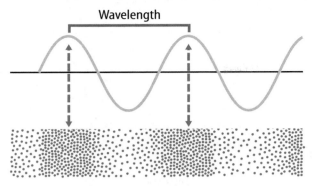

Wavelength — Low frequency; long wavelength

Wavelength — High frequency; short wavelength

Figure 11 If you compare two sound waves, the wave with the longer wavelength has a lower frequency.

Describing Sound Waves

Sounds depend on many properties of the sound waves that enter your ear. Loudness or softness depends on the amplitude of the wave. You also might describe sound according to how frequently the waves occur or how long the waves are.

Wavelength

One property of a sound wave is its wavelength. *The distance between a point on one wave and the nearest point just like it is called* **wavelength.** For example, you could measure a wavelength as the distance between the midpoint of one compression, or rarefaction, and the midpoint of the next compression, or rarefaction, as shown in **Figure 11.**

Frequency and Pitch

Suppose you could count sound waves produced by playing middle C on a piano. You would find that 262 wavelengths pass you each second. *The* **frequency** *of sound is the number of wavelengths that pass by a point each second.* Notice in **Figure 11** that as the wavelength of a sound wave decreases, its frequency increases. The frequency of one vibration, or wavelength, per second is called a hertz (Hz). The frequency of middle C on a piano is 262 Hz.

The perception of how high or low a sound seems is **pitch.** A higher frequency produces a higher pitch. For example, an adult male voice might range from 85 Hz to 155 Hz. An adult female voice might range from about 165 Hz to 255 Hz.

The human ear can detect sounds with frequencies between about 20 Hz and 20,000 Hz. Frequencies above this range are called ultrasound. The range of sounds heard by various animals is shown in **Figure 12.**

Active Reading **4. Connect** What is the relationship of wavelength to frequency and pitch?

Wavelength increases: _____

Wavelength decreases: _____

Figure 12 Many animals can hear sounds outside the range of human hearing.

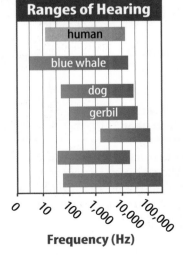

Ranges of Hearing

human
blue whale
dog
gerbil

Frequency (Hz)
0 10 100 1,000 10,000 100,000

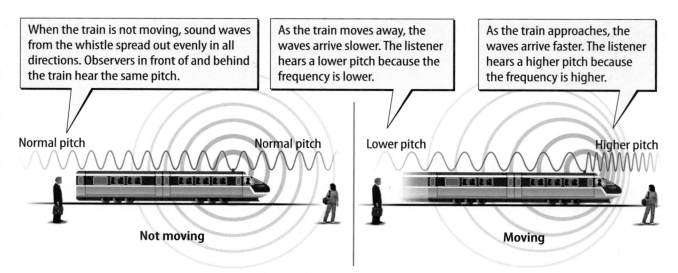

When the train is not moving, sound waves from the whistle spread out evenly in all directions. Observers in front of and behind the train hear the same pitch.

As the train moves away, the waves arrive slower. The listener hears a lower pitch because the frequency is lower.

As the train approaches, the waves arrive faster. The listener hears a higher pitch because the frequency is higher.

Normal pitch — Normal pitch — Lower pitch — Higher pitch

Not moving　　**Moving**

Figure 13 Sound waves bunch together ahead of a moving source.

The Doppler Effect

You might have heard the high pitch of a train whistle as the train approaches. As the train passes, the pitch drops. Sound frequency depends on the motions of the source and the listener. Compare the wave frequencies in front of and behind the moving train in **Figure 13.** The frequency increases if the distance between the listener and the source is decreasing. The frequency decreases if the distance between the listener and the source increases. *The change of pitch when a sound source is moving in relation to an observer is the* **Doppler effect.**

Sound Interference

If you walk through a room with stereo speakers at each end, the sound might seem louder in some places and softer in others. The waves from each speaker interact with one another. **Interference** *occurs when waves that overlap combine, forming a new wave.*

Figure 14 shows why this happens. When compressions meet, they join to form a wave with higher intensity and greater amplitude. This is called constructive interference. However, when a compression meets a rarefaction, the intensity and amplitude decrease. This is destructive interference.

Figure 14 Waves' interference can result in an increase or a decrease in amplitude.

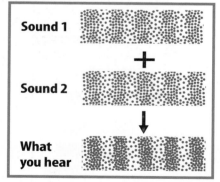

Constructive Interference
When the compressions and rarefactions of waves overlap, the combined compressions have greater intensity.

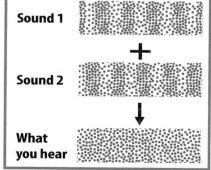

Destructive Interference
When the compressions of one wave overlap the rarefactions of another wave, the waves cancel and the result is no sound.

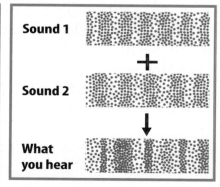

Beats
When the compressions of two waves are slightly offset, a pattern of increasing and decreasing compressions, called beats, occurs.

Inquiry **iLAB STATION** **Try It!** SC.7.N.1.6

MiniLab *How can you hear the beats?* at connectED.mcgraw-hill.com

Apply It!   After you complete the lab, answer these questions.

1. With two different sound frequencies, the rate at which the amplitude of the wave pulses is the beat frequency. Beat frequency is calculated with the formula $f_{beat} = f_1 - f_2$. Two tuning forks, one of 147 Hz and one of 152 Hz, are sounded together. What beat frequency will be observed?

2. A piano tuner uses beats to tune a piano. She will pluck a piano string and tap a tuning fork at the same time. She matches the piano string's frequency to the frequency of the tuning fork. Explain how the piano tuner knows when the piano string is "in tune."

Beats

Have you ever been to a concert where the musicians start by all playing the same note? Why do they do that? If the pitches of the notes are slightly different, the sounds will interfere. The audience might hear the notes get louder and softer several times a second. The repeating increases and decreases in amplitude are beats. Look back at **Figure 14.** The difference in frequencies determines how often beats occur. If one musician plays a note with a pitch of 392 Hz and another plays a note with a pitch of 395 Hz, the difference is 3 Hz. Beats will occur 3 times each second. Musicians can avoid beats by playing notes at the correct pitch on their instruments.

Fundamental and Overtones

When a musician plucks a guitar string, the string vibrates with a certain sound. If the musician plucks it again in the exact same way, the sound will be the same. All objects tend to vibrate with a certain frequency that depends on the object's properties.

The lowest frequency at which a material naturally vibrates is called its fundamental. Higher frequencies at which the material vibrates are called overtones. Objects vibrate with both a fundamental and overtones, as shown in **Figure 15.** The interference of these waves produces the sound you hear.

Active Reading   **5. Review** Highlight the difference between a fundamental and overtones.

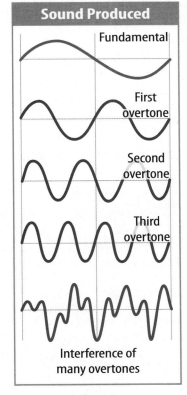

Sound Produced

Fundamental

First overtone

Second overtone

Third overtone

Interference of many overtones

Figure 15 🔑 A fundamental and overtones combine to produce an object's sound.

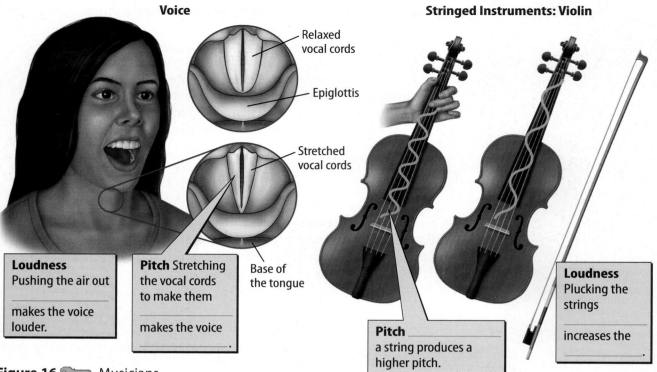

Voice

Relaxed vocal cords

Epiglottis

Stretched vocal cords

Base of the tongue

Loudness Pushing the air out _____ makes the voice louder.

Pitch Stretching the vocal cords to make them _____ makes the voice _____ .

Stringed Instruments: Violin

Pitch _____ a string produces a higher pitch.

Loudness Plucking the strings _____ increases the _____ .

Figure 16 🔑 Musicians change the pitch and loudness of sound in different ways.

Active Reading **6. Label** Fill in how pitch and loudness are changed for each music type.

Active Reading

FOLDABLES® LA.6.2.2.3

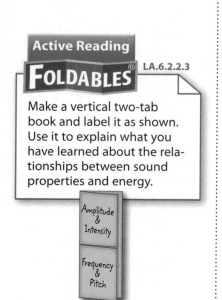

Make a vertical two-tab book and label it as shown. Use it to explain what you have learned about the relationships between sound properties and energy.

Amplitude & Intensity

Frequency & Pitch

Music

Have you ever been listening to your favorite music and someone tells you "Turn off that noise!" How are music and noise different? Unlike noise, music is sound with a pleasing pattern.

Sound Quality

The unique sound of a musical instrument is a mix of its fundamental and overtones. These waves interact to form a distinct sound, or timbre. Suppose, for example, that a clarinet player and a piano player both play the same note. The fundamental of both instruments is the same. The number and intensity of the overtones, however, differ. Overtones produce the complex sound waves that let you distinguish the unique sound quality of each instrument.

Active Reading **7. Identify** How can you recognize sounds from different sources?

_____

_____

Resonance

If you hold a guitar string by its ends and have a friend pluck the string, the sound is almost too low to hear. Instruments use resonance to amplify sound. **Resonance** *is an increase in amplitude that occurs when an object vibrating at its natural frequency absorbs energy from a nearby object vibrating at the same frequency.* The vibrating string causes the back of the guitar and the air inside to vibrate by resonance. The sound is then much louder.

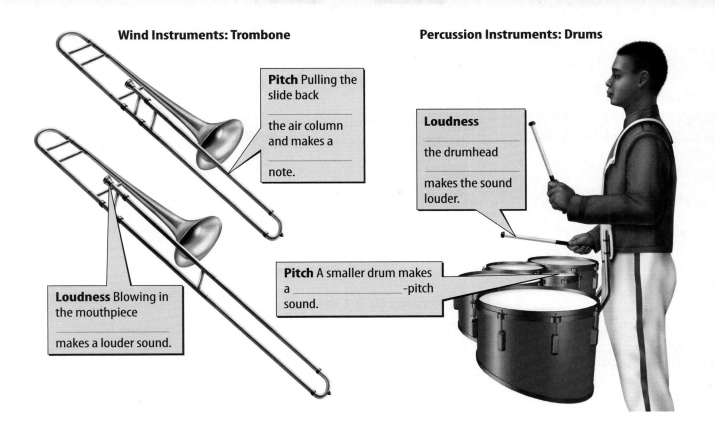

Wind Instruments: Trombone

Pitch Pulling the slide back _____ the air column and makes a _____ note.

Loudness Blowing in the mouthpiece _____ makes a louder sound.

Percussion Instruments: Drums

Loudness _____ the drumhead makes the sound louder.

Pitch A smaller drum makes a _____ -pitch sound.

Types of Musical Instruments

As shown in **Figure 16,** different types of instruments control the pitch and the loudness of sound in different ways.

String Instruments Musical instruments that have strings, such as a guitar, a violin, a harp, and a piano, produce sound when the string vibrates. A player plucks the strings of a guitar or a harp. The motion of a bow vibrates the strings of a violin. Pressing a piano key causes a felt-covered hammer to strike a particular string inside the piano. When the string inside the piano vibrates, you hear a tone. The pitch a string makes depends on its length and its thickness.

Pitch also depends on how tightly the string is stretched and the material it is made from. You can hear sounds a stringed instrument makes because of resonance between the string and the instrument's hollow body. Plucking or pressing the strings harder increases the volume.

Wind Instruments The vibrating medium in wind instruments, such as a saxophone or a trumpet, is air. Either your lips or a thin piece of wood, called a reed, vibrates. An air column then vibrates by resonance. The length of the air column determines pitch. Blowing harder increases the volume.

Percussion Instruments You make sound with a percussion instrument by striking it. Examples include a drum, cymbals, and a bell. The pitch depends on the instrument's size, its thickness, and the material from which it is made. Resonance can make the sound of a percussion instrument louder.

Voice The source of sound in your voice is vocal cords. Muscles in your throat allow you to increase the pitch of your voice by pulling the cords tighter. Other parts of your mouth and throat also affect the sounds your voice makes. You can make your voice louder by pushing out the air with a greater force.

Visual Summary

The intensity of sound decreases farther away from a source.

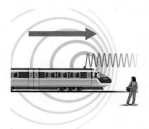

The pitch of sound increases if the distance between a sound source and a listener decreases.

An instrument's sound is a mix of its fundamental and overtones.

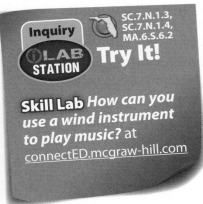

Inquiry SC.7.N.1.3, SC.7.N.1.4, MA.6.S.6.2

iLAB STATION Try It!

Skill Lab *How can you use a wind instrument to play music?* at connectED.mcgraw-hill.com

Use Vocabulary

1 As the _____ of a sound wave increases, the pitch of the sound gets lower. SC.7.N.1.4

2 **Describe** the Doppler effect in your own words.

_____

_____

3 **Use** the term *resonance* in a complete sentence.

_____

_____

Understand Key Concepts

4 Which increases if the amplitude of a sound wave increases? SC.7.N.1.4

   (A) intensity   (C) pitch

   (B) interference   (D) sound quality

5 **Identify** the vibrating media in three different types of musical instruments. SC.7.N.1.6

_____

_____

6 **Explain** how you can recognize the sound of a flute.

_____

_____

7 **Analyze** If the frequency of a sound wave increases, what happens to the wavelength?

_____

_____

Interpret Graphics

8 **Identify** Fill in the graphic organizer below. Use it to identify and briefly explain four properties that distinguish sounds. LA.6.2.2.3

Properties of Sound Waves

Connect Key Concepts 🔑

Evaluate What factors determine the speeds of sound?

1. Factor:

_____
of material

Gas particles are far apart; therefore, it takes longer to transfer sound energy from one particle to another.

2. Effect:

Factor:
Stiffness, or rigidity, of material

3.

4. Effect:

Factor:
Temperature increase in a gas

5.

6. Effect:

---

7. Factor:

Molecules slow down and move closer together, so they collide more often.

8. Effect:

Using Sound WAVES

Vocabulary

absorption p. 604

reflection p. 604

echo p. 604

reverberation p. 605

acoustics p. 605

echolocation p. 607

sonar p. 607

 Florida NGSSS

SC.7.N.1.1 Define a problem from the seventh grade curriculum, use appropriate reference materials to support scientific understanding, plan and carry out scientific investigation of various types, such as systematic observations or experiments, identify variables, collect and organize data, interpret data in charts, tables, and graphics, analyze information, make predictions, and defend conclusions.

SC.7.N.1.3 Distinguish between an experiment (which must involve the identification and control of variables) and other forms of scientific investigation and explain that not all scientific knowledge is derived from experimentation.

SC.7.N.1.6 Explain that empirical evidence is the cumulative body of observations of a natural phenomenon on which scientific explanations are based.

SC.7.P.10.3 Recognize that light waves, sound waves, and other waves move at different speeds in different materials.

LA.6.2.2.3 The student will organize information to show understanding or relationships among facts, ideas, and events (e.g., representing key points within text through charting, mapping, paraphrasing, summarizing, or comparing/contrasting);

MA.6.A.3.6 Construct and analyze tables, graphs, and equations to describe linear functions and other simple relations using both common language and algebraic notation.

 Launch Lab   SC.7.N.1.1, SC.7.P.10.3

15 minutes

Why didn't you hear the phone ringing?

A cell phone ringing on a countertop can be heard from far away, but the same phone in your jacket pocket is not so easy to hear. What explains this difference?

Procedure 🔬

1. Read and complete a lab safety form.

2. Set a **kitchen timer** for 1 s. Hold it about 15 cm from your ear, and listen to the sound.

3. Locate the exit hole for sound waves on the back of the timer. Set the timer to a 5-s delay. Hold a **foil pie pan** flat against the exit hole. Move your ear about 15 cm from the timer and pie pan. Observe the difference in the sound.

4. Set the timer with a 5-s delay. Place the timer in a **shoe box,** and cover it with several crumpled **paper towels.** Listen with your ear 15 cm from the box.

Think About This

1. Compare the movement of sound through the foil pan and through air.

2. What do you think changed the sound when you covered the timer with paper towels?

3. 🔑 **Key Concept** Based on your results, why do you think a cell phone is harder to hear in a jacket pocket than on a countertop?

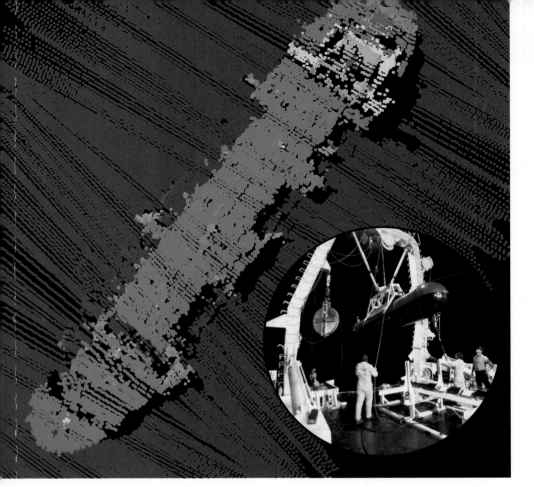

1. These workers are preparing to lower a special device into the water that uses sound waves to search for underwater objects. Sometimes they locate sunken ships, such as the oil tanker in the large image. What properties of sound waves do you think are useful for finding things under water?

_____

_____

_____

_____

Sound Waves and Matter

Why do the cheers of a crowd in an indoor gymnasium sound so different from a crowd yelling at an outdoor football game? Sound waves at the football game spread out with few barriers. What happens to sound waves when they strike a different medium, such as the walls of a building?

Transmission

Have you ever heard someone talking in the next room? This is possible because of transmission—the movement of sound waves through a medium. When sound waves move from air into a wall, the vibrations of air particles cause particles in the solid wall to vibrate. Even though solids transmit sound waves better than gases, most sound waves do not move easily from gases into a solid. However, loud sounds, which have a lot of energy, will move into and through the solid wall. Waves of quieter sounds, with less energy, may be partially or completely blocked. These waves don't carry enough energy to cause much vibration in the wall. As the vibrations reach the next room, they transfer the remaining energy to the air particles, and you hear the sound. The amplitude of the sound is lower because the wall could not transmit all of the energy.

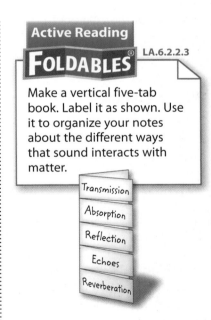

Active Reading

FOLDABLES® LA.6.2.2.3

Make a vertical five-tab book. Label it as shown. Use it to organize your notes about the different ways that sound interacts with matter.

Transmission

Absorption

Reflection

Echoes

Reverberation

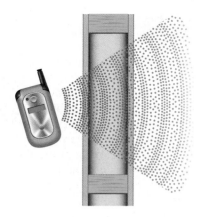

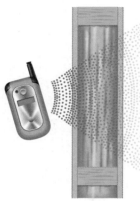

Figure 17 🔑

Insulation absorbs much of the energy in a sound wave and converts it to a small amount of thermal energy.

WORD ORIGIN

echo

from Greek *ekhe*, means "sound"

Absorption

If you throw a tennis ball at a pillow, it will not bounce back. Most of the ball's energy goes into the pillow. Some materials, such as the wall insulation in **Figure 17,** act like the pillow when sound waves strike them. *The transfer of energy by a wave to the medium through which it travels is called* **absorption.** How well a material absorbs the energy of a sound wave depends on various factors, such as its inner structure and the amount of air in it. Rather than passing from one particle to another, some of the sound energy changes to heat due to friction.

Reflection

What happens if you throw a tennis ball at a hard surface? The ball probably bounces back at you. Similarly, a sound wave might bounce back when it strikes a different medium. *The bouncing of a wave off a surface is called* **reflection.** The way in which sound waves reflect is shown in **Figure 18.** The angle at which a sound wave strikes a surface is always equal to the angle at which the sound wave is reflected off of the surface.

 2. NGSSS Check **Review** (Circle) some ways in which sound interacts with matter. SC.7.P.10.3

Echoes

Have you ever yelled a name in a gym and heard the same voice yell back? That was you, of course! As you yelled, you heard the original sound of your own voice. Then you heard the sound again after the sound waves reflected off the walls of the gymnasium and traveled back to your ears. *A reflected sound wave is an* **echo.**

Sound waves travel at about 343 m/s in air, or 34.3 m in 0.1 s. And, the brain holds onto a sound for about 0.1 s. When the reflecting surface is far enough away that the sound waves take more than 0.1 s to return, the listener hears the original sound followed by the reflected sound. So, if you clap your hands at one end of a long room, you hear an echo only if the sound wave returns more than 0.1 s later.

Figure 18 🔑 Sound waves reflect from a surface at the same angle at which they strike the surface.

Reverberation

In many closed spaces, sound waves reflect from surfaces that are different distances from the listener. Because some waves travel farther than other waves, reflected waves reach the listener at different times. *The collection of reflected sounds from the surfaces in a closed space is called* **reverberation.**

Sound waves that reach the listener directly are heard sooner than reflected sound waves. If each reflected wave reaches the ear before the previous sound fades, the original sound seems to last longer. However, too much reverberation can make words hard to understand because the echoes of old sounds can interfere with new ones.

Acoustics

In a room with no furniture, rugs, or drapes, the sound waves of footsteps and speech bounce around the room and sound loud. Soft or fuzzy materials, such as the rug, curtains, and padded furniture on the right in **Figure 19,** absorb much of the energy of sound waves. Footsteps are almost silent. Voices are softer. *The study of how sound interacts with structures is called* **acoustics.** Acoustical engineers use their knowledge of sound transmission, absorption, and reflection to control sounds.

SC.7.N.1.6, SC.7.P.10.3

Inquiry
LAB STATION **Try It!**

MiniLab *How fast is sound?* at connectED.mcgraw-hill.com

Apply It!

After you complete the lab, answer the questions below.

1. If sound travels at about 343 m/s, why does it take 2 s to hear the echo from a wall 343 m away?

2. The speed of sound in air depends on the temperature of the air. You can use the following formula to find the actual speed of sound at a given temperature: Velocity (in m/s) = 331 + 0.59 × Temperature (in °C). The highest temperature on record in Florida was 43°C in 1931 at Monticello. What was the speed of sound that day?

Figure 19 Soft materials reduce the reverberation in a room.

Figure 20 🔑 Noise-canceling earphones protect the worker from noises that otherwise would damage his hearing.

Active Reading **3. Describe** What are some of the ways that people can control sound?

Method	Description
Noise-canceling earphones	How they work • •
Designing spaces	Examples • •

Figure 21 🔑 Engineers design concert halls and recording studios to control sounds.

Noise Pollution and Control

One way acoustical engineers control sound is by developing methods to protect people from noise pollution. Think about the noises people might hear during a typical day. Trucks rumble by, cars honk their horns, and claps of thunder boom during a storm. At home, you might use noisy appliances such as a hair dryer, a vacuum cleaner, or a dishwasher. Severe noise pollution can result in hearing loss, stress, and other types of health problems. Laws limit the noise that can be produced by machinery or landing aircraft. The government requires ear protection for workers in many jobs.

Noise-canceling earphones, such as those worn by the person in **Figure 20,** work in several ways. They cover the ears and can block incoming sound waves. Other types of earphones reduce the sound by analyzing incoming sound waves and then producing waves that create destructive interference.

Designing Spaces

Acoustical engineers develop ways to control sound in buildings. When you enter a concert hall such as the one in **Figure 21,** you might think that the unusual appearance is just for looks. However, engineers have carefully chosen shapes and materials to control sound waves. The stage has a wooden floor to improve vibrations. The curved panels on the ceiling reflect sound waves in different directions to fill the space. However, the recording studio has foam panels on the walls. These soft materials absorb sound waves to **prevent** reverberation as the musician performs.

Active Reading **4. Offer** List 3 jobs that might require noise-canceling earphones.

_____

_____

_____

1 A dolphin sends out a series of high-pitched clicks.

2 When a sound wave strikes an object or another dolphin, some of the sound reflects back.

Figure 22 🔑 Organs in the dolphin's head send out ultrasonic waves and analyze the echoes.

Ultrasound

You have read that the waves of ultrasound have a higher frequency than humans can hear. Both animals and humans use these high-frequency sound waves.

Echolocation

Recall that many animals, such as bats and whales, can hear sounds above the range of human hearing. Animals use some of these ultrasound frequencies to communicate. They use other frequencies to locate and identify objects. *The process an animal uses to locate an object by means of reflected sounds is* **echolocation** (e koh loh KAY shun). The dolphins in **Figure 22** might use an echo to determine an object's distance, shape, and speed or to find other dolphins.

Sonar

If you go fishing often, you probably wish you could easily discover exactly where the fish are located. A device called a fish-finder does just that by using a technology similar to echolocation. A fish-finder is an example of **sonar**, *a system that uses the reflection of sound waves to find underwater objects.* The word *sonar* is an acronym for <u>So</u>und <u>Na</u>vigation and <u>R</u>anging. Ships use sonar to send a high-frequency sound wave into the water. As the sound wave moves deeper, it spreads out, forming a cone, or beam. When the sound wave strikes something within this beam, it bounces back to the ship. Sonar contrasts signals that strike the ocean floor with signals from other objects. It measures the amount of time between when the sound wave leaves and when it bounces back. The sonar system then calculates the distance and draws an image on a screen. The picture of the sunken oil tanker on the first page of this lesson shows what a sonar image might look like.

5. NGSSS Check
Summarize How can people "see" under water using sonar? SC.7.P.10.3

_____

_____

_____

Math Skills MA.6.A.3.6

Use a Formula

You can use a formula to calculate sonar distances. For example, sound travels at **1,530 m/s** in seawater. A ship sends out a signal to find the water's depth. The signal returns in **5.0 s**. How deep is the water?

1. Use the speed = distance/time formula to calculate distance.

   distance = *speed* × *time*

2. The signal traveled to the ocean's bottom and back to the ship. Divide by 2 to get the time it took to reach the bottom.

   5.0 s/2 = 2.5 s

3. Replace the terms with the values and multiply.

   d = (1,530 m/s)(2.5 s)
   = 3,825 m

Practice

6. A sonar signal traveling at 1,490 m/s returns to the ship in 4.0 s. What is the distance to the target?

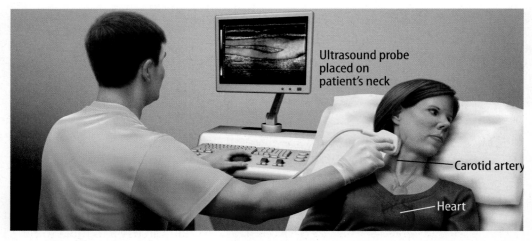

Figure 23 🔑 This ultrasound scanner produces an image of the patient's artery.

Medical Uses of Ultrasound

Suppose a doctor could see your heart beating or watch blood flowing through your arties. He or she could determine whether something was wrong without performing surgery. Believe it or not, as **Figure 23** shows, these things are possible using ultrasound.

Figure 24 🔑 Ultrasound can break large kidney stones into tiny pieces.

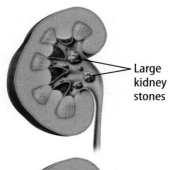

Large kidney stones

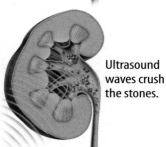

Ultrasound waves crush the stones.

Ultrasound Imaging Ultrasound scanners work much like sonar. A doctor moves the scanner over different parts of the body. The scanner emits safe, high-frequency sound waves. Body structures such as muscle, fat, blood, and bone reflect sound at different rates. Based on the reflected waves, the scanner produces an image called a sonogram. It is even possible to analyze motion with ultrasound. The scanner in **Figure 23** uses the Doppler effect. By determining how much the frequency of the reflected wave changes, the scanner can determine the blood's speed and direction. Doctors often use sonograms to check the health of unborn babies. Ultrasound is safer than X-rays because it doesn't damage cells.

Treating Medical Problems Has anyone ever massaged your neck or back when it was sore? It not only feels good, it also relaxes stiff muscles. Many physical therapists use ultrasound to treat joint and muscle sprains or to ease muscle spasms. The vibrations travel through the skin and soft tissue and act like hundreds of tiny fingers massaging the area. Short pulses of high-frequency sound waves can even break apart kidney stones, as shown in **Figure 24.**

Active Reading **7. Explain** What are some ways in which people use ultrasound? Fill in the graphic organizer at right.

Imaging ➡ Explanation:

Treatment ➡ Explanation:

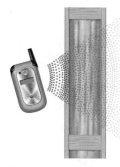

A medium might transmit, absorb, or reflect sound waves that strike it.

Sound waves reflect off hard surfaces in an enclosed space.

Acoustical engineers develop ways to control sound.

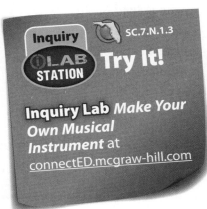

Inquiry SC.7.N.1.3

LAB STATION **Try It!**

Inquiry Lab *Make Your Own Musical Instrument* at connectED.mcgraw-hill.com

Use Vocabulary

1 **Describe** reverberation in your own words.

_____

_____

Understand Key Concepts 🗝

2 Which is the use of sound to identify the distances and positions of objects? SC.7.P.10.3

(A) echolocation   (C) reverberation

(B) reflection   (D) timbre

3 **Suggest** a way to prevent sounds from echoing in a large gymnasium.

_____

_____

4 **Describe** several ways to limit sound transmission through walls.

_____

_____

Interpret Graphics

5 **Summarize** Fill in the graphic organizer below. Use it to identify and briefly explain three ways sound can interact with matter. LA.6.2.2.3

Interactions of Sound and Matter

Critical Thinking

6 **Assess** the value of echolocation to animals such as bats and dolphins.

_____

_____

Math Skills   MA.6.A.3.6

7 A ship's sonar signal travels at 1,500 m/s and returns 0.40 s after it is sent. How deep is the shipwreck the ship locates?

Think About It! Sound is one of many forms of energy. Sound waves are one mode of transferring energy. People use the properties of sound and sound waves to perform many types of tasks. Each sound source creates a unique wave form that the ear can identify.

Key Concepts Summary

Vocabulary

LESSON 1 Producing and Detecting Sound

- **Vibrations** in a medium produce **sound waves.**

- Sound waves travel as **compressions** and **rarefactions.**

- Sound waves travel faster through materials in which the particles are closer together.

- The outer ear collects sound. The middle ear amplifies sound. The inner ear converts vibrations to nerve signals.

sound wave p. 585
longitudinal wave p. 585
vibration p. 585
medium p. 586
compression p. 586
rarefaction p. 586

LESSON 2 Properties of Sound Waves

- The greater a sound wave's energy, the larger the **amplitude** and the greater the wave's **intensity.**

- Sound waves with a longer **wavelength** have a lower **frequency** and a lower **pitch.**

- Different frequencies of sound waves combine and form a complex wave the brain recognizes.

- Strings, air columns, or surfaces of instruments vibrate and produce music.

amplitude p. 593
intensity p. 594
wavelength p. 595
frequency p. 595
pitch p. 595
Doppler effect p. 596
interference p. 596
resonance p. 598

LESSON 3 Using Sound Waves

- Sound waves can be transmitted, **reflected,** or **absorbed** by matter.

- Materials and shapes in a room can improve vibrations, absorb excess sound waves, and reflect sound waves to fill the room.

- Ultrasound is used for medical imaging and treatment.

absorption p. 604
reflection p. 604
echo p. 604
reverberation p. 605
acoustics p. 605
echolocation p. 607
sonar p. 607

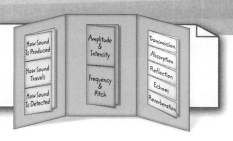

Active Reading
FOLDABLES® Chapter Project

Assemble your lesson Foldables as shown to make a Chapter Project. Use the project to review what you have learned in this chapter.

Use Vocabulary

1 Describe a longitudinal wave.

_____

_____

2 The part of a sound wave in which the particles are most spread out is called a(n)

_____ .

3 Explain the relationship between frequency and pitch.

_____

_____

4 A change in frequency called the

_____ depends on the motion of the sound source and the position of the listener.

5 Describe how resonance works.

_____

_____

6 The collection of sound reflections in a closed space is called

_____ .

7 An example of _____ is when a dolphin emits loud clicks and listens for the echo.

Link Vocabulary and Key Concepts

Use vocabulary terms from the previous page to complete the concept map.

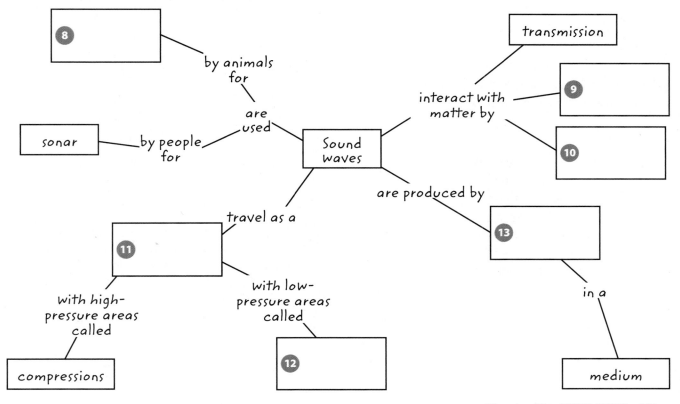

Chapter 15 Review

Fill in the correct answer choice.

🔑 Understand Key Concepts

1 Which type of matter would transmit sound waves fastest? SC.7.P.10.3

Ⓐ air at 5°C
Ⓑ air at 20°C
Ⓒ ice at 0°C
Ⓓ seawater at 20°C

2 A large windowpane on a storefront vibrates as a big truck drives by. Which type of interaction between sound and matter best explains this vibration?

Ⓐ absorption
Ⓑ resonance
Ⓒ reverberation
Ⓓ transmission

3 As _____ decreases, sound intensity decreases. SC.7.N.1.4

Ⓐ amplitude
Ⓑ quality
Ⓒ wave speed
Ⓓ wavelength

4 Wave frequency is measured in which unit?

Ⓐ decibel
Ⓑ hertz
Ⓒ meter
Ⓓ second

5 What causes the sound waves' intensity to decrease in the picture below? SC.7.N.1.4

Ⓐ insulation
Ⓑ interference
Ⓒ rarefaction
Ⓓ resonance

Critical Thinking

6 **Visualize** A bat in a dark cave sends out a high-frequency sound wave. The bat detects an increase in frequency after the sound bounces off its prey. Describe the possible motion of the bat and its prey.

_____

_____

_____

7 **Synthesize** An MP3 player at maximum volume produces sound at 110 dB. The table below shows the recommended time exposure before risk of damage to the ear. How many hours per day could you listen to your MP3 player at full volume without risking hearing loss? Explain.

_____

_____

_____

Recommended Noise Exposure Limits	
Sound Level (dB)	Time Permitted (hr)
90	8
95	4
100	2
105	1

8 **Construct** a diagram of two sound waves of equal amplitude and frequency that experience destructive interference.

9 **Create** a diagram that includes drawings of sound waves to show how the ear distinguishes among sounds, such as different voices or musical instruments.

10 **Compare** the effects on the ear of a single very loud sound of 150 dB with prolonged exposure to sounds of 90 dB.

_____

_____

_____

Writing in Science

11 **Write** A friend tells you that his family always complains about the noise when he practices his electric guitar in his room. Write at least four recommendations on a separate sheet of paper for how your friend might reduce the sound that escapes from his room. LA.6.2.2.3

Big Idea Review

12 Explain how sound is produced, travels from one place to another, and is detected by the ear. LA.6.2.2.3

_____

_____

_____

_____

_____

_____

13 Look at all the equipment in the music control room shown in the photograph. Think about what you have learned about sound in the chapter. How can musicians produce, describe, and use sound? LA.6.1.2.3

_____

_____

_____

_____

_____

_____

_____

Math Skills MA.6.A.3.6

Solve Equations

14 A ship sends out a sonar signal to locate other ships. Traveling at 1,530 m/s, a signal returns to the ship 3.6 s after it is sent. What is the distance from the other ship?

15 A sonar signal takes 3 s to return from a sunken ship directly below. Find the depth of the sunken ship if the speed of the signal is 1,440 m/s.

16 A ship is sailing in water that is 500 m deep. Sound travels at 1,520 m/s in the body of water. A sonar echo returns to the ship in 0.6 s. Did the signal bounce off the bottom? Explain.

Florida NGSSS

Benchmark Practice

Fill in the correct answer choice.

Multiple Choice

Use the figure below to answer questions 1–3.

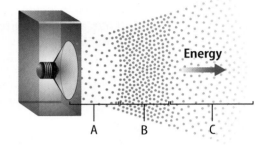

1 How does this speaker cause sound that can be heard? SC.7.P.10.3

- Ⓐ It produces echoes.
- Ⓑ It produces vibrations.
- Ⓒ It releases light.
- Ⓓ It releases heat.

2 What type of wave does the speaker produce for sound to be heard? SC.7.P.10.3

- Ⓕ electromagnetic
- Ⓖ longitudinal
- Ⓗ surface
- Ⓘ transverse

3 Point A in the figure is in a rarefaction. Point B is in a compression. Point C is a location in the air at normal air pressure before the sound wave has arrived. Which statement is true of the pressure at points A, B, and C? SC.7.P.10.3

- Ⓐ The pressure at point A is the greatest.
- Ⓑ The pressure at point B is the greatest.
- Ⓒ The pressure at point C is the greatest.
- Ⓓ The pressure at all three points is equal.

4 Which two factors can affect the speed of sound? SC.7.P.10.3

- Ⓕ size and density of the medium
- Ⓖ mass and density of the medium
- Ⓗ temperature and mass of the medium
- Ⓘ temperature and density of the medium

Use the table below to answer question 5.

Speed of Sound in Different Materials	
Material (at 20°C)	**Speed (m/s)**
Air	343
Glass	5,640
Iron	5,130
Water	1,481

5 A sound wave takes about 0.03 s to move through a material that is 10.3 m long. What is the material? MA.6.S.6.2

- Ⓐ air
- Ⓑ glass
- Ⓒ iron
- Ⓓ water

6 Which structure in the human ear transfers sound vibrations to the stirrup? SC.7.P.10.3

- Ⓕ anvil
- Ⓖ cochlea
- Ⓗ ear canal
- Ⓘ eardrum

7 Which type of sound wave has the greatest energy? SC.7.P.10.3

- Ⓐ a wave that has a very high intensity
- Ⓑ a wave that has a very low amplitude
- Ⓒ a wave with particles that do not vibrate from their rest positions
- Ⓓ a wave with particles that vibrate a very short distance from their rest positions

Use the diagram below to answer questions 8 and 9.

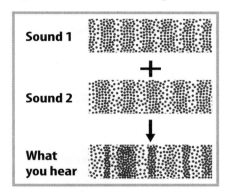

8 What can you tell about sounds 1 and 2 by looking at the figure? SC.7.P.10.3

  Ⓕ Sound 1 has a lower pitch.

  Ⓖ Sound 1 has a greater frequency.

  Ⓗ Sound 1 has a smaller amplitude.

  Ⓘ Sound 1 has a greater wavelength.

9 What does the bottom wave in the figure represent? SC.7.P.10.3

  Ⓐ silence

  Ⓑ interference

  Ⓒ the Doppler effect

  Ⓓ a fundamental frequency

10 What differs when the same note is played with the same amplitude by two different kinds of instruments? SC.7.P.10.3

  Ⓕ energy

  Ⓖ fundamentals

  Ⓗ intensity

  Ⓘ overtones

11 Which property describes the distance between two identical points on a sound wave? SC.7.P.10.3

  Ⓐ amplitude   Ⓒ pitch

  Ⓑ frequency   Ⓓ wavelength

12 When you hear a sound, its pitch is mostly influenced by which properties of a sound wave? SC.7.P.10.3

  Ⓕ amplitude and speed

  Ⓖ frequency and amplitude

  Ⓗ speed and frequency

  Ⓘ wavelength and frequency

13 Which shows overtones of a fundamental? SC.7.P.10.3

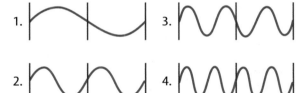

  Ⓐ only 1

  Ⓑ 1 and 2

  Ⓒ 2, 3, and 4

  Ⓓ 1, 2, 3, and 4

14 Which property of sound waves changes when you increase the volume on a car radio? SC.7.P.10.3

  Ⓕ amplitude   Ⓗ speed

  Ⓖ frequency   Ⓘ wavelength

15 Which is the region of a sound wave where particles of a medium are most spread out? SC.7.P.10.3

  Ⓐ compression   Ⓒ reverberation

  Ⓑ rarefaction   Ⓓ transmission

NEED EXTRA HELP?

If You Missed Question...	1	2	3	4	5	6	7	8	9	10	11	12	13	14	15
Go to Lesson...	1	1	2	1	2	2	1	2	1	2	2	2	2	2	1

Benchmark Mini-Assessment   Chapter 15 • Lesson 1

Multiple Choice *Bubble the correct answer.*

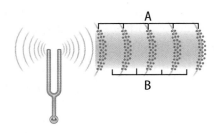

1. In the image above, which portion of a sound wave is represented by A? **SC.7.P.10.3**

- (A) compression
- (B) density
- (C) rarefaction
- (D) wavelength

2. Through which would sound move fastest? **SC.7.P.10.3**

- (F) air
- (G) ice
- (H) freshwater
- (I) seawater

3. Which part of the ear might become damaged from listening to loud music over a long period of time? **SC.7.P.10.3**

- (A) anvil
- (B) cochlea
- (C) hammer
- (D) stirrup

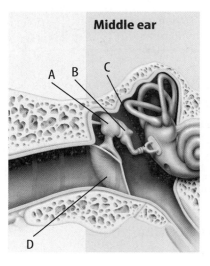

4. Which structure in the middle ear conducts sound directly to the cochlea? **SC.7.P.10.3**

- (F) Structure A
- (G) Structure B
- (H) Structure C
- (I) Structure D

Copyright © Glencoe/McGraw-Hill, a division of The McGraw-Hill Companies, Inc.

Multiple Choice *Bubble the correct answer.*

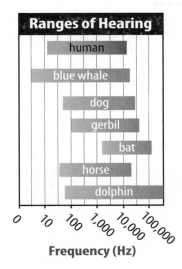

Ranges of Hearing

human
blue whale
dog
gerbil
bat
horse
dolphin

0   10   100   1,000   10,000   100,000

Frequency (Hz)

1. At which frequency could a sound be heard by a bat and a dog, but not by a human?

- (A) 9,000 Hz
- (B) 30,000 Hz
- (C) 75,000 Hz
- (D) 120,000 Hz

2. Which would NOT increase the resonance of a musical instrument?

- (F) the hollow body of a drum
- (G) the hollow chamber of a violin
- (H) the long tube of a clarinet
- (I) the thin reed of an oboe

3. Which of these instruments is likely to produce sounds with the longest wavelengths?

- (A) clarinet
- (B) flute
- (C) guitar
- (D) tuba

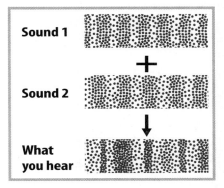

Sound 1

+

Sound 2

↓

What you hear

4. If Sound 1 and Sound 2 shown in the diagram were produced near you, what would you hear?

- (F) no sound produced
- (G) sound that fades in and out
- (H) sound that is less intense than either Sound 1 or Sound 2
- (I) sound that is more intense than either Sound 1 or Sound 2

Copyright © Glencoe/McGraw-Hill, a division of The McGraw-Hill Companies, Inc.

Name _____ Date _____

Benchmark Mini-Assessment　　Chapter 15 • Lesson 3

Multiple Choice *Bubble the correct answer.*

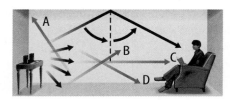

1. Which arrow indicates the way that a sound wave will be reflected in the room shown above?

- (A) Arrow A
- (B) Arrow B
- (C) Arrow C
- (D) Arrow D

2. Which of these rooms is likely to produce the most reverberation?

- (F) a den with carpet and couches
- (G) a kitchen with a carpeted floor
- (H) a bathroom with a tiled floor and tile-covered walls
- (I) a small bedroom with pillows and blankets on the bed

3. What characteristic of sound allows you to hear the noise from a television even if you are in a different room?

- (A) absorption
- (B) reflection
- (C) reverberation
- (D) transmission

4. The animals in the image above are communicating using

- (F) acoustics.
- (G) echolocation.
- (H) reverberation.
- (I) sonar.

Copyright © Glencoe/McGraw-Hill, a division of The McGraw-Hill Companies, Inc.

Notes

Name _____ Date _____

Red v. Blue

Three friends argued about the colors in the visible light spectrum. They disagreed on which color of the spectrum had the most energy of all the visible colors. This is what they said:

Kara: I think the red part of the spectrum has more energy than the blue part.

Jack: I think the blue part of the spectrum has more energy than the red part.

Li: I think the colors all have the same amount of energy.

Which friend do you agree with the most? Explain your thinking about visible light and energy.

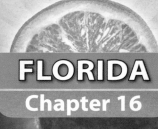
Electromagnetic
WAVES

FLORIDA BIG IDEAS

1 The Practice of Science
10 Forms of Energy
11 Energy Transfer and Transformations

Think About It!

How can you describe and use electromagnetic waves?

Each of these photos shows an image of the Sun taken with a camera that is sensitive to different temperatures. Color has been added with the help of a computer because the waves are invisible to the human eye. In this chapter, you will read that all objects emit electromagnetic waves. The wavelengths depend on the temperature of the object.

1 Why do you think all objects give off electromagnetic waves?

_____

_____

2 How do you think photographs such as these help scientists study the Sun?

_____

_____

3 How can you describe and use electromagnetic waves?

_____

_____

_____

Get Ready to Read

What do you think about electromagnetic waves?

Before you read, decide if you agree or disagree with each of these statements. As you read this chapter, see if you change your mind about any of the statements.

	AGREE	DISAGREE
1 Warm objects emit radiation, but cool objects do not.	☐	☐
2 Light always travels at a speed of 300,000 km/s.	☐	☐
3 Red light has the least amount of energy of all colors of light.	☐	☐
4 A television remote control emits radiation.	☐	☐
5 Thermal images show differences in the amount of energy people or objects give off.	☐	☐
6 When you call a friend on a cell phone, a signal travels directly from your phone to your friend's phone.	☐	☐

There's More Online!
Video • Audio • Review • ⓘLab Station • WebQuest • Assessment • Concepts in Motion • Multilingual eGlossary

Electromagnetic RADIATION

 How do electromagnetic waves form?

 What are some properties of electromagnetic waves?

Vocabulary

electromagnetic wave p. 625

radiant energy p. 625

 Inquiry Launch Lab

SC.7.P.10.1,
SC.7.N.1.6

15 minutes

How can you detect invisible waves?

You can see light from the Sun, but other forms of the Sun's energy are invisible. One way to detect invisible waves is to observe their effects on things.

Procedure

1. Read and complete a lab safety form.

2. With a **marker,** label a **clear plastic cup** *TAP* near the bottom of the cup. Fill it with tap water.

3. Label another cup *TONIC,* and fill it with **tonic water.**

4. Hold the cup with tap water near a **lamp.** Hold a sheet of **black construction paper** behind the cup. Observe the water as you slowly move the cup and paper away from and then closer to the lamp several times. Do you notice any change? Record observations.

5. Repeat step 4 with the tonic water. Record your observations.

6. Repeat steps 4 and 5, but this time move each cup and paper closer to and then away from bright sunlight instead of a lamp. Record your observations.

Data and Observations

Think About This

1. How was the effect of sunlight different from the effect of lamplight on each type of water?

2. **Key Concept** How did your results show that sunlight emits invisible waves?

 Florida NGSSS

SC.7.P.10.1 Illustrate that the sun's energy arrives as a radiation with a wide range of wavelengths, including infrared, visible, and ultraviolet, and that white light is made up of a spectrum of many different colors.

SC.7.N.1.6 Explain that empirical evidence is the cumulative body of observations of a natural phenomenon on which scientific explanations are based.

LA.6.2.2.3 The student will organize information to show understanding or relationships among facts, ideas, and events (e.g., representing key points within text through charting, mapping, paraphrasing, summarizing, or comparing/contrasting);

MA.6.A.3.6 Construct and analyze tables, graphs, and equations to describe linear functions and other simple relations using both common language and algebraic notation.

1. Camping in a remote area far from electric lines can be a challenge. Is it possible to get energy for things such as lights and heating? This camper can use these solar panels to capture energy from the Sun. Because electromagnetic waves travel through space, the Sun is Earth's most important source of energy, both in cities and in remote areas. Where do you think you might see solar cells similar to those on the panels in the photo?

_____

_____

_____

What are electromagnetic waves?

Suppose you live in a remote location. Would you be able to have electric lights? Could you watch television? Thanks to the Sun, the answer is yes! Like the camper you could use solar panels to capture the Sun's energy and transform it into electricity.

Energy from the Sun reaches Earth by traveling in waves. Many types of waves can travel only through a medium, or matter. Waves in a pond, for example, require water to travel. Waves from the Sun are different because they can travel through empty space. *A wave that can travel through empty space and through matter is called an* **electromagnetic wave.** These waves radiate, or spread out, in all directions from a source. *Energy carried by an electromagnetic wave is called* **radiant energy.** This energy also is known as electromagnetic radiation.

You will read in this lesson that the Sun is not the only source of radiant energy. However, the Sun is the source that provides most of Earth's radiant energy. You also will read about how electromagnetic waves form, and learn about some properties of electromagnetic waves.

Active Reading

3. Write Fill in information about electromagnetic waves in the boxes.

Electromagnetic Waves	
travel through	carry energy called
_____ and _____.	_____, also known as _____.

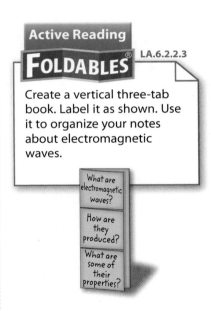

Active Reading

FOLDABLES LA.6.2.2.3

Create a vertical three-tab book. Label it as shown. Use it to organize your notes about electromagnetic waves.

What are electromagnetic waves?

How are they produced?

What are some of their properties?

Active Reading **2. Identify** What is electromagnetic radiation?

_____

_____

_____

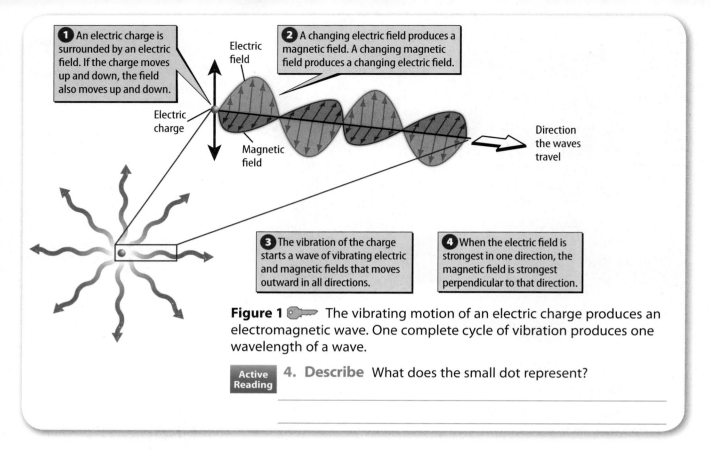

1 An electric charge is surrounded by an electric field. If the charge moves up and down, the field also moves up and down.

Electric field

2 A changing electric field produces a magnetic field. A changing magnetic field produces a changing electric field.

Electric charge

Magnetic field

Direction the waves travel

3 The vibration of the charge starts a wave of vibrating electric and magnetic fields that moves outward in all directions.

4 When the electric field is strongest in one direction, the magnetic field is strongest perpendicular to that direction.

Figure 1 🔑 The vibrating motion of an electric charge produces an electromagnetic wave. One complete cycle of vibration produces one wavelength of a wave.

Active Reading 4. **Describe** What does the small dot represent?

_____

_____

How Electromagnetic Waves Form

Electromagnetic waves form when an electric charge accelerates by either speeding up, slowing down, or changing direction. This happens when a charged particle vibrates, as shown in **Figure 1.**

Force Fields If you ever have played with a magnet, you know that it is surrounded by a field, or area, where the force of the magnet is present. The same is true for a charged particle, such as an electron. An electric field surrounds the charged particle.

Connected Fields Scientists have found that electric fields and magnetic fields are related.

• A changing electric field produces a magnetic field.

• A changing magnetic field produces a changing electric field.

As a charged particle vibrates, the electric field around it vibrates. This changing electric field produces a magnetic field. As the magnetic field changes, it produces a changing electric field. These connected fields spread out in all directions as electromagnetic waves. Just as shaking the end of the rope in **Figure 2** produces a wave, a vibrating charge produces a wave.

Active Reading 5. **Explain** How do electromagnetic waves form?

_____

_____

Figure 2 🔑 Shaking the rope up and down one time produces one wavelength. A charged particle also produces one wavelength when it moves up and down one time.

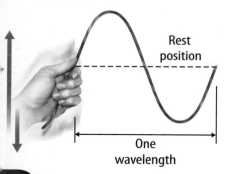

Rest position

One wavelength

Properties of Electromagnetic Waves

Electromagnetic waves usually are drawn as a single curve, like the rope in **Figure 2.** The waves are called transverse because the disturbance is perpendicular to the direction they travel.

Wavelength and Frequency As with all waves, the wavelength and frequency of electromagnetic waves are related. As shown in **Figure 2,** wavelength is the distance between one point on a wave to the nearest point just like it. Frequency is the number of wavelengths that pass by a point in a certain period of time, such as a second. As frequency decreases, wavelength increases. You will read in Lesson 2 that electromagnetic waves are grouped according to wavelength and frequency.

Wave Speed Electromagnetic waves travel through space at 300,000 km/s, or the speed of light (*c*). A wave's speed (*s*) is its frequency (*f*) multiplied by its wavelength (λ). To determine the wavelength of a wave moving through space, divide the speed of light (*c*) by the frequency of the wave, as shown in the Math Skills box.

What happens when electromagnetic waves move through matter? Suppose you run across a beach toward a lake or an ocean. You move quickly over the sand, but you slow down in the water. Electromagnetic waves behave similarly. When they encounter matter, electromagnetic waves slow down.

Active Reading 6. **Describe** What are some properties of electromagnetic waves?

_____

_____

_____

Math Skills MA.6.A.3.6

Solve One-Step Equations

Use inverse operations to keep sides of an equation equal. What is the wavelength of an electromagnetic wave in space that has a frequency of 500,000 Hz?

a. Use the wave-speed equation: **wave speed = frequency × wavelength:**

$$s = f\lambda$$

b. Divide by frequency. Then substitute known values. (Hint: Hz = 1/s)

$$\lambda = \frac{s}{f} = \frac{300{,}000 \text{ km/s}}{500{,}000 \text{ Hz}} = 0.6 \text{ km}$$

Practice

7. What is the wavelength of an electromagnetic wave with a frequency of 125,000 Hz?

Inquiry

LAB STATION **Try It!**

MiniLab *How are electric fields and magnetic field related?* at connectED.mcgraw-hill.com

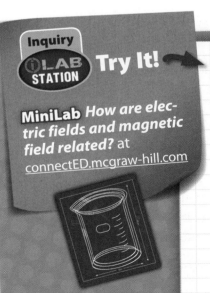

Apply It! After you complete the lab, answer these questions.

1. Recall that the wire did not touch the compass. Explain why the compass moved.

2. What was happening within the wires when they were touching?

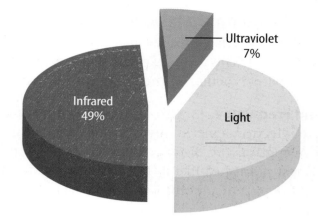

Figure 3 Most of the Sun's energy is carried by infrared waves.

Ultraviolet 7%

Infrared 49%

Light

Active Reading 8. **Calculate** What percentage of the Sun's energy is light? MA.6.A.3.6

REVIEW VOCABULARY

temperature
a measure of the average kinetic energy of the particles that make up an object

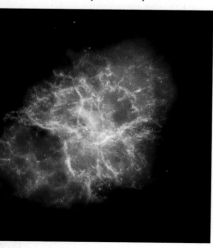

Figure 4 A supernova is the result of an exploding star. Colors were added with a computer to show the different wavelengths of electromagnetic waves emitted by this supernova.

Sources of Electromagnetic Waves

When you hear the term *electromagnetic radiation, or radiant energy,* you might imagine dangerous rays that you should avoid. It might surprise you to learn, however, that you are a source of electromagnetic waves! All matter contains charged particles that constantly vibrate. As a result, all matter—including you—produces electromagnetic waves, and therefore radiant energy. As an object's **temperature** increases, its particles vibrate faster. Therefore, the object produces electromagnetic waves with a greater frequency, or shorter wavelengths.

The Sun

Earth's most important energy source is the Sun. The Sun emits energy by giving off electromagnetic waves. Only a tiny amount of these waves reach Earth. The Sun has many areas with different temperatures and produces electromagnetic waves with many different wavelengths. As shown in **Figure 3,** almost all of the Sun's energy is carried by three types of waves—ultraviolet waves, light waves, and infrared waves. Light waves are the only type of electromagnetic waves you can see. Infrared waves have wavelengths longer than light. Ultraviolet waves have wavelengths shorter than light.

 9. **NGSSS Check** **Identify** Which electromagnetic wave from the Sun can you see? SC.7.P.10.1

Other Sources of Electromagnetic Waves

Even though the Sun is Earth's most important source of electromagnetic radiation, it is not the only one. All matter, both in space and on Earth, produces electromagnetic waves.

Sources in Space If you look up at the night sky, you see the Moon, stars, and planets. Telescopes on Earth and on satellites above Earth's atmosphere produce images of radiation emitted by these objects. Some of the radiation is visible, but most of it is not. **Figure 4** shows what radiation emitted by an exploding star might look like if your eyes could detect it.

Sources on Earth What do a campfire, a lightbulb, and a burner on an electric stove have in common? They all are hot enough to produce electromagnetic waves that carry energy you can feel. Look around you right now. Your book, the wall, people, and everything else you see produce electromagnetic waves, too. Some waves you detect, but others you do not. Telescopes produce visible images of radiation from space. Also, special cameras produce visible images of invisible waves on Earth. As shown in **Figure 5,** the ultraviolet waves from a flower produce an image very different from the waves you see with your eyes.

The Energy of Electromagnetic Waves

Have you ever seen someone with a terrible sunburn, such as the boy in **Figure 6?** Some of the Sun's waves have enough energy to damage your skin. The energy of electromagnetic waves is related to their frequency. Waves that have a higher frequency, such as ultraviolet waves, have higher energy. Waves that have a lower frequency, such as light waves and infrared waves, have lower energy. Light from the girl's flashlight in **Figure 6** can never damage her skin because the light waves do not have enough energy.

The relationship of energy to other wave properties is different for mechanical waves and electromagnetic waves. The energy of a mechanical wave is related to its amplitude. A water wave, for example, with a high amplitude has a lot of energy. The energy of electromagnetic waves is related to their frequency, not amplitude. As the frequency of an electromagnetic wave increases, the energy of the wave increases.

Active Reading **10. Explain** Why can the Sun's rays cause a burn, but the light from a flashlight cannot harm your skin?

Figure 5 The image on the bottom shows what the dandelions would look like if your eyes could detect the ultraviolet waves the flowers emit.

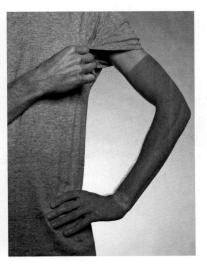

Figure 6 Ultraviolet waves from the Sun carry enough energy to damage skin cells. Light waves from a flashlight do not carry enough energy to cause damage.

Active Reading **11. Review** Highlight why ultraviolet waves can damage skin cells.

Visual Summary

An accelerating charge produces an electromagnetic wave similar to the wave produced by shaking the end of a rope.

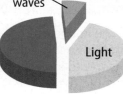

Almost all of the Sun's energy travels by infrared waves, light waves, or ultraviolet waves.

Electromagnetic waves transfer energy from one place to another, even through empty space.

Use Vocabulary

1 **Use the term** *radiant energy* in a sentence.

_____

_____

2 Electromagnetic waves carry _____.

Understand Key Concepts 🔑

3 What is the speed of electromagnetic waves in space?

(A) 30,000 m/s   (C) 300,000 m/s

(B) 30,000 km/s   (D) 300,000 km/s

4 **Identify** What must happen in order for an electromagnetic wave to form? SC.7.P.10.1

_____

_____

Interpret Graphics

5 **Sequence** Fill in the graphic organizer below to describe how an electromagnetic wave forms and travels away from a source. LA.6.2.2.3

electric charge accelerates ⇨ [ ] ⇨ [ ] ⇨ [ ]

Critical Thinking

6 **Create** a poster that explains how electromagnetic waves sometimes behave like a stream of particles. LA.6.2.2.3

Math Skills   MA.6.A.3.6

7 What is the wavelength of an electromagnetic wave with a frequency of 135,000 Hz?

Solar Sails

Using Light to Sail Through Space

How far and how fast can a spacecraft travel? Scientists are studying an exciting way to travel beyond the solar system. A technology called solar sails would allow a spacecraft to travel faster and farther than ever before possible.

Johannes Kepler first proposed solar sails about 400 years ago. He observed comets' tails facing away from the Sun and concluded that sunlight caused pressure—a force that might be harnessed with sails.

The Power of the Sun

Solar sails use light, not matter, to propel spacecraft. When reflected off a solar sail, light transfers its momentum to the surface of the sail, giving it a slight push. The solar sail moves a spacecraft very slowly at first. However, as long as there is a steady stream of light, the sail can eventually accelerate a spacecraft to incredibly high speeds.

The future of solar sail technology is unknown. So far, the use of solar sails has been successful only in a laboratory test chamber. The image below shows how scientists believe a solar sail could work if used in space.

Sun

Earth

Characteristics of a Solar Sail

- Large: at least as big as a football field

- Lightweight: 40–100 times thinner than a sheet of paper

- Very shiny: push provided by reflected light

- Rigid and durable: lasts many years in deep space

It's Your Turn

CALCULATE After traveling continuously for 3 years, a solar sail could reach a speed of 160,000 km/h. The speed of light is about 300,000 km/s. First calculate the kilometers per second a solar sail could reach. Then compare that number to the speed of light. What comparison can you make?

The Electromagnetic SPECTRUM

ESSENTIAL QUESTIONS

 What is the electromagnetic spectrum?

 How do electromagnetic waves differ?

Vocabulary

electromagnetic spectrum p. 633

radio wave p. 634

microwave p. 634

infrared wave p. 634

ultraviolet wave p. 635

X-ray p. 635

gamma ray p. 635

 Florida NGSSS

SC.7.P.10.1 Illustrate that the sun's energy arrives as radiation with a wide range of wavelengths, including infrared, visible, and ultraviolet, and that white light is made up of a spectrum of many different colors.

SC.7.N.1.1 Define a problem from the seventh grade curriculum, use appropriate reference materials to support scientific understanding, plan and carry out scientific investigation of various types, such as systematic observations or experiments, identify variables, collect and organize data, interpret data in charts, tables, and graphics, analyze information, make predictions, and defend conclusions.

SC.6.N.1.5 Recognize that science involves creativity, not just in designing experiments, but also in creating explanations that fit evidence.

SC.7.N.1.6 Explain that empirical evidence is the cumulative body of observations of a natural phenomenon on which scientific explanations are based.

SC.8.N.1.3 Use phrases such as "results support" or "fail to support" in science, understanding that science does not offer conclusive 'proof' of a knowledge claim.

MA.6.A.3.6 Construct and analyze tables, graphs, and equations to describe linear functions and other simple relations using both common language and algebraic notation.

Also covers: LA.6.2.2.3

SC.6.N.1.5

 Launch Lab

20 minutes

How do electromagnetic waves differ?

Procedure

1. Read and complete a lab safety form.

2. Obtain **four beads.** Place each in a separate **small, self-sealing plastic bag.**

3. Obtain **three types of sunscreen.** Each type should have a different sun protection factor (SPF).

4. Use a **permanent marker** to write the SPF rating on one side of each bag. On the other side, apply a thin layer of sunscreen that has that SPF. Label the fourth bag No Sunscreen.

5. Place the bags and beads near a **lamp** with the sunscreen side up. Observe the beads for several minutes.

6. Place the bags outside in sunlight. Observe the beads again for several minutes.

Think About This

1. Contrast what happened to the beads when you placed them near lamplight and in sunlight.

2. Describe the relationship between the SPF numbers and what happened to the beads.

3. **Key Concept** What do you think could have caused your results?

Inquiry **How Many Colors?**

1. Can you name the colors of the rainbow? You might know the familiar colors from red to violet, but no one can name all the colors. Rainbows are a continuous range of colors. Each color is an electromagnetic wave with a slightly different wavelength. What do you think are the various types of electromagnetic waves?

_____

_____

_____

What is the electromagnetic spectrum?

The waves that carry voices to your cell phone, the waves of energy that toast your bread, and the X-rays that a dentist uses to check the health of your teeth are all electromagnetic waves. The changing motion of an electric charge produces each type of electromagnetic wave. Each type of wave has a different frequency and wavelength, and each carries a different amount of energy. Electromagnetic waves might vibrate from a thousand times a second to trillions of times a second. They might be as large as a house or as small as an atom's nucleus. *The* **electromagnetic spectrum** *is the entire range of electromagnetic waves with different frequencies and wavelengths.*

Classifying Electromagnetic Waves

There are many shades among the familiar colors of a rainbow. Each color gradually becomes another. However, the electromagnetic spectrum is organized into groups based on the wavelengths and frequencies of the waves. Like the colors of a rainbow, each group blends into the next. You will read about how electromagnetic waves are grouped on the next pages.

Active Reading

FOLDABLES LA.6.2.2.3

Use four sheets of paper to make an eight-layer book. Label it as shown. Use it to compare and contrast the different types of electromagnetic waves.

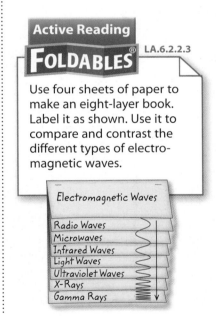

 2. NGSSS Check **Organize** As you read, fill in the table about electromagnetic waves of different wavelengths. SC.7.P.10.1

Waves with Longer Wavelengths Than Visible Light	Waves with Shorter Wavelengths Than Visible Light
1.	1.
2.	2.
3.	3.

Figure 7 Electromagnetic waves have a wide range of uses.

A **radio wave** *is a low-frequency, low-energy electromagnetic wave that has a wavelength longer than about 30 cm.* Some radio waves have wavelengths as long as a kilometer or more. Radio waves often are used for communication. The wavelengths are long enough to move around many objects, but the energy is low enough that they aren't harmful. On Earth, radio waves usually are produced by an electric charge moving in an antenna, but the Sun and other objects in space also produce radio waves.

① Radio waves   **② Microwaves**   **③ Infrared waves**

Longer wavelengths, Lower frequencies, Lower energy

A **microwave** *is a low-frequency, low-energy electromagnetic wave that has a wavelength between about 1 mm and 30 cm.* Like radio waves, microwaves are used for communication, such as cell phone signals. With shorter wavelengths than radio waves, microwaves are less often scattered by particles in the air. Microwaves are useful for satellite communications because they can pass through Earth's upper atmosphere. Because of the frequency range of microwaves, food molecules such as water and sugar can absorb their energy. This makes microwaves useful for cooking.

An **infrared wave** *is an electromagnetic wave that has a wavelength shorter than a microwave but longer than light.* Vibrating molecules in any matter emit infrared waves. Even your body emits infrared radiation. You cannot see infrared waves, but if you warm your hands near a campfire, you can feel them. Your skin senses infrared waves with longer wavelengths as warmth. Infrared waves with shorter wavelengths do not feel warm. Your television remote control, for example, sends out these waves.

Active Reading   **3. Name** Which type of electromagnetic waves have wavelengths too long for your eyes to see?

_____

_____

_____

4. Locate <u>Underline</u> the electromagnetic wave that is produced when the nucleus of an atom breaks apart or changes.

④ Light is electromagnetic waves that your eyes can see. You might describe light as red, orange, yellow, green, blue, indigo, and violet. Red light has the longest wavelength and the lowest frequency. Violet light has the shortest wavelength and the highest frequency. Each name represents a family of colors, each with a range of wavelengths.

④ **Light waves** ⑤ **Ultraviolet waves** ⑥ **X-rays** ⑦ **Gamma rays**

Shorter wavelengths, Higher frequencies, Higher energy

An **ultraviolet wave** *is an electromagnetic wave that has a slightly shorter wavelength and higher frequency than light and carries enough energy to cause chemical reactions.* Earth's atmosphere prevents most of the Sun's ultraviolet rays from reaching Earth. But did you know that you can get a sunburn on a cloudy day? This is because ultraviolet waves carry enough energy to move through clouds and to penetrate the skin. They can damage or kill cells, causing sunburn or even skin cancer.

⑤

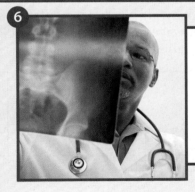

⑥

An **X-ray** *is a high-energy electromagnetic wave that has a slightly shorter wavelength and higher frequency than an ultraviolet wave.* Have you ever had an X-ray taken to see if you had a broken bone? X-rays have enough energy to pass through skin and muscle, but the calcium in bone can stop them. Scientists learn about objects and events in space, such as black holes and star explosions, by studying the X-rays they emit.

A **gamma ray** *is a high-energy electromagnetic wave with a shorter wavelength and higher frequency than all other types of electromagnetic waves.* Gamma rays are produced when the nucleus of an atom breaks apart or changes. As shown in the photo, gamma rays have enough energy that physicians can use them to destroy cancerous cells. Like X-rays, gamma rays form in space during violent events, such as the explosion of stars.

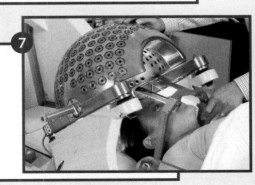

⑦

Light represents only a small portion of the total electromagnetic spectrum.

Because of differences in wavelength, frequency, and energy, electromagnetic waves have many common uses.

High-frequency electromagnetic waves carry so much energy that they often are used for medical imaging and treatment.

Inquiry SC.7.P.10.1,
SC.7.N.1.1,
SC.7.N.1.6,
SC.6.N.1.5,
SC.8.N.1.3,
MA.6.A.3.6

①LAB STATION

Try It!

Skill Lab *What's at the edge of a rainbow?* at connectED.mcgraw-hill.com

Use Vocabulary

1 **Explain** the difference between a microwave and a radio wave. SC.7.P.10.1

_____

_____

2 **Explain** the difference between an infrared wave and an ultraviolet wave. SC.7.P.10.1

_____

_____

Understand Key Concepts 🔑

3 Which electromagnetic waves have the highest energy? SC.7.P.10.1

  Ⓐ gamma rays   Ⓒ radio waves

  Ⓑ light waves   Ⓓ X-rays

4 **Name** the types of waves that make up the electromagnetic spectrum, from longest wavelength to shortest wavelength. SC.7.P.10.1

_____

_____

5 **Compare** the frequency of gamma waves with the frequency of light waves.

_____

_____

Interpret Graphics

6 **Sequence** Using the following electromagnetic waves; radio waves, microwaves, ultraviolet waves and gamma rays, fill in the boxes below from lowest to highest energy.

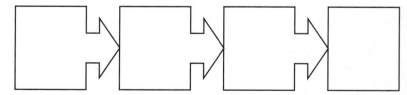

Critical Thinking

7 **Compare and contrast** infrared waves, light, and ultraviolet waves. SC.7.P.10.1

_____

_____

_____

8 **Infer** Why are X-rays and gamma rays able to penetrate the body, but radio waves are not?

_____

_____

Compile Differentiate wavelengths in the electromagnetic spectrum. The numbers in the boxes correspond to the art of the electromagnetic spectrum shown in your book.

A. Identify the type of wave. B. Describe the wave's wavelength, frequency, and energy. C. Determine a use or characteristic of the wave. D. Diagram of the wave	1. A. radio wave B. C. D.
2. A. B. C. D.	3. A. B. C. D.
4. A. B. C. D.	5. A. ultraviolet wave B. C. D.
6. A. X-ray B. C. D.	7. A. B. C. D.

Using Electromagnetic WAVES

ESSENTIAL QUESTIONS

 How are different types of electromagnetic waves used for communication?

 What are some everyday applications of electromagnetic waves?

 What are some medical uses of electromagnetic waves?

Vocabulary

broadcasting p. 639

carrier wave p. 640

amplitude modulation p. 640

frequency modulation p. 640

Global Positioning System p. 642

 Florida NGSSS

SC.7.N.1.3 Distinguish between an experiment (which must involve the identification and control of variables) and other forms of scientific investigation and explain that not all scientific knowledge is derived from experimentation.

SC.7.N.1.6 Explain that empirical evidence is the cumulative body of observations of a natural phenomenon on which scientific explanations are based.

SC.7.P.11.2 Investigate and describe the transformation of energy from one form to another.

SC.8.N.1.4 Explain how hypotheses are valuable if they lead to further investigations, even if they turn out not to be supported by the data.

LA.6.2.2.3 The student will organize information to show understanding or relationships among facts, ideas, and events (e.g., representing key points within text through charting, mapping, paraphrasing, summarizing, or comparing/contrasting);

LA.6.4.2.2 The student will record information (e.g., observations, notes, lists, charts, legends) related to a topic, including visual aids to organize and record information, as appropriate, and attribute sources of information;

 SC.7.N.1.6

(Inquiry) Launch Lab

15 minutes

How do X-rays see inside your teeth?

Dentists take X-rays of your mouth to see if you have any problems. How does an X-ray produce images of hard materials such as teeth?

Procedure

1 Read and complete a lab safety form.

2 Use **scissors** to cut an **index card** into four teeth shapes. Put the paper teeth in different places on a **piece of black paper** on the bottom of a **shoe box.**

3 Cover the shoebox opening with a piece of **screen.** Use **rubber bands** to attach the screen to the top of the shoe box.

4 Sprinkle **flour** evenly across the surface of the screen. Use a **toothbrush** to gently brush the flour on the screen until all of it falls through the screen.

5 Carefully remove the screen and remove the tooth shapes from the box.

6 Observe the paper "X-ray."

Think About This

1. What did you observe in the bottom of the box?

2. **Key Concept** If the screen represents your skin, muscle, and fatty tissues, what do you think this activity tells you about how X-ray machines form images?

1. A glowing fingerprint might look strange, but to a crime-scene detective, it's a clue. Detectives dust the scene with a powder that glows when ultraviolet waves strike it. If the powder sticks to a handprint, using ultraviolet waves might solve the crime. What do you think are some other uses of electromagnetic waves?

_____

_____

_____

Active Reading **2. Identify** Fold a sheet of paper into three columns. Label them *(K)* for what you already know about the uses of electromagnetic waves. *(W)* for what you want to learn, and *(L)* for the facts that you learned. Fill in the third column after you have read this lesson.

Active Reading **3. Find** <u>Underline</u> the reason why radio waves cannot harm humans.

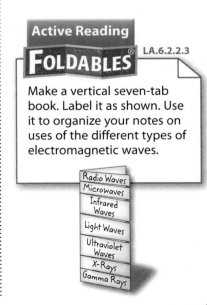

Active Reading

FOLDABLES® LA.6.2.2.3

Make a vertical seven-tab book. Label it as shown. Use it to organize your notes on uses of the different types of electromagnetic waves.

Radio Waves
Microwaves
Infrared Waves
Light Waves
Ultraviolet Waves
X-Rays
Gamma Rays

How do you use electromagnetic waves?

There are few things you do that don't involve some type of electromagnetic wave. Think about a typical day. Perhaps a clock radio wakes you in the morning with your favorite music. You are warm under your blanket because infrared radiation from your body changed into thermal energy. When you finally crawl out of bed, you turn on a lamp and use light to see which clothes to wear. These are just a few of the ways you use electromagnetic waves. In this lesson, you will read about common uses for all types of electromagnetic waves.

Radio Waves

It probably is no surprise to you that radio and television stations use radio waves to send out signals. Radio waves can pass through many buildings, yet they don't carry enough energy to harm humans. The wavelengths of radio waves are long enough to go around many obstacles. They travel through air at the same speed as all electromagnetic waves.

In the past, it might have taken minutes, hours, or even days for news to travel from one part of the United States to another. Today, you can watch and listen to events around the world as they happen! **Broadcasting** *is the use of electromagnetic waves to send information in all directions.* Broadcasting became possible when scientists learned to use radio waves.

Radio and Television

How do your radio and television receive information to produce sounds or images? A radio or TV station uses radio waves to carry information. Each radio and TV station sends out radio waves at a certain frequency. The station converts sounds or images into an electric signal. Then, the station produces a **carrier wave**—*an electromagnetic wave that a radio or television station uses to carry its sound or image signals.* The station **modulates,** or varies, the carrier wave to match the electric signal. The signal then gets converted back into images or sounds when it reaches your radio or television.

WORD ORIGIN

modulate

from Latin *modulari,* means "to play or sing"

Figure 8 Amplitude modulation and frequency modulation are two ways of changing a carrier wave to send information.

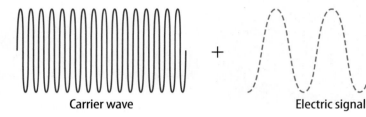

Carrier wave

+

Electric signal

=

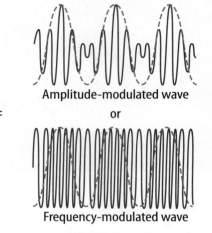

Amplitude-modulated wave

or

Frequency-modulated wave

Active Reading **4. Identify** Which type of modulation varies the strength of the carrier wave?

_____

_____

Two ways a station might change its carrier wave are shown in **Figure 8.** **Amplitude modulation** *(AM) is a change in the amplitude of a carrier wave.* Recall that amplitude is the maximum distance a wave varies from its rest position. **Frequency modulation** *(FM) is a change in the frequency of a carrier wave.* Changes in frequency match changes in sounds or images.

Digital Signals

Television stations and some radio stations broadcast digital signals. To understand types of signals, look at the amplitude-modulated wave in **Figure 8.** Notice how the wave changes smoothly from high to low. This type of change is called analog. A digital signal, however, changes in steps. Stations can produce a digital signal by changing different properties of a carrier wave. For example, the station might send the signal as pulses, or a pattern of starting and stopping. The wave might have a code of high and low amplitudes. Sounds and images sent by digital signals are usually clearer than analog signals.

Active Reading **5. Define** Highlight how an analog wave changes.

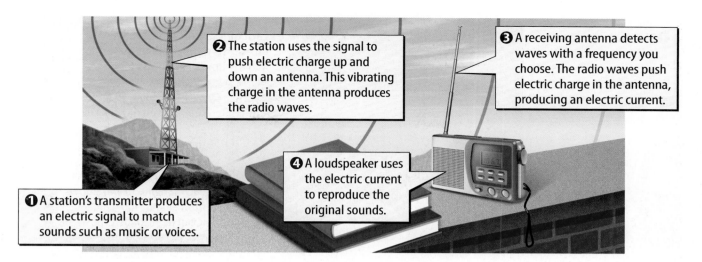

❷ The station uses the signal to push electric charge up and down an antenna. This vibrating charge in the antenna produces the radio waves.

❸ A receiving antenna detects waves with a frequency you choose. The radio waves push electric charge in the antenna, producing an electric current.

❹ A loudspeaker uses the electric current to reproduce the original sounds.

❶ A station's transmitter produces an electric signal to match sounds such as music or voices.

Transmission and Reception

The way a radio station broadcasts a signal is shown in **Figure 9**. A transmitter produces and sends out radio waves. An antenna receives the waves and changes them back to sound. Television transmission is similar, but for cable television, the waves travel through cables designed to carry radio waves.

The wavelength of an AM carrier wave is typically a lot longer than the wavelength of an FM carrier wave. This means AM signals can pass around obstacles and travel farther than FM waves. AM waves also are long enough to reach Earth's upper atmosphere without scattering. FM waves typically have higher frequencies and more energy than AM waves. Therefore, FM waves usually produce better sound quality than AM waves.

Active Reading **7. Explain** Why are AM waves able to reach Earth's atmosphere without scattering?

_____

_____

Microwaves

Did you know you use microwaves when you watch a television show that uses satellite transmission? Like radio waves, microwaves are useful for sending and receiving signals. However, microwaves can carry more information than radio waves because their wavelength is shorter. Microwaves also easily pass through smoke, light rain, snow, and clouds.

Cell Phones

A cell phone company sets up small regions of service called cells. Each cell has a base station with antennae, as shown in **Figure 10**. When you make a call, your phone sends and receives signals to and from the cell tower. Electric circuits in the base station direct your signal to other towers. A signal passes from cell to cell until it reaches the phone of the person you called.

Figure 9 🔑 Radio waves travel from a transmitting station to a radio antenna.

Active Reading **6. Describe** What type of energy changes take place during this process?

_____

_____

_____

Figure 10 🔑 Cell towers hold antennae from one or more service providers. Each provider positions its antennae toward a different direction to send and receive signals.

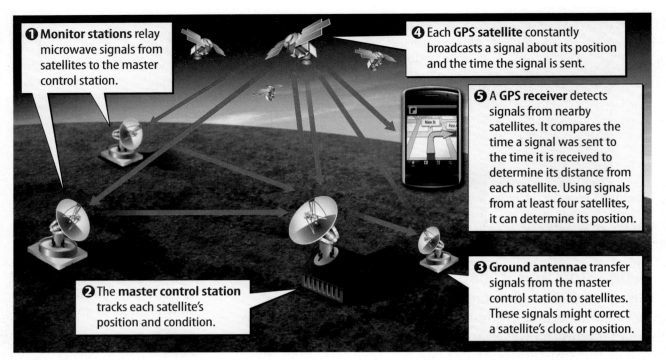

❶ Monitor stations relay microwave signals from satellites to the master control station.

❹ Each GPS satellite constantly broadcasts a signal about its position and the time the signal is sent.

❺ A GPS receiver detects signals from nearby satellites. It compares the time a signal was sent to the time it is received to determine its distance from each satellite. Using signals from at least four satellites, it can determine its position.

❷ The master control station tracks each satellite's position and condition.

❸ Ground antennae transfer signals from the master control station to satellites. These signals might correct a satellite's clock or position.

Figure 11 To calculate its position, a receiver needs signals from at least four satellites.

Active Reading
8. Point Out What is the purpose of ground antennae?

_____

_____

_____

SCIENCE USE V. COMMON USE

microwave

Science Use a low-frequency, low-energy electromagnetic wave

Common Use appliance used to cook food quickly

ACADEMIC VOCABULARY

analyze

(verb) to study how parts work together

Communication Satellites

Have you seen a sports event or heard a news story from around the world on your TV? Information about the sounds and the images probably traveled by **microwaves** from a satellite to your TV. Some homes have a satellite dish that receives signals directly from satellites. Satellites send and receive signals similar to the way antennae on Earth do. A transmitter sends microwave signals to the satellite. The satellite can pass the signal to other satellites or send it to another place on Earth.

You can use satellite signals to find directions. *The **Global Positioning System** (GPS) is a worldwide navigation system that uses satellite signals to determine a receiver's location.* As shown in **Figure 11,** at least 24 GPS satellites continually orbit Earth, sending out signals about their orbits and the time. Receivers **analyze** this information to calculate their location.

Active Reading
9. Review How are microwaves used for communication?

_____

_____

Infrared Waves

You read in Lesson 2 that vibrating molecules in any matter emit infrared waves. The wavelength of the infrared waves depends on the object's temperature. Hotter objects emit infrared waves with a greater frequency and shorter wavelength than do cooler objects. Scientists have developed technology that produces images showing how much thermal energy a person or thing emits.

Infrared Imaging

Thermal cameras take pictures by detecting infrared waves rather than light waves. They convert invisible infrared waves, or different temperatures, to different colors so that your eyes can interpret the information. Medical professionals use thermal imaging to locate areas of poor circulation in the body. Thermal imaging of a building can identify areas where excessive thermal energy loss occurs.

Can you see in the dark? Night-vision goggles, such as those in **Figure 12,** use infrared waves similar to the way a normal camera uses light. Using the small amount of infrared light present, the goggles produce enough light that objects can be seen easily in the dark.

Figure 12 Night-vision goggles make buildings and other objects visible at night.

Imaging Earth

Scientists launch satellites that detect and photograph infrared waves coming from Earth. These infrared images might show variations in vegetation or snowfall. Some images can even show a fire smoldering in a forest before it bursts into flame. The thermal image in **Figure 13** shows the lava flow from a volcano.

Active Reading **10. State** What are some ways in which infrared images help humans?

_____

_____

Figure 13 This infrared image of the Colima Volcano in Mexico shows recent lava flow as red.

Light

The Sun provides most of Earth's light, but light emitted on Earth also has important uses. Think about how difficult driving in a city would be without automobile headlights, street lights, and traffic signals, such as the one in **Figure 14.** Without electric lights, homes and businesses would be dark at night. Televisions, computers, and movie theaters also rely on light. Some forms of communication rely on light that travels through optical fibers—thin strands of glass or plastic that transmit a beam of light over long distances.

Active Reading **11. Identify** What technologies use visible light produced on Earth?

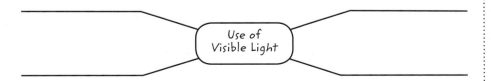

Use of Visible Light

Figure 14 🔑 A traffic signal is an example of how light is useful.

12. Apply What are some everyday applications of electromagnetic waves?

_____

_____

_____

_____

Ultraviolet Waves

Too much exposure to ultraviolet waves can damage your skin, but these same waves also can be useful. Certain materials glow when ultraviolet waves strike them. The materials absorb the energy of the waves and reemit it as light. Credit cards, for example, have invisible symbols stamped on them using this material. If a store clerk holds the card under an ultraviolet lamp, the symbol appears. Scientists also can identify some minerals in rocks by the way they glow under an ultraviolet lamp.

Ultraviolet waves also are useful for killing germs. Just as your skin can be damaged by absorbing the energy of ultraviolet waves, germs also can be damaged. Food manufacturers might use ultraviolet lamps to kill germs in some foods. Campers might use ultraviolet lamps to purify lake water.

Medical Uses

In hospitals and clinics, having a germ-free environment might be the difference between life and death. Medical facilities sometimes clean tools and surfaces by bringing them near an ultraviolet lamp. Some air and water purifiers use ultraviolet light to kill germs and reduce the spread of disease.

Ultraviolet light has other medical uses. For example, exposure to ultraviolet waves can help control or cure certain skin problems, such as psoriasis. Several times a week, the patient's skin is exposed to ultraviolet light. The waves carry enough energy to slow the growth of the diseased skin cells.

Another use of ultraviolet waves is shown in **Figure 15.** The dentist is using ultraviolet light to harden an adhesive in just a few seconds. Without ultraviolet light, adhesives might take several minutes to harden.

Figure 15 A dentist can use an ultraviolet wand to quickly harden adhesives.

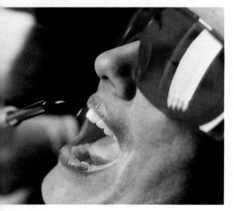

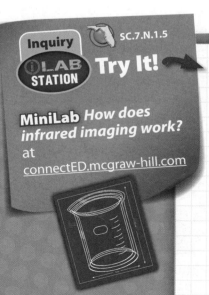

Inquiry SC.7.N.1.5

⚡LAB STATION **Try It!**

MiniLab *How does infrared imaging work?*
at
connectED.mcgraw-hill.com

Apply It! After you complete the lab, answer these questions.

1. Which parts of the false-color image show the least infrared light?

2. Why do thermal cameras convert different temperatures to different colors?

Fluorescent Lightbulbs

Lightbulbs like the one shown in **Figure 16** use the energy from ultraviolet waves to produce light. Similar to the way a symbol on a credit card glows under ultraviolet light, certain chemicals inside the lightbulb glow when they are exposed to ultraviolet light. When you flip a switch to provide electricity to the bulb, an electric current flows through a gas inside the bulb. The heated gas emits ultraviolet light. This light strikes a chemical that coats the inside of the bulb. The chemical absorbs the energy of the ultraviolet waves and reemits it as light.

Active Reading 13. **Recall** How does a fluorescent lightbulb use ultraviolet waves?

_____

_____

Figure 16 A fluorescent lightbulb uses ultraviolet waves to produce light.

X-Rays

Ultraviolet waves can be dangerous because they can pass through the upper layer of your skin. X-rays have even more energy than ultraviolet waves and can pass through skin and muscle. Although this property makes X-rays dangerous, it also makes them useful. Two important uses of X-rays are security and medical imaging.

Detection and Security

Have you ever had your luggage scanned at an airport, as shown in **Figure 17?** X-rays are useful for security scanning because they can pass through many materials, but metal objects block them. Computers can form images of the contents of luggage by measuring how different materials transmit the X-rays.

Active Reading 14. **Describe** What is shown on the screen when baggage is X-rayed?

_____

_____

Airport Security

Figure 17 Security workers use X-ray scanners to form images of the contents of passenger luggage.

Active Reading **15. Reason** Why do you think dentists cover your body in a blanket lined with lead when taking an X-ray of your mouth?

_____

_____

_____

_____

Medical Detection

You might have had an X-ray taken by a dentist or a doctor, similar to that shown in **Figure 18.** X-rays pass through soft parts of your body, but bone stops them. When X-rays strikes photographic film, the film turns dark. Light parts of the film show where bone absorbed the X-rays.

A doctor can obtain even more-detailed images using a computed tomography (CT) scanner. The scanner is an X-ray machine that rotates around a patient, producing a three-dimensional view of the body.

Active Reading **16. Differentiate** How are a CT scan and a normal X-ray different?

_____

_____

Gamma Rays

Because of their extremely high energy, gamma rays can be used to destroy diseased tissue in a patient. Gamma rays also can be used to diagnose medical conditions. Recall that gamma rays are produced when an atom's nucleus breaks apart or changes. In a procedure called a positron emission tomography (PET) scan, a detector monitors the breakdown of a chemical injected into a patient. The chemical is chosen because it is attracted to diseased parts of the body. The detector can find the location of the disease by detecting the gamma rays emitted by the chemical.

Active Reading **17. Give Examples** What are some medical uses of electromagnetic waves?

_____

_____

_____

Figure 18 Both X-rays and gamma rays are used in medicine to diagnose and treat diseases.

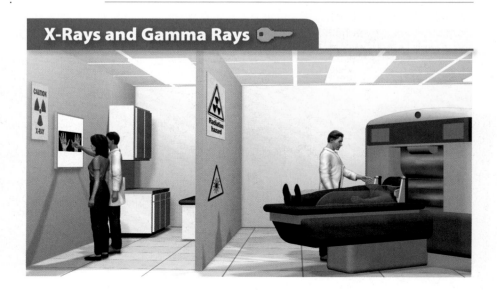

X-Rays and Gamma Rays

Visual Summary

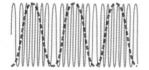

Radio and television stations can broadcast their programming with modulated radio waves.

Satellites send and receive microwave signals to communicate information over long distances.

Because warmer objects emit more infrared waves, scientists can study heat-related changes on Earth's surface, such as weather and volcanic activity.

Inquiry **iLAB STATION** **Try It!**

SC.7.N.1.3, SC.8.N.1.4, LA.6.4.2.2

Inquiry Lab *Design an Exhibit for a Science Museum* at connectED.mcgraw-hill.com

Use Vocabulary

1. The wave that is changed to carry a radio or television signal is a(n) _____ .

2. **Use the term** *broadcasting* in a complete sentence.

_____

_____

Understand Key Concepts 🔑

3. Which of the following changes in an AM radio wave?
   - Ⓐ amplitude
   - Ⓒ speed
   - Ⓑ frequency
   - Ⓓ wavelength

4. **Explain** why X-rays are less harmful than infrared waves in medical imaging.

_____

_____

5. **Describe** how microwaves are useful for communication.

_____

_____

6. **Identify** an everyday application of ultraviolet waves.

_____

_____

Interpret Graphics

7. **Identify** Fill in the graphic organizer below to identify several uses of electromagnetic waves. LA.6.2.2.3

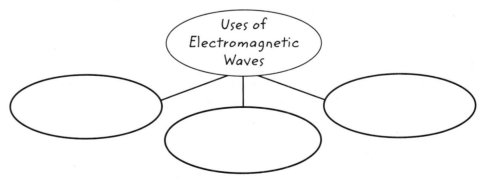

Uses of Electromagnetic Waves

Critical Thinking

8. **Predict** the effect on human life if the ozone layer, which blocks harmful ultraviolet waves, becomes thinner or develops more holes.

_____

_____

_____

Chapter 16 Study Guide

 Think About It! Electromagnetic waves transfer energy through matter and space. They exist in different forms, based on frequency and wavelength. Uses include communication and medical imaging.

Key Concepts Summary

Vocabulary

LESSON 1 Electromagnetic Radiation

- An accelerating electric charge produces an **electromagnetic wave**.
- Electromagnetic waves travel through space at 300,000 km/s. Properties include their wavelengths, their frequencies, and the amount of **radiant energy** each wave carries.

electromagnetic wave p. 625

radiant energy p. 625

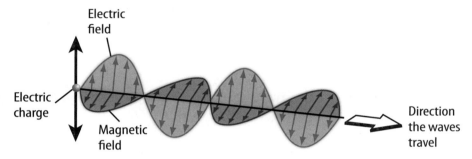

Electric field

Electric charge

Magnetic field

Direction the waves travel

LESSON 2 The Electromagnetic Spectrum

- The **electromagnetic spectrum** is the entire range of wavelengths and frequencies of electromagnetic waves.
- Electromagnetic waves differ in their wavelengths, frequencies, and energy. **Radio waves** have the longest wavelengths, the lowest frequencies, and the lowest energy. **Gamma rays** have the shortest wavelengths, the highest frequencies, and the highest energy.

electromagnetic spectrum p. 633

radio wave p. 634

microwave p. 634

infrared wave p. 634

ultraviolet wave p. 635

X-ray p. 635

gamma ray p. 635

LESSON 3 Using Electromagnetic Waves

- Electromagnetic waves are used in radio and television **broadcasting**, and in cell phone communication.
- Using light to see and microwaves to cook food are everyday applications of electromagnetic waves. Listening to radio and watching television require devices that use electromagnetic waves.
- Doctors use X-rays to identify broken bones or damaged teeth. Gamma rays can be used to diagnose or treat diseases.

broadcasting p. 639

carrier wave p. 640

amplitude modulation p. 640

frequency modulation p. 640

Global Positioning System p. 642

Assemble your lesson Foldables as shown to make a Chapter Project. Use the project to review what you have learned in this chapter.

Electromagnetic Waves

What are electromagnetic waves?
How are they produced?
What are some of their properties?

Radio Waves
Microwaves
Infrared Waves
Light Waves
Ultraviolet Waves
X-Rays
Gamma Rays

Radio Waves
Microwaves
Infrared Waves
Light Waves
Ultraviolet Waves
X-Rays
Gamma Rays

Use Vocabulary

1 A transverse wave that is able to move through space is a(n)

_____ .

2 The type of electromagnetic wave that causes sunburn is a(n)

_____ .

3 The electromagnetic wave that has the highest energy and is therefore most dangerous to humans is a(n)

_____ .

4 A television signal travels from one place to another by a(n) _____ that is modulated.

5 An FM station changes its carrier wave by

_____ .

Link Vocabulary and Key Concepts

Use vocabulary terms from the previous page to complete the concept map.

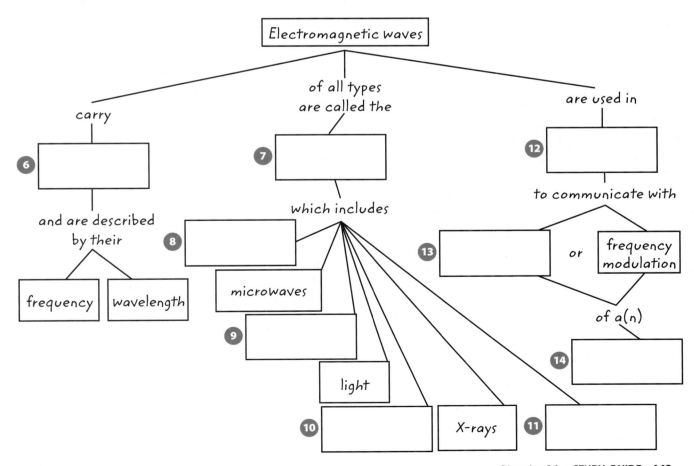

Fill in the correct answer choice.

🔑 Understand Key Concepts

1 What happens to a wave if its frequency decreases? SC.7.P.10.1

- Ⓐ Its amplitude increases.
- Ⓑ Its energy decreases.
- Ⓒ Its speed increases.
- Ⓓ Its wavelength decreases.

2 The image below most likely was produced using which type of electromagnetic wave? SC.7.P.10.1

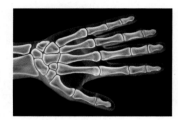

- Ⓐ gamma rays
- Ⓑ light
- Ⓒ microwaves
- Ⓓ X-rays

3 Which type of electromagnetic waves has the longest wavelengths? SC.7.P.10.1

- Ⓐ infrared waves
- Ⓑ microwaves
- Ⓒ radio waves
- Ⓓ ultraviolet waves

4 Which type of electromagnetic waves does your body emit? SC.7.P.10.1

- Ⓐ infrared waves
- Ⓑ microwaves
- Ⓒ radio waves
- Ⓓ ultraviolet waves

Critical Thinking

5 **Sequence** the types of electromagnetic waves according to their ability to penetrate matter. SC.7.P.10.1

_____

_____

_____

_____

6 **Synthesize** The waves in the diagram below represent the colors yellow, blue, and red. Describe which color is represented by waves A, B, and C, and explain how you made your choice. SC.7.P.10.1

_____

_____

_____

_____

_____

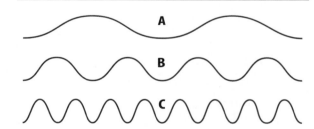

7 **Explain** how you can determine the frequency of an electromagnetic wave moving through space if you know the wavelength. MA.6.A.3.6

_____

_____

_____

_____

Writing in Science

8 **Write** A classmate can't understand what electromagnetic waves have to do with his or her everyday life. Write a note on a separate sheet of paper at least five sentences long to your friend explaining why understanding electromagnetic waves is important and how the information might apply to everyday life. LA.6.2.2.3

Big Idea Review

9 The photos below show that the Sun emits different types of electromagnetic waves. How can you describe the main types of electromagnetic waves the Sun emits? What are some ways in which people use electromagnetic waves? SC.7.P.10.1

_____

_____

_____

_____

_____

_____

_____

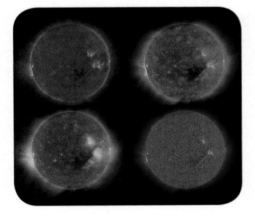

Math Skills   MA.6.A.3.6

Solve One-Step Equations

10 What is the wavelength of an electromagnetic wave with a frequency of 145,000 Hz?

11 What is the wavelength of an electromagnetic wave with a frequency of 165,000 Hz?

12 An electromagnetic wave has a wavelength of 0.65 km. What is its frequency?

13 An electromagnetic wave has a wavelength of 0.45 km. What is its frequency?

14 What is the wavelength of an electromagnetic wave with a frequency of 225,000 Hz?

Fill in the correct answer choice.

Multiple Choice

Use the table below to answer questions 1 and 2.

Regions of the Electromagnetic Spectrum		
Infrared waves	Radio waves	Gamma rays
X-rays	Visible light	Ultraviolet waves

1 If you arranged the list of electromagnetic waves shown above in order from shortest to longest wavelength, which would be first on the list? SC.7.P.10.1

Ⓐ gamma rays

Ⓑ radio waves

Ⓒ visible light

Ⓓ X-rays

2 Which region of the electromagnetic spectrum listed in the table includes microwaves? SC.7.P.10.1

Ⓕ ultraviolet waves

Ⓖ radio waves

Ⓗ infrared waves

Ⓘ gamma rays

3 Which three types of electromagnetic waves carry most of the Sun's energy that strikes Earth? SC.7.P.10.1

Ⓐ infrared waves, light, ultraviolet waves

Ⓑ ultraviolet waves, light, microwaves

Ⓒ light, ultraviolet waves, radio waves

Ⓓ X-rays, infrared waves, gamma waves

4 Which shows the correct order from lowest to highest energy for different electromagnetic waves? SC.7.P.10.1

Ⓕ light → X-ray → microwave

Ⓖ gamma ray → light → radio wave

Ⓗ X-ray → microwave → ultraviolet wave

Ⓘ radio wave → microwave → gamma ray

Use the figure to answer questions 5 and 6.

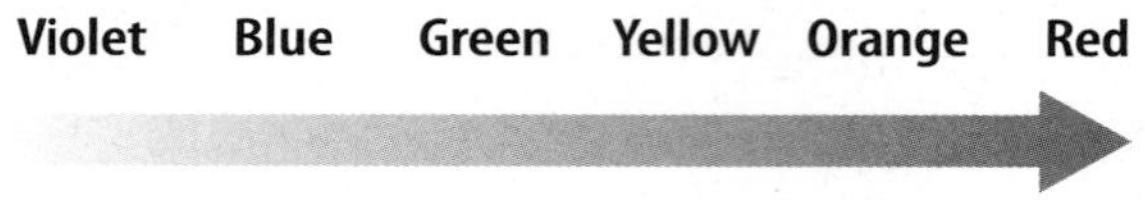

Violet	Blue	Green	Yellow	Orange	Red

5 The figure lists different colors of light. Based on the sequence shown, what could the arrow represent? SC.7.P.10.1

Ⓐ increasing energy

Ⓑ increasing frequency

Ⓒ increasing speed

Ⓓ increasing wavelength

6 What do all these colors of light have in common? SC.7.P.10.1

Ⓕ their energy

Ⓖ their frequency

Ⓗ their speed

Ⓘ their wavelength

Use the figure to answer questions 7 and 8.

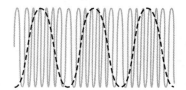

7 The figure shows a carrier wave used to broadcast radio signals. What is modulated in this wave in order to carry information? SC.7.P.10.1

Ⓐ amplitude

Ⓑ crest height

Ⓒ frequency

Ⓓ loudness

8 Which is true of this type of signal wave? SC.7.P.10.1

Ⓕ It can easily pass around obstacles.

Ⓖ It can be seen with the unaided eye.

Ⓗ It can travel far by bouncing off particles in the atmosphere.

Ⓘ It can have wavelengths that are less than a meter long.

9 Even on a cloudy day, you can get sunburned outside. However, inside a glass greenhouse, you won't get sunburned. Which type of electromagnetic wave will pass through clouds, but not glass? SC.7.P.10.1

Ⓐ ultraviolet waves

Ⓑ radio waves

Ⓒ microwaves

Ⓓ infrared waves

10 Which list of electromagnetic waves is in the correct order from highest to lowest energy? SC.7.P.10.1

Ⓕ gamma rays, radio waves, infrared waves, microwaves

Ⓖ ultraviolet waves, gamma rays, X-rays, light

Ⓗ light, infrared waves, microwaves, radio waves

Ⓘ X-rays, gamma rays, ultraviolet waves, light

11 Which type of electromagnetic wave causes a chemical to glow in a fluorescent lightbulb? SC.7.P.10.1

Ⓐ gamma waves

Ⓑ infrared waves

Ⓒ radio waves

Ⓓ ultraviolet waves

NEED EXTRA HELP?

If You Missed Question...	1	2	3	4	5	6	7	8	9	10	11
Go to Lesson...	2	2	2	2	2	2	3	3	2	2	3

Multiple Choice *Bubble the correct answer.*

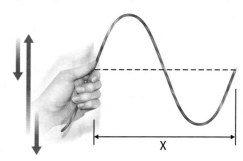

1. In the illustration of a wave shown above, what measurement is indicated by X? **SC.7.P.10.1**

   (A) crest

   (B) speed

   (C) frequency

   (D) wavelength

2. Where do electromagnetic waves move most quickly? **SC.7.P.10.1**

   (F) through metal

   (G) through space

   (H) through water

   (I) through wood

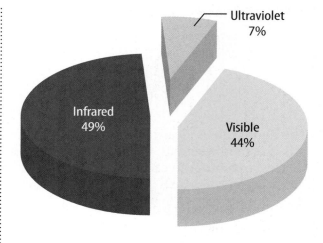

Ultraviolet 7%

Infrared 49%

Visible 44%

3. The pie chart above shows the three types of waves that carry nearly all of the Sun's energy to Earth. What can be inferred from the chart? **SC.7.P.10.1**

   (A) Visible light has fewer photons than infrared light.

   (B) Ultraviolet light has the shortest wavelength.

   (C) Most of the Sun's energy is carried by infrared light.

   (D) Infrared light has the highest frequency.

4. If the frequency of an electromagnetic wave decreases, which increases? **SC.7.P.10.1**

   (F) its energy

   (G) its photons

   (H) its speed

   (I) its wavelength

Copyright © Glencoe/McGraw-Hill, a division of The McGraw-Hill Companies, Inc.

Multiple Choice *Bubble the correct answer.*

Radio waves   Microwaves

X

1. The Venn diagram above can be used to compare radio waves and microwaves. Which can go where X is in the diagram? **LA.6.2.2.3**

(A) may be harmful

(B) not formed in space

(C) used for communication

(D) used for cooking

2. Which kind of electromagnetic wave can be sensed as heat? **SC.7.P.10.1**

(F) microwaves

(G) X-rays

(H) infrared waves

(I) ultraviolet waves

3. What do electromagnetic waves with frequencies higher than visible light have in common? **SC.7.P.10.1**

(A) They are used for communication.

(B) They can be extremely harmful.

(C) They have longer wavelengths than visible light.

(D) They have low amounts of energy.

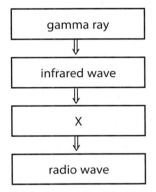

4. The flow chart above shows four kinds of electromagnetic waves, arranged from higher frequency to lower frequency. Which belongs at X? **LA.6.2.2.3**

(F) microwave

(G) X-ray

(H) ultraviolet wave

(I) visible light wave

Copyright © Glencoe/McGraw-Hill, a division of The McGraw-Hill Companies, Inc.

Multiple Choice *Bubble the correct answer.*

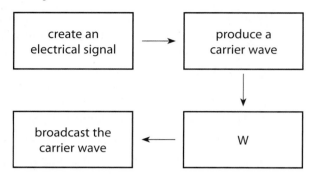

1. The flow chart above shows the steps needed to create and send signals with radio waves. Which goes in the box with W?
   LA.6.2.2.3

   (A) amplify the electrical signal

   (B) change the signal's frequency

   (C) modulate the carrier wave

   (D) send the carrier wave to an antenna

2. Which device uses the same kind of electromagnetic waves as a cell phone?
   SC.7.N.1.6

   (F) digital camera

   (G) microwave oven

   (H) night-vision binoculars

   (I) X-ray machine

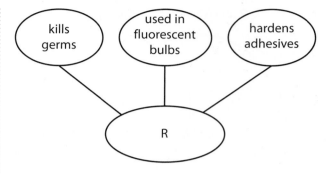

3. The graphic organizer above shows three different ways that one kind of electromagnetic wave might be used. Which goes in the oval containing R?
   LA.6.2.2.3

   (A) gamma rays

   (B) infrared waves

   (C) microwaves

   (D) ultraviolet waves

4. In a nuclear reactor, the nuclei of uranium atoms are split apart. This process releases large amounts of **SC.7.N.1.5**

   (F) gamma rays.

   (G) X-rays.

   (H) infrared waves.

   (I) ultraviolet waves.

Copyright © Glencoe/McGraw-Hill, a division of The McGraw-Hill Companies, Inc.

Notes

Name _____ Date _____

Seeing a Flower

When we look at flowers, we see that they come in many different shapes, sizes, and colors. What happens between a flower and our eyes that enables us to see it? (Circle) the answer that best matches your thinking.

A. The light in the room lights up the flower so our eyes can see it.

B. Something goes from our eyes to the flower so we can see it.

C. Something goes from the flower to our eyes so we can see it.

D. Particles of color travel to our brain so we can see the flower.

E. Something goes back and forth between our eye and the flower so we can see it.

Explain your thinking. Describe how it is possible for us to see the flower.

LIGHT

FLORIDA BIG IDEAS

1 The Practice of Science

3 The Role of Theories, Laws, Hypotheses, and Models

10 Forms of Energy

The Big Idea

Think About It!

How does matter affect the way you perceive and use light?

The tiny flowers seem real, but they are not. This is a close-up photo of a chrysanthemum flower covered with water drops. The tiny flower images are reflections in the water drops.

1 What do you think causes images of the flower to appear in the water drops?

_____

_____

_____

2 Why do you think reflections appear in the water but not on the flower?

_____

_____

3 How do you think matter affects the way you perceive light?

_____

_____

_____

Get Ready to Read

What do you think about light?

Before you read, decide if you agree or disagree with each of these statements. As you read this chapter, see if you change your mind about any of the statements.

AGREE DISAGREE

1 Both the Sun and the Moon produce their own light. ☐ ☐

2 The color of an object depends on the light that strikes it. ☐ ☐

3 Mirrors are the only surfaces that reflect light. ☐ ☐

4 The image in a mirror is always right side up. ☐ ☐

5 A prism adds color to light. ☐ ☐

6 When light moves from air into water or glass, it travels more slowly. ☐ ☐

7 A laser will burn a hole through your skin. ☐ ☐

8 Telephone conversations can travel long distances as light waves. ☐ ☐

 **There's More Online!**
Video • Audio • Review • ⓘLab Station • WebQuest • Assessment • Concepts in Motion • Multilingual eGlossary

Light, Matter, and COLOR

ESSENTIAL QUESTIONS

 What are some sources of light, and how does light travel?

 What can happen to light that strikes matter?

Why do objects appear to have different colors?

Vocabulary

light p. 663

reflection p. 663

transparent p. 664

translucent p. 664

opaque p. 664

transmission p. 665

absorption p. 665

 Florida NGSSS

SC.7.N.1.1 Define a problem from the seventh grade curriculum, use appropriate reference materials to support scientific understanding, plan and carry out scientific investigation of various types, such as systematic observations or experiments, identify variables, collect and organize data, interpret data in charts, tables, and graphics, analyze information, make predictions, and defend conclusions.

SC.7.N.1.6 Explain that empirical evidence is the cumulative body of observations of a natural phenomenon on which scientific explanations are based.

SC.7.P.10.1 Illustrate that the sun's energy arrives as radiation with a wide range of wavelengths, including infrared, visible, and ultraviolet, and that white light is made up of a spectrum of many different colors.

SC.7.P.10.2 Observe and explain that light can be reflected, refracted, and/or absorbed.

LA.6.2.2.3 The student will organize information to show understanding or relationships among facts, ideas, and events (e.g., representing key points within text through charting, mapping, paraphrasing, summarizing, or comparing/contrasting);

LA.6.4.2.2 The student will record information (e.g., observations, notes, lists, charts, legends) related to a topic, including visual aids to organize and record information, as appropriate, and attribute sources of information;

SC.7.N.1.6

Inquiry Launch Lab

10 minutes

How can you make a rainbow?

After it rains, you might see a rainbow stretching across the sky. How can sunlight change into so many different colors? How can you make a rainbow of colors by shining light through a prism?

Procedure

1. Read and complete a lab safety form.
2. Use **tape** to attach a sheet of **white paper** to a wall.
3. In a darkened room, stand about 1 m from the paper. Hold a **prism** in front of a **flashlight,** as shown in the photo. Turn on the flashlight. Shine the light through the prism onto the paper.
4. Adjust the prism until you see bands of color on the paper. Turn the prism, and have your partner trace the bands of color on the paper with **colored pencils.** Record your observations. Turn the prism two more times, and trace the colors.
5. Switch places with your partner, and repeat steps 3–4.

Data and Observations

Think About This

1. What is the sequence of colors in the bands of light in your drawings? Is it the same in all the drawings?

2. **Key Concept** How do you think the prism affected the white light from the flashlight?

 **Why So Colorful?**

1. On a clear day, it's exciting to watch colorful hot-air balloons launch into the blue sky. What do you think makes the balloons so colorful? Why are you able to see all the different colors?

_____

_____

_____

_____

_____

What is light?

Suppose you are taking a tour of a cave below Earth's surface and the lights go out. Would you see anything? No! You could not see anything because there would be no light for your eyes to perceive. **Light** *is electromagnetic radiation that you can see.* Electromagnetic radiation has wave properties and particle properties. A particle of electromagnetic radiation is called a photon. The frequency of a light wave depends on the amount of energy carried by a photon of light. Light waves can carry this energy through space and some matter.

Sources of Light

You can see an object if it is luminous or if it is illuminated. Luminous objects, such as the campfire in **Figure 1,** release, or emit, light. The Sun is luminous because it is made of hot, glowing gases. A traffic light and a firefly also are luminous objects. Luminous objects are sources of light. We also can see illuminated objects, such as the camper and trees in **Figure 1.** However, they do not emit light and are not sources of light. The Moon might appear bright, but it is just a rocky sphere. You see the Moon because it reflects light from the Sun. **Reflection** *is the bouncing of a wave off a surface.* In a dark cave, there is no light to reflect off objects so you cannot see anything.

2. NGSSS Check **Name** What are some sources of light? SC.7.P.10.1

_____

REVIEW VOCABULARY

electromagnetic radiation
energy carried through space or matter by electromagnetic waves

Figure 1 You see the campfire because it is luminous. You see other objects in the picture because light from the fire illuminates them.

Figure 2 Light spreads out as it moves away from a source. An object in the path of light forms shadows where there is less light.

Figure 3 Different amounts of light can pass through the transparent and translucent parts of this window. The opaque walls and table do not allow light to pass through them.

How Light Travels

Light travels as waves moving away from a source. Scientists often describe these waves as countless numbers of light rays spreading out in all directions from a source. You can model light rays by shining a light through a comb, as shown in **Figure 2.**

What are the spaces between the rays in **Figure 2?** They are shadows or places with less light. The shadows show that light normally travels in a straight line. However, objects in its path can cause light to change direction. It also can spread out slightly as it moves through a small opening. Sometimes, the effect is difficult to see because light waves are so small.

Active Reading 3. **Explain** How does light travel?

_____

_____

_____

Light and Matter

What can you see through your classroom window? Would you still be able to see anything if the blinds were closed? How do different types of matter affect light?

You can see objects clearly through air, clean water, plain glass, and some plastics. *A material that allows almost all the light that strikes it to pass through and form a clear image is* **transparent.** The unfrosted parts of the window in **Figure 3** are transparent.

Light also passes through the frosted parts of the window, but clear images do not form. *A material that allows most of the light that strikes it to pass through and form a blurry image is* **translucent.** Plastics with textured surfaces are examples of translucent materials.

A material through which light does not pass is **opaque.** No light passes through the chairs or the table in **Figure 3.** Wood and metal are examples of opaque materials.

 4. **NGSSS Check** Review Underline the three results of light interacting with matter.
SC.7.P.10.2

Figure 4 🔑 The window transmits, absorbs, and reflects light.

Active Reading 5. **Identify** Draw arrows and label materials in the photo as transparent or opaque.

Transmission of Light

You just read that light passes through transparent and translucent objects. *The passage of light through an object is called* **transmission.** The girls and you can see objects through the window in **Figure 4** because the window transmits light. A luminous object or an illuminated object on one side of the window is visible on the other side of the window. Also, the energy carried by the light waves from these objects can pass through the window.

Absorption of Light

Imagine standing near a window on a spring day. The transparent window transmits some sunlight, and it lands on you. If you touch the window, it might feel warm. Some of the energy in sunlight stays inside the window. *The transfer of energy by a wave to the medium through which it travels is called* **absorption.** The energy causes atoms in the material to vibrate faster, increasing the temperature of the material. All materials absorb some of the light that strikes them. The window feels warm because it absorbs some of the sunlight's energy.

Reflection of Light

When you look at a pane of glass, you sometimes can see an image of yourself. Light bounces off you, strikes the glass, and bounces back to your eye. Recall that the bouncing of a wave off a surface is called reflection.

Look again at the window in **Figure 4.** Notice that you can see the transmission and reflection of light. You cannot see it, but some of the light also is absorbed. Most types of matter interact with light in a combination of ways.

WORD ORIGIN

transmission
from Latin *transmissionem,* means "sending over or across"

Active Reading

FOLDABLES® LA.6.2.2.3

Make a vertical trifold book and label it as shown. Use it to organize your notes on the sources of light, light and matter, and the relationship between light and color.

Light Sources

Light & Matter

Light & Color

MiniLab *What color is that?* at connectED.mcgraw-hill.com

Apply It!

After you complete the lab, answer the questions below.

1. Imagine you are looking through a red filter at a green apple and a red apple that are both sitting in sunlight. The first apple appears black, and the second apple appears red. Explain how you can determine the true colors of the apples.

2. How would the apples appear if you viewed them both through a blue filter?

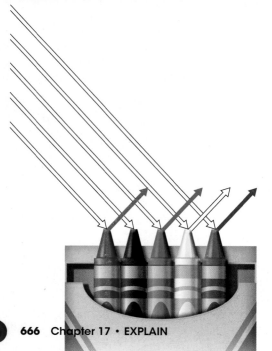

Light and Color

Recall that visible light is electromagnetic radiation with wavelengths of all colors of the rainbow. The longest wavelengths of light appear red. Violet has the shortest wavelengths. Other colors of light have different wavelengths. White light is a mixture of all wavelengths of light. How does this account for the colors you might see?

Colors you see result from wavelengths of light that enter your eyes. When you look at a luminous object such as a campfire, you see the colors emitted by the fire and glowing logs. What happens when you look at an illuminated object? That depends on whether the object is opaque, transparent, or translucent.

Opaque Objects

Suppose white light strikes a box of crayons, as shown in **Figure 5.** Each crayon absorbs all wavelengths of light except its color. For example, the green crayon absorbs all colors except green. The green wavelengths reflect back into your eyes, and you see green. The red crayon absorbs all colors except red, and it reflects red. The black crayon absorbs all colors. The color of an opaque object is the color of light it reflects.

What do you think would happen to the colors reflected by the crayons if red light, instead of white light, were shining on them? Would the green crayon still appear green? No. It would absorb the red light, but there would be no green light to reflect. The green crayon would appear black. The blue crayon also would appear black. The red crayon and the white crayon would appear red because they would reflect red light. The color you see always depends on the color of light that the object reflects.

Figure 5 ⬤➤ The color of an object is the color of light that reflects off the object.

Transparent and Translucent Objects

Absorption, transmission, and reflection also explain the color of transparent and translucent objects. For example, suppose white light, such as sunlight, shines through a piece of blue glass. The glass absorbs all wavelengths of light except blue. The blue wavelengths pass through the glass to your eyes. If the blue glass is translucent, it still only transmits blue light, but the image is blurry. The color of a transparent or translucent object is the color it transmits.

What color would you see if you held a piece of red cellophane, or a red filter, in front of each crayon in **Figure 5?** Remember that the color you see depends on the colors in the light source, the absorbed colors, and the colors that reach your eyes.

Combining Colors

Have you ever mixed several colors of watercolor paints to get just the shade you want? You know that you can make many different shades from a few basic colors. But if you mix too many colors, you get black! Why does that happen?

Combining Pigments Each color of paint in a set of watercolors contains different pigments, or dyes. Each pigment absorbs some colors of light and reflects other colors. Mixing pigments produces many different shades as certain wavelengths are absorbed and fewer colors are reflected to your eyes. As you add each color of pigment, the mixture gets darker and darker because more colors are absorbed. Cyan, magenta, and yellow are the primary pigments. Combining equal amounts of these pigments makes black, as shown in the center of the artist's palette in **Figure 6.**

Combining Colors of Light Red, green, and blue are the primary light colors. If you shine equal amounts of red light, green light, and blue light at a white screen, each color reflects to your eyes. Where two of the colors overlap, both wavelengths reflect to your eyes and you see a third color. Where the three colors overlap, all colors reflect and you see white light, also shown in **Figure 6**.

 6. **NGSSS Check** Summarize Why do objects appear to have different colors? SC.7.P.10.1

Figure 6 Colors of pigment combine by subtracting wavelengths. Colors of light combine by adding wavelengths.

Magenta + Yellow = _____

| Active Reading | 7. **Label** Complete the color equations. Magenta + Cyan = _____ |

Cyan + Yellow = _____

Combining Pigments

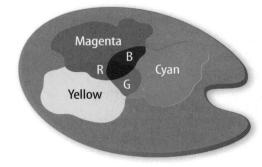

Each primary pigment subtracts color by absorption. Mixing all three pigments equally produces black.

Combining Light

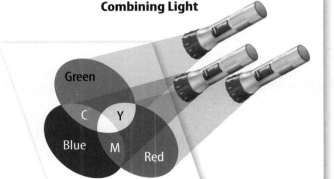

Adding the three primary colors of light produces the colors of the primary pigments and white.

Lesson Review 1

Visual Summary

The amount of light that can pass through a material determines whether it is transparent, translucent, or opaque.

Reflection is the bouncing of waves off a surface.

The color of an opaque object is the color of light that reflects off the object.

Use Vocabulary

1 **Compare and contrast** how an opaque object and a transparent object affect light.

_____

_____

2 The passage of light through an object is called

_____ .

3 An electromagnetic wave that you can see is

_____ .

Understand Key Concepts 🔑

4 Which indicates that light travels in straight lines? SC.7.P.10.1

  (A) colors   (C) shadows

  (B) pigments   (D) waves

5 **Contrast** how white light interacts with a blue book and a blue stained-glass window. SC.7.P.10.2

_____

_____

6 **Describe** three things that can happen when light strikes an object. SC.7.P.10.2

_____

_____

Interpret Graphics

7 **Summarize** Use the graphic organizer below. Fill in three ways you can describe a material according to whether light can pass through it. LA.6.2.2.3

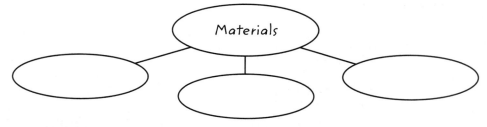

Critical Thinking

8 **Construct** a diagram showing what happens when rays of white light strike a yellow, opaque object. SC.7.P.10.1

Summarize each term below in your own words. Use your classroom to find an example of each and circle how light interacts with matter to produce the effect.

Translucent

3.

Definition

Example

Reflected

Transmitted

Absorbed

Opaque

2.

Definition

Example

Reflected

Transmitted

Absorbed

Transparent

1.

Definition

Example

Reflected

Transmitted

Absorbed

Reflection and MIRRORS

ESSENTIAL QUESTIONS

 How does light reflect from smooth surfaces and rough surfaces?

 What happens to light when it strikes a concave mirror?

 Which types of mirrors can produce a virtual image?

Vocabulary

law of reflection p. 671

regular reflection p. 672

diffuse reflection p. 672

concave mirror p. 673

focal point p. 673

focal length p. 673

convex mirror p. 674

 Florida NGSSS

SC.6.N.3.3 Give several examples of scientific laws.

SC.7.N.1.1 Define a problem from the seventh grade curriculum, use appropriate reference materials to support scientific understanding, plan and carry out scientific investigation of various types, such as systematic observations or experiments, identify variables, collect and organize data, interpret data in charts, tables, and graphics, analyze information, make predictions, and defend conclusions.

SC.7.N.1.6 Explain that empirical evidence is the cumulative body of observations of a natural phenomenon on which scientific explanations are based.

SC.7.P.10.2 Observe and explain that light can be reflected, refracted, and/or absorbed.

LA.6.2.2.3 The student will organize information to show understanding or relationships among facts, ideas, and events (e.g., representing key points within text through charting, mapping, paraphrasing, summarizing, or comparing/contrasting);

 Inquiry Launch Lab   SC.7.P.10.2, SC.7.N.1.1

10 minutes

How can you read a sign behind you?

Mirrors aren't just for looking at your reflection. You can use them to see around corners, over and under obstacles, and inside things.

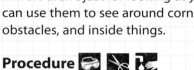

Procedure

1. Read and complete a lab safety form.
2. Use a **marker** to write *CAT* about 5 cm tall on an **index card.** Use **tape** to attach the card to a wall.
3. Stand with your back toward the wall. Use a **mirror** to look at the letters on the card. Sketch what you see.
4. Repeat step 3, but this time use **an additional mirror** to look at the letters in the first mirror.

Data and Observations

Think About This

1. Draw a simple diagram showing the path of light from the card to your eye for each setup.

2. **Key Concept** How did one reflection in a mirror affect the letters? How did two reflections affect the letters?

1. When you look in a mirror in your home, your image is very much like you. But these people are looking at their images in a carnival's fun-house mirror. Some parts of their images are much larger than they should be. Other parts are much smaller. What do you think causes the images to appear so strange?

_____

_____

_____

Reflection of Light

In Lesson 1, you read that reflection is the bouncing of a wave off a surface. Suppose you toss a tennis ball against a wall. If you throw the ball straight toward the wall, it will bounce back to you. Where on the wall would you throw the ball so that a friend standing to your left could catch it? You would throw it toward a point on the wall halfway between you and your friend.

Law of Reflection

Like the tennis ball, light behaves in predictable ways when it reflects. Scientists often model the path of light using straight arrows called rays. The rays in **Figure 7** show how light reflects. An imaginary line perpendicular to a reflecting surface is called the normal. The light ray moving toward the surface is the incident ray. The light ray moving away is the reflected ray.

Notice the angle formed where an incident ray meets a normal. This is the angle of incidence. A reflected ray forms an identical angle on the other side of the normal. This angle is the angle of reflection. According to the **law of reflection**, *when a wave is reflected from a surface, the angle of reflection is equal to the angle of incidence.*

2. **NGSSS Check Review** Underline how reflected and incident rays differ. SC.7.P.10.2

Figure 7 🔑 If you know a light ray's angle of incidence, you can predict its angle of reflection.

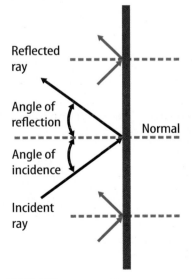

Active Reading 3. **Predict** If the angle of incidence is 40°, what is the angle of reflection?

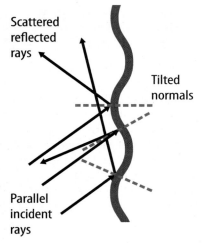

Figure 8 🔑 No clear image forms when light reflects off a rough surface.

Active Reading

FOLDABLES® LA.6.2.2.3

Make a horizontal two-tab concept map book. Label it as shown. Use it to compare the reflection of light on different types of surfaces.

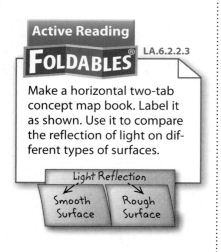

Light Reflection
Smooth Surface | Rough Surface

Figure 9 🔑 A plane mirror forms a virtual image.

Regular and Diffuse Reflection

You see objects when light reflects off them into your eyes. Why can you see your reflection in smooth, shiny surfaces but not in a piece of paper or on a painted wall? The law of reflection applies whether the surface is smooth or rough.

Reflection of light from a smooth, shiny surface is called **regular reflection.** Look again at **Figure 7** on the previous page. The three incident rays and the three reflected rays all are parallel. You see a sharp image when parallel rays reflect into your eyes. When light strikes an uneven surface, as in **Figure 8,** the angle of reflection still equals the angle of incidence at each point. However, rays reflect in different directions. *Reflection of light from a rough surface is called* **diffuse reflection.**

 4. NGSSS Check Contrast How do regular and diffuse reflection differ? SC.7.P.10.2

_____

_____

_____

Mirrors

Any surface that reflects light and forms images is a mirror. The type of image depends on whether the reflecting surface is flat or curved.

Plane Mirrors

The word *plane* means "flat," so a plane mirror has a flat reflecting surface. **Figure 9** shows how an image from a plane mirror forms. Only a few rays are shown, but the number of rays involved is infinite. Notice that the image formed is the same size as the object. However, it is a virtual image. A virtual image is an image of an object that your brain perceives to be in a place where the object is not.

If you look in a plane mirror and raise your right hand, your image raises its left hand. However, up and down are not reversed.

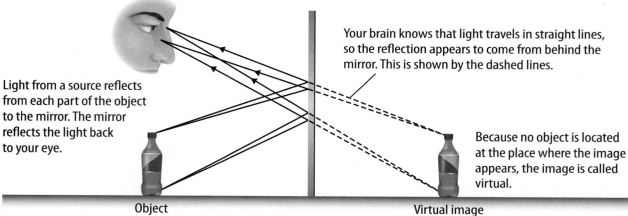

Your brain knows that light travels in straight lines, so the reflection appears to come from behind the mirror. This is shown by the dashed lines.

Light from a source reflects from each part of the object to the mirror. The mirror reflects the light back to your eye.

Because no object is located at the place where the image appears, the image is called virtual.

Object   Virtual image

Apply It! After you complete the lab, answer these questions.

1. **Analyze** As you stand in front of a plane mirror, you can see a virtual image of yourself. Why is the image called virtual?

2. **Classify** Which directions (left, right, top, bottom) are reversed when you view an object in a plane mirror? Which directions are not reversed?

Concave Mirrors

Not all mirrors are flat. *A mirror that curves inward is called a* **concave mirror,** like the mirror in **Figure 10.** A line perpendicular to the center of the mirror is the optical axis. The law of reflection determines the direction of reflected rays in **Figure 10.** When rays parallel to the optical axis strike a concave mirror, the reflected rays converge, or come together.

Focal Point Look again at **Figure 10.** Notice the point where the rays converge. *The point where light rays parallel to the optical axis converge after being reflected by a mirror or refracted by a lens is the* **focal point.** In the next lesson, you will read about lenses. Imagine that a concave mirror is part of a hollow sphere. The focal point is halfway between the mirror and the center of the sphere. *The distance along the optical axis from the mirror to the focal point is the* **focal length.** The lesser the curve of a mirror, the longer its focal length. The position of an object compared to the focal point determines the type of image formed by a concave mirror.

If you reverse the direction of the arrows in **Figure 10,** you can see how a flashlight works. The reflector behind the bulb is a concave mirror. The bulb is at the focal point. Light rays from the bulb strike the mirror and reflect as parallel rays.

 5. **NGSSS Check** Explain How does a concave mirror reflect light? SC.7.P.10.2

_____

_____

WORD ORIGIN

concave
from Latin *concavus,* means "hollow"

Figure 10 Light rays that strike a concave mirror converge.

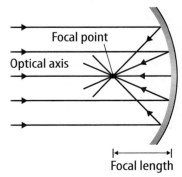

Focal point

Optical axis

|←——→|
Focal length

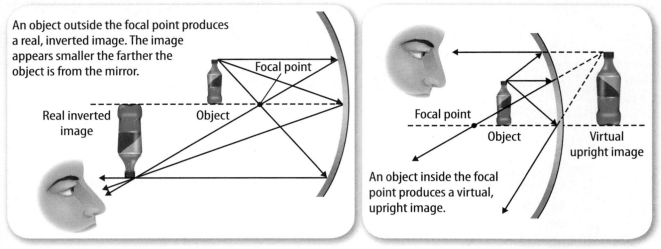

An object outside the focal point produces a real, inverted image. The image appears smaller the farther the object is from the mirror.

Focal point

Real inverted image

Object

Focal point

Object

Virtual upright image

An object inside the focal point produces a virtual, upright image.

Figure 11 🔑 A concave mirror can produce a real or a virtual image.

Figure 12 The image flips as the spoon moves away from the object.

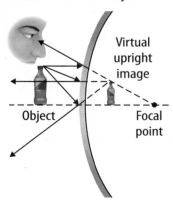

Figure 13 🔑 The image in a convex mirror looks smaller than the object.

Virtual upright image

Object

Focal point

Types of Images When you look at ray diagrams, think about what you would see if the rays reflected into your eyes. Remember that your brain only perceives where the rays appear to come from, not their actual paths. As shown in **Figure 11,** the image a concave mirror forms depends on the object's location relative to the focal point. The image is virtual if the object is between the focal point and the mirror. The image is real if the object is beyond the focal point. A real image is one that forms where rays converge. No image forms if the object is at the focal point.

Suppose you look at your reflection in the bowl of a shiny spoon. If your face is outside the spoon's focal point, your image is inverted, or upside down, as shown in **Figure 12.** Your image disappears when your face is at the focal point. When your face is inside the focal point, the image is upright.

Convex Mirrors

Have you ever seen a large, round mirror high in the corner of a store? The mirror enables store personnel to see places they cannot see with a plane mirror. *A mirror that curves outward, like the back of a spoon, is called a* **convex mirror.** Light rays diverge, or spread apart, after they strike the surface of a convex mirror. The dashed lines in **Figure 13** model how your brain interprets these rays as coming from a smaller object behind the mirror. Therefore, a convex mirror always produces a virtual image that is upright and smaller than the object.

As you just read, convex mirrors and plane mirrors form only virtual images. However, concave mirrors can form both virtual images and real images.

Active Reading 6. **List** Which types of mirrors can form virtual images?

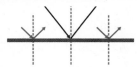

The angle of reflection equals the angle of incidence, even if a surface is rough.

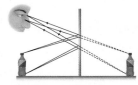

The reflection of an object in a plane mirror is a virtual image the same size as the object.

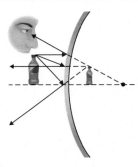

The image formed by a curved mirror might have a different size and might be inverted.

Inquiry SC.7.P.10.2, SC.7.N.1.1, SC.7.N.1.6

⊙LAB STATION Try It!

Skill Lab *How can you demonstrate the law of reflection?* at connectED.mcgraw-hill.com

Use Vocabulary

1 Distinguish between the focal point and the focal length of a mirror.

_____

_____

2 The reflection in a clear window of a store is a(n)

_____.

3 A mirror that converges light rays is a _____ mirror.

Understand Key Concepts 🔑

4 Which term describes reflection from a rough surface?

(A) diffuse   (C) regular

(B) irregular   (D) translucent

5 Which terms describe the image of an object located between a concave mirror and its focal point?

(A) real, inverted   (C) virtual, inverted

(B) real, upright   (D) virtual, upright

6 Explain why the image in a plane mirror is a virtual image.

_____

_____

Interpret Graphics

7 Organize Information Fill in the graphic organizer below describing how real and virtual images are produced by different types of mirrors. LA.6.2.2.3

Type of Mirror	How Images Form

Critical Thinking

8 Evaluate The rearview mirror on a car is often slightly convex. Explain why this is more practical than a plane mirror.

_____

_____

From Galileo to Hubble

How mirrors help see the universe

Since the days of Galileo in the 17th century, humans have always longed to see deeper into the unknown universe. People have had a desire to see stars and black holes and to know about what exists outside our small, blue planet. Until the late 20th century, however, astronomers and scientists have struggled to get a clear picture of stars. Earth's atmosphere creates a distortion of stars that we see even with the most powerful telescopes. In 1990, NASA launched a revolutionary new idea, the *Hubble Space Telescope*. Launched from Kennedy Space Center in Cape Canaveral, FL and deployed from the *Space Shuttle Discovery*, the *Hubble Space Telescope* (named after astronomer Edwin Hubble) was designed to orbit Earth, thus escaping the atmosphere.

After 20 years of service, the *Hubble Space Telescope* has provided an array of dazzling images and led to new understandings of our universe.

How does the *Hubble Telescope* work?

The way that *Hubble* works is quite amazing and simple. The telescope orbits the Earth every 97 minutes. As it travels, it captures light and directs it to several instruments.

Hubble is a reflector telescope that uses only mirrors to focus the light. Light enters the telescope, and hits the main, primary concave mirror. It bounces off the primary mirror, and then hits a secondary convex mirror. The secondary mirror focuses the light and directs it through a hole in the center of the primary mirror that leads to the telescope's instruments such as cameras and spectrographs (light measuring devices). The instruments then send the data back to the scientists at NASA.

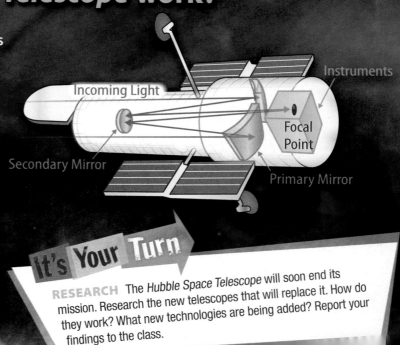

Incoming Light

Instruments

Focal Point

Secondary Mirror

Primary Mirror

It's Your Turn

RESEARCH The *Hubble Space Telescope* will soon end its mission. Research the new telescopes that will replace it. How do they work? What new technologies are being added? Report your findings to the class.

Refraction and LENSES

ESSENTIAL QUESTIONS

 What happens to light as it moves from one transparent substance to another?

 How do convex lenses and concave lenses affect light?

Vocabulary

refraction p. 678

lens p. 680

convex lens p. 680

concave lens p. 680

rod p. 685

cone p. 685

 Florida NGSSS

SC.7.N.1.3 Distinguish between an experiment (which must involve the identification and control of variables) and other forms of scientific investigation and explain that not all scientific knowledge is derived from experimentation.

SC.7.N.1.6 Explain that empirical evidence is the cumulative body of observations of a natural phenomenon on which scientific explanations are based.

SC.7.N.1.3 Distinguish between an experiment (which must involve the identification and control of variables) and other forms of scientific investigation and explain that not all scientific knowledge is derived from experimentation.

SC.7.N.1.6 Explain that empirical evidence is the cumulative body of observations of a natural phenomenon on which scientific explanations are based.

SC.7.P.10.1 Illustrate that the sun's energy arrives as radiation with a wide range of wavelengths, including infrared, visible, and ultraviolet, and that white light is made up of a spectrum of many different colors.

SC.7.P.10.2 Observe and explain that light can be reflected, refracted, and/or absorbed.

SC.7.P.10.3 Recognize that light waves, sound waves, and other waves move at different speeds in different materials.

Also covers: LA.6.2.2.3, MA.6.A.3.6 , SC.7.N.1.1

(inquiry) **Launch Lab**

SC.7.N.1.6, SC.7.P.10.2

15 minutes

What happens to light that passes from one transparent substance to another?

Objects look different in water than in air. Light that reflects off objects changes when it moves from one transparent material to another.

Procedure

1. Read and complete a lab safety form.

2. Pour about 150 mL of water into a **250-mL beaker.** Pour 150 mL of **vegetable oil** into a **second beaker.**

3. Place three **test tubes** in a **test-tube rack.** Leave one test tube empty. Half fill one with water. Half fill another with vegetable oil.

4. One at a time, place each test tube into the water beaker. Look through the side of the beaker at the test tube.

5. Remove and dry the test tubes with a **paper towel.**

6. Repeat steps 4–5 using the oil beaker.

Think About This

1. Draw a diagram showing the substances light passed through as it moved from the test tube to your eye for each setup. Remember to include the glass of the test tube and the beaker.

2. **Key Concept** Summarize your observations of the test tubes of air, water, and oil.

 **Why is its head so big?**

1. The cat's head might look gigantic compared to its body, but it is just an illusion. Light can produce interesting effects when it moves from one transparent material, such as glass, to another transparent material, such as water. Why do you think this happens?

_____

_____

_____

_____

Table 1 The index of refraction indicates how much a medium can change the direction of light.

Table 1	
Medium	**Index of Refraction**
Vacuum	1.0000
Air	1.0003
Ice	1.31
Water	1.333
Oil	1.47
Ovenproof glass	1.47
Diamond	2.417

Refraction of Light

Have you ever tried to pick up something from the bottom of a container of water and the object was deeper than you thought it was? This happens because light waves can change direction. Light always travels through empty space at the same speed—300,000 km/s. Light travels more slowly through a medium (plural, media), such as air, glass, or water. The atoms of the material interact with the light waves and slow them down. Some substances, such as air, only slow light a little. Others, such as glass, slow the light more.

As a light wave moves from one medium into another, its speed changes. If it enters the new medium at an angle, the wave will change direction. _The change in direction of a wave as it changes speed while moving from one medium to another is called_ **refraction.**

Each transparent material has a property called the index of refraction, as shown in **Table 1.** A medium that has a high index of refraction is sometimes called slow because light moves more slowly through it. A medium that has a relatively low index of refraction, such as air, is called fast.

 2. NGSSS Check **Classify** Based on **Table 1,** rank these four examples of transparent media from fastest to slowest according to their placement on the index of refraction. SC.7.P.10.3

	Air	**Diamond**	**Oil**	**Water**	
Fast					Slow

Moving Into a Slower Medium

Suppose you roll a toy car across a table straight at a piece of fabric. The front tires of the car slow down when they hit the fabric. The car continues to move in a straight line but more slowly. If you roll the car at an angle, one of the front tires will hit the fabric before the other. That side of the car will slow down, but the rest of the car will continue at the same speed until the other tire hits the fabric. This will cause the car to turn and change direction.

As shown in **Figure 14,** a light wave behaves in a similar way when it moves into a slower medium. Recall that a normal is a line perpendicular to a surface. As light moves into a slower medium at an angle, it changes direction toward the normal.

Moving Into a Faster Medium

What happens when light in the figure moves back into the air? Suppose you ride your bike from a sidewalk into a muddy field and then back onto a sidewalk. You use the same energy to pedal the whole time, but you move more slowly in the mud. When you move back onto the sidewalk, you speed up.

Similarly, as light moves into a medium with a lower index of refraction, it speeds up. The wave is still at an angle, so the part that leaves the slower medium first, will speed up sooner. This causes the wave to turn away from the normal, as shown on the right in **Figure 14.**

You see the boundaries between surfaces such as air, glass, and water because of refraction. If transparent substances have the same index of refraction, you do not see the surfaces. The Launch Lab at the start of this lesson shows this effect.

Math Skills   MA.6.A.3.6

Use Scientific Notation

You can calculate the index of refraction of a material by dividing the speed of light in a vacuum (3.0×10^8 m/s) by the speed of light in that material. When dividing numbers written in scientific notation, divide the coefficients and subtract the exponents. For example, what is the index of refraction of a substance in which light travels at 2.0×10^8 m/s?

a. Set up the problem.

Index of refraction
$$= \frac{3.0 \times 10^8 \text{ m/s}}{2.0 \times 10^8 \text{ m/s}}$$

Note that units cancel.

b. Divide the coefficients.

$$3.0 \div 2.0 = 1.5$$

c. Subtract the exponents.

$$8 - 8 = 0$$

The index of refraction is $1.5 \times 10^0 = 1.5 \times 1 = 1.5$.

Practice

3. The speed of light in a material is 1.56×10^8 m/s. What is its index of refraction?

Refraction 🔑

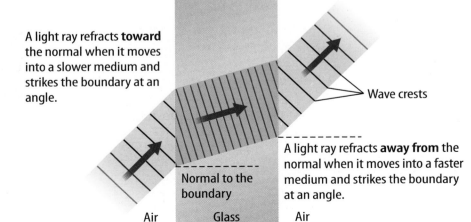

A light ray refracts **toward** the normal when it moves into a slower medium and strikes the boundary at an angle.

Wave crests

Normal to the boundary

A light ray refracts **away from** the normal when it moves into a faster medium and strikes the boundary at an angle.

Air   Glass   Air

Figure 14 The direction of refraction depends on each material's index of refraction.

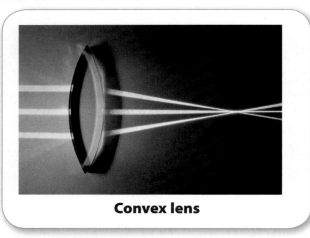

Convex lens

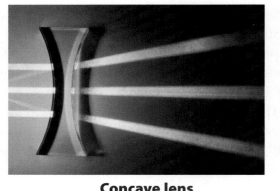

Concave lens

Figure 15 A convex lens often is called a converging lens. A concave lens often is called a diverging lens.

Lenses

What do binoculars, eyeglasses, your eyes, and a camera have in common? Each contains a lens. *A* **lens** *is a transparent object with at least one curved side that causes light to change direction.* Recall that most of the light that strikes a transparent material passes through it. Light refracts as it passes through a lens. The greater the curve of the lens, the more the light refracts. The direction of refraction depends on whether the lens is curved outward or inward.

As shown in **Figure 15,** light passes through two different lenses. *A lens that is thicker in the middle than at the edges is a* **convex lens.** Notice that the light rays that move through a convex lens come together, or converge. *A lens that is thicker at the edges than in the middle is a* **concave lens.** Notice that the light rays spread apart, or diverge, as they move through a concave lens.

Active Reading **4. Explain** Describe the shapes of a convex lens and a concave lens.

_____

_____

Inquiry

SC.7.P.10.2, SC.7.N.1.1

LAB STATION **Try It!**

MiniLab *How can water move light?* at connectED.mcgraw-hill.com

Apply It! After you complete the lab, answer these questions.

1. Read the section of text about lenses. Then, compare and contrast the clear plastic box of water and a curved lens.

2. In the MiniLab, which medium is faster? Which is slower? Explain your reasoning.

Convex Lenses

Have you ever used a magnifying lens to look at a tiny insect? A magnifying lens makes things appear larger. If you look closely, you can see that a magnifying lens is a convex lens because its center is thicker than its edges.

The refraction of light by a convex lens is shown in **Figure 16.** Notice that a normal to the curved surface slants toward the optical axis. Recall that light moving into a slower **medium** turns toward the normal. As a result, a convex lens refracts light inward and it converges.

Focal Point and Focal Length Similar to a mirror, the point where rays parallel to the optical axis converge after passing through a lens is the focal point. The distance along the optical axis between the lens and the focal point is the focal length of the lens. Because you can look through a lens from either side, a focal point is on both sides of the lens. For a lens with the same curve on both sides, the lens's two focal points are the same distance from it.

 5. NGSSS Check Recall <u>Underline</u> what happens to light as it passes through a convex lens. SC.7.P.10.2

Types of Images Like a concave mirror, the type of image a convex lens forms depends on the location of the object. A convex lens can form both real and virtual images, as shown in **Figure 17.** The diagrams show only two rays, but remember that in reality there are an infinite number of rays.

If you look through a magnifying lens at an object more than one focal length from the lens, the image you see is inverted and smaller. If you look at an object less than one focal length from the lens, the image you see is upright and larger. It is virtual because your brain interprets the rays as moving in a straight line, as shown by the dashed lines in **Figure 17.**

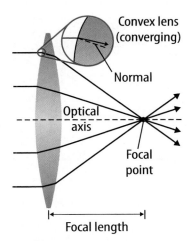

Figure 16 🔑 The outward curve of a convex lens refracts light rays inward.

SCIENCE USE V. COMMON USE

medium
Science Use matter through which a wave travels
Common Use middle or average

Figure 17 🔑 You can see either a virtual image or a real image if you look through a convex lens.

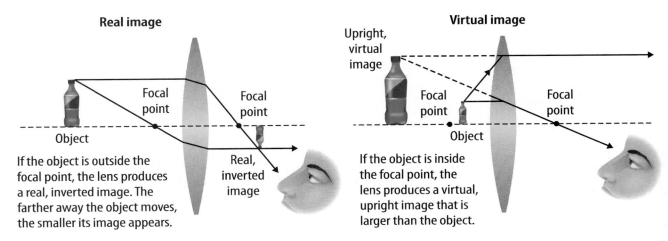

Real image

If the object is outside the focal point, the lens produces a real, inverted image. The farther away the object moves, the smaller its image appears.

Virtual image

If the object is inside the focal point, the lens produces a virtual, upright image that is larger than the object.

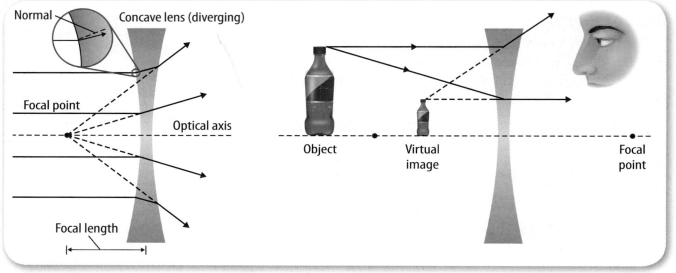

Normal Concave lens (diverging)

Focal point

Optical axis

Focal length

Object Virtual image Focal point

Figure 18 🔑 The inward curve of a concave lens refracts light rays outward. You see a virtual image if you look through a concave lens.

6. NGSSS Check
Describe What happens to light as it passes through a concave lens? SC.7.P.10.2

_____

_____

_____

Figure 19 Each color of light has a different wavelength and frequency.

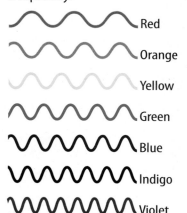

Red

Orange

Yellow

Green

Blue

Indigo

Violet

Concave Lenses

Light rays that pass through a concave lens diverge, or spread apart. The drawing on the left in **Figure 18** shows why. Notice that a normal to the curved surface slants away from the optical axis. Because light entering a slower medium changes direction toward the normal, the lens refracts light outward. The light diverges.

The drawing on the right in **Figure 18** shows the type of image a concave lens forms. Suppose you stand to the right of the lens and look at the object. The refracted rays reach your eyes, but your brain assumes the light traveled the straight dashed lines. You see a virtual image that is smaller than the object.

Refraction and Wavelength

You have read that the curved surfaces of lenses affect the refraction of light rays. The wavelength of light also affects the amount of refraction. You might have seen sparkling cut-glass ornaments hanging in a window. When sunlight strikes the glass, rainbows appear on the wall. These "sun catchers" work because different wavelengths of light refract different amounts.

You read in Lesson 1 that white light is made up of different colors. A few colors are shown in **Figure 19.** The speed of a wave in a material is related to its wavelength. Waves with longer wavelengths travel at greater speeds in a material than waves with shorter wavelengths. Therefore, when entering a material, light with longer wavelengths travels faster and refracts less than light with shorter wavelengths. As a result, violet refracts the most because its wavelength is the smallest. Red has the longest wavelength and refracts the least.

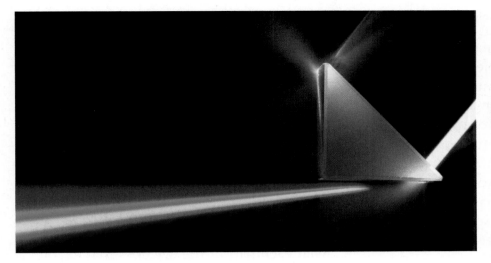

Figure 20 A prism spreads light into different wavelengths.

Active Reading **7. Review** Which color has a longer wavelength—yellow or blue?

Prisms

As white light passes through a piece of glass called a prism, each wavelength of light refracts when it enters and again when it leaves. Because each color refracts at a slightly different angle, the colors separate. They spread out into the familiar spectrum, as shown in **Figure 20.**

Rainbows

Why do rainbows appear in the sky only during or after a rain shower? Rainbows form when water droplets in the air refract light like a prism. Each wavelength of light refracts as it enters the droplet, reflects back into the droplet, and refracts again when it leaves the droplet. Notice in **Figure 21** that wavelengths of light near the blue end of the spectrum refract more than wavelengths near the red end of the spectrum. This effect produces the separate colors you see in a rainbow.

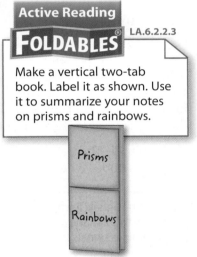

Active Reading

FOLDABLES® LA.6.2.2.3

Make a vertical two-tab book. Label it as shown. Use it to summarize your notes on prisms and rainbows.

Prisms

Rainbows

Figure 21 Sunlight refracts as it passes into and out of a raindrop.

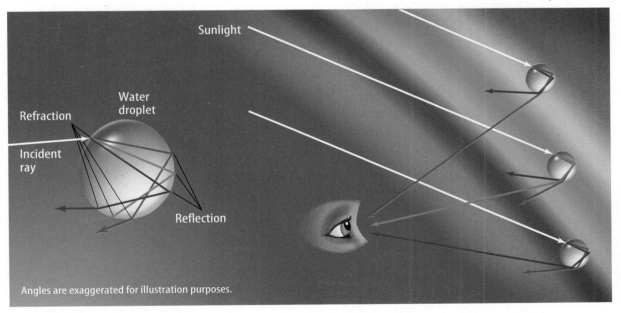

Sunlight

Water droplet

Refraction

Incident ray

Reflection

Angles are exaggerated for illustration purposes.

Active Reading **8. Hypothesize** Why do you have to stand with the Sun behind you to see a rainbow?

Detecting Light

You receive much information about the world around you when objects either emit light or reflect light into your eyes. You have read that the material that makes up an object reacts to various wavelengths of light in different ways. The material absorbs some wavelengths of light. Other wavelengths of light reflect to your eyes, providing information about shapes and colors. Your brain interprets this light energy so that you can recognize the images you see as people, places, and objects.

The Human Eye

Eyes respond quickly to changing conditions. As shown in **Figure 22,** the iris changes the size of the pupil, which controls the amount of light that enters an eye. The cornea acts as a convex lens and focuses light when it enters the eye. The shape of the cornea cannot change, but the shape of the eye's lens can. Changes to the shape of the lens can alter its focal point. This can enable a person to see objects clearly either far away or near. Follow the steps in **Figure 22** to learn what each part of the eye does that enables a person to see.

How the Eye Works 🔑

Figure 22 The parts of an eye work together to focus light and send signals about what you see to the brain.

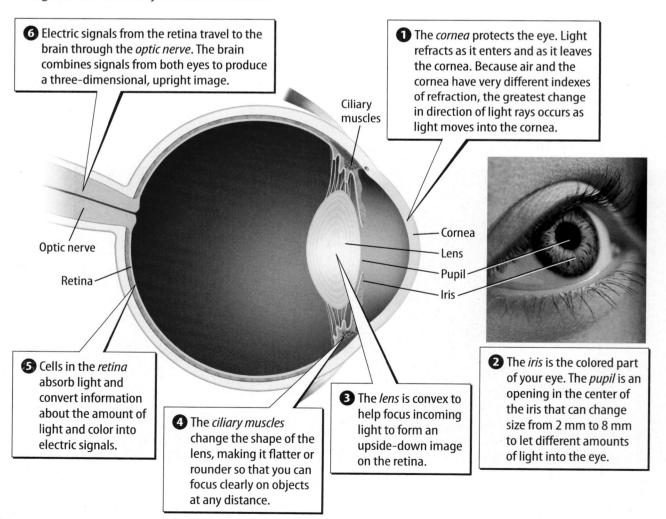

6 Electric signals from the retina travel to the brain through the *optic nerve.* The brain combines signals from both eyes to produce a three-dimensional, upright image.

1 The *cornea* protects the eye. Light refracts as it enters and as it leaves the cornea. Because air and the cornea have very different indexes of refraction, the greatest change in direction of light rays occurs as light moves into the cornea.

Ciliary muscles

Optic nerve

Retina

Cornea

Lens

Pupil

Iris

5 Cells in the *retina* absorb light and convert information about the amount of light and color into electric signals.

4 The *ciliary muscles* change the shape of the lens, making it flatter or rounder so that you can focus clearly on objects at any distance.

3 The *lens* is convex to help focus incoming light to form an upside-down image on the retina.

2 The *iris* is the colored part of your eye. The *pupil* is an opening in the center of the iris that can change size from 2 mm to 8 mm to let different amounts of light into the eye.

Seeing Color

You might have noticed that it is harder to see colors in dim light. Why does this happen? Different cells on the retina of an eye enable a person to see colors and shades of gray. *A* **rod** *is one of many cells in the retina of the eye that responds to low light.* There are about 120 million rods in the human retina. Rods enable you to see near the sides of your eyes rather than along the direct line of vision. This type of eyesight is called peripheral [puh RIH fuh rul] vision.

A **cone** *is one of many cells in the retina of the eye that respond to colors.* The human retina contains 6 million to 7 million cones that require more light to produce signals. These cones respond to red, green, and blue wavelengths of light. Recall that these primary colors of light can combine and form all the other colors. When light strikes a cone, it produces an electric signal that depends on the light's intensity and its wavelength. Your brain combines signals from all rods and cones and forms the colors, the shapes, and the brightness of the objects you see, as shown in **Figure 23.**

Correcting Vision

In a person with normal eyesight, the cornea and the lens focus light directly on the retina. This forms a sharp, clear image. Some eyes, however, have an irregular shape. If an eye is too long, light focuses in front of the retina. This person sees near objects clearly, but objects that are far away are blurred. In other instances, the eye can be too short, and light focuses behind the retina. People with this problem see faraway objects clearly, but nearby objects are blurred. **Figure 24** shows how manufactured lenses can correct these problems.

You have read that refraction and lenses enable you to see. Light can refract as it moves from one medium into another. Different colors refract at different angles. A lens causes light to change direction. An example is the focusing of light onto the retina by the cornea and lens of the eye. Eyeglasses or contact lenses may be needed to help focus this light.

Figure 23 The ability to distinguish colors allows you to see more detail in your surroundings.

Active Reading **9. Analyze** Compare and contrast rods and cones.

Rod
Both
Cone

Figure 24 Some people have eyes with irregular shapes that require lenses to correct their vision.

 Active Reading **10. Identify** Complete the sentences below.

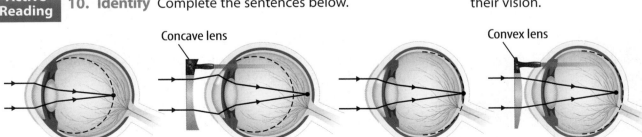

A person is **nearsighted** when light focuses in _____ of the retina. A **concave** lens corrects the problem.

A person is **farsighted** when light focuses _____ the retina. A **convex** lens corrects the problem.

Visual Summary

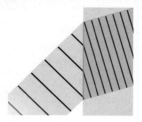

Refraction is the change in direction of a wave as it moves from one medium into another at an angle.

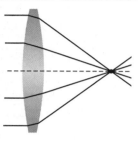

A lens changes the direction of light rays that pass through it.

Each color of light refracts differently as it passes through a prism.

Inquiry SC.7.P.10.2, SC.7.N.1.1, SC.7.N.1.3

iLAB STATION Try It!

Skill Lab *How does a lens affect light?* at connectED.mcgraw-hill.com

Use Vocabulary

1 **Identify** two differences in rods and cones.

_____

_____

2 **Describe** a convex lens in your own words.

_____

_____

Understand Key Concepts 🔑

3 Which terms describe the image produced by a concave lens? SC.7.P.10.2

(A) real, inverted   (C) virtual, inverted

(B) real, upright   (D) virtual, upright

4 **Contrast** refraction when light moves into an area of higher and lower index of refraction. SC.7.P.10.2

_____

_____

Interpret Graphics

5 **Sequence** Fill in the graphic organizer below to show what happens as light enters and interacts with parts of the eye and allows you to see an image. Add boxes as necessary. LA.6.2.2.3

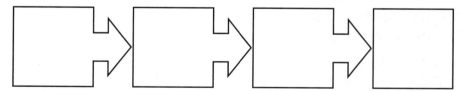

Critical Thinking

6 **Predict** How would an object appear if you looked at it through a concave lens and a convex lens at the same time? Why? SC.7.P.10.2

_____

_____

Math Skills   MA.6.A.3.6

7 Light travels through a sheet of plastic at 1.8×10^8 m/s. What is the index of refraction of the plastic?

Color!

There's more to it than meets the eye.

Do you have a favorite color? If you had to explain why it's your favorite, what would you say? You know that an object's color is due to the wavelengths of light it reflects, but you might be surprised to learn that color is more than something you see. Color can affect your mood, your emotions, and even your creativity at school or work!

Color psychology is the study of the effect of color on the way people think and behave. For example, the color blue is associated with peacefulness. It tends to make people feel calm. Studies have shown that blue rooms promote creativity.

▲ **This waiting room is designed to be soothing.**

What words would you use to describe a bright yellow room? Would you say the room is sunny and cheerful? People often connect happiness with yellow and nature with green. Green, like blue, tends to calm people. Hospitals often use green in waiting rooms to reduce people's anxiety. What about orange? Many people connect orange with energy and warmth.

Facts about how color affects moods can be fun to learn. However, for some people, it is their business. Industrial psychologists study ways to maximize workers' productivity and safety. Many industrial psychologists pay close attention to the colors used in different areas of the workplace. Those who work in advertising and merchandising choose specific colors for product packages that connect the product inside with certain ideas or emotions.

So, what does your favorite color mean to you? Perhaps it changes depending on your mood or experiences. Regardless of what your favorite color is, it is interesting to know that there is science behind colors chosen for many places and products that are used daily.

▲ **Do you think that blue calms people because it makes them think about the ocean or the sky?**

Red makes people think of energy and excitement. That's why it's a popular color for sports cars.

It's Your Turn

RESEARCH AND REPORT Find out more about how a person's culture might influence the way the person perceives colors. Make a poster to share what you learn with your classmates. LA.6.4.2.2

Optical TECHNOLOGY

ESSENTIAL QUESTIONS

 What do devices such as telescopes, microscopes, and cameras have in common?

 What is laser light, and how is it used?

 How do optical fibers work, and how are they used?

Vocabulary

optical device p. 689

refracting telescope p. 690

reflecting telescope p. 690

microscope p. 690

laser p. 692

hologram p. 693

 Florida NGSSS

SC.6.N.1.5 Recognize that science involves creativity, not just in designing experiments, but also in creating explanations that fit evidence.

SC.7.N.1.1 Define a problem from the seventh grade curriculum, use appropriate reference materials to support scientific understanding, plan and carry out scientific investigation of various types, such as systematic observations or experiments, identify variables, collect and organize data, interpret data in charts, tables, and graphics, analyze information, make predictions, and defend conclusions.

SC.7.N.1.6 Explain that empirical evidence is the cumulative body of observations of a natural phenomenon on which scientific explanations are based.

SC.7.P.10.2 Observe and explain that light can be reflected, refracted, and/or absorbed.

SC.8.N.1.4 Explain how hypotheses are valuable if they lead to further investigations, even if they turn out not to be supported by the data.

LA.6.2.2.3 The student will organize information to show understanding or relationships among facts, ideas, and events (e.g., representing key points within text through charting, mapping, paraphrasing, summarizing, or comparing/contrasting);

Inquiry Launch Lab

SC.7.N.1.1

10 minutes

How do long-distance phone lines work?

Procedure

1. Read and complete a lab safety form.

2. Fill a **clear plastic box** with water.

3. In a darkened room, shine a thin **flashlight** beam through the side of the box into the water. Stir **milk** into the water one drop at a time until you can see the beam in the water.

4. Position the beam of light through the side of the box so that it is almost perpendicular to the underside of the water's surface. Record your observations of the reflected beam and the water.

5. Hold an **index card** above the box so that any light transmitted through the surface shines on it.

6. Slowly change the angle of the beam until no light is transmitted onto the card. Draw your setup, and record your observations.

Data and Observations

Think About This

1. How did the angle of the incident beam affect the amount of light that appeared on the card?

2. **Key Concept** How do you think your observations show what happens to light in an optical fiber?

Inquiry How does the light do that?

1. A laser show is a spectacular site as the lights streak through the air. The lights in your home and school fill a room with brightness, but not the lights of a laser. How do you think laser lights travel in such a straight line? What do you think makes them different from other lights?

_____

_____

_____

_____

Active Reading

FOLDABLES LA.6.2.2.3

Use two sheets of paper to make a bound book and label it as shown. Use it to organize what you have learned about different types of optical devices.

Optical Devices

Optical Devices That Magnify

Do you wear sunglasses or contact lenses? Have you ever watched an animal through binoculars or looked at a planet through a telescope? If you answered yes to any of these questions, you have used an optical device. *Any instrument or object used to produce or control light is an* **optical device.** Some optical devices use lenses to help you see small or distant objects.

Telescopes

Hundreds of years ago, people could not see details of the Moon, such as those shown in **Figure 25.** However, by using telescopes, people have learned about objects in space. Telescopes are optical devices that produce magnified images of distant objects. A telescope uses a lens or a concave mirror to gather light. You might have used a small telescope to observe the Moon or planets, but some telescopes are much larger. The largest optical telescope in the United States has a mirror that is 10 m across. This enormous size means that the telescope can gather more light from a distant object than your eye can. The *Hubble Space Telescope* orbits about 600 km above Earth. It has a clearer view because it is above Earth's atmosphere. Astronomers use it to see objects billions of light-years away.

Figure 25 A telescope gives a close-up view of the Moon's surface.

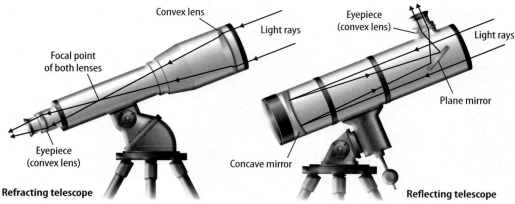

Refracting telescope

Reflecting telescope

Figure 26 🔑 Telescopes use either a lens or a mirror to gather light from a faraway source.

Refracting Telescope *A telescope that uses lenses to gather and focus light from distant objects is a* **refracting telescope.** The convex lens that gathers light is called the objective. As shown on the left in **Figure 26,** light rays that pass through the objective converge and form a real image. The convex lens that a viewer looks through is called the eyepiece. This lens magnifies the image. Recall that refraction differs slightly for different wavelengths of light. Eyepieces often include multiple lenses with different focal lengths so that all wavelengths of light focus clearly.

Reflecting Telescope *A telescope that uses a mirror to gather and focus light from distant objects is a* **reflecting telescope.** A simple reflecting telescope is shown on the right in **Figure 26.** Light enters the tube and reflects off a concave mirror at the far end. The light begins to converge, but before it converges completely, it reflects off a plane mirror toward the eyepiece. A real image forms, and the eyepiece lens magnifies the image.

Light Microscopes

Telescopes are useful for observing faraway objects, but to observe tiny objects up close, you need a different instrument. *A* **microscope** *is an optical device that forms magnified images of very small, close objects.* A microscope works like a refracting telescope, but its lenses have shorter focal lengths. As shown in **Figure 27,** light from a mirror or a light source near the microscope's base passes through a thin sample of an organism or object. The light then travels through a convex lens, called an objective, that focuses it. The eyepiece magnifies the image. Microscopes often have several objective lenses with different focal lengths for different magnifications.

WORD ORIGIN

microscope
from Latin *microscopium,* means "an instrument used to view small things"

Figure 27 🔑 The main difference in microscopes and refracting telescopes is focal length.

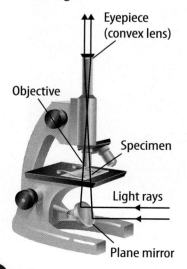

Eyepiece (convex lens)

Objective

Specimen

Light rays

Plane mirror

Active Reading **2. Contrast** How are refracting telescopes and microscopes different?

Device	Function
Refracting Telescope	
Microscope	

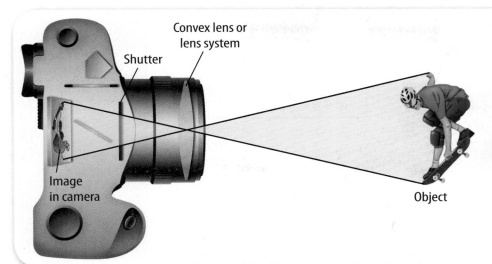

Convex lens or lens system

Shutter

Image in camera

Object

Figure 28 🗝 A digital camera uses one or more convex lenses to focus light onto a sensor.

Active Reading
3. Predict Draw the lens system's location on the camera if it were preparing to take a photograph of a more distant object.

Cameras

Cameras come in many different designs that use lenses to focus images. The design of a typical digital camera is illustrated in **Figure 28.** Suppose you want to take a photo of your friend on a skateboard. Light reflects off your friend and enters the camera. The light passes through a convex lens or, more often, a lens system. A shutter normally blocks light from entering the camera. However, pressing a button briefly opens the shutter. Light enters and is detected by an electronic sensor in the back of the camera. Recall that a convex lens is a converging lens. Inside the camera, a lens or system of lenses converges the light rays toward the sensor.

The boy you photograph is farther away from the lens than the focal point. For a convex lens, this means the image produced is real, inverted, and smaller than the object. Notice in the figure that this is the type of image that appears on the electronic sensor.

With some cameras, you can zoom, or move the lenses and increase the size of an image. Zooming changes the lenses' focal lengths. This means a camera can take enlarged images of distant or close objects.

Active Reading
4. Explain What does a camera have in common with a telescope and a microscope?

_____

_____

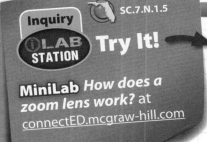

Inquiry SC.7.N.1.5
①LAB STATION **Try It!**

MiniLab *How does a zoom lens work?* at connectED.mcgraw-hill.com

Apply It!

After you complete the lab, answer the questions below.

1. What variable determines how much a zoom lens enlarges an image?

2. Contrast the distance between the object and the camera lens with the distance between the camera lens and the image produced on the camera's sensor.

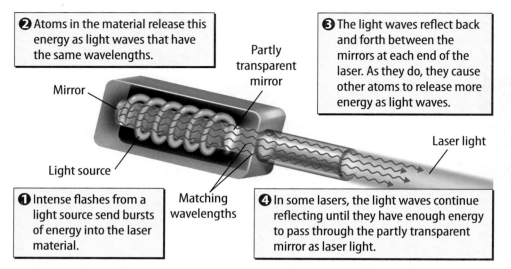

❷ Atoms in the material release this energy as light waves that have the same wavelengths.

Mirror

Light source

Partly transparent mirror

❸ The light waves reflect back and forth between the mirrors at each end of the laser. As they do, they cause other atoms to release more energy as light waves.

Laser light

❶ Intense flashes from a light source send bursts of energy into the laser material.

Matching wavelengths

❹ In some lasers, the light waves continue reflecting until they have enough energy to pass through the partly transparent mirror as laser light.

Figure 29 Light from the laser spreads much less than light from other sources because the light waves have the same wavelength.

Active Reading
5. Describe
What is a laser light?

_____

_____

_____

_____

Figure 30 Most DVDs hold over ten times more information than a CD. For example, you can record an entire movie on one DVD.

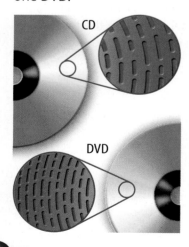

CD

DVD

Lasers

Some lasers are gentle enough to operate on the human eye, but some are powerful enough to cut through steel. *A **laser** is an optical device that produces a narrow beam of coherent light.* Recall that light from a luminous source, such as a lightbulb, has many wavelengths and spreads out in all directions. Coherent light is different. The light waves of coherent light all have the same wavelength and travel together. The beam is narrow because the waves do not spread apart, unlike light from other sources. Also, the intensity of laser light does not decrease quickly as it moves away from its source.

The word *laser* stands for <u>l</u>ight <u>a</u>mplification through <u>s</u>timulated <u>e</u>mission of <u>r</u>adiation. One type of laser is shown in **Figure 29.** An energy source causes atoms of a material in the laser to emit light. This is the stimulated emission. This light travels back and forth within a tube, causing other atoms to emit similar waves. In this way, the light intensity increases. This is the light amplification.

CDs and DVDs

Whenever you use a CD player, you use a laser. To make a CD, electric signals represent sounds. A laser burns tiny pits in the surface of the CD that correspond to the signals. In your CD player, another laser passes over the pits as the CD spins. Different amounts of light reflect back to a sensor that converts the light back into an electric signal. The signal causes the speakers to produce sound.

DVDs are similar to CDs except the DVD's pits are much smaller and closer together. Therefore, more information fits on a DVD than on a CD, as shown in **Figure 30.**

Holograms

Perhaps you've seen images on a book cover or on a credit card that seem to float in space. You can see different sides of the object, just as you would with a real object. *A **hologram** is a three-dimensional photograph of an object.*

Figure 31 shows how one type of hologram forms. First, laser light splits into two beams. One beam passes through a convex lens and then reflects from the object onto photographic film. The other beam passes through a convex lens and then travels to the film. The combined light from the two beams produces a pattern on the film that shows both the brightness of light and its direction. The pattern appears as swirls, but when a laser shines on the film, a holographic image appears.

Some paper bills and credit cards have a different type of hologram printed on them to prevent counterfeiting. The holograms are made with lasers but can be viewed under regular light. Other uses of lasers are shown in **Figure 32.**

Active Reading 6. **Review** Highlight some ways lasers are used.

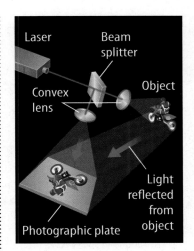

Figure 31 Interference of two laser-light beams produces a hologram.

Figure 32 🔑 Laser light is useful because its intensity does not decrease quickly.

▲ Lasers scan barcodes at stores and libraries.

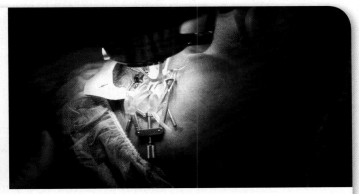

▲ Doctors use low-power lasers to correct problems in the retina.

▲ Lasers guide tunnel-boring machines to keep tunnels straight.

◄ Powerful lasers cut through metal to form machine parts or intricate designs.

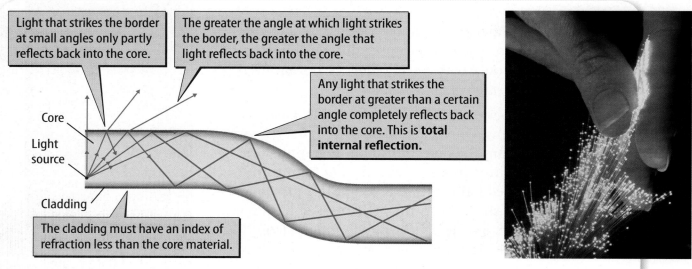

Light that strikes the border at small angles only partly reflects back into the core.

The greater the angle at which light strikes the border, the greater the angle that light reflects back into the core.

Any light that strikes the border at greater than a certain angle completely reflects back into the core. This is **total internal reflection.**

Core

Light source

Cladding

The cladding must have an index of refraction less than the core material.

Figure 33 🔑 Light moves through an optical fiber because of a phenomenon called total internal reflection.

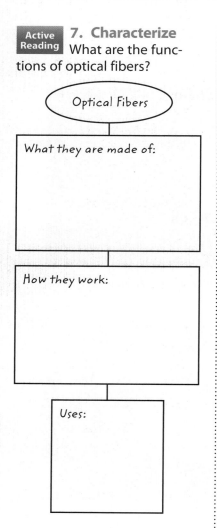

Active Reading **7. Characterize** What are the functions of optical fibers?

Optical Fibers

What they are made of:

How they work:

Uses:

Optical Fibers

Almost every time you make a phone call, use a computer, or watch a television show, you are using light technology. The electronic signals for these devices travel at least part of the way through optical fibers. An optical fiber is a thin strand of glass or plastic that can transmit a beam of light over long distances.

Total Internal Reflection Just as water moves through a pipe, light moves through an optical fiber. **Figure 33** shows how this happens. A core material surrounded by another material called the cladding makes up an optical fiber. As light moves through the core, it strikes the cladding at an angle greater than a certain angle and reflects back into the core. This process is called total internal reflection. Because light stays in the core, a fiber can carry light signals long distances without losing strength.

Uses of Optical Fibers Phone conversations, computer data, and TV signals travel as pulses of light through optical fibers. Using different wavelengths of light, one fiber can carry thousands of phone conversations at a time. Optical fibers also have medical uses, such as providing light for a physician to perform surgery through a small incision in the body.

Relating Optical Technology Each of the optical devices that you have read about uses lenses or mirrors to produce or control light. Telescopes and light microscopes produce magnified images. Cameras use lenses to focus light on an image sensor. Lasers produce coherent light. Optical fibers are also a type of optical device because they transmit and reflect light.

Visual Summary

A telescope gathers and focuses light and produces a magnified image of an object that is far away.

Lasers have many uses because they produce an intense beam of light.

Almost no light is lost as light passes through an optical fiber.

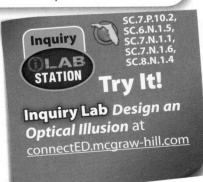

Inquiry

SC.7.P.10.2,
SC.6.N.1.5,
SC.7.N.1.1,
SC.7.N.1.6,
SC.8.N.1.4

iLAB STATION **Try It!**

Inquiry Lab *Design an Optical Illusion* at connectED.mcgraw-hill.com

Use Vocabulary

1 **Use the term** *microscope* in a sentence.

_____

2 **Define** *hologram* in your own words.

_____

_____

3 An optical device that produces a narrow beam of coherent light is a(n) _____ .

Understand Key Concepts

4 Which collects light in a reflecting telescope? SC.7.P.10.2

(A) concave lens   (C) convex lens

(B) concave mirror   (D) convex mirror

5 **Contrast** a microscope and a camera.

_____

_____

6 **Describe** how coherent light is produced in a laser. SC.7.P.10.2

_____

_____

Interpret Graphics

7 **Compare and contrast** Fill in the graphic organizer below. How are the lenses in a refracting telescope and a microscope alike and different? LA.6.2.2.3

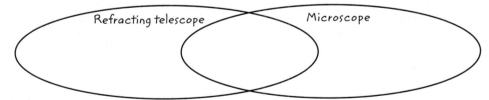

Refracting telescope   Microscope

Critical Thinking

8 **Compare and contrast** a hologram and a regular image taken by a camera. SC.7.P.10.2

_____

_____

Think About It! Energy from the Sun has a wide range of wavelengths. Visible light causes changes in the retina of the eye. Objects are perceived as light from the object is reflected, refracted, or absorbed.

Key Concepts Summary

LESSON 1 Light, Matter, and Color

- Luminous objects, such as a flashlight, are sources that produce **light** that spreads out in straight lines.
- Matter can emit, absorb, reflect, or transmit light.
- An object's color is determined by the wavelengths of light the object reflects or transmits to you.

Vocabulary

light p. 663
reflection p. 663
transparent p. 664
translucent p. 664
opaque p. 664
transmission p. 665
absorption p. 665

LESSON 2 Reflection and Mirrors

- Light rays reflect parallel from smooth surfaces and in many directions from rough surfaces.
- The shape of a mirror affects the way it reflects light. **Concave mirrors** converge light rays. **Convex mirrors** diverge light rays.
- All mirrors can produce virtual images, but only concave mirrors can produce real images.

law of reflection p. 671
regular reflection p. 672
diffuse reflection p. 672
concave mirror p. 673
focal point p. 673
focal length p. 673
convex mirror p. 674

LESSON 3 Refraction and Lenses

- Light changes direction if it moves into a medium with a different index of **refraction.**
- **Convex lenses** converge light rays. **Concave lenses** diverge light rays.
- **Rods** in the retina of the eye detect the intensity of light. **Cones** detect the colors of objects.

refraction p. 678
lens p. 680
convex lens p. 680
concave lens p. 680
rod p. 685
cone p. 685

LESSON 4 Optical Technology

- Telescopes, **microscopes,** and cameras use lenses to focus light.
- **Lasers** produce a narrow beam of coherent light. Uses of laser light include detecting information on DVDs, making **holograms,** and cutting metal.
- Light completely reflects back into optical fibers. These fibers can carry light signals over long distances.

optical device p. 689
refracting telescope p. 690
reflecting telescope p. 690
microscope p. 690
laser p. 692
hologram p. 693

Active Reading
FOLDABLES® Chapter Project

Assemble your lesson Foldables as shown to make a Chapter Project. Use the project to review what you have learned in this chapter.

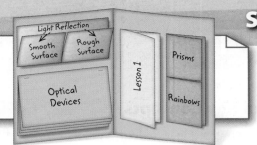

Use Vocabulary

1 You can see light but not objects clearly through a(n) _____ object.

2 Contrast opaque and transparent objects.

_____

_____

3 A mirror that causes light waves to converge is a(n)

_____ .

4 The place where rays parallel to the optical axis cross after reflecting from a mirror is called the

_____ .

5 Describe refraction in your own words.

_____

_____

6 A three-dimensional image of an object is a(n)

_____ .

7 Which optical instrument uses at least two convex lenses to magnify small, close objects?

_____

_____

Link Vocabulary and Key Concepts

Use vocabulary terms from the previous page to complete the concept map.

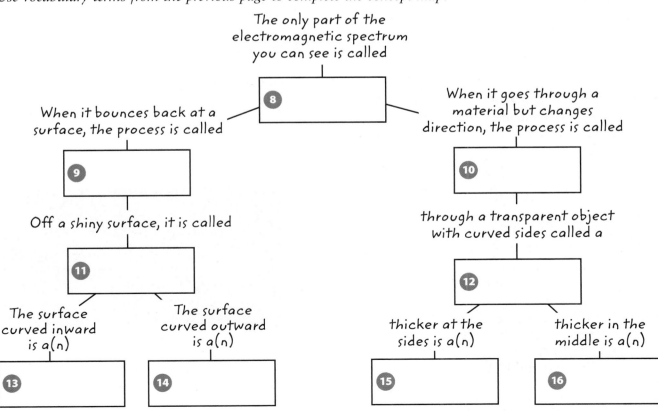

Chapter 17 Review

Fill in the correct answer choice.

🔑 Understand Key Concepts

1 Which best describes the image formed in a plane mirror? SC.7.P.10.2

(A) a real image behind the mirror
(B) a real image in front of the mirror
(C) a virtual image behind the mirror
(D) a virtual image in front of the mirror

2 Which MOST determines the color of an opaque object? SC.7.P.10.1

(A) diffraction
(B) reflection
(C) refraction
(D) transmission

3 Which describes a material that does NOT transmit any light? SC.7.P.10.2

(A) opaque
(B) reflective
(C) translucent
(D) transparent

4 If the angle of incidence of a ray striking a plane mirror is 50°, what is the angle of reflection? SC.7.P.10.2

(A) 40°
(B) 50°
(C) 100°
(D) 140°

5 Which terms describe the image that forms as light passes through the lens shown below? SC.7.P.10.2

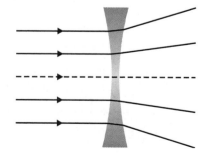

(A) real image; larger than the object
(B) real image; smaller than the object
(C) virtual image; larger than the object
(D) virtual image; smaller than the object

Critical Thinking

6 **Analyze** A girl wants to take a picture of her reflection in a plane mirror. She stands 1.3 m in front of the mirror. At what distance should she focus the camera to get a clear image of her reflection? SC.7.P.10.2

_____

_____

_____

7 **Hypothesize** How might a magician make an object seem to disappear using a concave mirror? SC.7.P.10.2

_____

_____

_____

8 **Infer** A solid, transparent object is placed in a container of clear liquid. The object is no longer visible. What could explain this effect? SC.7.P.10.2

_____

_____

_____

9 **Deduce** At sunset, the Sun is at position A, but the observer sees the Sun at position B. Why does this happen? SC.7.P.10.1

_____

_____

_____

_____

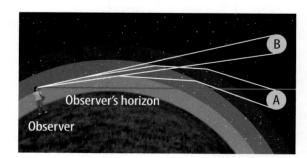

10 **Construct** a diagram showing how light forms an image in a concave mirror when the object is farther from the mirror than the focal point is. Explain the type of image that forms. SC.7.P.10.2

11 **Predict** what color a white shirt would appear to be if the light reflected from the shirt passes through a red filter and then a green filter before reaching your eye. SC.7.P.10.1

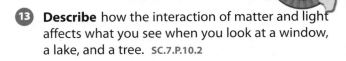

Writing in Science

12 **Write** You are an eye doctor with a patient who has problems seeing things up close. The patient's distance vision is good. On a separate sheet of paper, describe the patient's problem in detail. Identify how you would recommend solving the patient's problem. LA.6.2.2.3

Big Idea Review

13 **Describe** how the interaction of matter and light affects what you see when you look at a window, a lake, and a tree. SC.7.P.10.2

14 How does matter affect the way you perceive light in the picture on page 660? SC.7.P.10.2

Math Skills   MA.6.A.3.6

Use Scientific Notation

15 Light travels through a manufactured transparent material at about 1.0×10^8 m/s. What is the index of refraction of the material?

16 The speed of light is 1.83×10^8 m/s through a material. What is the material's index of refraction?

17 Light travels through a material at a speed of 1.42×10^8 m/s. What is the index of refraction of the material?

18 The speed of light in a sodium chloride crystal is 1.95×10^8 m/s. What is the index of refraction of sodium chloride?

19 A diamond's index of refraction is about 2.4. What is the speed of light through a diamond? [Hint: The speed of light through a substance equals the speed of light through a vacuum divided by the index of refraction.]

Fill in the correct answer choice.

Multiple Choice

1 Which describes one way concave mirrors and convex mirrors differ? SC.7.P.10.2

Ⓐ Concave mirrors are curved, and convex mirrors are flat.

Ⓑ Concave mirrors always produce a virtual image, and convex mirrors can produce a real image.

Ⓒ Convex mirrors are curved, and concave mirrors are flat.

Ⓓ Convex mirrors always produce a virtual image, and concave mirrors can produce a real image.

2 What happens to most of the light that strikes a flat transparent object? SC.7.P.10.2

Ⓕ It is absorbed by the object.

Ⓖ It is magnified by the object.

Ⓗ It is reflected by the object.

Ⓘ It is transmitted by the object.

Use the figure below to answer question 3.

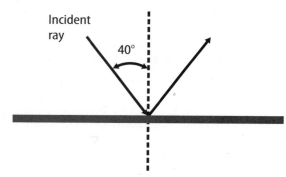

3 The figure shows how light is affected by a plane mirror. What angle does the reflected ray form with the normal? SC.7.P.10.2

Ⓐ 20°

Ⓑ 40°

Ⓒ 80°

Ⓓ 90°

Use the figure below to answer question 4.

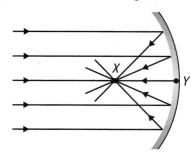

4 The figure shows how light rays are affected by a concave mirror. What is between points X and Y? SC.7.P.10.2

Ⓕ the angle of incidence

Ⓖ the focal length

Ⓗ the focal point

Ⓘ a virtual image

5 Which correctly describes an optical fiber? SC.7.P.10.2

Ⓐ The core is opaque and luminous.

Ⓑ The core is opaque and illuminates other objects.

Ⓒ The core is transparent and luminous.

Ⓓ The core is transparent and transmits light.

6 Which describes how a refracting telescope and a reflecting telescope differ? SC.7.P.10.2

Ⓕ A refracting telescope gathers light using a concave lens, and a reflecting telescope uses a convex lens.

Ⓖ A refracting telescope gathers light using a concave mirror, and a reflecting telescope uses a convex lens.

Ⓗ A refracting telescope gathers light using a convex lens, and a reflecting telescope uses a concave mirror.

Ⓘ A refracting telescope gathers light using a concave mirror, and a reflecting telescope uses a convex mirror.

7 Which describes how concave lenses and convex lenses differ? SC.7.P.10.2

(A) Concave lenses are flat on both sides. Convex lenses are curved on both sides.

(B) Concave lenses are curved on both sides. Convex lenses are flat on both sides.

(C) Concave lenses are thicker in the middle. Convex lenses are thinner in the middle.

(D) Concave lenses are thinner in the middle. Convex lenses are thicker in the middle.

Use the figure below to answer question 8.

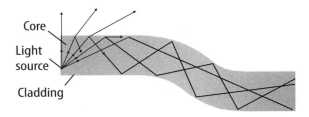

8 What determines whether light that enters the fiber will reflect back into the core? SC.7.P.10.2

(F) the angle at which the light strikes

(G) the intensity of the light

(H) the thickness of the fiber

(I) the wavelength of the light

9 What is unique about light from a laser? SC.7.P.10.2

(A) It is more spread out.

(B) It is more coherent.

(C) It has more wavelengths.

(D) Its intensity decreases faster.

10 What color of light is produced when the three primary colors are combined in equal amounts? SC.7.P.10.1

(F) black   (H) red

(G) blue   (I) white

Use the diagram below to answer questions 11–13.

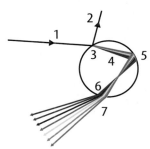

11 The diagram above shows light entering a raindrop. What produced the light represented by ray 2? SC.7.P.10.2

(A) It was absorbed by the drop.

(B) It was reflected from the drop.

(C) It was refracted by the drop.

(D) It was transmitted by the drop.

12 What caused the spread of the colors at point 7? SC.7.P.10.2

(F) absorption

(G) reflection

(H) refraction

(I) transmission

13 What process took place at position 5? SC.7.P.10.2

(A) absorption

(B) holography

(C) coherent light refraction

(D) total internal reflection

14 What term describes the degree to which a material causes light to move more slowly? SC.7.P.10.2

(A) focal length

(B) focal point

(C) index of refraction

(D) reverse

NEED EXTRA HELP?

If You Missed Question...	1	2	3	4	5	6	7	8	9	10	11	12	13	14
Go to Lesson...	2	1	3	3	4	3	3	4	4	1	2	1	4	3

Multiple Choice *Bubble the correct answer.*

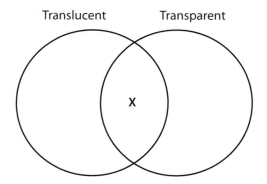

Translucent   Transparent

X

1. The Venn diagram above can be used to compare a translucent object with a transparent object. Which is X? **LA.6.2.2.3**

   (A) Most light is blocked.

   (B) Most light is reflected.

   (C) No light is transmitted.

   (D) Some light is absorbed.

2. Shadows are created when light waves are **SC.7.P.10.2**

   (F) absorbed.

   (G) blocked.

   (H) reflected.

   (I) transmitted.

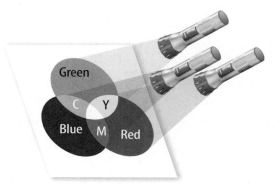

Green

C   Y

Blue   M   Red

3. Which is shown in the image above? **SC.7.P.10.2**

   (A) Adding wavelengths of the primary colors of light produces white.

   (B) Combining the primary pigments will produce white.

   (C) Combining red, blue, and green creates the primary colors of light.

   (D) Subtracting wavelengths of the primary pigments produces white.

4. An opaque object is placed in front of a spotlight. Which does NOT occur? **SC.7.P.10.2**

   (F) absorption

   (G) illumination

   (H) reflection

   (I) transmission

Copyright © Glencoe/McGraw-Hill, a division of The McGraw-Hill Companies, Inc.

Benchmark Mini-Assessment　　Chapter 17 • Lesson 2

Multiple Choice *Bubble the correct answer.*

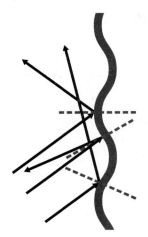

1. Which object has a surface that reflects light rays in the manner shown in the diagram above? **SC.7.P.10.2**

(A) convex mirror

(B) glass window

(C) polished silver spoon

(D) pond with ripples

2. Which can create a real, inverted image of an object? **SC.7.P.10.2**

(F) concave mirror

(G) convex mirror

(H) diffuse reflection

(I) regular reflection

3. The law of reflection states that the angle of reflection is equal to the angle of **SC.7.P.10.2**

(A) diffusion.

(B) incidence.

(C) refraction.

(D) transmission.

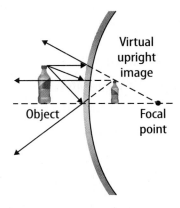

4. In the diagram above, the light rays hit the convex mirror, and then they **SC.7.P.10.2**

(F) converge.

(G) diffuse.

(H) diverge.

(I) focus.

Copyright © Glencoe/McGraw-Hill, a division of The McGraw-Hill Companies, Inc.

Multiple Choice *Bubble the correct answer.*

Medium	Index of Refraction
Vacuum	1.0000
Air	1.0003
Water	1.333
Oil	1.47

1. Using the information in the table above, which change in medium will cause a beam of light to refract away from the normal? **SC.7.P.10.2**

(A) from air into oil

(B) from air into water

(C) from a vacuum into water

(D) from water into a vacuum

2. You are looking through a magnifying glass at a tack. The tack sits beyond the focal point of the magnifying glass. Which describes what you see? **SC.7.P.10.2**

(F) larger, inverted tack

(G) larger, upright tack

(H) smaller, inverted tack

(I) smaller, upright tack

Real image

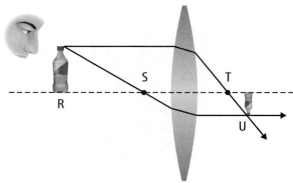

3. In the diagram above, which are the focal points? **SC.7.P.10.2**

(A) R and S

(B) R and U

(C) S and T

(D) T and U

4. Which is the result when someone wears glasses with concave lenses? **SC.7.P.10.2**

(F) The focal point moves away from the cornea.

(G) The focal point moves away from the retina.

(H) The focal point moves toward the cornea.

(I) The focal point moves toward the iris.

Copyright © Glencoe/McGraw-Hill, a division of The McGraw-Hill Companies, Inc.

Multiple Choice *Bubble the correct answer.*

1. In the illustration above, which is the objective? **SC.7.P.10.2**

(A) J

(B) K

(C) L

(D) M

2. A refracting telescope gathers and focuses light from distant objects using **SC.7.P.10.2**

(F) beam splitters.

(G) concave mirrors.

(H) convex lenses.

(I) plane mirrors.

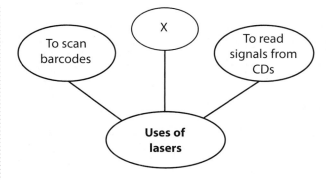

3. The graphic organizer above shows three common uses of laser light. Which does NOT go in X? **SC.7.P.10.2**

(A) to create holograms

(B) to create images in a camera

(C) to cut through metal

(D) to perform eye surgery

4. The light emitted by a laser is a single color because it **SC.7.P.10.2**

(F) has waves that are all the same length.

(G) is created by a colored bulb.

(H) passes through a colored filter.

(I) reflects off a partly transparent mirror.

Copyright © Glencoe/McGraw-Hill, a division of The McGraw-Hill Companies, Inc.

Name _____ Date _____

Electric Charge

Electric charges surround you all the time. Some electric charges, such a lightning, are quite strong. Others are barely noticeable. What happens when two electrically charged objects get near each other? (Circle) the response that best matches your thinking.

A. The two objects will be pulled toward each other.

B. The two objects will be pushed away from each other.

C. The two objects will either be pulled together or pushed apart.

D. The charges will cancel out and neither object will move toward or away from the other.

Explain your thinking. Describe your ideas about electric charge.

ELECTRICITY

FLORIDA BIG IDEAS

1   The Practice of Science

3   The Role of Theories, Laws, Hypotheses, and Models

11   Energy Transfer and Transformations

13   Forces and Changes in Motion

Think About It!

How do electric circuits and devices transform energy?

You might not know the name Nikola Tesla, despite his great achievements in electrical science. Tesla made this photograph to demonstrate that his inventions in electricity were safe for everyday use. His work set the stage for the power and lighting systems now used around the world.

1   What might happen if any of the electric streaks touch the man?

_____

_____

2   What different ways do you think electricity is used in the room?

_____

_____

3   How do you this energy is transformed in this scene?

_____

_____

_____

Get Ready to Read

What do you think about electricity?

Before you read, decide if you agree or disagree with each of these statements. As you read this chapter, see if you change your mind about any of the statements.

	AGREE	DISAGREE
1 Protons and electrons have opposite electric charges.	☐	☐
2 Objects must be touching to exert a force on each other.	☐	☐
3 When electric current flows in a wire, the number of electrons in the wire increases.	☐	☐
4 Electrons flow more easily in metals than in other materials.	☐	☐
5 In any electric circuit, current stops flowing in all parts of the circuit if a connecting wire is removed or cut.	☐	☐
6 The light energy given off by a flashlight comes from the flashlight's batteries.	☐	☐

 **ConnectED**   **There's More Online!**
Video • Audio • Review • ⓘLab Station • WebQuest • Assessment • Concepts in Motion • Multilingual eGlossary

Electric Charge and ELECTRIC FORCES

ESSENTIAL QUESTIONS

 How do electrically charged objects interact?

How can objects become electrically charged?

What is an electric discharge?

Vocabulary

static charge p. 712

electric insulator p. 714

electric conductor p. 714

polarized p. 715

electric discharge p. 717

grounding p. 717

 Florida NGSSS

SC.6.N.1.1 Define a problem from the sixth grade curriculum, use appropriate reference materials to support scientific understanding, plan and carry out scientific investigation of various types, such as systematic observations or experiments, identify variables, collect and organize data, interpret data in charts, tables, and graphics, analyze information, make predictions, and defend conclusions.

SC.6.P.13.1 Investigate and describe types of forces including contact forces and forces acting at a distance, such as electrical, magnetic, and gravitational.

LA.6.2.2.3 The student will organize information to show understanding or relationships among facts, ideas, and events (e.g., representing key points within text through charting, mapping, paraphrasing, summarizing, or comparing/contrasting);

LA.6.4.2.2 The student will record information (e.g., observations, notes, lists, charts, legends) related to a topic, including visual aids to organize and record information, as appropriate, and attribute sources of information;

 Launch Lab

10 minutes

How can you bend water?

Procedure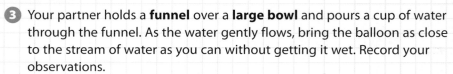

1. Read and complete a lab safety form.

2. Inflate a **balloon**, and tie the end. With a **permanent marker**, draw an *X* on one side of the balloon.

3. Your partner holds a **funnel** over a **large bowl** and pours a cup of water through the funnel. As the water gently flows, bring the balloon as close to the stream of water as you can without getting it wet. Record your observations.

4. Next, rub the X side of the balloon on your **sweater**, and then hold the balloon next to the spot where you rubbed. Observe the interaction between your sweater and the balloon. Record your observations.

5. Rub the X side of the balloon against your sweater again. Immediately repeat step 3.

Data and Observations

Think About This

1. **Infer** Why did you rub the balloon on your sweater? Predict what might have happened if you simply touched the balloon to your sweater instead of rubbing it.

2. **Key Concept** **Assess** Why do you think the balloon interacted the way it did with your sweater and with the stream of water?

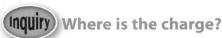

1. You are surrounded by electric charges at all times. Without them, you could not exist. However, some electric charges, such as those in lightning, can be dangerous. How can you keep yourself safe from dangerous amounts of electric charge? How can you control electric charges to make your life better?

_____

_____

_____

_____

_____

_____

_____

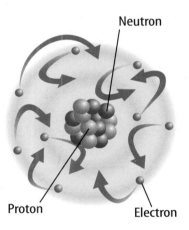

Figure 1 Protons, neutrons, and electrons make up an atom.

Active Reading

2. Classify Fill in the table. List the atomic particles that are described by each heading.

Have Electric Charge	Have No Electric Charge
1.	1.
2.	

Electric Charges

Imagine a hot summer afternoon. Dark clouds fill the sky. Suddenly, a bolt of lightning streaks across the sky. Seconds later, you hear thunder in the distance. The lightning released a tremendous amount of energy. Some of the energy was released as the light that flashed through the sky. Some of the energy was released as the sound you heard as thunder. And, some of the energy was released as thermal energy that heated the air. Where did the energy of the lightning come from? The answer has to do with electric charge.

Charged Particles

Recall that all matter is made of particles called atoms. Also recall that atoms are made of even smaller particles—protons, neutrons, and electrons. Protons and neutrons make up the nucleus of an atom, as shown in **Figure 1**. Electrons move around the nucleus. Protons and electrons have the property of electric charge, neutrons do not. As you read further, you will learn how charged particles interact to affect your everyday life.

Positive Charge and Negative Charge

There are two types of electric charge—positive charge and negative charge. Protons in the nuclei of atoms have positive charge. Electrons moving around a nucleus have negative charge. The amount of positive charge of one proton is equal to the amount of negative charge of one electron.

Recall from previous chapters that oppositely charged particles attract each other. Similarly charged particles repel each other. Therefore, a positively charged proton and a negatively charged electron attract each other. Two protons, or two electrons, repel, or push away from each other.

An atom is electrically neutral—it has equal amounts of positive charge and negative charge. This is because an atom has equal numbers of protons and electrons. Electrons can move from one atom to another. When an atom gains one or more electrons, it becomes negatively charged. If an atom loses one or more electrons, it becomes positively charged. Electrically charged atoms are called ions. It is important to understand that any object can become electrically charged.

Neutral Objects

Similar to electrically neutral atoms, larger electrically neutral objects have equal amounts of positive and negative charge. Electrically neutral objects do not attract or repel each other.

Charged Objects

Just as atoms sometimes gain or lose electrons, larger objects can gain or lose electrons, too. Some materials hold electrons more loosely than other materials. As a result, electrons often move from one object to another. When this happens, the positive charge and negative charge on the objects are unbalanced. *An unbalanced negative or positive electric charge on an object is sometimes referred to as a* **static charge**.

Like an atom, an object that gains eletrons has more negative charge than positive charge. It is said to be negatively charged. Likewise, an object that loses electrons has more positive charge than negative charge. It is said to be positively charged.

Electric Forces

The region surrounding a charged object is called an electric field. An electric field applies a force, called an electric force, to other charged objects, even if the objects are not touching.

The electric force applied by an object's electric field will either attract or repel other charged objects. **Figure 2** shows that objects with opposite electric charges attract each other. On the other hand, objects with similar electric charges repel each other.

Active Reading   **3. Show** Fill in the blanks in **Figure 2** to indicate whether the charged particles attract or repel each other.

Figure 2 Charged objects can attract or repel each other.

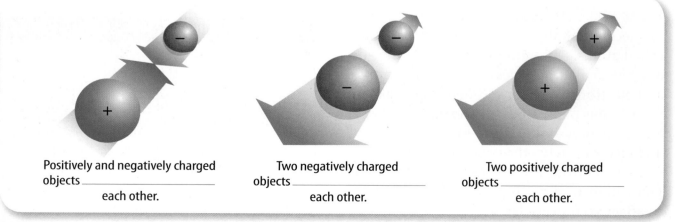

Positively and negatively charged objects _____ each other.

Two negatively charged objects _____ each other.

Two positively charged objects _____ each other.

Electric Force and Amount of Charge

The strength of the electric force between two charged objects depends on two variables—the total amount of charge on both objects and the distance between the objects. For example, when you brush your hair, electrons move from the brush to your hair. This causes the brush and your hair to each have a static charge. The brush is positively charged. Your hair is negatively charged. Because the brush and your hair have opposite electric charges, they attract each other.

The left portion of **Figure 3** illustrates how the amount of charge affects electric force. If you brush your hair once or twice, some electrons transfer from the brush to your hair. The force of attraction is not very strong. However, if you continue brushing your hair, more electrons transfer from the brush to your hair. This increases the strength of the electric force of attraction between the brush and your hair.

Electric Force and Distance

Distance also affects the strength of the electric force between electric charges. As described above, brushing your hair produces an unbalanced positive charge on the brush and an unbalanced negative charge on your hair. Electric fields surround these electric charges. The fields are more intense near the charges. A more intense electric field applies a stronger electric force to other objects. Therefore, when the brush is close to your hair, the force of attraction is very noticeable. The electric force is weaker when the charged objects are far from each other, as shown on the right side of **Figure 3**.

Active Reading 4. **Summarize** Fill in the blanks below to summarize the factors that affect electric forces.

Active Reading

FOLDABLES® LA.6.2.2.3

Make a vertical three-tab book, and label it as shown. Use it to organize your notes about electrically charged objects.

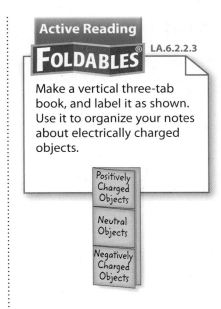

Positively Charged Objects

Neutral Objects

Negatively Charged Objects

Figure 3 The strength of the electric force between two charged objects depends on the distance between the objects and the total amount of electric charge of the objects.

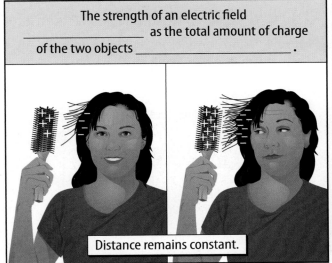

The strength of an electric field _____ as the total amount of charge of the two objects _____.

Distance remains constant.

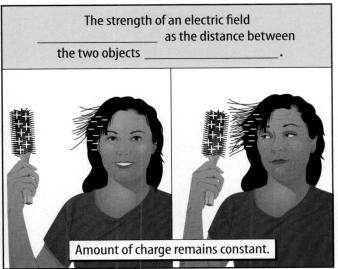

The strength of an electric field _____ as the distance between the two objects _____.

Amount of charge remains constant.

Active Reading

5. Distinguish Fill in the table below to compare insulators and conductors.

Insulator	Conductor
Definition:	Definition:
Examples:	Examples:
Common use:	Common use:

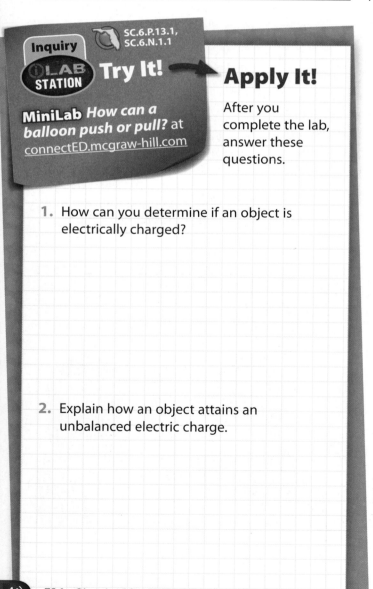

Inquiry

SC.6.P.13.1, SC.6.N.1.1

iLAB STATION Try It!

MiniLab *How can a balloon push or pull?* at connectED.mcgraw-hill.com

Apply It!

After you complete the lab, answer these questions.

1. How can you determine if an object is electrically charged?

2. Explain how an object attains an unbalanced electric charge.

Transferring Electrons

You read that a hair brush and strands of hair become electrically charged when electrons transfer from the brush to your hair. How do electrons move from one object to another?

Insulators and Conductors

To understand how electrons move from one object to another, you need to know about two basic types of materials—electric insulators and electric conductors. *A material in which electrons cannot easily move is an* **electric insulator**. Glass, rubber, wood, and even air are good electric insulators. *A material in which electrons can easily move is an* **electric conductor**. Most metals, such as copper and aluminum, are good electric conductors.

Electric insulators and electric conductors are all around you. Electric conductors and electric insulators are used in electrical power cords around your house. In **Figure 4**, copper wire is the conductor of electrons in an extension cord. The copper allows electrons to easily move through the wire. Plastic and rubber are used as protective electric insulators around the metal wire. Electrons cannot easily move in the plastic and rubber.

Electrons transfer between objects by contact, induction, or conduction. As you will read, electric insulators and conductors play an important role in these processes.

Figure 4 Using conductors and insulators can safely control the flow of electric charges.

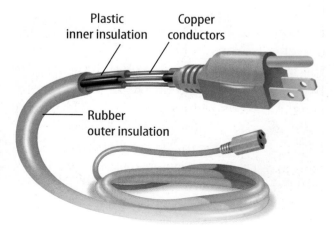

Plastic inner insulation

Copper conductors

Rubber outer insulation

Transferring Charge by Contact

Recall that some materials hold their electrons more loosely than other materials. When objects made of different materials touch, electrons tend to collect on the object that holds electrons more tightly. This is called transferring charge by contact.

The wool sweater in **Figure 5** holds electrons more loosely than the rubber balloon. When the balloon comes in contact with the sweater, electrons from the surface of the sweater transfer to the surface of the balloon, creating a static charge on both objects. Because the balloon gained electrons, it has an unbalanced negative charge. Because the sweater lost electrons, it has an unbalanced positive charge. Both insulators and conductors can be charged by contact.

Transferring Charge by Induction

Transferring charge by induction is a process by which one object causes two other objects that are conductors to become charged without touching them. **Figure 6** illustrates how this works.

Part 1 of **Figure 6** shows a negatively charged balloon repelling electrons in a metal soda can. Because the can is aluminum and, therefore, a conductor, electrons in the can easily move toward the far end of the can. The can is not charged because it has not gained or lost any electrons. Instead, the can is polarized. *When electrons concentrate at one end of an object, the object is* **polarized**.

In part 2, when a charged balloon is brought near two cans that are touching, the balloon polarizes the cans as if they are one object. Electrons in both cans move toward the far end of the can on the right. Then, the two cans separate. As shown in part 3, the cans that were originally polarized as a group are now individually charged. The can on the right has an unbalanced negative electric charge, and the can on the left has an unbalanced positive electric charge.

Figure 5 🔑 More loosely held electrons on wool easily transfer onto the surface of a rubber balloon.

Figure 6 Objects that are electric conductors can be charged by induction.

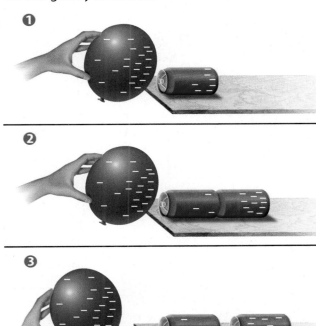

 6. Explain Why do negative charges in the aluminum cans tend to move away from the balloon?

_____

_____

_____

7. Evaluate Fill in the table to explain how materials transfer charge. (Circle) the way in which insulators can be charged.

Method	Description
Contact	
	Two conducting objects are polarized when they touch.
Conduction	

Figure 7 Charges flow between objects until the charge on both objects is equal.

Transferring Charge by Conduction

Another way that electrons transfer between two conductors is called transferring charge by conduction. As shown in **Figure 7**, when conducting objects with unequal charges touch, electrons flow from the object with a greater concentration of negative charge to the object with a lower concentration of negative charge. This is similar to water flowing from a container with a higher water level to a container with a lower water level. The flow of electrons continues until the concentration of charge of both objects is equal.

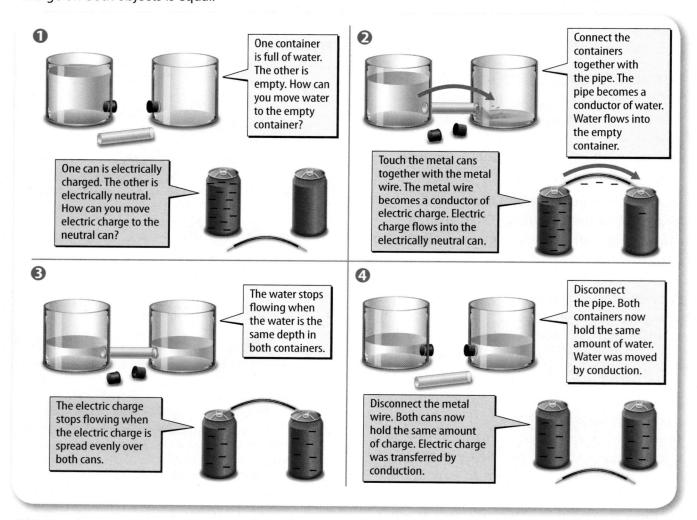

❶ One container is full of water. The other is empty. How can you move water to the empty container?

One can is electrically charged. The other is electrically neutral. How can you move electric charge to the neutral can?

❷ Connect the containers together with the pipe. The pipe becomes a conductor of water. Water flows into the empty container.

Touch the metal cans together with the metal wire. The metal wire becomes a conductor of electric charge. Electric charge flows into the electrically neutral can.

❸ The water stops flowing when the water is the same depth in both containers.

The electric charge stops flowing when the electric charge is spread evenly over both cans.

❹ Disconnect the pipe. Both containers now hold the same amount of water. Water was moved by conduction.

Disconnect the metal wire. Both cans now hold the same amount of charge. Electric charge was transferred by conduction.

8. Compare How are the pipe and the wire in **Figure 7** similar?

_____

_____

_____

Electric Discharge

When you brush your hair, electrons transfer from the brush to your hair. What happens to these excess electrons in your hair? They transfer to objects that come in contact with your hair, such as a hat or your pillow. Electrons even transfer to the air. Gradually, your hair loses its charge. *The process of an unbalanced electric charge becoming balanced is an* **electric discharge**.

Lightning Rods and Grounding

An electric discharge can occur slowly, such as when your hair loses its negative charge and is no longer attracted to a brush. Or, an electric discharge can occur as quickly as a flash. For example, lightning is a powerful electric discharge that occurs in an instant.

A lightning strike can severely damage a building or injure people. Lighting rods help protect people against these dangers. **Figure 8** shows a metal lightning rod attached to the roof of a building. A thick wire connects the lightning rod to the ground. The wire provides a path for the electrons released in a lightning strike to travel into the ground. *Providing a path for electric charges to flow safely into the ground is called* **grounding**. **Table 1** provides tips on how to protect yourself from a lightning strike.

What causes lightning?

Scientists are not entirely clear on what causes lightning. However, many scientists believe that lightning is related to the electric charges that separate within storm clouds. The large amounts of ice, hail, and partially frozen water droplets that thunderstorms create seem to play a role, too.

Forecasting exactly where and when lightning will strike might never be possible. **Figure 9** on the next page summarizes what is known about how lightning forms.

Figure 8 Lightning rods help protect tall buildings from the damaging effects of lightning.

Active Reading **9. Organize** Complete the graphic organizer below on electric discharge.

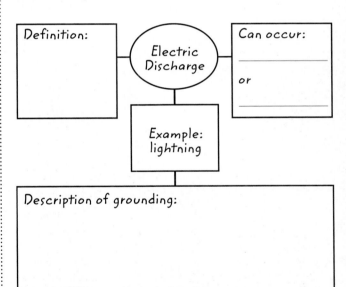

Definition:

Electric Discharge

Can occur: _____ or _____

Example: lightning

Description of grounding:

Table 1 Lightning Safety Tips

If the time between a lightning flash and thunder is less than 30 seconds, the storm is dangerously close. Protect yourself in the following ways:

- Seek shelter in an enclosed building or a car with a metal top. Never stand under a tree.
- Do not touch metal surfaces.
- Get away from water if swimming, boating, or bathing.
- Wait 30 minutes after the last flash of lightning before leaving the shelter, even if the Sun comes out.

1

Within a storm cloud, warm air rises past falling cold air. The cold air is filled with hail, ice, and partially frozen water droplets that pick up electrons from the rising air. This causes the bottom of the cloud to become negatively charged.

2

The negatively charged cloud polarizes Earth's surface by repelling negative charges in the ground. Thus, the surface of the ground becomes positively charged.

3

When the bottom of the cloud accumulates enough negative charge, the attraction of the positively charged ground causes electrons in the cloud to begin moving toward the ground.

4

As electrons approach the ground, positive charges quickly flow upward, making an electric connection between the cloud and the ground. You see this electric discharge as lightning.

5

Accumulation of charge does not only occur between clouds and the ground. Charges also separate within or between storm clouds, causing cloud-to-cloud lightning.

Visual Summary

Any object can be positively charged, negatively charged, or neutral.

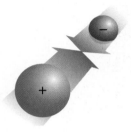

Charged objects exert forces on each other.

Lightning is one form of electric discharge.

Use Vocabulary

1 Providing a path for electric charges to flow safely into the ground is called _____.

2 **Use the term** *static charge* in a sentence. SC.6.P.13.1

_____

_____

Understand Key Concepts

3 Particles that attract each other are SC.6.P.13.1

Ⓐ two electrons.　Ⓒ one proton and one electron.

Ⓑ two protons.　Ⓓ one proton and one neutron.

4 **Summarize** How does the electric force between charged objects change as the objects move away from each other? SC.6.P.13.1

_____

_____

_____

5 **Describe** Imagine you have to design a safe, useful device that uses an electric discharge to make your life better. How would you describe it to others?

_____

_____

_____

Interpret Graphics

6 **Organize** Fill in the type of charge on protons, neutrons, and electrons. LA.6.2.2.3

Particle	Type of Electric Charge
Proton	
Neutron	
Electron	

Critical Thinking

7 **Analyze** How can objects become positively charged?

_____

_____

_____

Van de Graaff Generator

How can a machine move electrons?

Have you ever seen a Van de Graaff generator? Originally, scientists studied the nuclei of atoms with these machines. Now, they are often seen in science museums, where they are used to demonstrate the effects electric charge.

How does this device generate electric charge? Recall that some materials hold electrons more loosely than other materials. In a Van de Graaff generator, the metal dome holds electrons more loosely than the belt. As the belt travels over the top roller past the top comb, electrons move from the metal dome through the upper comb and onto the belt. This leaves the dome positively charged.

The excess electrons on the belt move from the belt onto the lower comb as the belt moves over the roller at the bottom of the generator. The electrons then travel through a wire out of the machine and into the ground. This process of electrons moving from the dome to the ground continues as long as the generator is running.

Soon, the dome loses so many electrons that it acquires a very large positive charge. The positive charge on the dome becomes so great that electrons create an electric spark as they jump back onto the dome from any object that will release them. That object could be you if you stand close enough.

▼ The Van de Graaff generator causes the girl's body, including her hair, to become electrically charged. Because all the strands of her hair acquire the same charge and repel each other, her hair stands on end.

Upper comb

Metal dome

Belt

Electric motor

Lower comb

Ground

It's Your Turn

REPORT Find out how to build a simple Van de Graaff generator using everyday materials. Share your findings with your classmates. LA.6.4.2.2

Electric Current and
SIMPLE CIRCUITS

ESSENTIAL QUESTIONS

🔑 What is the relationship between electric charge and electric current?

🔑 What are voltage, current, and resistance? How do they affect each other?

Vocabulary

electric current p. 722

electric circuit p. 722

electric resistance p. 724

voltage p. 725

Ohm's law p. 726

 Florida NGSSS

SC.6.N.3.3 Give several examples of scientific laws.

SC.7.N.1.4 Identify test variables (independent variables) and outcome variables (dependent variables) in an experiment.

SC.7.N.1.6 Explain that empirical evidence is the cumulative body of observations of a natural phenomenon on which scientific explanations are based.

SC.7.P.11.2 Investigate and describe the transformation of energy from one form to another.

LA.6.2.2.3 The student will organize information to show understanding or relationships among facts, ideas, and events (e.g., representing key points within text through charting, mapping, paraphrasing, summarizing, or comparing/contrasting);

MA.6.A.3.6 Construct and analyze tables, graphs, and equations to describe linear functions and other simple relations using both common language and algebraic notation.

Inquiry Launch Lab   SC.7.N.1.6

20 minutes

What is the path to turn on a light?

How can you make a lightbulb light up?

Procedure

1. Read and complete a lab safety form.

2. Examine a **D-cell battery.** Notice the differences between the two ends. Record your observations.

3. Using a **paper clip, small lightbulbs,** and the battery, design a way to light the bulb. Draw a diagram of your plan.
   ⚠ The paperclip can get hot!

4. Find two other ways to light the bulb using the paper clip and the battery. Draw your plans.

5. Record three configurations that do not light the bulb.

Data and Observations

Think About This

1. **Infer** Why do some arrangements of the materials light the bulb, but other arrangements do not?

1. Across the country, over 300,000 km of wires carry electric energy from about 500 power companies to homes and businesses. A flashlight operates with only a few centimeters of wires. How do you think these vastly different systems are similar?

_____

_____

_____

_____

_____

Electric Current and Electric Circuits

In Lesson 1, you read about ways to transfer, or move electric charges. The transferring of electric charges allows you to power the electrical devices you use every day. For example, conduction of electric charges occurs every time you turn on a flashlight or a television. Induction is used for wireless charging of devices, such as electrical toothbrushes. And, as you read in the last lesson, transferring charge by contact occurs every time lightning strikes. How are a flash of lightning and a TV similar? They both transform the energy of moving electrons to light, sound, and thermal energy. *The movement of electrically charged particles is an* **electric current**.

A Simple Electric Circuit

The movement of electrons in a lightning strike lasts only a fraction of a second. In a TV, electrons continue moving as long as the TV remains on. An electric current flows in a closed path to and from a source of electric energy. *A closed, or complete, path in which an electric current travels is an* **electric circuit**.

All electric circuits have one thing in common—they transform electric energy to other forms of energy. For example, electric circuits in a microwave oven transform electric energy to the thermal energy that quickly cooks your food. Circuits in your television transform electric energy to the light and sound that entertains and informs you. How do electric energy transformations improve your life?

Active Reading **2. Recognize** What do all electric circuits have in common? Highlight your answer in the text.

Figure 10 An electric current flows in a circuit if the circuit is complete, or closed. Current will not flow if the circuit is broken, or open.

Figure 11 🔑 When current flows in a wire, the number of electrons in the wire does not change.

How Electric Charges Flow in a Circuit

Figure 10 shows a battery, wires, and a lightbulb connected in a circuit. The lightbulb glows as electrons flow from of one end of the battery, through the wires and lightbulb, and back into the other end of the battery. If the circuit is broken, or open, the electrons stop flowing, and the bulb stops glowing.

A current of electrons in a wire is somewhat like water being pumped through a hose. The amount of water flowing into one end of a hose is the same as the amount of water flowing out the other end. As shown in **Figure 11**, the number of electrons flowing into a wire from a power source equals the number of electrons flowing out of the wire back into the source.

The Unit For Electric Current

Electric current is approximately measured as the number of electrons that flow past a point every second. However, electrons are so tiny and there are so many in a circuit that you could never count them one at a time. Therefore, just like you count eggs by the dozen, scientists count electrons by a quantity called the coulomb (KEW lahm). A coulomb is a very large number—about 6×10^{18}, or 6,000,000,000,000,000,000! That is 6 quintillion.

The SI unit for electric current is the ampere (AM pihr), commonly called an amp. Its symbol is A. One ampere of current equals about one coulomb of electrons flowing past a point in a circuit every second. The electric current through a 120-W lightbulb is about 1 A. A typical hair dryer uses a current of about 10 A, or 60,000,000,000,000,000,000 electrons per second.

Active Reading **3. List** What 3 way electric circuits are used to transform electric energy in your home?

_____

_____

_____

Active Reading **4. Summarize** Fill in the blanks below to help you understand electric current.

Electric current is the movement of _____ , _____ or the number of _____ that flow past a point every second. Scientists count electrons by the _____ , which is about 6×10^{18} _____. The SI unit for electric current is the _____. One ampere of electric current equals _____ coulomb of electric charge flowing past a given point in _____ second.

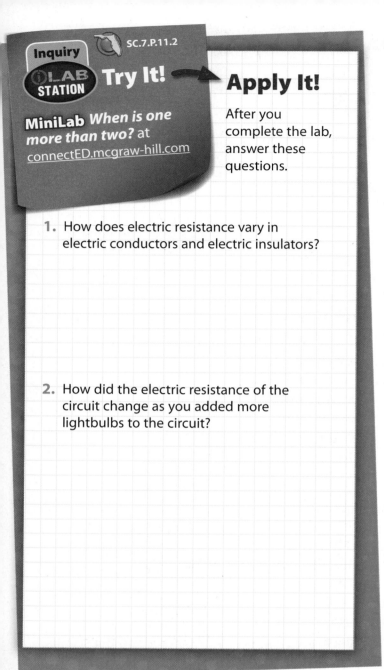

Inquiry **Try It!**

MiniLab *When is one more than two?* at connectED.mcgraw-hill.com

Apply It!

After you complete the lab, answer these questions.

1. How does electric resistance vary in electric conductors and electric insulators?

2. How did the electric resistance of the circuit change as you added more lightbulbs to the circuit?

Table 2 Electric Resistance of Different Materials	
Material (20 cm x 1 mm)	Resistance (ohms)
Copper	0.004
Gold	0.006
Iron	0.025
Carbon	8.9
Rubber	2,500,000,000,000,000,000,000.0

What is electric resistance?

See **Figure 12** on the next page. Suppose you replaced one of the green wires in the circuit on the left side with a piece of string. The lightbulb would not glow because there would be no current in the circuit. Why do electrons flow easily in metal wire but not in string? The answer is that wire has much less electric resistance than string. **Electric resistance** *is a measure of how difficult it is for an electric current to flow in a material.*

The unit of electric resistance is the ohm (OHM), which is symbolized by the Greek letter Ω (omega). An electric resistance of 20 ohms is written 20 Ω. **Table 2** lists the electric resistances of different materials.

Electric Resistance of Conductors and Insulators

You read that electric conductors are materials, such as copper and aluminum, in which electrons easily move. A good conductor has low electric resistance. Usually electric wires are made of copper because copper is one of the best conductors.

Recall that electrons cannot easily move through insulators such as plastic, wood, or string. A good electric insulator has high electric resistance. Atoms of an insulator hold electrons tightly. This prevents electric charges from easily moving through the material. Therefore, replacing a wire in a circuit with string prevents electrons from flowing in a circuit.

Resistance—Length and Thickness

A material's electric resistance also depends on the material's length and thickness. A thick copper wire has less electric resistance than a thin copper wire of the same length. Because the thick wire has less resistance, it will conduct better. Increasing the length of a conductor also increases its electric resistance.

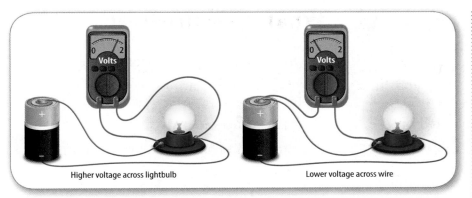

Higher voltage across lightbulb   Lower voltage across wire

Figure 12 🔑 • Voltage can be different in different parts of a circuit. The voltmeter shows where most of the battery's energy is used.

Active Reading **6. Infer** Why is the voltage reading across the lightbulb higher than across the segment of wire?

_____

_____

_____

_____

_____

What is voltage?

You probably heard the term *volt*. You use 1.5-V batteries in a flashlight. You plug a hair dryer into a 120-V outlet. But, what does this mean?

Battery Voltage

In **Figure 12**, a battery creates an electric current in a closed circuit. Energy stored in the battery moves electrons in the circuit. As the electrons move through the circuit, the amount of energy transformed by the circuit depends on the battery's voltage. **Voltage** *is the amount of energy the source uses to move one coulomb of electrons through the circuit.* A circuit with a high voltage source transforms more electric energy to other energy forms than a circuit with a low voltage source. For example, a lightbulb connected to a 9-V battery produces about six times more light and thermal energy than the same lightbulb connected to a 1.5-V battery.

Voltage in Different Parts of a Circuit

Electric energy transforms to other forms of energy in all parts of a circuit. For example, the lightbulbs in **Figure 12** transform electric energy to light and thermal energy. Even the wires and batteries produce a small amount of thermal energy. In other words, different amounts of energy transform in different parts of a circuit. The voltage measured across a portion of a circuit indicates how much energy transforms in that portion of the circuit. For example, the voltage is greater across the lightbulb in **Figure 12** than across the wire. Therefore, the lightbulb transforms more energy than the wire.

Active Reading **5. Describe** What happens to the energy flowing in an electric circuit?

_____

_____

_____

Active Reading

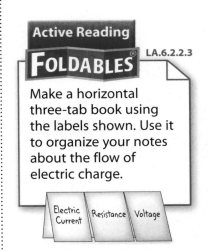

FOLDABLES LA.6.2.2.3

Make a horizontal three-tab book using the labels shown. Use it to organize your notes about the flow of electric charge.

Electric Current | Resistance | Voltage

Ohm's Law

When designing electrical devices, engineers choose materials based on their electric resistance. For example, the heating coils in a toaster must be made of a metal with very high electric resistance. As explained on the next page, this allows the coils to transform most of the circuit's energy to thermal energy. But, how much resistance should a conductor have? The answer is found with Ohm's law.

Using Ohm's Law

Named after German physicist Georg Ohm, **Ohm's law** *is a mathematical equation that describes the relationships among voltage, current, and resistance.* The law states that as the voltage of a circuit's electric energy source increases, the current in the circuit increases, too. Also, as the resistance of a circuit increases, the current decreases.

Ohm's Law can be written as the following equation:

> ### Ohm's Law Equation
> **voltage (V) = current (I) × resistance (R)**
> $$V = IR$$

V is the symbol for voltage, measured in volts (V). I is the symbol for electric current, which is measured in amperes (A). And, R is the symbol for electric resistance, measured in ohms (Ω). If you know the value of two of the variables in the equation, you can determine the third, as described in **Figure 13**. You can measure any of these variables with a multimeter, as shown in **Figure 14**.

Figure 13 Ohm's law can be used to calculate unknown quantities in a circuit.

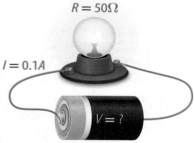

Calculate the voltage across the lightbulb.

$R = 50\Omega$

$I = 0.1A$

$V = ?$

To find the voltage, start with Ohm's law:

$$V = IR$$

Substitute the known values into the equation:

$$V = 0.1A \times 50\Omega$$
$$V = 5V$$

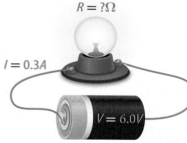

Calculate the resistance of the lightbulb.

$R = ?\Omega$

$I = 0.3A$

$V = 6.0V$

To find the resistance, start with this form of Ohm's law:

$$R = \frac{V}{I}$$

Substitute the known values into the equation:

$$R = \frac{6V}{0.3A}$$
$$R = 20\Omega$$

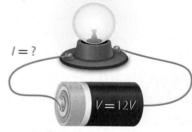

Calculate the current flowing through the lightbulb.

$R = 50\Omega$

$I = ?$

$V = 12V$

To find the current, start with this form of Ohm's law:

$$I = \frac{V}{R}$$

Substitute the known values into the equation:

$$I = \frac{12V}{50\Omega}$$
$$I = 0.2A$$

Active Reading **7. Explain** What are voltage, current, and resistance, and how do they affect each other?

8. Identify In the table below, fill in the correct vocabulary word, definition, abbreviation, unit, and unit symbol.

Vocabulary word	Definition	Abbreviation	Unit	Unit symbol
Electric current		I	ampere	
Electric resistance				Ω
Voltage				

Voltage, Resistance, and Energy Transformation

Figure 14 shows two lightbulbs and a battery connected in a circuit. Both bulbs are connected one-after-another in a single loop. The currents through both bulbs are equal. However, the two lightbulbs are not identical. One has greater electric resistance than the other.

You determine which lightbulb has more electric resistance with a voltmeter and an understanding of Ohm's law. Ohm's law states that, with equal current, the voltage is greater across the device with the greater resistance. **Figure 14** also shows that the lightbulb with the greater electric resistance has the greater voltage across it. The higher-resistance lightbulb on the left transforms more electric energy to light.

Active Reading **9. Explain** Look at **Figure 14**. What is the relationship between the voltages shown on the meters and the voltage of the battery?

_____

_____

Figure 14 When two devices are in the same circuit, the voltage is greater across the device with higher resistance. The device with higher resistance transforms more electric energy to other forms of energy.

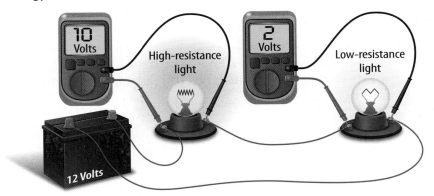

10 Volts — High-resistance light

2 Volts — Low-resistance light

12 Volts

Math Skills   MA.6.A.3.6

Solve for Voltage

The current through a lightbulb is **0.5 A**. The resistance of the lightbulb is **220 Ω**. What is the voltage across the lightbulb?

Use Ohm's law to solve this problem.

$$V = IR$$

Substitute the values for **I** and **R** and multiply:

$$V = 0.5\,A \times 220\,\Omega = 110\,V$$

The answer is **110 V**

Practice

10. What is the voltage across the ends of a wire coil in a circuit if the current in the wire is 0.1 A and the resistance is 30 Ω?

Lesson Review 2

Visual Summary

An electric current is the flow of negative electric charges through a conductor.

Electrical devices function only when they are connected in a closed circuit.

Ohm's law shows the relationship among voltage, current, and resistance.

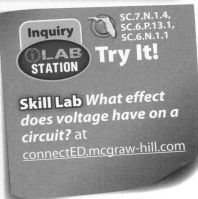

Inquiry

SC.7.N.1.4,
SC.6.P.13.1,
SC.6.N.1.1

iLAB STATION Try It!

Skill Lab *What effect does voltage have on a circuit?* at connectED.mcgraw-hill.com

Use Vocabulary

1. Electrons flow more easily in a material with lower _____.

2. A closed path in which electric charges can flow is a(n) _____.

3. The flow of electric charges is a(n) _____.

Understand Key Concepts 🔑

4. Ohm's law is NOT related to SC6.N.3.3
   - (A) current.
   - (C) resistance.
   - (B) mass.
   - (D) voltage.

5. **Describe** in your own words the relationship between electric charge and electric current.

Interpret Graphics

6. **Explain** Fill in the effect of changing voltage and resistance on electric current. SC.6.2.2.3

Variable	Effect of Increase in Variable on Current
Voltage	
Resistance	

Critical Thinking

7. **Evaluate** How does the total electric charge of a wire in a circuit change when current stops flowing in the circuit?

Math Skills   MA.6.A.3.6

8. The current in a hair dryer is 10 A. The hair dryer is plugged into a 110-V outlet. How much electric resistance is there in the hair dryer's circuit?

Working with Amusement Parks and Electricity

Hard Work Can be Fun

For many people, amusement parks are a popular destination for recreation and entertainment. In fact, amusement parks in Florida have become some of the most visited parks around the world. With high-tech electric systems, today's amusement parks consume huge amounts of electric energy. It takes thousands of electrical workers and electrical engineers to keep a park running.

Electricians are problem solvers. Each electrical job has its own set of challenges. For example, at one Florida park, electricians built an underground system of lights to create the illusion of campfires on the African plain. At parks with live animals, electricians must have knowledge of animal behavior. When working in elephant enclosures, electricians bury wires deeply. If elephants feel or hear vibrations, they might dig up the wires. At another park in Florida, a series of lights is installed on a 50-meter high tower that arches over a major attraction. When bulbs burn out, electricians trained in high-access rappelling change the light bulbs.

At top theme parks, electricians work with lighting to create the proper mood. For instance, when lights are installed in a gift shop, electricians design the lighting to make souvenirs more appealing to guests.

Working with electricity in amusement parks can be fun. Could this be a career for you?

It's Your Turn

RESEARCH AND REPORT Write or talk to a person in the Human Resources department of a major amusement park in Florida. Research the training necessary to become a park electrician. What salary can you expect to earn in this field? Write a short report to share what you learned with your class.

Describing CIRCUITS

 What are the basic parts of an electric circuit?

How do the two types of electric circuits differ?

Vocabulary

series circuit p. 733

parallel circuit p. 734

 Florida NGSSS

SC.6.N.1.1 Define a problem from the sixth grade curriculum, use appropriate reference materials to support scientific understanding, plan and carry out scientific investigation of various types, such as systematic observations or experiments, identify variables, collect and organize data, interpret data in charts, tables, and graphics, analyze information, make predictions, and defend conclusions.

SC.6.N.1.4 Discuss, compare, and negotiate methods used, results obtained, and explanations among groups of students conducting the same investigation.

SC.6.N.1.5 Recognize that science involves creativity, not just in designing experiments, but also in creating explanations that fit evidence.

SC.7.N.1.1 Define a problem from the seventh grade curriculum, use appropriate reference materials to support scientific understanding, plan and carry out scientific investigation of various types, such as systematic observations or experiments, identify variables, collect and organize data, interpret data in charts, tables, and graphics, analyze information, make predictions, and defend conclusions.

SC.7.N.1.3 Distinguish between an experiment (which must involve the identification and control of variables) and other forms of scientific investigation and explain that not all scientific knowledge is derived from experimentation.

SC.7.P.11.2 Investigate and describe the transformation of energy from one form to another.

LA.6.2.2.3 The student will organize information to show understanding or relationships among facts, ideas, and events (e.g., representing key points within text through charting, mapping, paraphrasing, summarizing, or comparing/contrasting);

Inquiry Launch Lab

SC.6.N.1.5

20 minutes

How would you wire a house?

If you turn off the kitchen light, why does the refrigerator keep running?

Procedure

1. Read and complete a lab safety form.

2. Using **alligator clip wires**, light a **small bulb in a base** with a **D-cell battery in a base**. Draw the setup.

3. Disconnect the wires between the battery and the bulb. Add a **second bulb** between the battery and the first bulb. Remove one of the bulbs from its base. Record your observations.

4. Take apart the setup, and reassemble the setup in step 2. Now, add a second bulb by connecting one terminal of the second bulb to one of the terminals of the lit bulb. Connect the other terminal of the second bulb to the remaining terminal of the lit bulb. Both bulbs should be lit. Remove one of the bulbs from its base. Record your observations.

Data and Observations

Think About This

1. **Describe** Which of your assembled circuits would be the best to use when wiring a house?

2. **Key Concept** **Recommend** Which setup is the best way to connect lights in your home? Explain.

inquiry Can wires fly?

1. Modern jets may contain over 160 km of electric circuits. This wiring can be compared to the body's nervous system. However, aircraft wiring does not last for the life of an airplane. Wiring systems in airplanes can wear out in 10 to 15 years. Why do you think it is important for airplane mechanics to understand electric circuits?

_____

_____

_____

_____

ACADEMIC VOCABULARY
device
(noun) a piece of equipment

Parts of an Electric Circuit

Do you study by an electric lamp? Is there a computer on your desk? These **devices** contain electric circuits.

Three common parts of most electric circuits are a source of electric energy, electrical devices that transform the electric energy, and conductors, such as wires, that connect the other components.

Electric Energy to Kinetic Energy

An energy source, such as the battery in **Figure 15**, produces an electric current in a circuit. Some electrical devices are designed to transform the electric energy of the current to kinetic energy—the energy of motion. For example, the motor in an electric fan transforms electric energy to the kinetic energy of moving air particles that keep you cool.

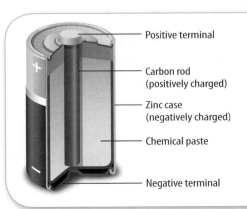

Figure 15 In a battery, chemical reactions in the moist paste cause the carbon rod to become positively charged. The zinc case becomes negatively charged.

Active Reading

2. Name What are the three basic parts of an electric circuit?

_____

_____

_____

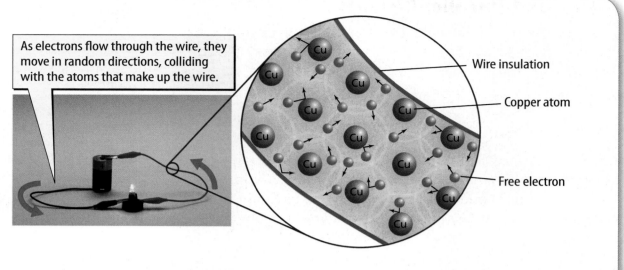

As electrons flow through the wire, they move in random directions, colliding with the atoms that make up the wire.

Wire insulation

Copper atom

Free electron

Figure 16 🔑 The battery produces a current of electrons in a closed circuit. The kinetic energy of the electrons transforms to other forms of energy.

Batteries supply electric energy.

A battery is often used as a source of electric energy for a circuit. **Figure 15** on the previous page shows a cross section of a common battery. A battery is just a can of chemicals. As the chemicals react, electrons in the battery concentrate at the battery's negative terminal. When a closed circuit connects the terminals, electrons flow in the circuit from the battery's negative terminal to the positive terminal. If the circuit is closed and the chemicals in the battery keep reacting, an electric current continues.

Electric circuits transform energy.

Energy transformations occur in all parts of an electric circuit. For example, a battery transforms stored chemical energy to the electric energy of electrons moving as an electric current. Electrical devices in the circuit transform most of the electric energy to other useful forms of energy. For example, a lightbulb transforms electric energy to light, and stereo speakers transform electric energy to sound. In addition, all parts of a circuit, including the energy source and the connecting wires, transform some of the electric energy to wasted thermal energy.

Figure 16 shows how these energy transformations occur. For example, electrons flowing in the wire filament of a lightbulb collide with the atoms that make up the filament. The collisions transfer the electrons' energy to the atoms. The atoms immediately release the energy in other forms, such as light and thermal energy.

You read that circuits and devices release wasted thermal energy. However, many modern electrical devices, such as compact fluorescent lamps (CFLs), are designed to waste less energy. CFLs conduct an electric current through a gas and do not use wire filaments. More energy transforms to light, and less is wasted as thermal energy.

Wires connect parts of a circuit.

Recall that an electric current flows only in a closed circuit. Therefore, a circuit's energy source and device must be connected with some form of conducting material. Metal wires often are used to complete circuits. Because of their low electric resistance, wires transform only a small amount of electric energy to wasted thermal energy. This leaves more energy available for useful devices in the circuit.

Series and Parallel Circuits

Some strings of holiday lights will not light if any bulbs are missing. Other strings of lights work with or without missing bulbs. These two examples of holiday lights represent the two types of electric circuits—series circuits and parallel circuits.

Series Circuit—A Single Current Path

A **series circuit** *is an electric circuit with only one path for an electric current to follow.* Many strings of holiday lights are series circuits. In these series circuits all the lightbulbs connect end-to-end in a single conducting loop. As shown in **Figure 17**, breaking the loop at any point stops the current of electrons throughout the entire circuit. If a wire loop in a series circuit is broken, or open, all devices in the circuit will turn off.

Active Reading
3. **Explain** Why do all devices in a series circuit stop working if the circuit is broken at any point?

_____

_____

_____

The amount of current in a circuit depends on the number of devices in the circuit. **Figure 17** shows that adding devices to a series circuit adds resistance to the total circuit. According to Ohm's law, when resistance increases and voltage remains the same, there is less current in the circuit. Because a string of holiday lights contains so many devices, the current through the circuit is very low. The low current produces very little thermal energy, making these lights safe to use.

Active Reading

FOLDABLES LA.6.2.2.3

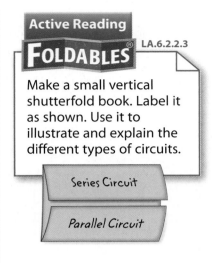

Make a small vertical shutterfold book. Label it as shown. Use it to illustrate and explain the different types of circuits.

Series Circuit

Parallel Circuit

Active Reading
4. **Write** Using the following words, fill in the boxes in **Figure 17**:
- open
- circuit
- current
- energy
- resistance

Figure 17 🔑 A series circuit has only one path for current. The current stops if that path is broken.

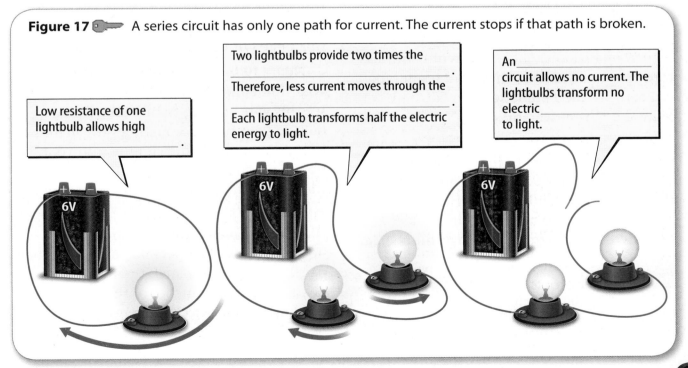

Low resistance of one lightbulb allows high _____.

Two lightbulbs provide two times the _____.
Therefore, less current moves through the _____.
Each lightbulb transforms half the electric energy to light.

An_____ circuit allows no current. The lightbulbs transform no electric_____ to light.

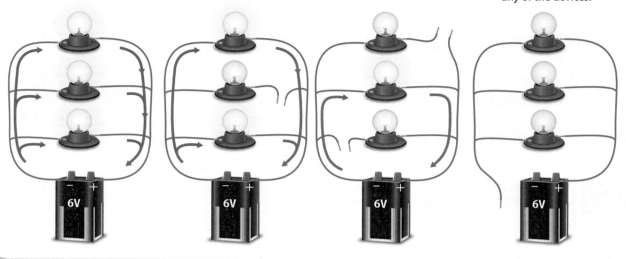

Figure 18 Parallel circuits have more than one path through which electric current can flow.

In a parallel circuit, each device has its own path through which current can flow.

If any of the paths of a parallel circuit is broken, current still can flow through the other devices.

However, if the circuit is open at the source of electric energy, no current can flow through any of the devices.

Parallel Circuit— Multiple Current Paths

If the electrical devices in your home were connected as a series circuit, you would need to turn on every electrical device in the house just to watch TV! Luckily, devices in your home are connected to an electric source as a **parallel circuit**—*an electric circuit with more than one path, or branch, for an electric current to follow.*

As shown in **Figure 18**, if you open one branch of a parallel circuit, current continues through the other branches. As a result, you can turn off the TV but keep the kitchen light and the refrigerator on.

You read that a device in a parallel circuit connects to the source with its own branch. The current in one branch has no effect on the current in other branches. However, adding branches to a parallel circuit does increase the total current through the source.

Active Reading **5. Contrast** How do the two types of electric circuits differ? Highlight the definition of each.

Electric Circuits in the Home

Most people use electrical devices every day without thinking much about them. However, where does the electric energy that we use come from?

Electric energy used in most homes and businesses is generated at large power plants. These power plants may be many kilometers from your home. A complex system of transmission wires carries the electric energy to all parts of the country.

Figure 19 shows that electric energy enters your home through a main wire. You might see this wire coming from a utility pole outside your home. Sometimes the main wire is underground. Before coming into your house, the main wire travels through an electric meter. The meter measures the energy used in the electric circuits of your home.

From the meter, the wire enters your house and goes to a steel box called the main panel. At the main panel, the main wire divides into the branches of the parallel circuit that carry electric energy to all parts of your house.

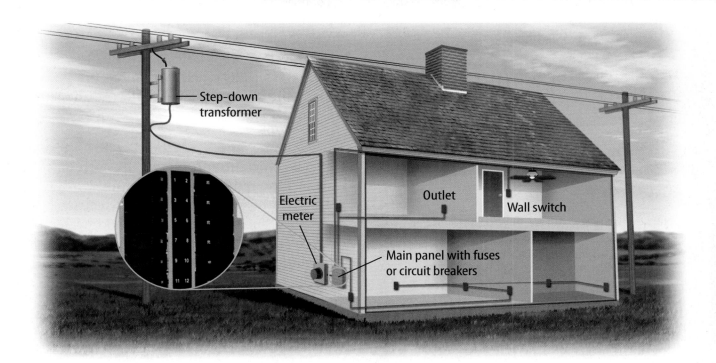

Step-down transformer

Electric meter

Outlet

Wall switch

Main panel with fuses or circuit breakers

Fuses, Circuit Breakers, and GFCI Devices

Recall that adding branches and devices to a parallel circuit increases the current through the source. Too many branches or devices in a circuit can create dangerously high current in the circuit. Excess current can cause the circuit to become hot enough to cause a fire. The branches of your home's parallel circuit connect to the source with a safety mechanism, such as a fuse or a circuit breaker, to help prevent such disasters.

Fuses and circuit breakers automatically open a circuit when the current becomes dangerously high. A fuse is a piece of metal that breaks a circuit by melting from the thermal energy produced by a high current. A circuit breaker is a switch that automatically opens a circuit when the current is too great.

There is another type of automatic safety mechanism found in some circuits. Are there electric outlets in your home with two small buttons labeled *test* and *reset*? Those special outlets are ground-fault circuit interrupters (GFCI). A GFCI is used where an outlet is found near a source of water, such as a sink. A GFCI protects you from a dangerous electric shock.

For example, imagine you are in the bathroom using a hair dryer. Then, accidentally, water is splashed on you and the hair dryer. Since tap water is an electric conductor, some of the electric current from the outlet flows through you and not through the hair dryer. If a current flows through you, it could be fatal. As soon as the GFCI senses that not all of the current flows through the hair dryer, it opens the circuit and stops the current. It is able to react as quickly as 1/30 of a second.

Figure 19 The electric devices in a house are connected in parallel circuit. Each outlet or fixture is connected to a separate branch.

Active Reading **6. Paraphrase** In the chart below, contrast fuses and circuit breakers with ground fault circuit interrupters.

| **Fuse and Circuit breaker** |
| Function: |
| |
| Location: |
| |
| |

| **GFCI** |
| Function: |
| |
| Location: |
| |
| |

MiniLab *What else can a circuit do?* at
connectED.mcgraw-hill.com

Apply It!

After you complete the lab, answer the question below.

1. Name an electric device in your kitchen whose function could be accomplished without using electric energy. Explain how you could replace this using electric energy.

Electric Safety

An electric shock can be painful, and sometimes deadly. Each year, more than 500 people die by accidental electric shock in the United States.

What causes an electric shock?

An electric current follows the path of least electric resistance to the ground. That path could be through any good electric conductor, such as metal, wet wood, water, or even you! An electric shock occurs when an electric current passes through the human body. If you touch a bare electric wire or faulty appliance while you are grounded, an electric current could pass through you to the ground, resulting in a dangerous shock.

Current as small as 0.01 A can produce a painful shock. More than 0.1 A of electric current can cause death. The voltage of household electrical devices can cause dangerous amounts of current to pass through the body.

How can you be safe?

Listed below are some ways you can help protect yourself from a deadly electric shock:

- Never use electrical devices with damaged power cords.

- Stay away from water when using electrical devices plugged into an outlet.

- Avoid using extension cords and never plug more than two home appliances into an outlet at once.

- Never allow any object that you are touching to contact electric power lines, such as a kite string or a ladder.

- Do not touch anyone or anything that is touching a downed electric wire.

- Never climb utility poles or play on fences surrounding electricity substations.

A battery can be the source of electric energy in a circuit.

A series circuit has only one path for all devices in the circuit.

A parallel circuit has a separate path for each device in the circuit.

SC.6.N.1.4,
SC.6.N.1.5,
SC.7.N.1.3,
SC.7.P.11.2,
SC.7.N.1.1

Inquiry

⊕LAB STATION Try It!

Inquiry Lab *Design an Elevator* at connectED.mcgraw-hill.com

Use Vocabulary

1. Electric current decreases as more devices are added to a(n) _____ circuit.

2. Different amounts of current can flow through each device in a(n) _____ circuit.

Understand Key Concepts 🔑

3. One source of electric energy for an electric circuit is a SC.7.P.11.2
   (A) battery.   (C) switch.
   (B) lightbulb.   (D) wire.

4. **List** the basic components of a simple electric circuit. Explain why each of the components is necessary.

   _____

   _____

   _____

5. **Contrast** How does the electric current change when more devices are added to a series circuit? To a parallel circuit?

   _____

   _____

   _____

Interpret Graphics

6. **Organize Information** Fill in the graphic organizer below to show how electric energy is transformed to thermal energy by an electric current. Add additional boxes, if necessary. LA.6.2.2.3

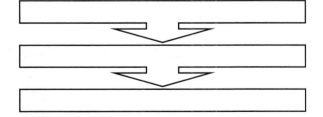

Critical Thinking

7. **Hypothesize** If a lightbulb in a circuit becomes dimmer, how did the energy supplied to the lightbulb change? Explain your answer.

   _____

   _____

   _____

Think About It! An electric energy source changes the motion of the electrons of a circuit, creating an electric current. Energy is conserved as a circuit transforms the energy of moving electrons to light, sound, and thermal energy.

🔑 Key Concepts Summary

Vocabulary

LESSON 1 Electric Charge and Electric Forces

- Particles that have the same type of electric charge repel each other. Particles that have different types of electric charge attract each other.
- Objects become negatively charged when they gain electrons, and become positively charged when they lose electrons.
- An **electric discharge** is the loss of **static charge.**

static charge p. 712
electric insulator p. 714
electric conductor p. 714
polarized p. 715
electric discharge p. 717
grounding p. 717

LESSON 2 Electric Current and Simple Circuits

- An **electric current** is the flow of electrically charged particles through a conductor.
- According to Ohm's law, across any portion of an **electric circuit, voltage** (V), current (I), and **electric resistance** (R) are related by the equation $V = IR$.

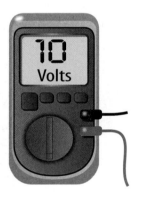

electric current p. 722
electric circuit p. 722
electric resistance p. 724
voltage p. 725
Ohm's law p. 726

LESSON 3 Describing Circuits

- Electric circuits have a source of electric energy to produce an electric current, one or more electric devices to transform electric energy to useful forms of energy, and wires to connect the circuit's device(s) to the energy source.
- A **series circuit** has one path in which current flows. A **parallel circuit** has more than one path in which current flows.

series circuit p. 733
parallel circuit p. 734

Assemble your lesson Foldables as shown to make a Chapter Project. Use the project to review what you have learned in this chapter.

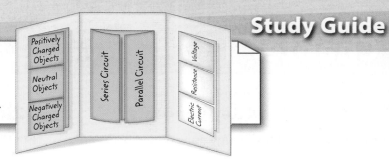

Use Vocabulary

1 An unbalanced electric charge is sometimes called a(n) _____.

2 Ohm's law states that as the voltage of a circuit increases, the current _____.

3 Electric charges flow easily in an electric _____.

4 A measure of the energy transformed in a portion of an electric circuit is _____.

5 Electric current decreases as more devices are added to a(n) _____ circuit.

6 Lightning is a(n) _____ that occurs in a fraction of a second.

7 A closed path in which electric charges flow is an electric _____.

8 Disconnecting one device in a _____ circuit will not cause other devices in the circuit to stop working.

Link Vocabulary and Key Concepts

Use vocabulary terms from the previous page and other terms from the chapter to complete the concept map.

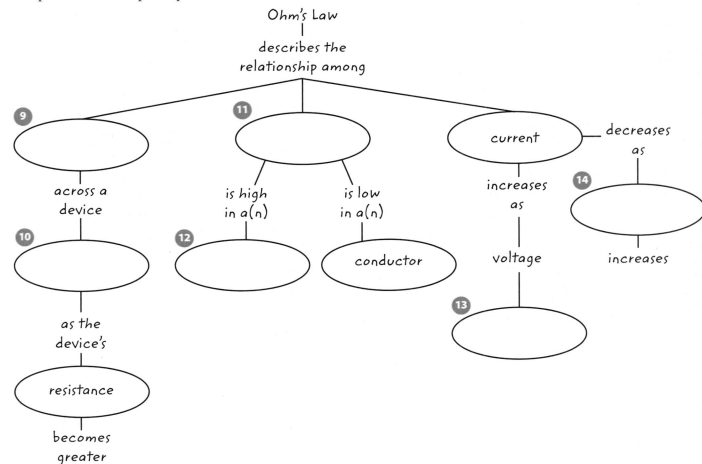

Fill in the correct answer choice.

🔑 Understand Key Concepts

1 The measure of the energy transformed between two points in an electric circuit is SC.7.P.11.2

Ⓐ resistance.
Ⓑ voltage.
Ⓒ electric current.
Ⓓ electric force.

2 The switch in a circuit breaker opens when which of the following in the circuit becomes too high? SC.7.P.11.2

Ⓐ current
Ⓑ resistance
Ⓒ static charge
Ⓓ total charge

3 The electric force between two charges increases when SC.6.P.13.1

Ⓐ both charges become negative.
Ⓑ both charges become positive.
Ⓒ the charges get closer together.
Ⓓ the charges get farther apart.

4 Which best describes an electrically charged object? SC.6.P.13.1

Ⓐ an electron without a static charge
Ⓑ an object that has gained or lost electrons
Ⓒ an object that has gained or lost neutrons
Ⓓ a proton without a static charge

5 In the figure below, what form of energy is being converted to electric energy? SC.7.P.11.2

Ⓐ chemical
Ⓑ light
Ⓒ nuclear
Ⓓ thermal

Critical Thinking

6 **Suggest** What are two ways the electric force between two charged objects can be increased? SC.7.N.1.4

_____

_____

7 **Evaluate** Dry air is a better electric insulator than humid air. Would the electric discharge from a charged balloon happen more slowly in dry air or humid air? Explain your answer. SC.7.N.1.4

_____

_____

8 **Recommend** If a metal wire is made thinner, how would you change the length of the wire to keep the electric resistance the same? SC.7.N.1.4

_____

_____

9 **Suggest** What are two ways to increase electric current in a simple circuit? SC.7.N.1.4

_____

_____

_____

10 **Evaluate** The current in a lightbulb stays the same, but the voltage across the lightbulb decreases. How does the electric energy to the lightbulb change? SC.7.N.1.4

_____

_____

_____

11 **Recommend** Show on the diagram below where a switch could be placed that would turn only lightbulb B off and on. SC.7.N.1.4

12 **Assess** Lightbulb A and lightbulb B are connected in a circuit. The voltage across lightbulb A is higher than the voltage across lightbulb B. Which lightbulb is brighter? Explain your reasoning. SC.7.N.1.4

_____

_____

_____

_____

Writing in Science

13 **Write** a short essay describing some of the electrical devices you use every day. Include in your essay the energy transformations that occur in each device. Use a separate sheet of paper. LA.6.2.2.3

Big Idea Review

14 How do electric charges flowing in a circuit regain some of the energy they transfer to atoms in a circuit? SC.6.P.13.1

_____

_____

_____

_____

Math Skills   MA.6.A.3.6

Use a Simple Equation

15 A current of 20 A flows into a hot water heater. If the electric resistance of the hot water heater is 12 Ω, what is the voltage across the hot water heater?

16 A refrigerator is plugged into an electrical outlet. If the voltage across the refrigerator is 120 V and the current flowing into the refrigerator is 6 A, what is the total electric resistance of the refrigerator?

17 What is the current in a flashlight bulb if the bulb has an electric resistance of 60 Ω and the voltage across the bulb is 6 V?

Fill in the correct answer choice.

Multiple Choice

Use the figures to answer questions 1 and 2.

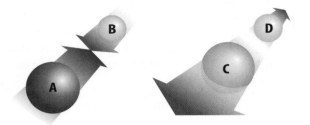

1 What must be true of particles A and B? SC.6.P.13.1

Ⓐ A and B must be like charges.

Ⓑ A and B must be opposite charges.

Ⓒ A must be positive, and B must be negative.

Ⓓ A must be negative, and B must be positive.

2 If the amount of charge on particles A, B C, and D are equal, what must be true of particle pairs AB and CD? SC.6.P.13.1

Ⓕ The attractive force of AB is greater than the repulsive force of CD.

Ⓖ The attractive force of AB is less than the repulsive force of CD.

Ⓗ The attractive force of AB equals the repulsive force of CD.

Ⓘ The repulsive force of AB is less than the attractive force of CD.

3 The movement of negative charges between nonconducting objects that are touching is called transferring charge by SC.6.P.13.1

Ⓐ conduction.

Ⓑ contact.

Ⓒ induction.

Ⓓ polarization.

4 Which phenomenon is an example of electric discharge? SC.7.P.11.2

Ⓕ friction

Ⓖ lightning

Ⓗ magnetism

Ⓘ static cling

5 The charged particles that move as an electric current are SC.7.P.11.2

Ⓐ electrons.

Ⓑ light particles.

Ⓒ metal atoms.

Ⓓ neutrons.

Use the table below to answer question 6.

Circuit	Current (A)	Resistance (Ω)
1	0.25	220
2	0.5	220
3	0.5	110

6 Based on the data, use Ohm's law to determine which statement is true. MA.6.A.3.6

Ⓕ Circuits 1 and 2 have the same voltage.

Ⓖ Circuits 2 and 3 have the same voltage.

Ⓗ Circuits 1 and 3 have the same voltage.

Ⓘ Each circuit has a different voltage.

7 When negative charges concentrate at one end of an object that is made of a conducting material, the object is SC.6.P.13.1

Ⓐ inducted.

Ⓑ insulated.

Ⓒ polarized.

Ⓓ undergoing friction.

Use the figure below to answer questions 8-10.

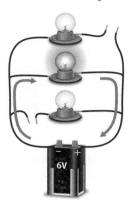

8 When the wires in this circuit are connected and all bulbs are lit, what type of circuit is shown? SC.7.P.11.2

Ⓕ negative

Ⓖ parallel

Ⓗ positive

Ⓘ series

9 If the broken wire of the top branch is connected, what will be true of the lightbulbs? SC.7.P.11.2

Ⓐ All of the bulbs would be lit.

Ⓑ None of the bulbs would be lit.

Ⓒ Only the top bulb would be lit.

Ⓓ Only the top two bulbs would be lit.

10 Connecting then disconnecting the broken ends of the top wire is similar to SC.7.N.1.4

Ⓕ adding then removing a fourth lightbulb.

Ⓖ adding then removing a second battery.

Ⓗ turning a switch on and off.

Ⓘ turning the middle lightbulb on and off.

11 What property of a wire changes when the wire is made thicker? SC.7.N.1.4

Ⓐ current

Ⓑ resistance

Ⓒ static charge

Ⓓ total electric charge

12 An electric field surrounds SC.6.P.13.1

Ⓕ an electron.

Ⓖ a neutron.

Ⓗ both an electron and a neutron.

Ⓘ neither an electron nor a neutron.

13 Which is a good conductor of electricity? SC.6.P.13.1

Ⓐ glass

Ⓑ gold

Ⓒ plastic

Ⓓ wood

14 Which lightbulbs in the diagram below will remain lit if the wire is disconnected at point B? SC.7.P.11.2

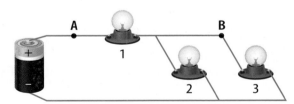

Ⓐ 1 and 2

Ⓑ 1 and 3

Ⓒ 2 and 3

Ⓓ 1, 2, and 3

Need Extra Help?

If You Missed Question...	1	2	3	4	5	6	7	8	9	10	11	12	13	14
Go to Lesson...	1	1	1	1	2	2	1	2	2	2	2	3	2	3

Benchmark Mini-Assessment   Chapter 18 • Lesson 1

Multiple Choice *Bubble the correct answer.*

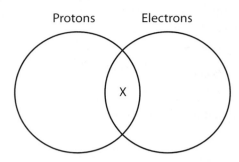

Protons   Electrons

X

1. In the Venn diagram above, which goes in X? **LA.6.2.2.3**

    (A) has the property of electric charge

    (B) is electrically neutral

    (C) makes an atom positively charged

    (D) moves from one object to another object

2. As two objects with oppositely charged electric fields are brought closer together, which will happen? **SC.6.P.13.1**

    (F) The electric fields will gain protons.

    (G) The objects will repel each other.

    (H) The electric fields will become negatively charged.

    (I) The strength of the electric field will increase.

Use the image below to answers questions 3 and 4.

Can X   Can Y

3. When a metal wire is placed between the two cans, which happens? **SC.6.P.13.1**

    (A) Negative charges flow from Can X to Can Y.

    (B) Negative charges flow from Can Y to Can X.

    (C) Positive charges flow from Can X to Can Y.

    (D) Positive charges flow from Can Y to Can X.

4. Which type of transferring of electrons is shown? **SC.6.P.13.1**

    (F) conduction

    (G) contact

    (H) grounding

    (I) induction

Copyright © Glencoe/McGraw-Hill, a division of The McGraw-Hill Companies, Inc.

Multiple Choice *Bubble the correct answer.*

1. Which cannot be inferred from the image above? **SC.7.N.1.6**

   (A) The battery and lightbulb form a closed circuit.

   (B) The connecting wires are electrically neutral.

   (C) The electric current is flowing clockwise through the circuit.

   (D) The resistance is lower in the battery than in the lightbulb.

2. Which does NOT affect an object's electrical resistance? **SC.7.N.1.3**

   (F) its length

   (G) its material

   (H) its temperature

   (I) its thickness

Use the image below to answer questions 3 and 4.

Circuit	Voltage of Battery	Current in Wire (in hundredths of an amp)
Lamp 1	9V	1
Lamp 2	12V	3
Lamp 3	25V	5
Lamp 4	36V	12

3. Each circuit in the table consists of a battery, wire, and a lamp. According to Ohm's law and the information in the table, which lamp has the greatest resistance? **MA.6.A.3.6**

   (A) Lamp 1

   (B) Lamp 2

   (C) Lamp 3

   (D) Lamp 4

4. If the lamps are identical, which lamp is transforming the least amount of electric energy to light energy? **MA.6.A.3.6**

   (F) Lamp 1

   (G) Lamp 2

   (H) Lamp 3

   (I) Lamp 4

Copyright © Glencoe/McGraw-Hill, a division of The McGraw-Hill Companies, Inc.

Multiple Choice *Bubble the correct answer.*

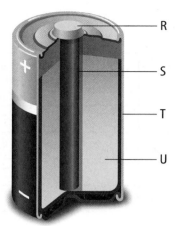

1. In the image of a battery shown above, in which part do chemical reactions occur? **SC.7.P.11.2**

   (A) R

   (B) S

   (C) T

   (D) U

2. Which is an example of electric energy transformed into kinetic energy? **SC.7.P.11.2**

   (F) an electric car being powered by a battery

   (G) an electric current powering a space heater

   (H) a computer monitor using electricity to produce images

   (I) a wind turbine's blades rotating to generate electric current

3. Which comparison between compact fluorescent lamps and regular lightbulbs is true? **SC.7.P.11.2**

   (A) CFLs are more efficient.

   (B) CFLs do not produce heat.

   (C) CFLs do not use electricity.

   (D) CFLs produce more light.

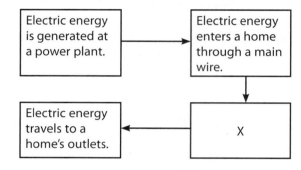

4. The flow chart above shows how electric energy can travel. Which belongs in the box containing X? **LA.6.2.2.3**

   (F) The current flows through a step-down transformer.

   (G) The current travels along a utility pole.

   (H) The main wire travels through an electric meter.

   (I) A series circuit allows the current to reach a home's wall switches.

Copyright © Glencoe/McGraw-Hill, a division of The McGraw-Hill Companies, Inc.

Notes

Name _____ Date _____

which pole is it?

The top picture shows a bar magnet with the poles labeled N and S. The magnet was cut into two pieces—one short piece and one long piece. How should the cut end of the short piece of magnet be labeled? Circle the answer that best matches your thinking.

A. N

B. S

C. No label—it no longer has a N or S pole on the cut end.

Explain your thinking. What rule or reasoning did you use to decide how to label the cut end of the short piece of magnet?

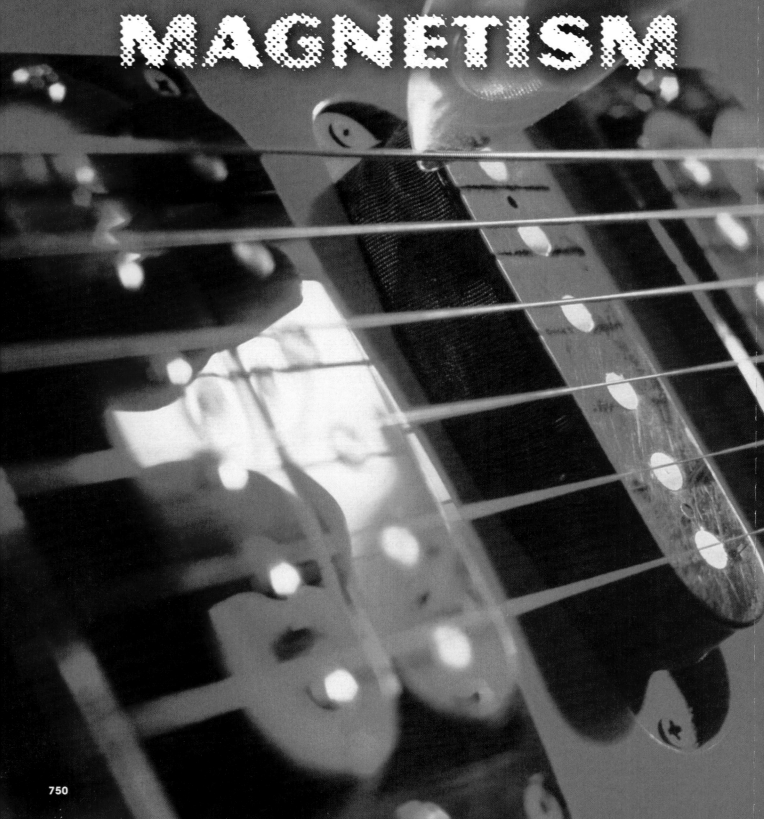

MAGNETISM

FLORIDA BIG IDEAS

1   The Practice of Science

11   Energy Transfer and Transformations

13   Forces and Changes in Motion

Think About It!

How are electric charges and magnetic fields related?

Since it was first manufactured in 1932, the electric guitar has inspired many new types of music. It is a prominent instrument in rock music and one of the most famous instruments to originate in the United States.

1 How do you think an acoustic guitar produces sound?

_____

_____

_____

2 Where do you think the magnets are in the picture?

_____

_____

_____

3 How are electric charges and magnetic fields related?

_____

_____

_____

_____

Get Ready to Read

What do you think about magnetism?

Before you read, decide if you agree or disagree with each of these statements. As you read this chapter, see if you change your mind about any of the statements.

		AGREE	DISAGREE
1	All metal objects are attracted to a magnet.	☐	☐
2	Two magnets can attract or repel each other.	☐	☐
3	A magnetic field surrounds a moving electron.	☐	☐
4	Unlike a permanent magnet, an electromagnet has two north magnetic poles or two south magnetic poles.	☐	☐
5	A battery produces an electric current that reverses direction in a regular pattern.	☐	☐
6	An electric generator transforms thermal energy into electric energy.	☐	☐

There's More Online!
Video • Audio • Review • ⓘLab Station • WebQuest • Assessment • Concepts in Motion • Multilingual eGlossary

Magnets and MAGNETIC FIELDS

ESSENTIAL QUESTIONS

 What types of forces do magnets apply to other magnets?

 Why are some materials magnetic?

 Why are some magnets temporary while others are permanent?

Vocabulary

magnet p. 753

magnetic pole p. 754

magnetic force p. 754

magnetic material p. 757

ferromagnetic element p. 757

magnetic domain p. 758

temporary magnet p. 759

permanent magnet p. 759

 Florida NGSSS

SC.6.N.1.1 Define a problem from the sixth grade curriculum, use appropriate reference materials to support scientific understanding, plan and carry out scientific investigation of various types, such as systematic observations or experiments, identify variables, collect and organize data, interpret data in charts, tables, and graphics, analyze information, make predictions, and defend conclusions.

SC.6.P.13.1 Investigate and describe types of forces including contact forces and forces acting at a distance, such as electrical, magnetic, and gravitational.

SC.7.N.1.6 Explain that empirical evidence is the cumulative body of observations of a natural phenomenon on which scientific explanations are based.

LA.6.2.2.3 The student will organize information to show understanding or relationships among facts, ideas, and events (e.g., representing key points within text through charting, mapping, paraphrasing, summarizing, or comparing/contrasting);

LA.6.4.2.2 The student will record information (e.g., observations, notes, lists, charts, legends) related to a topic, including visual aids to organize and record information, as appropriate, and attribute sources of information;

 Inquiry Launch Lab

SC.6.P.13.1

10 minutes

What does *magnetic* mean?

	Magnet		Nail		Rubbed Nail		Dropped Nail	
	"N" End	"S" End	Point	Head	Point	Head	Point	Head
Paper clip								
Paper								
Aluminum foil								
Choice #1								
Choice #2								

Procedure

1. Read and complete a lab safety form.

2. Touch each end of a **magnet** to a **paper clip**, **a piece of paper**, **a piece of aluminum foil**, and **two objects of your choosing**. Record your observations.

3. Repeat step 3, this time touching the ends of a **nail** to each object.

4. Rub the nail 25 times in the same direction across one end of the magnet. Repeat step 3, using the rubbed nail.

5. Drop the nail several times onto a hard surface. Repeat step 3 using the dropped nail. Record your observations.

Data and Observations

Think About This

1. **Point Out** When does the nail behave like the magnet?

2. **Explain** How do the two ends of the magnet interact with the various materials?

1. If you ever play with magnets you will probably notice that they seem to have magical powers. What makes magnets so special and necessary for everyone on Earth?

_____

_____

_____

_____

_____

_____

Magnets

Did you use a computer or a hair dryer today? Did you listen to a stereo system? It might surprise you to know that all these devices contain magnets. Magnets also are used to produce the electric energy that makes these familiar devices work. *A* **magnet** *is any object that attracts the metal iron.*

As you just read, magnets are used in many ways. Therefore, magnets are manufactured in many shapes and sizes, as shown in **Figure 1**. You might be familiar with common bar magnets and horseshoe magnets. Some of the magnets holding papers on your refrigerator might be disc-shaped or flat and flexible. However, all magnets have certain things in common, regardless of their shape and size.

 2. NGSSS Check Define Explain what a magnet is and list two examples of devices that use magnets. SC.6.P.13.1

_____

_____

_____

Figure 1 Magnets come in many sizes and shapes.

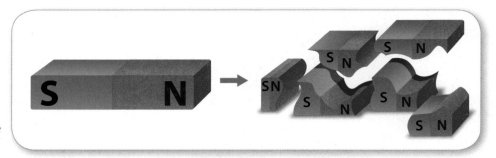

Figure 2 When a magnet is broken into pieces, each piece is still a magnet.

Magnetic Poles

You might have noticed that the ends of some magnets are different colors or labeled *N* and *S*. These colors and labels identify the magnet's magnetic poles. *A* **magnetic pole** *is a place on a magnet where the force it applies is strongest.* There are two magnetic poles on all magnets—a north pole and a south pole. As shown in **Figure 2**, if you break a magnet into pieces, each piece will have a north pole and a south pole.

The Forces Between Magnetic Poles

A force exists between the poles of any two magnets. If similar poles of two magnets, such as north and north or south and south, are brought near each other, the magnets repel. This means that the magnets will push away from each other. If the north pole of one magnet is brought near the south pole of another magnet, the two magnets attract. This means the magnets will pull together. In other words, as shown in **Figure 3**, similar poles repel, and opposite poles attract.

A force of attraction or repulsion between the poles of two magnets is a **magnetic force**. A magnetic force becomes stronger as magnets move closer together and becomes weaker as the magnets move farther apart.

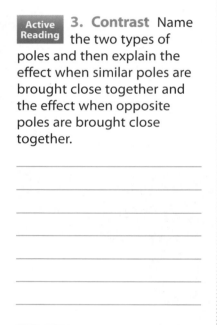

Active Reading **3. Contrast** Name the two types of poles and then explain the effect when similar poles are brought close together and the effect when opposite poles are brought close together.

_____

_____

_____

_____

_____

_____

_____

Figure 3 🔑 A magnetic force can be either a force of attraction or a force of repulsion.

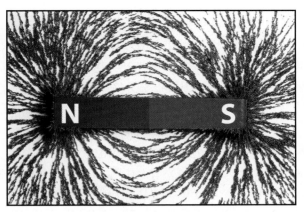
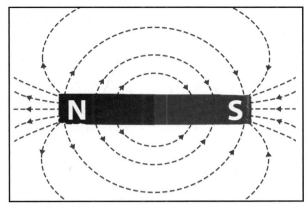

Figure 4 The magnetic field around a magnet can be shown with iron filings. Drawings of magnetic field lines include arrows to show the north-to-south direction of the magnetic field.

Magnetic Fields

Recall from a previous chapter that charged objects repel or attract each other even when they are not touching. Similarly, magnets can repel or attract each other even when they are not touching. An invisible magnetic field surrounds all magnets. It is this magnetic field that applies forces on other magnets.

Magnetic Field Lines

In **Figure 4**, iron filings have been sprinkled around a bar magnet. The iron filings form a pattern of curved lines that reveal the magnet's magnetic field.

A magnet's magnetic field can be represented by lines, called magnetic field lines. Magnetic field lines have a direction, which is shown on the right in **Figure 4**. Notice that the lines are closest together at the magnet's poles. This is where the magnetic force is strongest. As the field lines become farther apart, the field and the force become weaker.

Combining Magnetic Fields

What happens to the magnetic fields around two bar magnets that are brought together? The two fields combine and form one new magnetic field. The pattern of the new magnetic field lines depends on whether two like poles or two unlike poles are near each other, as shown in **Figure 5**.

Figure 5 When magnetic fields combine, they form field lines with different patterns.

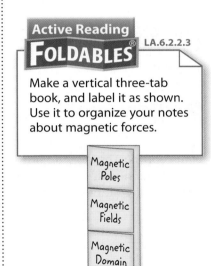

Active Reading
FOLDABLES® LA.6.2.2.3

Make a vertical three-tab book, and label it as shown. Use it to organize your notes about magnetic forces.

> Magnetic Poles
> Magnetic Fields
> Magnetic Domain

Active Reading **4. Point Out** Fill in the blanks below to identify which types of poles cause either an attraction or a repulsion.

Attraction

When _____ poles of two magnets are near each other, the resulting magnetic field applies a force of attraction.

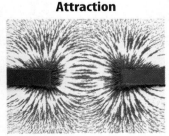

Repulsion

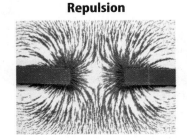

When _____ poles of two magnets are near each other, the resulting magnetic field applies a force of repulsion.

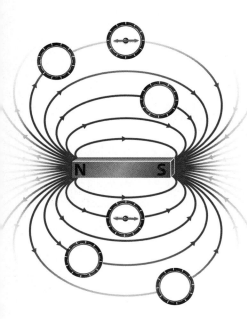

Figure 6 A compass needle will align with a magnet's magnetic field lines.

✓ **5. Visual Check**

Draw In the figure above, draw the needles in the four blank compasses. Why do the compasses' needles point in different directions?

_____

_____

_____

_____

Earth's Magnetic Field

A magnetic field surrounds Earth similar to the way a magnetic field surrounds a bar magnet. Earth has a magnetic field due to molten iron and nickel in its outer core. Like all magnets, Earth has north and south magnetic poles.

Compasses

Have you ever wondered why a compass needle points toward north? The needle of a compass is a small magnet. Like other magnets, a compass needle has a north pole and a south pole. Earth's magnetic field exerts a force on the needle, causing it to rotate. **Figure 6** shows that if a compass needle is within any magnetic field, including Earth's, it will line up with the magnet's field lines. A compass needle does not point directly toward the poles of a magnet. Instead, the needle aligns with the field lines and points in the direction of the field lines. Earth's magnetic poles and geographic poles are not in the same spot, as shown in **Figure 7**. Therefore, you cannot find your way to the geographic poles with only a compass.

Auroras

Earth's magnetic field protects Earth from charged particles from by the Sun. These particles can damage living organisms if they reach the surface of Earth. Earth's magnetic field deflects most of these particles. Sometimes, large numbers of particles from the Sun travel along Earth's magnetic field lines and concentrate near the magnetic poles. There, the particles collide with atoms of gases in the atmosphere, causing the atmosphere to glow. The light forms shimmering sheets of color known as auroras. An aurora is shown in **Figure 7**.

Figure 7 A magnetic field surrounds Earth. This magnetic field can cause auroras to occur near Earth's magnetic north pole and south pole.

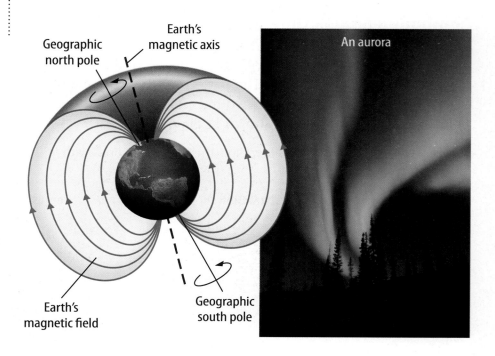

Geographic north pole

Earth's magnetic axis

An aurora

Earth's magnetic field

Geographic south pole

Magnetic Materials

You have read that magnets attract other magnets. But magnets also attract objects, such as nails, which are not magnets. *Any material that is strongly attracted to a magnet is a* **magnetic material**. Magnetic materials often contain ferromagnetic (fer oh mag NEH tik) elements. **Ferromagnetic elements** *are elements, including iron, nickel, and cobalt, that have an especially strong attraction to magnets.*

It is important to understand that while not all materials are magnetic, a magnetic field does surround every atom that makes up all materials. The field is created by the atom's constantly moving electrons. The interactions of these individual fields determine whether a material is magnetic. Next, you will read about why magnetic materials make good magnets.

REVIEW VOCABULARY

element
a substance made up of only one kind of atom

Active Reading **6. Analyze** In the table below, read each statement about magnetic materials. Decide whether each statement is true or false and explain your reasoning.

Statement	T or F	Explanation
All magnets are magnetic materials.		
All magnetic materials are magnets.		

Inquiry **Try It!**

SC.6.P.13.1,
SC.6.N.1.1,
SC.7.N.1.6

iLAB STATION

MiniLab *Where is magnetic north?* at connectED.mcgraw-hill.com

Apply It! After you complete the lab, answer these questions.

1. Why is the compass's needle a magnet and not simply a piece of magnetic material that is not a magnet?

2. Why did you move your compass away from the metal object in step 6 of the MiniLab?

7. Explain What makes a material a magnetic material? What do magnetic domains look like in a magnet?

_____

_____

_____

_____

_____

ACADEMIC VOCABULARY

align

(verb) to arrange in a straight line; adjust according to a line

Figure 8 🔑 In many materials, atoms act like tiny magnets. Most materials are nonmagnetic materials (a), but some are magnetic with magnetic domains (b). In a magnet, the poles of the domains line up (c).

Magnetic Domains

If all atoms act like tiny magnets, why do magnets not attract most materials, such as glass and plastic? The answer is that in most materials, the magnetic fields of the atoms point in different directions, as in part *a* of **Figure 8**. As the fields around the atoms combine, they cancel each other. Thus, most types of matter are nonmagnetic materials and are not attracted to magnets.

In a magnetic material, atoms form groups called magnetic domains. *A* **magnetic domain** *is a region in a magnetic material in which the magnetic fields of the atoms all point in the same direction,* as shown in part *b* of **Figure 8**. The magnetic fields of the atoms in a domain combine, forming a single field around the domain. Each domain is a tiny magnet with a north pole and a south pole.

When Domains Don't Line Up

Many magnetic materials are not magnets. This is because the magnetic fields of their domains point in random directions, as shown in part *b* of **Figure 8**. Similar to the atoms of a nonmagnetic material, the random magnetic fields of the domains cancel each other. Even though the individual domains are magnets, the entire object has no effective magnetic field.

When Domains Line Up

How do a bar magnet and a steel nail that is not a magnet differ? Both are magnetic materials. Both have atoms grouped into magnetic domains. However, for an object to be a magnet, its magnetic domains must **align** as in part *c* of **Figure 8**. When the domains align, their magnetic fields combine, forming a single magnetic field around the entire material. This causes the object to become a magnet.

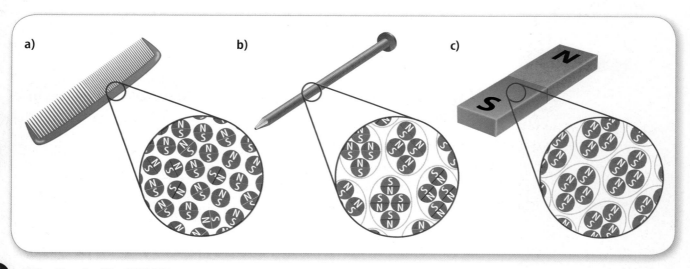

a)   b)   c)

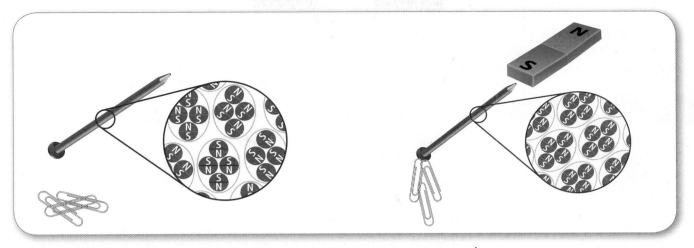

Figure 9 🔑 A nail is made of a magnetic material.

How Magnets Attract Magnetic Materials

Figure 9 shows a nail coming close to one of the poles of a bar magnet. Remember, even though the nail itself is not a magnet, each magnetic domain in the nail is a small magnet. The magnetic field around the bar magnet applies a force to each of the nail's magnetic domains. This force causes the domains in the nail to align along the bar magnet's magnetic field lines. As the poles of the domains of the nail point in the same direction, the nail becomes a magnet. Now the nail can attract other magnetic materials, such as paper clips.

Temporary Magnets

In **Figure 9**, the nail becomes a temporary magnet. *A magnet that quickly loses its magnetic field after being removed from a magnetic field is a* **temporary magnet**. The nail is a magnet only when it is close to the bar magnet. There, the magnetic field of the bar magnet is strong enough to cause the nail's magnetic domains to line up. However, when you move the nail away from the bar magnet, the poles of the domains in the nail return to pointing in different directions. The nail no longer is a magnet and no longer attracts other magnetic materials.

Permanent Magnets

A magnet that remains a magnet after being removed from another magnetic field is a **permanent magnet**. In a permanent magnet, the magnetic domains remain lined up. Some magnetic materials can be made into permanent magnets by placing them in a very strong magnetic field. This causes the magnetic domains to align and stay aligned. The material then remains a magnet after it is removed from the field.

✓ **8. Visual Check**
Describe What happens to the magnetic domains of the nail as it becomes a temporary magnet?

_____

_____

_____

Active Reading **9. Differentiate** How do a temporary magnet and a permanent magnet differ? Explain the reasons for the difference.

Temporary Magnet
Definition:
Reason:

Permanent Magnet
Definition:
Reason:

Visual Summary

When the magnetic fields of two magnets combine, a magnetic force exists between the magnets.

Objects made of magnetic materials are attracted to magnets.

The magnetic domains of a permanent magnet are aligned even when the magnet is not near another magnetic field.

Use Vocabulary

1. Elements that are strongly attracted to a magnet are _____.

2. The magnetic poles of the atoms in a(n) _____ point in the same direction.

3. An object that attracts other objects made of magnetic materials is a(n) _____. SC.6.P.13.1

Understand Key Concepts 🔑

4. **Identify** Where around a magnet is the magnetic field strongest?

   _____

5. Which does NOT contain magnetic domains?
   - (A) bar magnet
   - (B) disc magnet
   - (C) paper airplane
   - (D) steel nail

6. **Contrast** the magnetic domains in a steel nail that is far from a bar magnet and one that is close to a bar magnet. SC.6.P.13.1

   _____

   _____

Interpret Graphics

7. **Organize** Fill in the table below to describe the type of force between the magnetic poles. LA.6.2.2.3

Poles	Type of Force
South pole and south pole	
North pole and north pole	
South pole and north pole	

Critical Thinking

8. **Predict** Where would the north pole of a compass needle point if Earth's magnetic field switched direction?

   _____

   _____

9. **Evaluate** Two nails have identical sizes and shapes. In one nail, 20 percent of the domains are lined up. In the other nail, 80 percent of the domains are lined up. Which has a stronger magnetic field?

   _____

   _____

Roller Coasters

How Magnets Make a Wild Ride a Safe Ride

Roller coasters certainly have changed throughout their long history. Four hundred years ago, the ancestors of roller coasters appeared in Russia. These rides were long, steep wooden slides covered in ice. Some were over 20 m high. Riders slid down the slope in sleds made of wood or blocks of ice, crashing into a sand pile. Today's roller coasters can reach speeds over 160 km/h! Crashing into a sand pile at this speed would be no way to stop a ride safely.

Due to their safety, magnetic brakes are a technology that is gaining popularity with roller-coaster designers. Magnetic brakes rely on the interaction of magnetic fields—these brakes never come in contact with the cars. One design of magnetic brake uses rows of strong magnets built at the side of the track. A metal fin on the car passing between the rows of magnets creates a magnetic field that opposes the fin's motion. Magnetic braking is virtually fail-safe because it relies on the basic properties of magnetism and requires no electricity.

Engineers are designing future roller coasters with magnets to hold the cars to the rails during death-defying drops at extreme speeds never before possible. Computerized sensors and automatic detectors on new generation roller coasters will control the entire roller coaster experience from start to finish. Will science and technology soon make having fun on a roller coaster a process as sophisticated as a rocket launch?

Metal fin: typically copper or a copper/aluminum alloy.

Neodymium magnets: an alloy of neodymium, iron, and boron, currently the strongest type of permanent magnet.

Hydraulic system: High-pressure fluids push against a piston to lower the magnets when braking is not needed.

It's Your Turn

RESEARCH AND REPORT Roller-coaster cars are not the only kind of vehicles that make use of magnets. Research maglev trains to find out more about how these trains use magnets and magnetic fields to provide a safe, smooth ride. Write a short report to share what you learn with your class.

Making Magnets with an ELECTRIC CURRENT

Vocabulary

electromagnet p. 764
electric motor p. 766

 Florida NGSSS

SC.6.N.1.1 Define a problem from the sixth grade curriculum, use appropriate reference materials to support scientific understanding, plan and carry out scientific investigation of various types, such as systematic observations or experiments, identify variables, collect and organize data, interpret data in charts, tables, and graphics, analyze information, make predictions, and defend conclusions.

SC.6.P.11.2 Investigate and describe the transformation of energy from one form to another.

SC.6.P.13.1 Investigate and describe types of forces including contact forces and forces acting at a distance, such as electrical, magnetic, and gravitational.

LA.6.2.2.3 The student will organize information to show understanding or relationships among facts, ideas, and events (e.g., representing key points within text through charting, mapping, paraphrasing, summarizing, or comparing/contrasting);

 Launch Lab   SC.6.P.13.1
20 minutes

When is a wire a magnet?

In 1820, Hans Øersted made a remarkable discovery about how magnetism and electricity are parts of the same thing. How do magnetism and electricity relate?

Procedure

1. Read and complete a lab safety form.
2. With a **clamp attachment**, hang a length of **insulated wire** through the center of an iron ring on a **ring stand**. Make a small hole in the center of a 10-cm square piece of **cardboard**. Slide the wire through the hole. Rest the card on the iron ring. Secure the cardboard to the iron ring with **tape**.
3. Place four **small compasses** on the card in a circle around the wire.
4. Draw your setup. Include the wire through the card and the circle of compasses in your drawing. Indicate the direction that each compass points.
5. Use **alligator clip wires** to connect two **D-cell batteries in holders** in series with the ends of the wire. Draw a second diagram. Show the direction the compasses point with the batteries connected. Use another sheet of paper if necessary.
⚠ *Unhook the wire after a few seconds to prevent it from overheating!*

Think About This

1. **Infer** What would happen if you reversed the batteries?

2. **Conclude** Why do you think the compass needles behaved as they did after the batteries were connected?

 **What can magnets find?**

1. What does *metal detector* mean to you? Do you think of searching for buried treasure off Florida's Treasure Coast? Or, maybe a metal detector will help you find the lost gold of the Calusa tribes near Ft. Myers. Metal-detector technology is a huge part of our lives. Can you think of other ways these electrical and magnetic devices are used today?

_____

_____

_____

_____

_____

_____

Moving Charges and Magnetic Fields

Recall that a magnetic field surrounds a magnet. In addition, a magnetic field surrounds an electric current. This is why a compass needle moves when placed near a current-carrying wire. The needle moves because the magnetic field around the wire applies a force to the compass needle.

The Magnetic Field Around a Current

A magnetic field surrounds all moving charged particles. Remember that an electric current is the flow of electric charge. In a current-carrying wire, the magnetic fields of the flowing charges combine to produce a magnetic field around the wire, as shown in **Figure 10**. The field around the wire becomes stronger as the current in the wire increases, or as more electrons flow in the wire.

Figure 10 The magnetic field around a current-carrying wire forms closed circles.

| Active Reading | **2. Express** What effect does a magnet have on a wire carrying electric current? |

| A magnetic field surrounds moving charged particles. | A current-carrying wire contains moving negative charges. | Therefore, |

Figure 11 🔑 A magnet applies a force to a current-carrying wire. The direction of the force is related to the direction of the current.

Active Reading

FOLDABLES® LA.6.2.2.3

Make a vertical three-tab Venn book. Use it to compare and contrast the properties of permanent magnets and electromagnets.

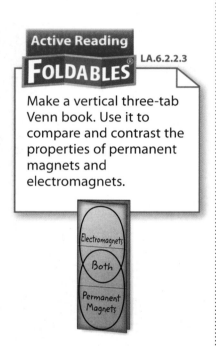

Figure 12 🔑 A ferromagnetic core becomes an electromagnet when surrounded by a current-carrying wire.

Magnets and Electric Currents

What happens when a magnet comes near a current-carrying wire? Because negatively charged electrons are moving in the wire, the magnet applies a force to the moving charges. This force causes the wire to move, as shown in **Figure 11**. The direction of the force on the wire depends on the direction of the current and the direction of the magnet's field.

Active Reading

3. Explain Why does a magnet exert a force on an electric current?

_____

_____

_____

Electromagnets

What do a stereo speaker and a hair dryer have in common? Both use electromagnets. *An* **electromagnet** *is a magnet created by wrapping a current-carrying wire around a ferromagnetic core.* The core becomes a magnet when an electric current flows through the coil. If a device uses electricity and has moving parts, chances are an electromagnet is inside.

Making an Electromagnet

Recall that an electric current in a wire produces a magnetic field around the wire. The strength of the magnetic field around a current-carrying wire increases when the wire is wound into a coil, as shown in the center of **Figure 12**. The magnetic fields around the individual loops of the coil combine, making the magnetic field around the entire coil stronger.

An electromagnet is made by placing a ferromagnetic material, such as iron, as a core within the wire coil, as shown on the right in **Figure 12**. The magnetic field of the coil causes the ferromagnetic core to become a magnet. The core greatly increases the strength of the coil's magnetic field. This makes the electromagnet's force stronger. And as the number of loops in the coil increases, the magnetic field of the electromagnet becomes stronger.

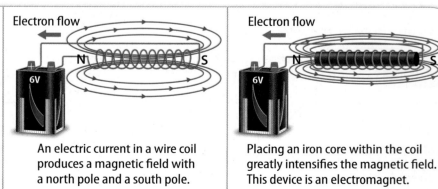

An electric current in a wire produces a magnetic field around the wire.

An electric current in a wire coil produces a magnetic field with a north pole and a south pole.

Placing an iron core within the coil greatly intensifies the magnetic field. This device is an electromagnet.

Properties of Electromagnets

An electromagnet differs from other magnets in several ways. First, the magnetic field around an electromagnet can be turned off by turning off the current. Without an electric current in the coil, an electromagnet is just a piece of iron wrapped with wire. But with the current on, it becomes a magnet.

Second, the strength of an electromagnet can be controlled. You read that the strength of the magnetic force of an electromagnet depends on the number of loops in the electromagnet's wire coil. Also, increasing the electric current in the coil strengthens the magnetic force.

Finally, the poles of an electromagnet can be reversed. Changing the direction of the current in the wire coil reverses an electromagnet's north and south magnetic poles.

Using Electromagnets

Because an electromagnet has reversible magnetic poles, other magnets can attract and then repel an electromagnet. For example, a loudspeaker produces sound waves when the direction of the electric current in its electromagnetic coil rapidly and repeatedly reverses. **Figure 13** shows the electromagnet in a loudspeaker that connects to a paper or plastic drive cone. The changing direction of the current causes the electromagnet to attract and then repel a permanent magnet. This makes the cone vibrate, producing sound waves.

Active Reading **4. Explain**
Complete the table to explain the function of each part of an electromagnet.

Part	Description
Battery	provides current
Wire coil	
Magnetic core	

Figure 13 A rapidly changing electric current in the coil of an electromagnet causes the drive cone of a loudspeaker to produce sound waves.

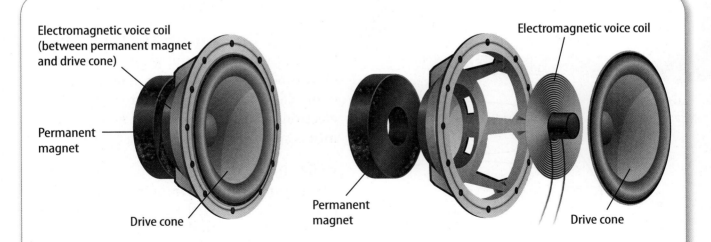

Electromagnetic voice coil (between permanent magnet and drive cone)

Permanent magnet

Drive cone

Electromagnetic voice coil

Permanent magnet

Drive cone

As the electric current in the voice coil changes direction (up to thousands of times per second), the coil's magnetic field changes with the current. The interaction of the coil's changing magnetic field and the permanent magnet causes the drive cone to move with a back-and-forth motion. The vibrating cone produces sound waves in the surrounding air.

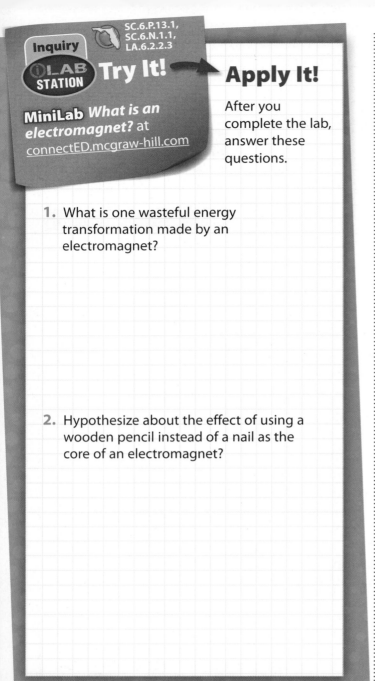

Inquiry

⊙LAB STATION Try It!

Apply It!

MiniLab *What is an electromagnet?* at connectED.mcgraw-hill.com

After you complete the lab, answer these questions.

1. What is one wasteful energy transformation made by an electromagnet?

2. Hypothesize about the effect of using a wooden pencil instead of a nail as the core of an electromagnet?

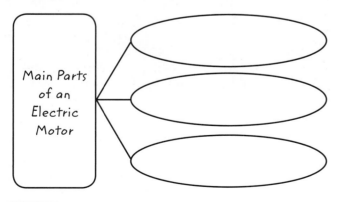

Main Parts of an Electric Motor

Active Reading **5. Identify** What are the 3 main parts of an electric motor?

Magnets and Electric Motors

Power tools, electric fans, hair dryers, computers, and even microwave ovens use electric motors. *An* **electric motor** *is a device that uses an electric current to produce motion.* How many devices can you think of that use electric motors?

A Simple Electric Motor

A simple electric motor is shown in **Figure 14**. The main parts of an electric motor are a coil of wire connected to a rotating shaft, a permanent magnet, and a source of electric energy, such as a battery.

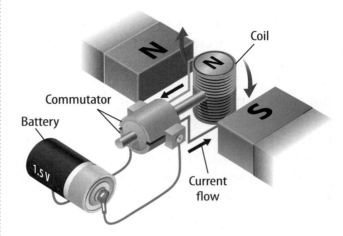

Coil

Commutator

Battery

1.5 V

Current flow

Figure 14 A rotating electromagnet in an electric motor is mounted between the poles of a permanent magnet. Here, a battery supplies a current to the electromagnet.

Making the Motor Spin

Recall that one useful property of electromagnets is that their magnetic poles easily can be reversed. This property of electromagnets is what makes an electric motor spin.

Unlike Poles Attract When an electric current is supplied to the motor, the unlike poles of the permanent magnet and electromagnet attract each other, causing the motor to begin to turn.

Reversing the Electric Current As the unlike poles of the motor's electromagnet and permanent magnet line up, the attraction between them will stop the motor's spinning. To keep the motor turning, the poles of the electromagnet must reverse.

The poles reverse when the direction of the current in the electromagnet changes, as shown in **Figure 15**. The commutator is the device in the motor that reverses the direction of the current in the electromagnet.

Similar Poles Repel Now, the similar poles of the electromagnet and the permanent magnet are close to each other. Thus, the poles repel each other and the electromagnet keeps spinning.

To keep the electromagnet spinning, the direction of the current in the coiled wire must continue changing direction. The commutator reverses the current when the poles of the electromagnet come near the poles of the permanent magnet. The permanent magnet and the electromagnet continuously attract, then repel each other. This keeps the electromagnet spinning.

Using Electric Motors

You read that an electric motor includes an electromagnet mounted on a shaft. As the electromagnet spins, the shaft spins, too. This spinning motion can be used to create other motions. Can you think of any parts powered with an electric motor in a car?

A system of levers connected to an electric motor produces the back-and-forth motion of the windshield wipers. Electric motors make the power windows go up and down. A compact disk player uses an electric motor to spin a CD and move a small laser across the disk. And now, electric motors are replacing gasoline engines in many cars.

Figure 15 🔑 The controlled forces of attraction and repulsion between an electromagnet and a permanent magnet make an electric motor spin.

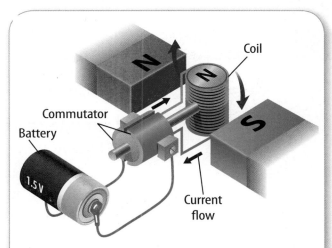

The opposite magnetic poles of the electromagnet and the permanent magnet attract each other and make the electromagnet rotate.

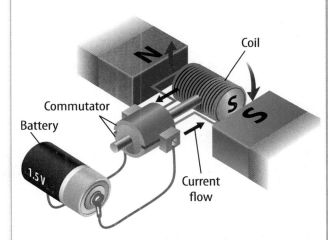

The commutator causes the current in the electromagnet to change direction, which reverses the poles of the electromagnet. The magnets now repel each other and the motor keeps rotating.

Active Reading **6. Describe** What happens to the magnetic poles of the permanent magnets in an electric motor as the electromagnet rotates?

_____

_____

_____

_____

Visual Summary

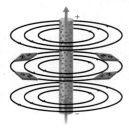

A current-carrying wire is surrounded by a magnetic field.

A magnetic core within a curent-carrying wire coil is an electromagnet.

Alternating attraction and repulsion between a permanent magnet and an electromagnet causes an electric motor to rotate.

Inquiry SC.6.P.13.1,
SC.6.N.1.1

LAB STATION Try It!

Skill Lab *How can you measure an electric current?* at connectED.mcgraw-hill.com

Use Vocabulary

1 A current-carrying wire coil around a ferromagnetic core is a(n) _____ .

2 An electric current causes a shaft in a(n) _____ to rotate. SC.6.P.13.1

Understand Key Concepts 🔑

3 **Identify** Why does a magnet apply a force to a current-carrying wire?

_____

_____

4 **Summarize** How do electromagnets and permanent magnets differ?

_____

_____

_____

5 In an electric motor, a force is applied to the electromagnet by
Ⓐ a battery.   Ⓒ an electric current.
Ⓑ a commutator.   Ⓓ a permanent magnet.

Interpret Graphics

6 **Compare** How would the magnetic field around the wire change if the electrons flowed in the opposite direction?

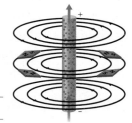

_____

_____

7 **Sequence Events** Fill in the graphic organizer below to show the sequence of events that occur during one revolution of an electric motor. LA.6.2.2.3

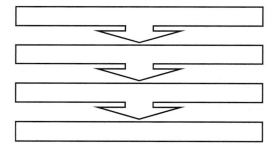

Critical Thinking

8 **Predict** A current-carrying wire made of nonmagnetic copper is attracted to a bar magnet. How will the force between the wire and the magnet change if more current flows in the wire?

_____

_____

_____

🔑 **Model** magnetic domains in groups of atoms as described below. Draw and label each material. Use these labels:

- magnet
- non-magnet
- magnetic material

16 atoms that are not arranged in magnetic domains	16 atoms arranged in 4 magnetic domains that do not line up	16 atoms arranged in 4 magnetic domains that line up
1.	2.	3.

4. 🔑 **Analyze It** Think about materials you encounter every day at home and at school. Which ones are generally magnetic? How does your observation relate to why Earth has a magnetic field?

_____

_____

_____

_____

_____

Making an Electric Current with MAGNETS

ESSENTIAL QUESTIONS

 How can a wire and a magnet produce an electric current?

 How do electric generators create an electric current?

 How are transformers used to bring an electric current into your home?

Vocabulary

electric generator p. 772

direct current p. 773

alternating current p. 773

turbine p. 774

transformer p. 775

 Florida NGSSS

SC.6.N.1.1 Define a problem from the sixth grade curriculum, use appropriate reference materials to support scientific understanding, plan and carry out scientific investigation of various types, such as systematic observations or experiments, identify variables, collect and organize data, interpret data in charts, tables, and graphics, analyze information, make predictions, and defend conclusions.

SC.6.N.1.5 Recognize that science involves creativity, not just in designing experiments, but also in creating explanations that fit evidence.

SC.6.P.13.1 Investigate and describe types of forces including contact forces and forces acting at a distance, such as electrical, magnetic, and gravitational.

SC.7.P.11.2 Investigate and describe the transformation of energy from one form to another.

SC.7.P.11.3 Cite evidence to explain that energy cannot be created nor destroyed, only changed from one form to another.

LA.6.2.2.3 The student will organize information to show understanding or relationships among facts, ideas, and events (e.g., representing key points within text through charting, mapping, paraphrasing, summarizing, or comparing/contrasting);

MA.6.A.3.6 Construct and analyze tables, graphs, and equations to describe linear functions and other simple relations using both common language and algebraic notation.

 inquiry Launch Lab

SC.6.P.13.1

20 minutes

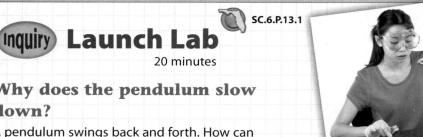

Why does the pendulum slow down?

A pendulum swings back and forth. How can magnetism be used to stop a pendulum?

Procedure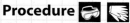

1. Read and complete a lab safety form.

2. Using a **thread**, suspend a **strong magnet** from a **ring stand**. Adjust the height of the ring so that the magnet hangs 0.5 cm off the table.

3. Pull the pendulum back 5 cm, and gently release it. Count the swings until the pendulum comes to a stop. Record your result.

4. Fold a **foil square** into a smooth 3-cm-wide strip. With **tape**, secure the foil under the pendulum, parallel to the direction of swing. The magnet should not touch the foil.

5. Repeat step 3. Record your results.

Data and Observations

Think About This

1. **Differentiate** How might your results differ if you used a marble or a baseball as the pendulum?

2. **Explain** Describe the interaction of the foil and the magnet.

Inquiry How bright is a magnet?

1. Riding a bicycle safely at night means being visible to other vehicles. A small generator mounted against the bicycle's wheel provides the electric current that powers the lights, which help keep your ride safe. How do you think this simple device changes the motion of your bicycle into a bright light at night?

_____

_____

_____

_____

_____

_____

Magnets and Wire Loops

In the previous lesson, you read that an electric current is surrounded by a magnetic field. You experience a magnetic field surrounding an electric current every time you use a device that has an electric motor. Now, you will read about how magnetic fields can produce electric currents. Electric power plants use powerful magnetic fields to produce the electric energy supplied to homes and businesses.

Generating Electric Current

How can a magnetic field produce an electric current? **Figure 16** shows one way. A magnet is moved through a wire coil that is part of a closed electric circuit. As the magnetic field moves over the wire coil, an electric current is produced in the circuit. When the magnet stops moving, there is no current in the circuit. The direction of the current depends on the direction in which the magnet moves.

Active Reading	**2. Relate** How can a wire and a magnet produce an electric current? Highlight your answer in the text above.

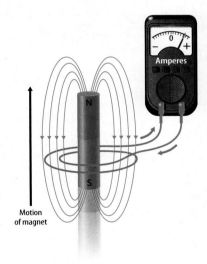

Figure 16 🔑 An electric current is produced in a wire coil when a magnet moves within the coil.

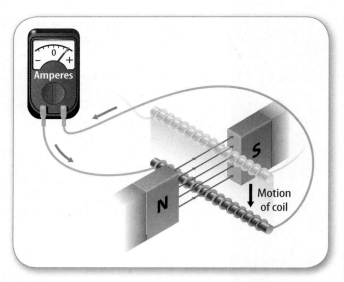

Figure 17 🔑 When a wire moves through a magnetic field, an electric current is produced in the circuit. There is no current when the coil is not moving.

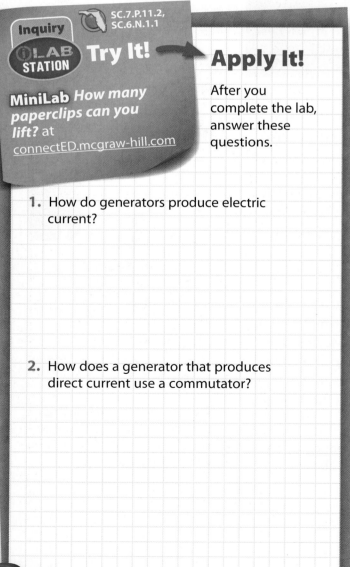

Inquiry

SC.7.P.11.2, SC.6.N.1.1

🔵 **LAB STATION** Try It!

Apply It!

MiniLab *How many paperclips can you lift?* at connectED.mcgraw-hill.com

After you complete the lab, answer these questions.

1. How do generators produce electric current?

2. How does a generator that produces direct current use a commutator?

A Moving Wire and an Electric Current.

You just read that a magnetic field passing over a wire coil produces an electric current. **Figure 17** shows moving a wire coil through a magnetic field as another way to generate an electric current. The magnetic force between the magnet and the electrons in the wire causes the electrons in the wire to move as an electric current in the wire.

Either a magnetic field passing over a wire coil or a wire coil passing through a magnetic field produces an electric current. You will now read how this relationship powers the many electric devices you use every day.

Active Reading **3. Describe** How can a magnetic field produce an electric current in a wire coil?

_____

_____

_____

_____

Electric Generators

When you turn on a flashlight, batteries produce an electric current in the lightbulb. However, when you turn on a TV, large electric generators in distant power plants provide the electric current to the TV. *An* **electric generator** *is a device that uses a magnetic field to transform mechanical energy to electric energy.*

A Simple Electric Generator

Figure 18 on the next page shows a hand generator in a circuit. The crank rotates a wire coil through the magnetic field of a small permanent magnet. This produces an electric current in the circuit. The current continues as the crank continues to rotate the coil within the magnetic field.

Unlike the hand-cranked generator that uses the magnetic field of a small permanent magnet, larger generators often use powerful electromagnets to produce a magnetic field.

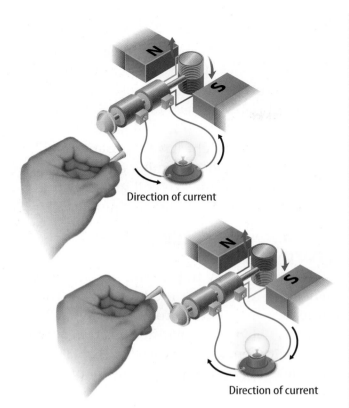

Direction of current

Direction of current

Figure 18 A generator produces an electric current when a wire coil rotates between the poles of two magnets.

Direct Current and Alternating Current

The electric current produced by a battery differs from the current produced by the generator in **Figure 18.** The current produced by a battery flows in a circuit in only one direction. *An electric current that flows in one direction is* **direct current**.

The current produced by a generator changes direction as the poles of the rotating coil line up with the poles of the magnet. *An electric current that changes direction in a regular pattern is* **alternating current**.

Some generators do produce direct current. They use a commutator, such as the one in **Figure 15** in Lesson 2. A commutator is a switch that rotates with the wire coil and prevents the current from changing directions. This results in direct current.

Active Reading

FOLDABLES LA.6.2.2.3

Make a vertical three-tab book. Label it as shown. Use it to organize notes about the electric current in each device.

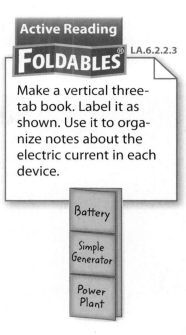

Battery

Simple Generator

Power Plant

Active Reading

4. Sequence Fill in the table to explain how a simple electric generator, such as a hand generator, produces an electric current.

A wire loop inside the generator is connected in a _____. The loop is between _____.

⬇

As the crank is turned, the _____ rotates through the _____ producing _____ through the circuit and through any device in the circuit.

⬇

The current continues as long as the _____ is turned, rotating the _____ within the _____.

Figure 19 Enormous electric generators at Hoover Dam on the Colorado River generate the electric energy used by more than 1 million households in the region.

Figure 20 This turbine will connect to an electric generator. There, it will spin as steam is forced through its blades.

Active Reading

5. Recall How is a turbine used to produce an electric current?

_____

_____

_____

_____

_____

_____

Generators and Power Plants

Electric power plants use huge generators, such as the ones shown in **Figure 19**. They generate the electric current used in thousands of homes. Instead of one wire coil, these generators have several large coils of wire. Each coil might have thousands of loops. Increasing the number of coils and the number of loops in each coil increases the amount of current a generator produces.

Mechanical Energy to Electric Energy

A useful energy transformation occurs in a generator. A generator produces an electric current as its wire coils rotate through a magnetic field. The rotating coils transform mechanical energy to electric energy. The electric energy is the kinetic energy of the current of electrons in the coil.

Supplying Mechanical Energy

As you turn a hand generator, you supply mechanical energy to the generator. The generators in electric power plants also need a source of mechanical energy to keep the coils rotating. Some power plants use the mechanical energy of high-pressure jets of steam from boiling water. Hydroelectric power plants use the mechanical energy of falling water.

In electric power plants, a generator usually is connected to a turbine (TUR bine), as shown in **Figure 20**. *A **turbine** is a shaft with a set of blades that spins when a stream of pressurized fluid strikes the blades.* A turbine transfers the mechanical energy of a stream of water, steam, or air to the generator.

Modern turbines that generate electric power are very efficient. They waste little of the mechanical energy used to rotate them.

Transformers—Changing Voltage

Recall that voltage is a measure of the energy transformed by a circuit. A high-voltage circuit transforms more energy than a low-voltage circuit with the same electric current.

Household electrical outlets provide an electric current at 120 V—great enough to cause a dangerous electric shock. However, in large transmission wires, voltage can exceed 500,000 V! At electric power plants, transformers raise the voltage for long-distance transmission. Then, other transformers lower the voltage for household use. A **transformer** *is a device that changes the voltage of an alternating current.*

How a Transformer Works

A transformer consists of two wire coils wrapped around a single iron core, as shown in **Figure 21**. Alternating current in the primary coil produces a continually reversing magnetic field around the core. This changing magnetic field produces alternating current in the secondary coil. If the input voltage of the primary coil increases, the output voltage of the secondary coil increases. And, if the input voltage of the primary coil decreases, the output voltage of the secondary coil decreases.

Step-Down Transformer The output voltage from a transformer also depends on the number of loops in the transformer's coils. If there are fewer loops in the secondary coil than in the primary coil, the transformer's output voltage is less than the input voltage. This is a step-down transformer.

Step-Up Transformer If the number of loops in the secondary coil is greater than the number of loops in the primary coil, the output voltage is greater than the input voltage. This type of transformer is called a step-up transformer.

Figure 21 A transformer's output voltage can be more or less than its input voltage.

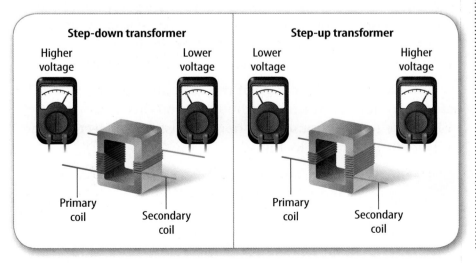

Step-down transformer
Higher voltage — Lower voltage
Primary coil — Secondary coil

Step-up transformer
Lower voltage — Higher voltage
Primary coil — Secondary coil

Math Skills   MA.6.A.3.6

Solve an Equation

You can use math to find the output voltage in a transformer.

Variable	Notation
Number of Loops on Primary Coil	N_p
Number of Loops on Secondary Coil	N_s
Input Voltage to Primary Coil	V_p
Output Voltage from Secondary Coil	V_s

A transformer's primary coil has **200 loops**. Its secondary coil has **3,000 loops**. If its input voltage is **90 V**, what is its output voltage?

$$V_s = V_p \times (N_s \div N_p)$$

Replace the terms in the equation and solve.

$$V_s = 90.0\text{ V} \times (3{,}000\text{ loops}/200\text{ loops}) = 1{,}350\text{ V}$$

Practice

6. A transformer's primary coil has 50 loops. Its secondary coil has 1,500 loops. If its input voltage is 120 V, what is its output voltage?

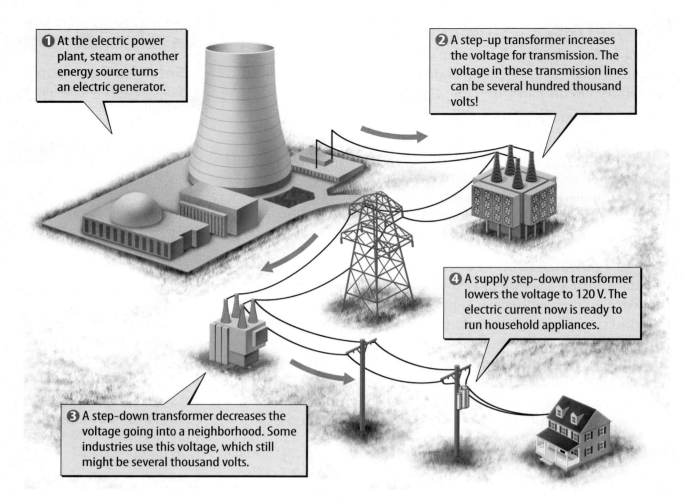

① At the electric power plant, steam or another energy source turns an electric generator.

② A step-up transformer increases the voltage for transmission. The voltage in these transmission lines can be several hundred thousand volts!

③ A step-down transformer decreases the voltage going into a neighborhood. Some industries use this voltage, which still might be several thousand volts.

④ A supply step-down transformer lowers the voltage to 120 V. The electric current now is ready to run household appliances.

Figure 22 Power plants move electric energy through long-distance wires. Transformers raise and lower the voltage of the current as needed by homes and businesses.

☑ **7. Visual Check** Point Out What type of transformer is used to make electric current safe to enter a home?

_____

_____

_____

_____

_____

Electric Energy—From a Power Plant to Your Home

Electric energy is transmitted from an electric power plant to homes, as **Figure 22** shows. Transformers control the voltage of the current as it travels from the power plant to homes and businesses. Step-up transformers increase the voltage of the alternating current produced at electric power plants.

High-voltage current is transmitted through transmission wires, or lines. The electric resistance of the wires causes some electric energy to transform to thermal energy. High voltage current in transmission wires reduces the amount of electric energy released as thermal energy to the environment.

However, the high voltage of the current must be reduced for safe use in homes. At neighborhood substations and on utility poles close to your house, step-down transformers decrease the voltage of the current to 120 V.

Active Reading **8. Explain** What is the function of transformers as electric power is delivered to your home?

_____

_____

Visual Summary

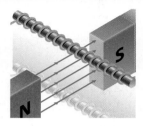

An electric current is produced in a wire coil that passes through a magnetic field.

Electric generators transform mechanical energy to electric energy.

A magnetic field can produce an electric current, and an electric current produces a magnetic field.

SC.6.P.13.1, SC.7.P.11.2, SC.6.N.1.1, SC.6.N.1.5

Inquiry

①LAB STATION Try It!

Inquiry Lab *Design a Wind-Powered Generator* at connectED.mcgraw-hill.com

Use Vocabulary

1. A device that transforms mechanical energy into electric energy is a(n) _____.

2. Electric current that reverses direction in a regular pattern is _____.

3. A device that changes the voltage of an alternating current is a(n) _____.

Understand Key Concepts 🔑

4. What do step-up transformers do?
   (A) control resistance   (C) increase current
   (B) decrease voltage   (D) increase voltage

5. **Explain** how an electric generator produces an electric current. SC.7.P.11.2

_____

_____

_____

Interpret Graphics

6. **Sequence Events** Show how transformers are used as electric energy is transmitted from a power plant to a home. LA.6.2.2.3

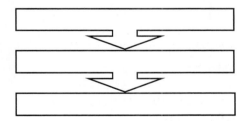

Critical Thinking

7. **Assess** Describe the electric current flowing in the primary coil of a transformer if there is no current flowing in the secondary coil. Explain your answer. SC.6.P.11.1

_____

_____

_____

Math Skills   MA.6.A.3.6

8. A transformer's primary coil has 10,000 loops. The secondary coil has 100 loops. The input voltage is 120 V What is the output voltage?

Think About It! A magnetic field applies a non-contact force to an electrically charged object. A magnetic push or pull causes a useful change in the motion of the shaft of an electric motor.

Key Concepts Summary

Vocabulary

LESSON 1 Magnets and Magnetic Fields

- A **magnet** is surrounded by a magnetic field that exerts forces on other magnets.
- **Magnetic materials** have **magnetic domains** that point in the same direction.
- The domains of a **temporary magnet** do not remain aligned when removed from a magnetic field. The domains of a **permanent magnet** remain in alignment, even when they are not near another magnet.

magnet p. 753
magnetic pole p. 754
magnetic force p. 754
magnetic material p. 757
ferromagnetic element p. 757
magnetic domain p. 758
temporary magnet p. 759
permanent magnet p. 759

LESSON 2 Making Magnets with an Electric Current

- A magnetic field exerts a force on any moving, electrically charged particle, including the charged particles in an electric current.
- The magnetic field around an **electromagnet** can be turned on and off, can reverse direction, and can be made stronger or weaker.
- An **electric motor** contains a permanent magnet and an electromagnet. When an electric current flows in the electromagnet, the forces between the two magnets cause the electromagnet to rotate.

electromagnet p. 764
electric motor p. 766

LESSON 3 Making an Electric Current with Magnets

- An electric current is produced when a magnet and a closed wire loop move past each other.
- An **electric generator** has wire loops that rotate within a magnetic field. The magnetic field causes a current of electric charges to move in the wire.
- A **transformer** changes the voltage of an **alternating current**. Step-up transformers raise voltage for cross-country transmission. Step-down transformers lower voltage for household use.

electric generator p. 772
direct current p. 773
alternating current p. 773
turbine p. 774
transformer p. 775

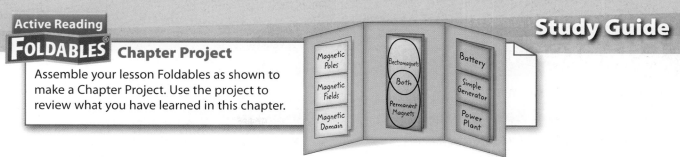

FOLDABLES Chapter Project

Assemble your lesson Foldables as shown to make a Chapter Project. Use the project to review what you have learned in this chapter.

Use Vocabulary

1 A(n) _____ can change the mechanical energy of a turbine into electric energy.

2 Similar poles of the _____ of a magnet point in the same direction.

3 An electric generator operates like a(n) _____ in reverse.

4 The magnetic field around a wire reverses direction in a regular pattern if a(n) _____ flows in the wire.

5 A(n) _____ remains a magnet for a long period of time.

6 A core of a ferromagnetic material placed inside the coil of a wire is a(n) _____.

Link Vocabulary and Key Concepts

Use vocabulary terms from the previous page and other terms from the chapter to complete the concept map.

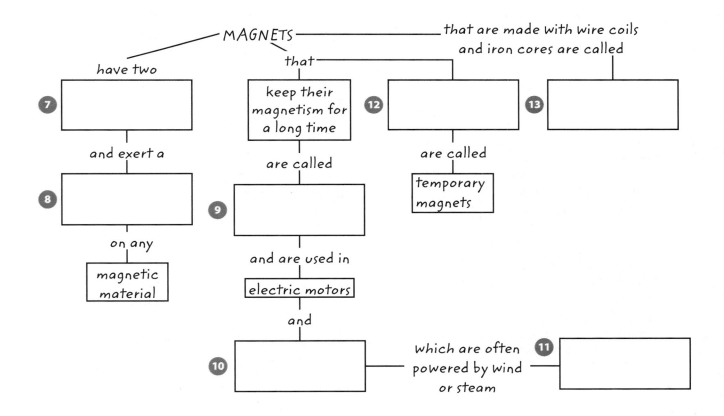

Fill in the correct answer choice.

🔑 Understand Key Concepts

1 When the current in an electromagnet increases, what does its magnetic field do? SC.6.P.13.1
- Ⓐ changes direction
- Ⓑ does not change
- Ⓒ gets stronger
- Ⓓ gets weaker

2 What is the function of the commutator in an electric motor? SC.6.P.13.1
- Ⓐ It decreases the current in the coil.
- Ⓑ It reverses the current in the coil.
- Ⓒ It reverses the poles of the permanent magnet.
- Ⓓ It weakens the field of the permanent magnet.

3 When a bar magnet moves in a wire loop, what does the wire loop produce? SC.7.P.11.2
- Ⓐ an electric current
- Ⓑ electrical resistance
- Ⓒ ferromagnetic elements
- Ⓓ magnetic domains

4 Which is NOT a ferromagnetic element? SC.6.P.13.1
- Ⓐ aluminum
- Ⓑ cobalt
- Ⓒ iron
- Ⓓ nickel

5 Which best describes the force between the two magnets shown below? SC.6.P.13.1

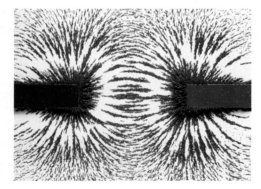

- Ⓐ alternating
- Ⓑ attractive
- Ⓒ direct
- Ⓓ repulsive

Critical Thinking

6 **Predict** A step-up transformer is used to connect a lightbulb to a wall outlet. What happens to the brightness of the lightbulb if the transformer is replaced with a step-down transformer? SC.6.P.11.1

_____

_____

_____

7 **Defend** One electromagnet has a wood core. Another electromagnet has an iron core. Which has the stronger magnetic field? Explain your answer. SC.7.P.11.2

_____

_____

_____

_____

8 **Assess** Compare and contrast the ways magnetic poles interact with the ways electric charges interact. SC.6.P.13.1

_____

_____

_____

_____

9 **Suggest** How could you determine a car door is made of a nonmagnetic material, such as plastic? SC.6.P.13.1

_____

_____

_____

_____

10 **Suggest** What would you do to a wire to make it attract to a magnet? SC.7.P.11.1

_____

_____

_____

11 **Evaluate** The figure below shows the field lines of a magnet. Where is the magnet's north pole? Explain your answer. SC.6.P.13.1

_____

_____

_____

12 **Recommend** How could you increase the number of times per second that an alternating current from a generator changes direction? SC.7.P.11.2

_____

_____

_____

13 **Hypothesize** An alternating current flows in an electromagnet. If the electromagnet is placed in a wire coil, will there be an electric current in the wire coil? Explain your answer. SC.7.P.11.2

_____

_____

_____

_____

Writing in Science

14 **Write** a short essay describing the different devices you use in a typical day that have electric motors. Include how the electric motors are used to make something move. Use a separate sheet of paper. LA.6.2.2.3

Big Idea Review

15 How can a magnetic field cause a current to flow in a wire coil? SC.7.P.11.2

_____

_____

_____

16 How are electric charges and magnetic fields related? SC.7.P.11.2

_____

_____

_____

Math Skills MA.6.A.3.6

17 A transformer has 300 loops on its primary coil and 90,000 loops on its secondary coil. The input voltage is 60 V. What is the output voltage?

18 A transformer has 80 loops on its primary coil and 1,200 loops on its secondary coil. The input voltage is 60 V. What is the output voltage?

19 The primary coil of a transformer has 150 loops and has a 120 V primary source. How many loops should the secondary coil have to produce an output voltage of 600 V?

Florida NGSSS

Fill in the correct answer choice.

Multiple Choice

1 Which statement about magnets is true? SC.6.P.13.1

Ⓐ Two north poles attract each other.

Ⓑ Two south poles attract each other.

Ⓒ A north pole and a south pole attract each other.

Ⓓ A north pole and a south pole repel each other.

Use the images below to answer question 2.

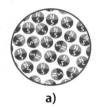 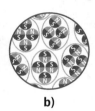 

a)   b)   c)

2 Which statement best describes these models? LA.6.2.2.3

Ⓕ Only A is a magnet.

Ⓖ Only C is a magnet.

Ⓗ A and B are magnets.

Ⓘ A and C are magnets.

3 Why does a compass needle move if held near a current-carrying wire? SC.6.P.13.1

Ⓐ A current-carrying wire will cause any object to move.

Ⓑ A compass needle moves when it is near any piece of metal wire.

Ⓒ Both the compass needle and the current-carrying wire have magnetic fields.

Ⓓ The compass needle and the current-carrying wire have the same electric charge.

Use the image below to answer question 4.

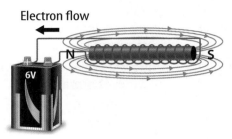

Electron flow

4 Which describes what will happen if the iron core is removed from the electromagnet? SC.6.P.13.1

Ⓕ It is no longer a magnet.

Ⓖ It is a stronger magnet.

Ⓗ It is a weaker magnet.

Ⓘ It loses its north and south poles.

5 What causes the poles of the electromagnet in an electric motor to reverse? SC.6.P.13.1

Ⓐ a change in the direction of the current

Ⓑ a change in the strength of the magnetic field

Ⓒ a change in the shape of the current-carrying wire

Ⓓ a change in the position of the permanent magnet

6 What happens when a magnet passes through a closed loop of wire? SC.7.P.11.2

Ⓕ An electric current is generated in the wire.

Ⓖ Any nearby magnets gain an electric charge.

Ⓗ The looped part of the wire no longer is attracted to magnets.

Ⓘ Any device in the circuit becomes a permanent magnet.

7 How do an electric motor and an electric generator differ? SC.6.P.13.1

(A) One uses a permanent magnet, while the other uses temporary magnets.

(B) One produces electric energy, while the other produces thermal energy.

(C) One contains moving parts, while the other has only parts that do not move.

(D) One uses an electric current to produce motion, while the other uses motion to produce an electric current.

Use the figure below to answer questions 8 and 9.

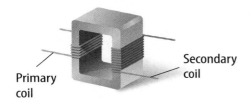

Primary coil

Secondary coil

8 What would be the use for this device? SC.7.P.11.1

(F) to generate electric energy

(G) to increase the voltage

(H) to make an electric motor

(I) to reduce the voltage

9 For the device, which statement is true? SC.7.P.11.1

(A) The primary voltage is greater than the secondary voltage.

(B) The primary voltage is less than the secondary voltage.

(C) The primary voltage is the same secondary voltage.

(D) The voltages alternate high to low.

10 Where on a magnet is the magnetic field the strongest? SC.6.P.13.1

(F) both magnetic poles

(G) center of the magnet

(H) magnetic north pole

(I) magnetic south pole

11 In the figure below, how does the magnetic force on the wire change if the current in the wire changes direction? SC.6.P.13.1

(A) reverses

(B) strengthens

(C) weakens

(D) remains constant

12 Which energy transformation occurs in a generator? SC.7.P.11.2

(F) electric energy to light energy

(G) electric energy to mechanical energy

(H) mechanical energy to electric energy

(I) thermal energy to electric energy

13 Which describes the direction of the magnetic field lines around a magnet? SC.6.P.13.1

(A) start at center, end at each pole

(B) start at each pole, end at center

(C) start at north pole, end at south pole

(D) start at south pole, end at north pole

NEED EXTRA HELP?

If You Missed Question...	1	2	3	4	5	6	7	8	9	10	11	12	13
Go to Lesson...	1	1	2	2	2	3	3	3	3	1	2	3	1

Multiple Choice *Bubble the correct answer.*

1. In the image above, what is true about the magnetic fields of the two magnets?
SC.6.P.13.1

Ⓐ They form an attraction.

Ⓑ They form a compass.

Ⓒ They form a domain.

Ⓓ They form a repulsion.

2. Which statement is NOT true about Earth's magnetic field? SC.7.N.1.6

Ⓕ It can cause auroras near Earth's geographic poles.

Ⓖ It is caused by molten metals in Earth's core.

Ⓗ It helps protect Earth from charged particles emitted by the Sun.

Ⓘ It forms magnetic poles that are in the same places as Earth's geographic poles.

3. Gemma holds a magnet next to her refrigerator. The magnet sticks. She then holds a steel pin next to the refrigerator. The pin does not stick. Gemma's observations show that the refrigerator is
SC.6.P.13.1

Ⓐ a compass.

Ⓑ a magnet.

Ⓒ a non-magnet made of magnetic material.

Ⓓ a non-magnet made of non-magnetic material.

4. The groups of atoms shown above are most likely found in a SC.6.N.1.5

Ⓕ glass cup.

Ⓖ steel nail.

Ⓗ bar magnet.

Ⓘ plastic spoon.

Copyright © Glencoe/McGraw-Hill, a division of The McGraw-Hill Companies, Inc.

Multiple Choice *Bubble the correct answer.*

Electron flow

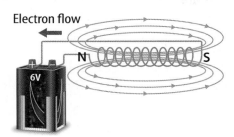

6V

1. Jim constructs the electromagnet shown above. How could Jim increase the strength of this magnet? **SC.6.N.1.4**

   (A) He could decrease the number of wire loops used.

   (B) He could connect another wire to the battery.

   (C) He could place an iron core within the wire coil.

   (D) He could spin the wire coil several times.

2. Which device does NOT use an electric motor to operate? **SC.7.P.11.2**

   (F) lamp

   (G) refrigerator

   (H) CD player

   (I) hair dryer

3. Which statement is NOT true about electromagnets? **SC.6.P.13.1**

   (A) The magnetic field around an electromagnet can be turned off and on.

   (B) Other magnets can attract and then repel electromagnets.

   (C) The poles of an electromagnet cannot be reversed.

   (D) The strength of an electromagnet can be controlled.

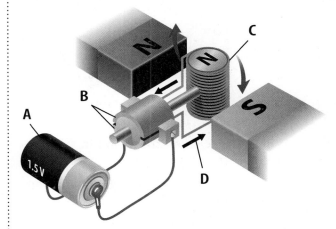

4. Which part of the motor shown above reverses the direction of the current in the motor's electromagnet? **SC.6.P.13.1**

   (F) Part A

   (G) Part B

   (H) Part C

   (I) Part D

Copyright © Glencoe/McGraw-Hill, a division of The McGraw-Hill Companies, Inc.

Multiple Choice *Bubble the correct answer.*

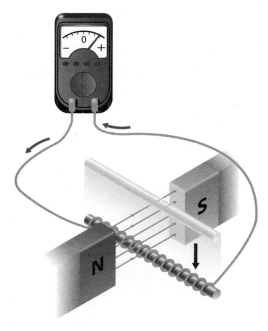

1. Which is true about the action in the diagram above? **SC.6.P.13.1**

(A) Moving a magnet through a wire loop generates a current.

(B) Moving a permanent magnet through a magnetic field increases voltage.

(C) Moving a wire loop between magnets produces a magnetic field.

(D) Moving a wire loop through a magnetic field generates a current.

2. A transformer with 50,000 loops on its primary coil and 100,000 loops on its secondary coil would **SC.7.P.11.2**

(F) increase voltage.

(G) reduce voltage.

(H) change current direction.

(I) generate electric current.

3. Which device uses direct current? **SC.7.P.11.2**

(A) flashlight

(B) lamp

(C) microwave oven

(D) toaster oven

4. What is the function of the device on the telephone pole in the picture? **SC.7.P.11.2**

(F) to increase the voltage of the current

(G) to lower the voltage of the current

(H) to convert alternating current to direct current

(I) to convert direct current to alternating current

Copyright © Glencoe/McGraw-Hill, a division of The McGraw-Hill Companies, Inc.

Notes

Glossary/Glosario

g Multilingual eGlossary

A science multilingual glossary is available on the science website. The glossary includes the following languages.

Arabic	Hmong	Tagalog
Bengali	Korean	Urdu
Chinese	Portuguese	Vietnamese
English	Russian	
Haitian Creole	Spanish	

Cómo usar el glosario en español:
1. Busca el término en inglés que desees encontrar.
2. El término en español, junto con la definición, se encuentran en la columna de la derecha.

Pronunciation Key

Use the following key to help you sound out words in the glossary.

a . . . . back (BAK)	ew. . . . . food (FEWD)
ay . . . . day (DAY)	yoo . . . . pure (PYOOR)
ah. . . . . father (FAH thur)	yew. . . . . . . few (FYEW)
ow . . . . flower (FLOW ur)	uh. . . . . comma (CAH muh)
ar . . . . car (CAR)	u (+ con) . . . . rub (RUB)
e . . . . less (LES)	sh . . . . shelf (SHELF)
ee . . . . leaf (LEEF)	ch . . . . nature (NAY chur)
ih . . . . trip (TRIHP)	g . . . . gift (GIHFT)
i (i + com + e) idea (i DEE uh)	j . . . . gem (JEM)
oh. . . . . go (GOH)	ing . . . . sing (SING)
aw . . . . soft (SAWFT)	zh . . . . vision (VIH zhun)
or . . . . orbit (OR buht)	k. . . . . cake (KAYK)
oy . . . . coin (COYN)	s . . . . seed, cent (SEED, SENT)
oo . . . . foot (FOOT)	z . . . . zone, raise (ZOHN, RAYZ)

English — A — **Español**

absorption/alkaline earth metal

absorption: the transfer of energy from a wave to the medium through which it travels.

acceleration: a measure of the change in velocity during a period of time.

acid: a substance that produces a hydronium ion (H_3O^+) when dissolved in water.

acoustics: the study of how sound interacts with structures.

activation energy: the minimum amount of energy needed to start a chemical reaction.

alkali (AL kuh li) metal: an element in group 1 on the periodic table.

alkaline (AL kuh lun) earth metal: an element in group 2 on the periodic table.

absorción/metal alcalinotérreo

absorción: transferencia de energía desde una onda hacia el medio a través del cual viaja.

aceleración: medida del cambio de velocidad durante un período de tiempo.

ácido: sustancia que produce ión hidronio (H_3O^+) cuando se disuelve en agua.

acústica: estudio de cómo interactúa el sonido con las estructuras.

energía de activación: cantidad mínima de energía necesaria para iniciar una reacción química.

metal alcalino: elemento del grupo 1 de la tabla periódica.)

metal alcalinotérreo: elemento del grupo 2 de la tabla periódica.

alternating current: an electric current that changes direction in a regular pattern.

amino acid: a carbon compound that contains the two functional groups amino and carboxyl.

amino group: a group consisting of a nitrogen atom covalently bonded to two hydrogen atoms ($-NH_2$).

amplitude: the maximum distance a wave varies from its rest position.

amplitude modulation (AM): a change in the amplitude of a carrier wave.

Archimedes' (ar kuh MEE deez) principle: principle that states that the buoyant force on an object is equal to the weight of the fluid that the object displaces.

atmospheric pressure: the ratio of the weight of all the air above you to your surface area.

atom: a small particle that is the building block of matter.; the smallest piece of an element that still represents that element.

atomic number: the number of protons in an atom of an element.

average acceleration: a change in velocity during a time interval divided by the time interval during which the velocity changes.

average atomic mass: the average mass of the element's isotopes, weighted according to the abundance of each isotope.

average speed: the total distance traveled divided by the total time taken to travel that distance.

corriente alterna: corriente eléctrica que cambia la dirección en un patrón regular.

aminoácido: compuesto de carbono que contiene los dos grupos funcionales amino y carboxilo.

grupo amino: grupo que consiste en un átomo de nitrógeno unido de manera covalente a dos átomos de hidrógeno ($-NH_2$).

amplitud: distancia máxima que varía una onda desde su posición de reposo.

amplitud modulada (AM): cambio en la amplitud de una onda portadora.

principio de Arquímedes: principio que establece que la fuerza de empuje ejercida sobre un objeto es igual al peso del fluido que el objeto desplaza.

presión atmosférica: peso del aire sobre una superficie.

átomo: partícula pequeña que es el componente básico de la materia.; parte más pequeña de un elemento que mantiene la identidad de dicho elemento.

número atómico: número de protones en el átomo de un elemento.

aceleración promedio: cambio de velocidad durante un intervalo de tiempo dividido por el intervalo de tiempo durante el cual cambia la velocidad.

masa atómica promedio: masa atómica promedio de los isótopos de un elemento, ponderado según la abundancia de cada isótopo.

rapidez promedio: distancia total recorrida dividida por el tiempo usado para recorrerla.

B

balanced forces: forces acting on an object that combine and form a net force of zero.

base: a substance that produces hydroxide ions (OH^-) when dissolved in water.

Bernoulli's (ber NEW leez) principle: principle that states that the pressure of a fluid decreases when the speed of that fluid increases.

biological molecule: a large organic molecule in any living organism.

fuerzas en equilibrio: fuerzas que actúan sobre un objeto, se combinan y forman una fuerza neta de cero.

base: sustancia que produce iones hidróxido (OH^-) cuando se disuelve en agua.

principio de Bernoulli: principio que establece que la presión de un fluido disminuye cuando la rapidez de dicho fluido aumenta.

molécula biológica: molécula orgánica grande de cualquier organismo vivo.

Boyle's Law: the law that pressure of a gas increases if the volume decreases and pressure of a gas decreases if the volume increases, when temperature is constant.

broadcasting: the use of electromagnetic waves to send information in all directions.

buoyant (BOY unt) force: an upward force applied by a fluid on an object in the fluid.

Ley de Boyle: ley que afirma que la presión de un gas aumenta si el volumen disminuye y que la presión de un gas disminuye si el volumen aumenta, cuando la temperatura es constante.

radiodifusión: uso de ondas electromagnéticas para enviar información en todas las direcciones.

fuerza de empuje: fuerza ascendente que un fluido aplica a un objeto que se encuentra en él.

carbohydrate: a group of organic molecules that includes sugars, starches and cellulose.

carboxyl group: a group consisting of a carbon atom with a single bond to a hydroxyl group and a double bond to an oxygen atom (−COOH).

carrier wave: an electromagnetic wave that a radio or television station uses to carry its sound or image signals.

catalyst: a substance that increases reaction rate by lowering the activation energy of a reaction.

centripetal (sen TRIH puh tuhl) force: in circular motion, a force that acts perpendicular to the direction of motion, toward the center of the curve.

Charles's Law: the law the volume of a gas increases with increasing temperature, if the pressure is constant.

chemical bond: a force that holds two or more atoms together.

chemical change: a change in matter in which the substances that make up the matter change into other substances with different chemical and physical properties.

chemical equation: a description of a reaction using element symbols and chemical formulas.

chemical formula: a group of chemical symbols and numbers that represent the elements and the number of atoms of each element that make up a compound.

carbohidrato: grupo de moléculas orgánicas que incluye azúcares, almidones y celulosa.

grupo carboxilo: grupo que consiste en un átomo de carbono unido a un grupo hidroxilo mediante un enlace sencillo y a un átomo de oxígeno mediante un enlace doble (−COOH).

onda portadora: onda electromagnética que una estación de radio o televisión usa para transportar señales de sonido o imagen.

catalizador: sustancia que aumenta la velocidad de reacción al disminuir la energía de activación de una reacción.

fuerza centrípeta: en movimiento circular, la fuerza que actúa de manera perpendicular a la dirección del movimiento, hacia el centro de la curva.

Ley de Charles: ley que afirma que el volumen de un gas aumenta cuando la temperatura aumenta, si la presión es constante.

enlace químico: fuerza que mantiene unidos dos o más átomos.

cambio químico: cambio de la materia en el cual las sustancias que componen la materia se transforman en otras sustancias con propiedades químicas y físicas diferentes.

ecuación química: descripción de una reacción con símbolos de los elementos y fórmulas químicas.

fórmula química: grupo de símbolos químicos y números que representan los elementos y el número de átomos de cada elemento que forman un compuesto.

chemical property: the ability or inability of a substance to combine with or change into one or more new substances.

chemical reaction: a process in which atoms of one or more substances rearrange to form one or more new substances.

circular motion: any motion in which an object is moving along a curved path.

coefficient: a number placed in front of an element symbol or chemical formula in an equation.

combustion: a chemical reaction in which a substance combines with oxygen and releases energy.

compound: a substance containing atoms of two or more different elements chemically bonded together.

compression: region of a longitudinal wave where the particles of the medium are closest together.

concave lens: a lens that is thicker at the edges than in the middle.

concave mirror: a mirror that curves inward.

concentration: the amount of a particular solute in a given amount of solution.

condensation: the change of state from a gas to a liquid.

conduction: the transfer of thermal energy due to collisions between particles.

cone: one of many cells in the retina of the eye that respond to colors.

constants: the factors in an experiment that remain the same.

constant speed: the rate of change of position in which the same distance is traveled each second.

contact force: a push or a pull on one object by another object that is touching it.

control group: the part of a controlled experiment that contains the same factors as the experimental group, but the independent variable is not changed.

convection: the transfer of thermal energy by the movement of particles from one part of a material to another.

propiedad química: capacidad o incapacidad de una sustancia para combinarse con o transformarse en una o más sustancias.

reacción química: proceso en el cual átomos de una o más sustancias se acomodan para formar una o más sustancias nuevas.

movimiento circular: cualquier movimiento en el cual un objeto se mueve a lo largo de una trayectoria curva.

coeficiente: número colocado en frente del símbolo de un elemento o de una fórmula química en una ecuación.

combustión: reacción química en la cual una sustancia se combina con oxígeno y libera energía.

compuesto: sustancia que contiene átomos de dos o más elementos diferentes unidos químicamente.

compresión: región de una onda longitudinal donde las partículas del medio están más cerca.

lente cóncavo: lente que es más grueso en los extremos que en el centro.

espejo cóncavo: espejo que dobla hacia adentro.

concentración: cantidad de cierto soluto en una cantidad dada de solución.

condensación: cambio de estado gaseoso a líquido.

conducción: transferencia de energía térmica debido a colisiones entre partículas.

cono: una de muchas células en la retina del ojo que es sensible a los colores.

constantes: factores que no cambian en un experimento.

velocidad constante: velocidad a la que se cambia de posición, en la cual se recorre la misma distancia por segundo.

fuerza de contacto: empuje o arrastre ejercido sobre un objeto por otro que lo está tocando.

grupo de control: parte de un experimento controlado que contiene los mismos factores que el grupo experimental, pero la variable independiente no se cambia.

convección: transferencia de energía térmica por el movimiento de partículas de una parte de la materia a otra.

convection current: the movement of fluids in a cycle because of convection.

convex lens: a lens that is thicker in the middle than at the edges.

convex mirror: a mirror that curves outward.

covalent bond: a chemical bond formed when two atoms share one or more pairs of valence electrons.

crest: the highest point on a transverse wave.

critical thinking: comparing what you already know with information you are given in order to decide whether you agree with it.

corriente de convección: movimiento de fluidos en un ciclo debido a la convección.

lente convexo: lente que es más grueso en el centro que en los extremos.

espejo convexo: espejo que curva hacia afuera.

enlace covalente: enlace químico formado cuando dos átomos comparten uno o más pares de electrones de valencia.

cresta: punto más alto en una onda transversal.

pensamiento crítico: comparación que se hace cuando se sabe algo acerca de información nueva, y se decide si se está o no de acuerdo con ella.

decomposition: a type of chemical reaction in which one compound breaks down and forms two or more substances.

density: the mass per unit volume of a substance.

dependent variable: the factor a scientist observes or measures during an experiment.

deposition: the process of changing directly from a gas to a solid.

description: a spoken or written summary of an observation.

diffraction: the change in direction of a wave when it travels by the edge of an object or through an opening.

diffuse reflection: reflection of light from a rough surface.

direct current: an electric current that continually flows in one direction.

displacement: the difference between the initial, or starting, position and the final position of an object that has moved.

dissolve: to form a solution by mixing evenly.

Doppler effect: the change of pitch when a sound source is moving in relation to an observer.

double-replacement reaction: a type of chemical reaction in which the negative ions in two compounds switch places, forming two new compounds.

descomposición: tipo de reacción química en la que un compuesto se descompone y forma dos o más sustancias.

densidad: cantidad de masa por unidad de volumen de una sustancia.

variable dependiente: factor que el científico observa o mide durante un experimento.

deposición: proceso de cambiar directamente de gas a sólido.

descripción: resumen oral o escrito de una observación.

difracción: cambio en la dirección de una onda cuando ésta viaja por el borde de un objeto o a través de una abertura.

reflexión difusa: reflexión de la luz en una superficie rugosa.

corriente directa: corriente eléctrica que fluye de manera continua en una dirección.

desplazamiento: diferencia entre la posición inicial, o salida, y la final de un objeto que se ha movido.

disolver: preparar una solución mezclando de manera homogénea.

efecto Doppler: cambio de tono cuando una fuente sonora se mueve con relación a un observador.

reacción de sustitución doble: tipo de reacción química en la que los iones negativos de dos compuestos intercambian lugares, para formar dos compuestos nuevos.

drag force: a force that opposes the motion of an object through a fluid.

ductility (duk TIH luh tee): the ability to be pulled into thin wires.

fuerza de arrastre: fuerza que se opone al movimiento de un objeto a través de un fluido.

ductilidad: capacidad para formar alambres delgados.

echo: a reflected sound wave.

echolocation (e koh loh KAY shun): the process an animal uses to locate an object by means of reflected sounds.

electric circuit: a closed, or complete, path in which an electric current flows.

electric conductor: a material through which electrons easily move.

electric current: the movement of electrically charged particles.

electric discharge: the process of an unbalanced electric charge becoming balanced.

electric energy: energy carried by an electric current.

electric generator: a device that uses a magnetic field to transform mechanical energy to electric energy.

electric insulator: a material through which electrons cannot easily move.

electric motor: a device that uses an electric current to produce motion.

electric resistance: a measure of how difficult it is for an electric current to flow in a material.

electromagnet: a magnet created by wrapping a current-carrying wire around a ferromagnetic core.

electromagnetic spectrum: the entire range of electromagnetic waves with different frequencies and wavelengths.

electromagnetic wave: a transverse wave that can travel through empty space and through matter.

electron: a negatively charged particle that occupies the space in an atom outside the nucleus.

electron cloud: the region surrounding an atom's nucleus where one or more electrons are most likely to be found.

eco: onda sonora reflejada.

ecocolocación: proceso que un animal hace para ubicar un objeto por medio de sonidos reflejados.

circuito eléctrico: trayectoria cerrada, o completa, por la que fluye corriente eléctrica.

conductor eléctrico: material a través del cual se mueven los electrones con facilidad.

corriente eléctrica: movimiento de partículas cargadas eléctricamente.

descarga eléctrica: proceso por el cual una carga eléctrica no balanceada se vuelve balanceada.

energía eléctrica: energía transportada por una corriente eléctrica.

generador eléctrico: aparato que usa un campo magnético para transformar energía mecánica en energía eléctrica.

aislante eléctrico: material por el cual los electrones no pueden fluir con facilidad.

motor eléctrico: aparato que usa corriente eléctrica para producir movimiento.

resistencia eléctrica: medida de qué tan difícil es para una corriente eléctrica fluir en un material.

electroimán: imán fabricado al enrollar un alambre que transporta corriente alrededor de un núcleo ferromagnético.

espectro electromagnético: rango completo de ondas electromagnéticas con frecuencias y longitudes de onda diferentes.

onda electromagnética: onda transversal que puede viajar a través del espacio vacío y de la materia.

electrón: partícula cargada negativamente que ocupa el espacio por fuera del núcleo de un átomo.

nube de electrones: región que rodea el núcleo de un átomo en donde es más probable encontrar uno o más electrones.

electron dot diagram: a model that represents valence electrons in an atom as dots around the element's chemical symbol.

element: a substance that consists of only one type of atom.

endothermic reaction: a chemical reaction that absorbs thermal energy.

energy: the ability to cause change.

enzyme: a catalyst that speeds up chemical reactions in living cells.

evaporation: the process of a liquid changing to a gas at the surface of the liquid.

exothermic reaction: a chemical reaction that releases thermal energy.

experimental group: the part of the controlled experiment used to study relationships among variables.

explanation: an interpretation of observations.

diagrama de puntos de Lewis: modelo que representa electrones de valencia en un átomo a manera de puntos alrededor del símbolo químico del elemento.

elemento: sustancia que consiste de un sólo tipo de átomo.

reacción endotérmica: reacción química que absorbe energía térmica.

energía: capacidad de causar cambio.

enzima: catalizador que acelera reacciones químicas en las células vivas.

evaporación: proceso por el cual un líquido cambia a gas en la superficie de dicho líquido.

reacción exotérmica: reacción química que libera energía térmica.

grupo experimental: parte del experimento controlado que se usa para estudiar las relaciones entre las variables.

explicación: interpretación de las observaciones.

ferromagnetic (fer ro mag NEH tik) element: elements, including iron, nickel, and cobalt, that have an especially strong attraction to magnets.

fluid: any substance that can flow and take the shape of the container that holds it.

focal length: the distance along the optical axis from the mirror to the focal point.

focal point: the point where light rays parallel to the optical axis converge after being reflected by a mirror or refracted by a lens.

force: a push or a pull on an object.

force pair: the forces two objects apply to each other.

fossil fuel: the remains of ancient organisms that can be burned as an energy source.

frequency: the number of wavelengths that pass by a point each second.

frequency modulation (FM): a change in the frequency of a carrier wave.

elemento ferromagnético: elementos, incluidos hierro, níquel y cobalto, que tienen una atracción especialmente fuerte hacia los imanes.

fluido: cualquier sustancia que puede fluir y toma la forma del recipiente que lo contiene.

distancia focal: distancia a lo largo del eje óptico desde el espejo hasta el punto focal.

punto focal: punto donde rayos de luz paralelos al eje óptico convergen después de ser reflejados por un espejo o refractados por un lente.

fuerza: empuje o arrastre ejercido sobre un objeto.

par de fuerzas: fuerzas que dos objetos se aplican entre sí.

combustible fósil: restos de organismos antiguos que pueden quemarse como fuente de energía.

frecuencia: número de longitudes de onda que pasan por un punto cada segundo.

frecuencia modulada (FM): cambio en la frecuencia de una onda portadora.

friction: a contact force that resists the sliding motion of two surfaces that are touching.

functional group: an atom or group of atoms that determine the function or properties of the compound.

fricción: fuerza que resiste el movimiento de dos superficies que están en contacto.

grupo funcional: átomo o grupo de átomos que determina la función o las propiedades de un compuesto.

gamma ray: a high-energy electromagnetic wave with a shorter wavelength and higher frequency than all other types of electromagnetic waves.

gas: matter that has no definite volume and no definite shape.

Global Positioning System (GPS): a worldwide navigation system that uses satellite signals to determine a receiver's location.

gravity: an attractive force that exists between all objects that have mass.

grounding: providing a path for electric charges to flow safely into the ground.

group: a column on the periodic table.

rayo gamma: onda electromagnética de alta energía con longitud de onda más corta y frecuencia más alta que los demás tipos de ondas electromagnéticas.

gas: materia que no tiene volumen ni forma definidos.

sistema de posicionamiento global (SPG): sistema mundial de navegación que usa señales satelitales para determinar la ubicación de un receptor.

gravedad: fuerza de atracción que existe entre todos los objetos que tienen masa.

polo a tierra: suministrar una trayectoria para que las cargas eléctricas fluyan con seguridad hacia el suelo.

grupo: columna en la tabla periódica.

halide group: a functional group which contains group 17 halogens—fluorine, chlorine, bromine, and iodine.

halogen (HA luh jun): an element in group 17 on the periodic table.

heat: the movement of thermal energy from a region of higher temperature to a region of lower temperature.

heat engine: a machine that converts thermal energy into mechanical energy.

heating appliance: a device that converts electric energy into thermal energy.

heterogeneous mixture: a mixture in which substances are not evenly mixed.

hologram: a three-dimensional photograph of an object.

homogeneous mixture: a mixture in which two or more substances are evenly mixed but not bonded together.

grupo halide: grupo funcional que contiene 17 halógenos—flúor, cloro, bromo y yodo.

halógeno: elemento del grupo 17 de la tabla periódica.

calor: movimiento de energía térmica de una región de alta temperatura a una región de baja temperatura.

motor térmico: máquina que convierte energía térmica en energía mecánica.

calentador: aparato que convierte energía eléctrica en energía térmica.

mezcla heterogénea: mezcla en la cual las sustancias no están mezcladas de manera uniforme.

holograma: fotografía tridimensional de un objeto.

mezcla homogénea: mezcla en la cual dos o más sustancias están mezcladas de manera uniforme, pero no están unidas químicamente.

hydrocarbon: a compound that contains only carbon and hydrogen atoms.

hydronium ion (H_3O^+): a positively charged ion formed when an acid dissolves in water.

hydroxyl group: the chemical group consisting of one atom of hydrogen and one atom of oxygen (−OH).

hypothesis: a possible explanation for an observation that can be tested by scientific investigations.

hidrocarburo: compuesto que contiene solamente átomos de carbono e hidrógeno.

ión hidronio (H_3O^+): ión cargado positivamente que se forma cuando un ácido se disuelve en agua.

grupo hidroxilo: grupo químico que consiste en un átomo de hidrógeno y un átomo de oxígeno (−OH).

hipótesis: explicación posible para una observación que puede ponerse a prueba en investigaciones científicas.

I

independent variable: the factor that is changed by the investigator to observe how it affects a dependent variable.

indicator: a compound that changes color at different pH values when it reacts with acidic or basic solutions.

inertia (ihn UR shuh): the tendency of an object to resist a change in its motion.

inexhaustible energy resource: an energy resource that cannot be used up.

inference: a logical explanation of an observation that is drawn from prior knowledge or experience.

infrared wave: an electromagnetic wave that has a wavelength shorter than a microwave but longer than visible light.

inhibitor: a substance that slows, or even stops, a chemical reaction.

instantaneous speed: an object's speed at a specific instant in time.

intensity: the amount of energy that passes through a square meter of space in one second.

interference: occurs when waves overlap and combine to form a new wave.

International System of Units (SI): the internationally accepted system of measurement.

ion (I ahn): an atom that is no longer neutral because it has lost or gained electrons.

ionic bond: the attraction between positively and negatively charged ions in an ionic compound.

variable independiente: factor que el investigador cambia para observar cómo afecta la variable dependiente.

indicador: compuesto que cambia de color a diferentes valores de pH cuando reacciona con soluciones ácidas o básicas.

inercia: tendencia de un objeto a resistirse al cambio en su movimiento.

recurso energético inagotable: recurso energético que no puede agotarse.

inferencia: explicación lógica de una observación que se obtiene a partir de conocimiento previo o experiencia.

onda infrarroja: onda electromagnética que tiene una longitud de onda más corta que la de una microonda, pero más larga que la de la luz visible.

inhibidor: sustancia que disminuye, o incluso detiene, una reacción química.

velocidad instantánea: velocidad de un objeto en un instante específico en el tiempo.

intensidad: cantidad de energía que atraviesa un metro cuadrado de espacio en un segundo.

interferencia: ocurre cuando las ondas coinciden y combinan para forma una onda nueva.

Sistema Internacional de Unidades (SI): sistema de medidas aceptado internacionalmente.

ión: átomo que no es neutro porque ha ganado o perdido electrones.

enlace iónico: atracción entre iones cargados positiva y negativamente en un compuesto iónico.

isomer: one of two or more compounds that have the same molecular formula but different structural arrangements.

isotopes: atoms of the same element that have different numbers of neutrons.

isómero: uno de dos o más compuestos que tienen la misma fórmula molecular, pero diferentes arreglos estructurales.

isótopos: átomos del mismo elemento que tienen diferente número de neutrones.

K

kinetic (kuh NEH tik) energy: energy due to movement

kinetic molecular theory: an explanation of how particles in matter behave.

energía cinética: energía debida al movimiento.

teoría cinética molecular: explicación de cómo se comportan las partículas en la materia.

L

laser: an optical device that produces a narrow beam of coherent light.

law of conservation of energy: law that states that energy can be transformed from one form to another, but it cannot be created or destroyed.

law of conservation of mass: law that states that the total mass of the reactants before a chemical reaction is the same as the total mass of the products after the chemical reaction.

law of conservation of momentum: a principle stating that the total momentum of a group of objects stays the same unless outside forces act on the objects.

law of reflection: law that states that when a wave is reflected from a surface, the angle of reflection is equal to the angle of incidence.

lens: a transparent object with at least one curved side that causes light to change direction.

light: electromagnetic radiation that you can see.

lipid: a type of biological molecule that includes fats, oils, hormones, waxes, and components of cellular membranes.

liquid: matter with a definite volume but no definite shape.

láser: aparato óptico que produce un haz angosto y coherente de luz.

ley de la conservación de la energía: ley que plantea que la energía puede transformarse de una forma a otra, pero no puede crearse ni destruirse.

ley de la conservación de la masa: ley que plantea que la masa total de los reactivos antes de una reacción química es la misma que la masa total de los productos después de la reacción química.

ley de la conservación del momentum: principio que establece que el momentum total de un grupo de objetos permanece constante a menos que fuerzas externas actúen sobre los objetos.

ley de la reflexión: ley que establece que cuando una onda se refleja desde una superficie, el ángulo de reflexión es igual al ángulo de incidencia.

lente: objeto transparente que tiene, al menos, un lado curvo que hace que la luz cambrie de dirección.

luz: radiación electromagnética que puede verse.

lípido: tipo de molécula biológica que incluye grasas, aceites, hormonas, ceras y componentes de las membranas celulares.

líquido: materia con volumen definido y forma indefinida.

longitudinal (lahn juh TEWD nul) wave: a wave in which the disturbance is parallel to the direction the wave travels.

luster: the way a mineral reflects or absorbs light at its surface.

onda longitudinal: onda en la que la perturbación es paralela a la dirección en que viaja la onda.

brillo: forma en que un mineral refleja o absorbe la luz en su superficie.

magnet: an object that attracts iron.

magnetic domain: region in a magnetic material in which the magnetic fields of the atoms all point in the same direction.

magnetic force: the force that a magnet applies to another magnet.

magnetic material: any material that a magnet attracts.

magnetic pole: the place on a magnet where the force it exerts is the strongest.

malleability (ma lee uh BIH luh tee): the ability of a substance to be hammered or rolled into sheets.

mass: the amount of matter in an object.

mass number: the sum of the number of protons and neutrons in an atom.

matter: anything that has mass and takes up space.

mechanical energy: sum of the potential energy and the kinetic energy in a system.

mechanical wave: a wave that can travel only through matter.

medium: a material in which a wave travels.

metal: an element that is generally shiny, is easily pulled into wires or hammered into thin sheets, and is a good conductor of electricity and thermal energy.

metallic bond: a bond formed when many metal atoms share their pooled valence electrons.

metalloid (MEH tul oyd): an element that has physical and chemical properties of both metals and nonmetals.

microscope: an optical device that forms magnified images of very small, close objects.

imán: objeto que atrae al hierro.

dominio magnético: región en un material magnético en el que los campos magnéticos de los átomos apuntan en la misma dirección.

fuerza magnética: fuerza que un imán aplica a otro imán.

material magnético: cualquier material que un imán atrae.

polo magnético: lugar en un imán donde la fuerza que éste ejerce es la mayor.

maleabilidad: capacidad de una sustancia de martillarse o laminarse para formar hojas.

masa: cantidad de materia en un objeto.

número de masa: suma del número de protones y neutrones de un átomo.

materia: cualquier cosa que tiene masa y ocupa espacio.

energía mecánica: suma de la energía potencial y la energía cinética en un sistema.

onda mecánica: onda que puede viajar sólo a través de la materia.

medio: material en el cual viaja una onda.

metal: elemento que generalmente es brillante, fácilmente puede estirarse para formar alambres o martillarse para formar hojas delgadas y es buen conductor de electricidad y energía térmica.

enlace metálico: enlace formado cuando muchos átomos metálicos comparten su banco de electrones de valencia.

metaloide: elemento que tiene las propiedades físicas y químicas de metales y no metales.

microscopio: aparato óptico que forma imágenes aumentadas de objetos cercanos muy pequeños.

microwave: a low-frequency, low-energy electromagnetic wave that has a wavelength between about 1 mm and 30 cm.

mixture: matter that can vary in composition.

molecule (MAH lih kyewl): two or more atoms that are held together by covalent bonds and act as a unit.

momentum: a measure of how hard it is to stop a moving object.

monomer: one of the small organic molecules that make up the long chain of a polymer.

motion: the process of changing position.

microonda: onda electromagnética de baja frecuencia y baja energía que tiene una longitud de onda entre 1 mm y 30 cm.

mezcla: materia cuya composición puede variar.

molécula: dos o más átomos que están unidos mediante enlaces covalentes y actúan como una unidad.

momentum: medida de qué tan difícil es detener un objeto en movimiento.

monómero: una de las moléculas orgánicas pequeñas que forman la larga cadena de un polímero.

movimiento: proceso de cambiar de posición.

net force: the combination of all the forces acting on an object.

neutron: a neutral particle in the nucleus of an atom.

Newton's first law of motion: law that states that if the net force acting on an object is zero, the motion of the object does not change.

Newton's second law of motion: law that states that the acceleration of an object is equal to the net force exerted on the object divided by the object's mass.

Newton's third law of motion: law that states that for every action there is an equal and opposite reaction.

noble gas: an element in group 18 on the periodic table.

noncontact force: a force that one object applies to another object without touching it.

nonmetal: an element that has no metallic properties.

nonrenewable energy resource: an energy resource that is available in limited amounts or that is used faster than it is replaced in nature.

nuclear decay: a process that occurs when an unstable atomic nucleus changes into another more stable nucleus by emitting radiation.

nuclear energy: energy stored in and released from the nucleus of an atom.

fuerza neta: combinación de todas las fuerzas que actúan sobre un objeto.

neutrón: partícula neutra en el núcleo de un átomo.

primera ley del movimiento de Newton: ley que establece que si la fuerza neta ejercida sobre un objeto es cero, el movimiento de dicho objeto no cambia.

segunda ley del movimiento de Newton: ley que establece que la aceleración de un objeto es igual a la fuerza neta que actúa sobre él divida por su masa.

tercera ley del movimiento de Newton: ley que establece que para cada acción hay una reacción igual en dirección opuesta.

gas noble: elemento del grupo 18 de la tabla periódica.

fuerza de no contacto: fuerza que un objeto puede aplicar sobre otro sin tocarlo.

no metal: elemento que tiene propiedades no metálicas.

recurso energético no renovable: recurso energético disponible en cantidades limitadas o que se usa más rápido de lo que se repone en la naturaleza.

desintegración nuclear: proceso que ocurre cuando un núcleo atómico inestable cambia a otro núcleo atómico más estable mediante emisión de radiación.

energía nuclear: energía almacenada en y liberada por el núcleo de un átomo.

nucleic acid (new KLEE ihk • A sud): a biological polymer that stores and transmits genetic information.

nucleus: the region in the center of an atom where most of an atom's mass and positive charge are concentrated.

ácido nucleico: polímero biológico que almacena y transmite información genética.

núcleo: región en el centro de un átomo donde se concentra la mayor cantidad de masa y las cargas positivas.

O

observation: the act of using one or more of your senses to gather information and take note of what occurs.

Ohm's law: a mathematical equation that describes the relationship between voltage, current, and resistance.

opaque: a material through which light does not pass.

optical device: any instrument or object used to produce or control light.

organic compound: chemical compound that contains carbon atoms usually bonded to at least one hydrogen atom.

observación: acción de usar uno o más sentidos para reunir información y tomar notar de lo que ocurre.

Ley de Ohm: ecuación matemática que describe la relación entre voltaje, corriente y resistencia.

opaco: material por el que no pasa la luz.

aparato óptico: cualquier instrumento usado para producir o controlar luz.

compuesto orgánico: compuesto químico que contienen átomos de carbono generalmente unidos a, al menos, un átomo de hidrógeno.

P

parallel circuit: an electric circuit with more than one path, or branch, for an electric current to follow.

Pascal's (pas KALZ) principle: principle that states that when pressure is applied to a fluid in a closed container, the pressure increases by the same amount everywhere in the container.

percent error: the expression of error as a percentage of the accepted value.

period: a row on the periodic table.

periodic table: a chart of the elements arranged into rows and columns according to their physical and chemical properties.

permanent magnet: a magnet that retains its magnetism even after being removed from a magnetic field.

pH: an inverse measure of the concentration of hydronium ions (H_3O^+) in a solution.

physical change: a change in the size, shape, form, or state of matter that does not change the matter's identity.

circuito en paralelo: circuito eléctrico con más de una trayectoria, o rama, para que fluya una corriente eléctrica.

principio de Pascal: principio que establece que cuando se aplica presión a un fluido en un recipiente cerrado, la presión aumenta el mismo valor en todas las partes del recipiente.

error porcentual: expresión del error como porcentaje del valor aceptado.

periodo: hilera en la tabla periódica.

tabla periódica: cuadro en que los elementos están organizados en hileras y columnas según sus propiedades físicas y químicas.

imán permanente: imán que retiene su magnetismo incluso después de haberlo retirado de un campo magnético.

pH: medida inversa de la concentración de iones hidronio (H_3O^+) en una solución.

cambio físico: cambio en el tamaño, la forma o el estado de la materia en el que no cambia la identidad de la materia.

physical property: a characteristic of matter that you can observe or measure without changing the identity of the matter.

pitch: the perception of how high or low a sound is; related to the frequency of a sound wave.

polarized: the condition of having electrons concentrated at one end of an object.

polar molecule: a molecule with a slight negative charge in one area and a slight positive charge in another area.

polymer: a molecule made up of many of the same small organic molecules covalently bonded together, forming a long chain.

polymerization: the chemical process in which small organic molecules, or monomers, bond together to form a chain.

position: an object's distance and direction from a reference point.

potential (puh TEN chul) energy: stored energy due to the interactions between objects or particles.

prediction: a statement of what will happen next in a sequence of events.

pressure: the amount of force per unit area applied to an object's surface.

product: a substance produced by a chemical reaction.

protein: a biological polymer made of amino acid monomers.

proton: positively charged particle in the nucleus of an atom.

propiedad física: característica de la materia que puede observarse o medirse sin cambiar la identidad de la materia.

tono: percepción de qué tan alto o bajo es el sonido; relacionado con la frecuencia de la onda sonora.

polarizado: condición de tener electrones concentrados en un extremo de un objeto.

molécula polar: molécula con carga ligeramente negativa en una parte y ligeramente positiva en otra.

polímero: molécula hecha de muchas de las mismas moléculas orgánicas pequeñas unidas por enlaces covalentes, y que forman una cadena larga.

polimerización: proceso químico en el que moléculas orgánicas pequeñas, o monómeros, se unen para formar una cadena.

posición: distancia y dirección de un objeto según un punto de referencia.

energía potencial: energía almacenada debido a las interacciones entre objetos o partículas.

predicción: afirmación de lo que ocurrirá después en una secuencia de eventos.

presión: cantidad de fuerza por unidad de área aplicada a la superficie de un objeto.

producto: sustancia producida por una reacción química.

proteína: polímero biológico hecho de monómeros de aminoácidos.

protón: partícula cargada positivamente en el núcleo de un átomo.

Q

qualitative data: the use of words to describe what is observed in an experiment.

quantitative data: the use of numbers to describe what is observed in an experiment.

datos cualitativos: uso de palabras para describir lo que se observa en un experimento.

datos cuantitativos: uso de números para describir lo que se observa en un experimento.

R

radiant energy: energy carried by an electromagnetic wave.

radiation: the transfer of thermal energy by electromagnetic waves.

energía radiante: energía que transporta una onda electromagnética.

radiación: transferencia de energía térmica por ondas electromagnéticas.

radioactive: any element that spontaneously emits radiation.

radio wave: a low-frequency, low-energy electromagnetic wave that has a wavelength longer than about 30 cm.

rarefaction (rayr uh FAK shun): the regions of a longitudinal wave where the particles of the medium are farthest apart.

reactant: a starting substance in a chemical reaction.

reference point: the starting point you use to describe the motion or the position of an object.

reflecting telescope: a telescope that uses a mirror to gather and focus light from distant objects.

reflection: the bouncing of a wave off a surface.

refracting telescope: a telescope that uses lenses to gather and focus light from distant objects.

refraction: the change in direction of a wave as it changes speed in moving from one medium to another.

refrigerator: a device that uses electric energy to pump thermal energy from a cooler location to a warmer location.

regular reflection: reflection of light from a smooth, shiny surface.

renewable energy resource: an energy resource that is replaced as fast as, or faster than, it is used.

resonance: an increase in amplitude that occurs when an object vibrating at its natural frequency absorbs energy from a nearby object vibrating at the same frequency.

reverberation: the collection of reflected sounds from the surfaces in a closed space.

rod: one of many cells in the retina of the eye that responds to low light.

radiactivo: cualquier elemento que emite radiación de manera espontánea.

onda de radio: onda electromagnética de baja frecuencia y baja energía que tiene una longitud de onda mayor de más o menos 30 cm.

rarefacción: regiones de una onda longitudinal donde las partículas del medio están más alejadas.

reactivo: sustancia inicial en una reacción química.

punto de referencia: punto que se escoge para describir la ubicación, o posición, de un objeto.

telescopio de reflexión: telescopio que tiene un espejo para reunir y enfocar luz de objetos lejanos.

reflexión: rebote de una onda desde una superficie.

telescopio de refracción: telescopio que tiene lentes para reunir y enfocar luz de objetos lejanos.

refracción: cambio en la dirección de una onda a medida que cambia de velocidad al moverse de un medio a otro.

refrigerador: aparato que usa energía eléctrica para bombear energía térmica desde un lugar más frío hacia uno más caliente.

reflexión especular: reflexión de la luz desde una superficie lisa y brillante.

recurso energético renovable: recurso energético que se repone tan rápido, o más rápido, de lo que se consume.

resonancia: aumento en la amplitud que ocurre cuando un objeto que vibra a su frecuencia natural absorbe energía de un objeto cercano y vibran a la misma frecuencia.

reverberación: colección de sonidos reflejados de superficies en un espacio cerrado.

bastón: una de muchas células en la retina del ojo sensible a la luminosidad baja.

S

saturated hydrocarbon: hydrocarbon that contains only single bonds.

saturated solution: a solution that contains the maximum amount of solute the solu-tion can hold at a given temperature and pressure.

science: the investigation and exploration of natural events and of the new information that results from those investigations.

scientific law: a rule that describes a pattern in nature.

scientific notation: a method of writing or displaying very small or very large numbers.

scientific theory: an explanation of observations or events that is based on knowledge gained from many observations and investigations.

semiconductor: a substance that conducts electricity at high temperatures but not at low temperatures.

series circuit: an electric circuit with only one closed path for an electric current to follow.

single-replacement reaction: a type of chemical reaction in which one element replaces another element in a compound.

solid: matter that has a definite shape and a definite volume.

solubility (sahl yuh BIH luh tee): the maximum amount of solute that can dissolve in a given amount of solvent at a given temperature and pressure.

solute: any substance in a solution other than the solvent.

solution: another name for a homogeneous mixture.

solvent: the substance that exists in the greatest quantity in a solution.

sonar: a system that uses the reflection of sound waves to find underwater objects.

sound energy: energy carried by sound waves.

hidrocarburos saturados: hidrocarburo que solamente contiene enlaces sencillos.

solución saturada: solución que contiene la cantidad máxima de soluto que la solución puede sostener a cierta temperatura y presión.

ciencia: investigación y exploración de eventos naturales y la información nueva que resulta de dichas investigaciones.

ley científica: regla que describe un patrón en la naturaleza.

notación científica: método para escribir o expresar números muy pequeños o muy grandes.

teoría científica: explicación de las observaciones y los eventos basada en conocimiento obtenido en muchas observaciones e investigaciones.

semiconductor: sustancia que conduce electricidad a altas temperaturas, pero no a bajas temperaturas.

circuito en serie: circuito eléctrico con sólo una trayectoria cerrada para que fluya una corriente eléctrica.

reacción de sustitución sencilla: tipo de reacción química en la que un elemento reemplaza a otro en un compuesto.

sólido: materia con forma y volumen definidos.

solubilidad: cantidad máxima de soluto que puede disolverse en una cantidad dada de solvente a temperatura y presión dadas.

soluto: cualquier sustancia en una solución diferente del solvente.

solución: otro nombre para una mezcla homogénea.

solvente: sustancia que existe en mayor cantidad en una solución.

sonar: sistema que usa la reflexión de ondas sonoras para encontrar objetos bajo el agua.

energía sonora: energía que transportan las ondas sonoras.

sound wave: a longitudinal wave that can travel only through matter.

specific heat: the amount of thermal energy it takes to increase the temperature of 1 kg of a material by 1°C.

speed: the distance an object moves divided by the time it takes to move that distance.

static charge: an unbalanced electric charge on an object.

sublimation: the process of changing directly from a solid to a gas.

substance: matter with a composition that is always the same.

substituted hydrocarbon: an organic compound in which a carbon atom is bonded to an atom, or group of atoms, other than hydrogen.

surface tension: the uneven forces acting on the particles on the surface of a liquid.

synthesis (SIHN thuh sus): a type of chemical reaction in which two or more substances combine and form one compound.

onda sonora: onda longitudinal que sólo viaja a través de la materia.

calor específico: cantidad de energía térmica necesaria para aumentar la temperatura de 1 Kg de un material en 1°C.

rapidez: distancia que un objeto recorre dividida por el tiempo que éste tarda en recorrer dicha distancia.

carga estática: carga eléctrica no balanceada en un objeto.

sublimación: proceso de cambiar directamente de sólido a gas.

sustancia: materia cuya composición es siempre la misma.

hidrocarburo de sustitución: compuesto orgánico en el cual un átomo de carbono está unido a un átomo, o grupo de átomos, diferente del hidrógeno.

tensión superficial: fuerzas desiguales que actúan sobre las partículas en la superficie de un líquido.

síntesis: tipo de reacción química en el que dos o más sustancias se combinan y forman un compuesto.

technology: the practical use of scientific knowledge, especially for industrial or commercial use.

temperature: the measure of the average kinetic energy of the particles in a material.

temporary magnet: a magnet that quickly loses its magnetism after being removed from a magnetic field.

thermal conductor: a material through which thermal energy flows quickly.

thermal contraction: a decrease in a material's volume when the temperature is decreased.

thermal energy: the sum of the kinetic energy and the potential energy of the particles that make up an object.

thermal expansion: an increase in a material's volume when the temperature is increased.

tecnología: uso práctico del conocimiento científico, especialmente para empleo industrial o comercial.

temperatura: medida de la energía cinética promedio de las partículas de un material.

imán temporal: imán que rápidamente pierde su magnetismo después de haberlo retirado de un campo magnético.

conductor térmico: material mediante el cual la energía térmica se mueve con rapidez.

contracción térmica: disminución del volumen de un material cuando disminuye la temperatura.

energía térmica: suma de la energía cinética y potencial de las partículas que componen un objeto.

expansión térmica: aumento en el volumen de un material cuando aumenta la temperatura.

thermal insulator: a material through which thermal energy flows slowly.

thermostat: a device that regulates the temperature of a system.

transformer: a device that changes the voltage of an alternating current.

transition element: an element in groups 3–12 on the periodic table.

translucent: a material that allows most of the light that strikes it to pass through, but through which objects appear blurry.

transmission: the passage of light through an object.

transparent: a material that allows almost all of the light striking it to pass through, and through which objects can be seen clearly.

transverse wave: a wave in which the disturbance is perpendicular to the direction the wave travels.

trough: the lowest point on a transverse wave.

turbine (TUR bine): a shaft with a set of blades that spins when a stream of pressurized fluid strikes the blades.

aislante térmico: material en el cual la energía térmica se mueve con lentitud.

termostato: aparato que regula la temperatura de un sistema.

transformador: aparato que cambia el voltaje de una corriente alterna.

elemento de transición: elemento de los grupos 3–12 de la tabla periódica.

translúcido: material que permite el paso de la mayor cantidad de luz que lo toca, pero a través del cual los objetos se ven borrosos.

transmisión: paso de la luz a través de un objeto.

transparente: material que permite el paso de la mayor cantidad de luz que lo toca, y a través del cual los objetos pueden verse con nitidez.

onda transversal: onda en la que la perturbación es perpendicular a la dirección en que viaja la onda.

seno: punto más bajo en una onda transversal.

turbina: eje con una serie de paletas que gira cuando un chorro de fluido a presión golpea las paletas.

ultraviolet wave: an electromagnetic wave that has a slightly shorter wavelength and higher frequency than visible light.

unbalanced forces: forces acting on an object that combine and form a net force that is not zero.

unsaturated hydrocarbon: a hydrocarbon that contains one or more double or triple bonds.

unsaturated solution: a solution that can still dissolve more solute at a given temperature and pressure.

onda ultravioleta: onda electromagnética que tiene una longitud de onda ligeramente menor y mayor frecuencia que la luz visible.

fuerzas no balanceadas: fuerzas que actúan sobre un objeto, se combinan y forman una fuerza neta diferente de cero.

hidrocarburos insaturados: hidrocarburo que contiene uno o más enlaces dobles o triples.

solución insaturada: solución que aún puede disolver más soluto a cierta temperatura y presión.

valence electron: the outermost electron of an atom that participates in chemical bonding.

vapor: the gas state of a substance that is normally a solid or a liquid at room temperature.

vaporization: the change in state from a liquid to a gas.

electrón de valencia: electrón más externo de un átomo que participa en el enlace químico.

vapor: estado gaseoso de una sustancia que normalmente es sólida o líquida a temperatura ambiente.

vaporización: cambio de estado líquido a gaseoso.

variable: any factor that can have more than one value.

velocity: the speed and the direction of a moving object.

vibration: a rapid back-and-forth motion that can occur in solids, liquids, or gases.

viscosity (vihs KAW sih tee): a measurement of a liquid's resistance to flow.

voltage: the amount of energy used to move one coulomb of electrons through an electric circuit.

variable: cualquier factor que tenga más de un valor.

velocidad: rapidez y dirección de un objeto en movimiento.

vibración: movimiento rápido de atrás hacia adelante que puede ocurrir en sólidos, líquidos o gases.

viscosidad: medida de la resistencia de un líquido a fluir.

voltaje: cantidad de energía usada para mover un culombio de electrones por un circuito eléctrico.

wave: a disturbance that transfers energy from one place to another without transferring matter.

wavelength: the distance between one point on a wave and the nearest point just like it.

weight: the gravitational force exerted on an object.

work: the amount of energy used as a force moves an object over a distance.

onda: perturbación que transfiere energía de un lugar a otro sin transferir materia.

longitud de onda: distancia entre un punto de una onda y el punto más cercano similar al primero.

peso: fuerza gravitacional ejercida sobre un objeto.

trabajo: cantidad de energía usada como fuerza que mueve un objeto a cierta distancia.

X-ray: a high-energy electromagnetic wave that has a slightly shorter wavelength and higher frequency than an ultraviolet wave.

rayo X: onda electromagnética de alta energía que tiene una longitud de onda ligeramente más corta y frecuencia más alta que una onda ultravioleta.

Index

Absorption

Italic numbers = illustration/photo   **Bold numbers** = vocabulary term
lab = indicates entry is used in a lab on this page

Boyle's law

INDEX

INDEX

INDEX

INDEX

INDEX

INDEX

INDEX

Volts, 726
Volume
 explanation of, 229, *230,* 263
 gas, 281 *lab,* 283–285
 liquid, 263

W

Waste energy, 151, *151*
Water
 bases in, 480, *480*
 boiling point of, 277
 chemical formula for, 429, *429,* 430, *430*
 composition of, 222
 as electric conductor, 735
 explanation of, 423
 freezing point of, 280
 heating curve of, 277, *277*
 melting point of, 277
 polarity of, 470, *470*
 as solvent, 470, *471*
 specific heat of, 189
 states of, 277, *277*
Water molecules, 393, *393*
Water waves. *See also* **Waves**
 energy transfer by, 545, 546, *546*
 explanation of, *550*
 raindrops as source of energy for, *544*
 reflection of, 565
Wavelength
 color and, 666, 667, *667,* 682, *682*
 detection of light and, *684,* 684–685, *685*
 of electromagnetic waves, 627, 633, *634–635,* 640, *640*

 explanation of, 557, **557,** *558,* 595, *595*
 frequency and, 558, *558*
 refraction and, 682–683
Waves
 amplitude and, 555–556, *556,* 561
 collision of, 562 *lab*
 diffraction of, *566,* 566–567
 electromechanic, 551, *551*
 as energy source, 545, *545*
 energy transfer by, 546, *546*
 explanation of, **545**
 frequency of, 558, *558*
 interaction between matter and, *563,* 563–565, *564*
 interference of, *567,* 567–568, *568*
 longitudinal, 548, *548*
 mechanical, 547–550, *549, 550*
 method to create, 544 *lab*
 reflection of, 564–565, *565*
 refraction of, 566, *566*
 speed of, 559
 standing, 568, *568*
 transverse, 547, *547*
 tsunamis as, 553
Wave speed equation, 559
Weight
 buoyant force and, 109, *109,* 112
 explanation of, **54**
 mass and, 54, *54,* 228
White light, 666, 667, 682
Wind, 117, *117*
Wind energy
 advantages and disadvantages of, *162*
 explanation of, 160, *160*
Wind instruments, 599, 601

Wind turbines
 explanation of, 160, *160*
Word Origin, NOS 5, NOS 14, 14, 15, 25, 32, 51, 64, 72, 79, 97, 106, 118, 139, 149, 156, 181, 192, 198, 217, 229, 235, 243, 263, 276, 311, 322, 343, 352, 361, 363, 384, 393, 428, 435, 441, 463, 474, 480, 503, 513, 551, 555, 586, 598, 625, 633, 640, 665, 673, 685, 717, 722, 733, 754, 767
Work
 energy and, 142, *142*
 explanation of, 115
Writing In Science, NOS 29, 39, 85, 123, 205, 251, 291, 331, 407, 449, 489, 509, 529, 573, 613, 651, 699, 741, 781

X

x-axis, 32
Xenon, 362
X-rays
 for detection and security purposes, 645, *645*
 medical uses for, 638 *lab,* 646, *646*

Y

y-axis, 32

Z

Zinc, 102, 343, *343*
Zoom lens, 691

Credits

Photo Credits

COVER Matt Meadows/Peter Arnold, Inc.; **vii** The McGraw-Hill Companies; **NOS 2** AP Photo/NTSB; **NOS 2-3** photo by John Kaplan, NASA; **NOS 4** (inset)SMC Images/Getty Images; **NOS 4** Popperfoto/Getty Images; **NOS 5** (c)Stephen Alvarez/Getty Images; **NOS 5** (b)Science Source/Photo Researchers, Inc.; **NOS 5** (t)Maria Stenzel/Getty; **NOS 6** Martyn Chillmaid/photolibrary.com; **NOS 7** NASA; **NOS 9** (t)Andy Sacks/Getty Images, (c)NASA, H. Ford (JHU), G. Illingworth (UCSC/LO), M.Clampin (STScI), (b)Brand X Pictures/PunchStock; **NOS 10** StockShot/Alamy; **NOS 11** Michael Newman/PhotoEdit; **NOS 12** Tim Wright/CORBIS; **NOS 14** (bkgd)Hutchings Photography/Digital Light Source; **NOS 16** (tl)Matt Meadows, (tr)ASP/YPP/age footstock, (bl)Blair Seitz/photolibrary.com, (br)The McGraw-Hill Companies, Inc./Louis Rosenstock, photographer; **NOS 17** (t)The McGraw-Hill Comapanies, (c)photostock1/Alamy, (b)Blend Images/Alamy; **NOS 18** The McGraw-Hill Companies; **NOS 20** MANDEL NGAN/AFP/Getty Images; **NOS 21** Brian Stevenson/Getty Images; **NOS 22** (t)Hisham Ibrahim/Getty Images, (bl)Bordner Aerials; **NOS 23** AP Photo/The Minnesota Daily, Stacy Bengs; **NOS 24** LARRY DOWNING/Reuters /Landov; **NOS 25** Plus Pix/age fotostock; **NOS 26** ASSOCIATED PRESS; Hartig (STScI), the ACS Science Team, and ESA; **4** (t0Wildscape/Alamy, (b) Andy Crawford/BBC Visual Effects - modelmaker/Dorling Kindersley; **5** (t) Miriam Maslo/Science Photo Library/CORBIS, (b) The Gravity Group, LLC; **8** (t)Jasper James/Getty Images, (b)Picturenet/Getty Images; **9** (t)Scott Smith/photolibrary, (c)image100/CORBIS, (b)Simon Stuart-Miller/Alamy; **10** Jeremy Horner/Getty Images; **11** Camille Moirenc/Getty Images; **13** Hutchings Photography/Digital Light Source; **14** Aerial Photos of New Jersey, sciencephotos/Alamy; **16** Camille Moirenc/Getty Images; **17** David J. Green - technology/Alamy; **18** Hutchings Photography/Digital Light Source; **19** Heinrich van den Berg/Getty Images; **20** (tl)Don Farrall, (tr)Don Farrall/Getty Images, (b)Hutchings Photography/Digital Light Source; **23** Getty Images; **24** Michael Dunning/Photo Researchers; **25** MIXA/Getty Images; **26** Getty Images; **27** Hutchings Photography/Digital Light Source; **28** MICHAEL KAPPELER/AFP/Getty Images; **29** (t)Hutchings Photography/Digital Light Source, (b)StockTrek/Getty Images, Leo Dennis Productions/Brand X/CORBIS; **31** Hutchings Photography/Digital Light Source; **32** (b)Robert Holmes/CORBIS, Getty Images; **33** (t)Comstock/PunchStock, (tl)Phil Degginger/Getty Images, (tr)James Leynse/CORBIS, (b)Eyecon Images/Alamy Images, (bl)AP Photo/Thomas Kienzle, (br)Construction Photography/CORBIS; **34** (t)NASA, (b)Bruce Ando/Getty Images; **35** (t)Phil Degginger/Getty Images, (b)Agence Zoom/Getty Images; **36** Steve Cole/Getty Images; **37** Hutchings Photography/Digital Light Source; **38** (t to b) Hutchings Photography/ Digital Light Source, (2)sciencephotos/Alamy, (3)Don Farrall, (4)Bruce Ando/Getty Images; **41** Charles Krebs/Getty Images; **48-49** Jason Horowitz/zefa/CORBIS; **50** (t)Hutchings Photography/Digital Light Source; **51** (t)NASA, (b)Terje Rakke/Getty Images; **52** (t)Hutchings Photography/Digital Light Source,(c)Daniel Smith/Getty Images, (b)Ryuhei Shindo/CORBIS/Jupiter Images; **54** NASA; **55** (tl)Horizons Companies, (b)Hutchings Photography/Digital Light Source; **57** (t)Terje Rakke/Getty Images, (b)Hutchings Photography/Digital Light Source, (c)VisionsofAmerica/Joe Sohm/Getty Images; **58** (b)Ursula Gahwiler/photolibrary.com, (bkgd)StockTrek/Getty Images; **60** (t)Charles Krebs, (b)Arthur Morris/CORBIS; **61 62** (l)Hutchings Photography/Digital Light Source; **63** Tim Keatley/Alamy; **64** AP Photo/Insurance Institute for Highway Safety; **65** (t)Hutchings Photography/Digital Light Source, (c)Tim Keatley/Alamy, (b)AP Photo/Insurance Institute for Highway Safety; **67** Hutchings Photography/Digital Light Source; **68** Michael Steele/Getty Images; **69** david sanger photography/Alamy; **73** Tim Garcha/CORBIS; **76** (t) Patrik Giardino/CORBIS, b)Hutchings Photography/Digital Light Source; **77** (t)LIN HUI/Xinhua /Landov, (b)Duncan Soar/Alamy; **78** (l)David Madison/Getty Images, (r)Design Pics Inc./Alamy; **80** (tl)Richard Megna, Fundamental Photographs, NYC, (bl)Richard Megna, Fundamental Photographs, NYC; **81** (t)Hutchings Photography/Digital Light Source, (c)David Madison/Getty Images, (b)Richard Megna, Fundamental Photographs, NYC; **82** (t to b) Hutchings Photography/Digital Light Source, (2)Arthur Morris/CORBIS, (3)david sanger photography/Alamy; **85** Jason Horowitz/zefa/CORBIS; **94-95** Jeff Greenberg/Alamy; **96** Hutchings Photography/Digital Light Source; **97** Angelo Cavalli/Getty Images; **99 102 103** Hutchings Photography/Digital Light Source; **104** (t)Harbor Branch Oceanographic Institute_at Florida Atlantic University, (b)Check Six/Getty Images; **105** Hutchings Photography/Digital Light Source; **106** (t)Guillen Photography/Travel/USA/Florida/Alamy, (b)Photo and Co/Getty Images; **113** Hutchings Photography/Digital Light Source; **114** Auto Imagery, Inc.; **115** Justin Sullivan/Getty Images; **117** Stephen Morton/Getty Images); **118** (b)Michael Blann/Getty Images; **119** Justin Sullivan/Getty Images; **120** Hutchings Photography/Digital Light Source, (b)Michael Blann/Getty Images; **132** (t)Jochen Tack/photolibrary, (bl)Biophoto Associates/Photo Researchers, Inc., (br)Oleksiy Maksymenko/Alamy; **132** (t) Dennis Kunkel Microscopy, Inc./Visuals Unlimited, Inc.,(c) PhotoAlto/PunchStock, (b)Amanda Hall/Robert Harding World Imagery/Getty Images; **136-137** David R. Frazier Photolibrary, Inc./Alamy; **138** Hutchings Photography/Digital Light Source; **139** (t)Edward Kinsman/Photo Researchers, Inc., (bl)Thinkstock/

CORBIS, (bc)Sharon Dominick/Getty Images, (br) Dimitri Vervitsiotis/Getty Images; **141** Hutchings Photography/Digital Light Source, (b)Image Source/ CORBIS; **143** (t to b)David Madison/Getty Images, (2)Andrew Lambert Photography/Photo Researchers, Inc., (3)Design Pics/age footstock, (4) VStock/Alamy, (5)Matthias Kulka/zefa/CORBIS, (6) Goodshoot/PunchStock; **145** (t)Ivy Close Images/ Alamy, (b)Lawrence Manning/CORBIS, (bkgd) Royalty-Free/CORBIS; **146** Hutchings Photography/ Digital Light Source; **147** Alaska Stock LLC/Alamy; **149** Hutchings Photography/ Digital Light Source; **150** Peter Cade/Getty Images; **151** Lorcan/Getty Images; **153** (l)AP Photo/Hugh Scoggins, (r)AP Photo/Phil Coale, Bill Bachmann/First Light/ CORBIS; **154** Hutchings Photography/Digital Light Source; **155** (t)Phil Noble/Reuters/CORBIS, (bl) Jupiterimages, (bc)BrandX/Jupiterimages, (br)Pixtal/ age fotostock; **159** Chinch Gryniewicz; Ecoscene/ CORBIS; **160** Frank Whitney/Getty Images; **161** Bill Bachmann/Alamy; **162** (t to b)Bill Ross/CORBIS, (2) Royalty-Free/CORBIS, (3)Courtesy USDA/NRCS, photo by Lynn Betts, (5)Ingram Publishing/ SuperStock, (6)Arctic-Images/CORBIS, (7)Joel Sartore/National Geographic/Getty Images; **176-177** Tyrone Turner/National Geographic Stock; **178** (t)Hutchings Photography/Digital Light Source; **179** (t) Philip Scalia/Alamy, (b)Jamie Sabau/Getty Images; **181** Johner/Getty Images; **183** Summer Jones/Alamy; **186** Hutchings Photography/Digital Light Source; **187** (t)Alaska Stock LLC/Alamy, (b) The McGraw-Hill Companies; **188** Anthony- Masterson/Getty Images; **190** (t)Matt Meadows, (b) The McGraw-Hill Companies; **191** (t)CORBIS (Royalty-Free); **192 194** (t)Matt Meadows; **195** (tl) Rachael Bowes/Alamy, (tr)CHRISTOPHER WEDDLE/UPI /Landov, (b)Paul Glendell/Alamy, (bkgd)David Papazian/Beateworks/CORBIS; **197** Kevin Foy/Alamy; **198** (bkgd)steven langerman/ Alamy, (inset)Thomas Northcut/Getty Images; **201** (l)steven langerman/Alamy; **202** (t)Summer Jones/Alamy, (b)CORBIS (Royalty-Free); **204** (l) Anthony-Masterson/Getty Images; **205** (l)Tom Uhlman/Alamy, (r)Tyrone Turner/National Geographic Stock; **207** Thomas Northcut/Getty Images; **214-215** Gregory G. Dimijian/Photo Researchers,Inc.; **216** Hutchings Photography/ Digital Light Source; **217** (t)VEER Alison Shaw/Getty Images, (b)The McGraw-Hill Companies; **221** (l) Mark Steinmetz, (cl)The McGraw-Hill Companies, (cr)Foodfolio/age footstock, (r)Photo Spin/Getty Images; **225** Hutchings Photography/Digital Light Source; **226** (bkgd)Alaska Stock Images/age footstock, (inset)Mike Perry/Alamy; **227** Rob Rae/ age fotostock; **228** Royalty-Free/CORBIS; **230** (l) Royalty-Free/CORBIS, (c)Getty Images, (r)ULTRA.F/ Getty Images; **231** (l to r)Brand X/CORBIS, (2) Studio/age footstock, (3)Steve Shott/Steve Shott/ Getty Images, (4)Dorling Kindersley/Getty Images, (5)Crawford/Dorling Kindersley/Getty Images; **234** Hutchings Photography/Digital Light Source;

235 John Terence Turner/Taxi/Getty Images; **236** Perennou Nuridsany/Photo Researchers, Inc.; **237** (l)Digital Vision/Getty Images, (c)BananaStock/ AGEfotostock, (r)IT Stock/ age fotostock; **238** GK Hart/Vikki Hart/Getty Images; **241** Hutchings Photography/Digital Light Source; **242** Luis Calabor/Getty Images; **243** (l)Brand X Pictures/ Punchstock, (cl)IMAGEMORE Co.,Ltd./Getty Images, (cr)Martin Hospach/Getty Images, (r)Siede Preis/Getty Images; **246** (t b)The McGraw-Hill Companies, (cl)The Print Collector/age footstock, (cr)Ambient Images Inc./Alamy; **260-261** Gregor M. Schmid/CORBIS; **262** Hutchings Photography/ Digital Light Source; **263** Atlantide Phototravel/ CORBIS; **265** (t)Royalty Free/CORBIS, (c)Steve Hamblin/Alamy, (b)The McGraw-Hill Companies; **266** (t)Vito Palmisano/Getty Images, (b)creativ collection/age fotostock; **267** (l)Mauritius/ SuperStock; **268** Alberto Coto/Getty Images; **270** (t)Finley - StockFood Munich, (b)The McGraw-Hill Companies, (bkgd)Michael Rosenfeld/Getty Images; **271** Hutchings Photography/Digital Light Source; **272** Robert Fried/Alamy; **275** Hutchings Photography/Digital Light Source; **276** (l)Charles D. Winters/Photo Researchers, Inc.; **276** (r)Jean du Boisberranger/Getty Images; **279** (t)Alaska Stock/ age footstock, (c)Hutchings Photography/Digital Light Source, (b)Jean du Boisberranger/Getty Images; **281** Hutchings Photography/Digital Light Source; **282** Check Six/Getty Images; **284** (c) Summer Jones/Alamy; **303** The McGraw-Hill Companies Inc./Ken Cavanagh Photographer; **300** (t)Roger Ressmeyer/CORBIS, (b)American Museum of Natural History; **304-305** (bkgd)JAMES BRITTAIN/Photolibrary; **306** (t)Hutchings Photography/Digital Light Source; **307** (l)DAJ/Getty Images; **307** (t)Royalty-Free/CORBIS, (r)Creatas/ PunchStock, (b)Royalty-Free/CORBIS, (inset)Drs. Ali Yazdani & Daniel J. Hornbaker/Photo Researchers, Inc., (bkgd)Alessandro Della Bella/Keystone/ CORBIS; **308** (l)Scala/Art Resource, NY, (r)The Royal Institution, London, UK /The Bridgeman Art Library; **309** (t)Horizons Companies, (b)Drs. Ali Yazdani & Daniel J. Hornbaker/Photo Researchers, Inc.; **317** (t)Horizons Companies, (b)Drs. Ali Yazdani & Daniel J. Hornbaker/Photo Researchers, Inc.; **318** Royalty-Free/CORBIS; **319** Hutchings Photography/Digital Light Source; **320** (bkgd) Derrick Alderman/Alamy, (inset)Derrick Alderman/ Alamy; **324** (t)SPL/Photo Researchers, Inc., (b)Time Life Pictures/Mansell/Time Life Pictures/Getty Images; **338-339** Nick Caloyianis/National Geographic/Getty Images; **340** Hutchings Photography/Digital Light Source; **341** (t)P.J. Stewart/Photo Researchers, Inc, Hutchings Photography/Digital Light Source; **343** (t)Hutchings Photography/Digital Light Source, (tl)DEA/A.RIZZI/ De Agostini Picture Library/Getty Images, (tr)Astrid & Hanns-Frieder Michler/Photo Researchers, Inc., (cl)Visuals Unlimited/Ken Lucas/Getty Images, (cr) Richard Treptow/Photo Researchers, Inc.,(bl)

Credits

CORBIS, (br)ImageState/Alamy; **346** (l)David J. Green/Alamy, (c)WILDLIFE/Peter Arnold Inc, (r)Mark Schneider/Visuals Unlimited/Getty Images; **347** (l) LBNL/Photo Researchers, Inc, (c)Boyer/Roger Viollet/Getty Images, (r)ullstein bild/Peter Arnold, Inc.; **350** Hutchings Photography/Digital Light Source; **351** Paul Katz/photolibrary.com; **352** (tl)The McGraw-Hill Companies, (tc)Paul Katz/Getty Images,(tr)Egyptian National Museum, Cairo, Egypt, Photo © Boltin Picture Library/The Bridgeman Art Library International; **352** (bl)NASA, (bc)Hutchings Photography/Digital Light Source, (br)Charles Stirling/Alamy; **353** (l)The McGraw-Hill Companies, Inc./Stephen Frisch, photographer, (c) sciencephotos/Alamy, (r)Martyn Chillmaid/Oxford Scientific (OSF)/photolibrary.com; **354** (l)Royalty-Free/CORBIS, (cl)Dr. Parvinder Sethi; **354** (cr)Joel Arem/Photo Researchers, Inc., (r)Ingram Publishing/ SuperStock; **356** (t)Egyptian National Museum, Cairo, Egypt, Photo © Boltin Picture Library / The Bridgeman Art Library International, (c)The McGraw-Hill Companies, Inc./Stephen Frisch, photographer, (b)Paul Katz/Getty Images; **357** Jeff Hunter/Getty Images; **358** Hutchings Photography/ Digital Light Source; **359** E.O. lawrence Berkely National Laboratory, University of California, U.S. Department of Energy; **360** (tl)Ted Foxx/Alamy, (tr) Richard Treptow/Photo Researchers, Inc., (c) Hutchings Photography/Digital Light Source, (bl) Photodisc/Getty Images, (br)Charles D. Winters/ Photo Researchers, Inc.; **361** (l)sciencephotos/ Alamy; **362** NASA-JPL; **363** (l)Ingemar Aourell/Getty Images,(cl)Don Farrall/Getty Images, (cr)Gabe Palmer/Alamy, (r)Henrik Sorensen/Getty Images; **364** (t)PhotoLink/Getty Images; **365** (t)Richard Treptow/Photo Researchers, Inc., (c)sciencephotos/ Alamy; **365** (b)PhotoLink/Getty Images, (t)David J. Green/Alamy; **366** (b)Mark Schneider/Visuals Unlimited/Getty Images; **378-379** altrendo images/ Getty Images; **380** Hutchings Photography/Digital Light Source; **381** Douglas Fisher/Alamy; **388** (t) Popperfoto/Getty Images, (c)Underwood & Underwood/, (bl)John Meyer, (br)Ilene MacDonald/ Alamy; **389** Hutchings Photography/Digital Light Source; **390** Gazimal/Getty Images; **397** Hutchings Photography/Digital Light Source; **398** Brent Winebrenner/Photolibrary.com; **403** (t)Photodisc/ Getty Images, (c)C Squared Studios/Getty Images, (b)Jennifer Martine/Jupiter Images; **419** The McGraw-Hill Companies, Inc./Stephen Frisch, photographer; **416** (t) PARIS PIERCE/Alamy, (bl) British Library, London, UK/© British Library Board. All Rights Reserved/The Bridgeman Art Library, (br) University Library, Leipzig, Germany/Archives Charmet/The Bridgeman Art Library; **417** ACE STOCK LIMITED/Alamy; **420-421** Anton Luhr/ Photolibrary; **422** Hutchings Photography/Digital Light Source; **423** Darwin Dale/Photo Researchers, Inc.C5454; **424** (tl)Royalty-Free/CORBIS, (tr)Charles D. Winters/Photo Researchers, Inc., (cl)London Scientific Films/photolibrary.com, (bl)Brand X

Pictures, (br)Dante Fenolio/Photo Researchers, Inc.; **428** Hutchings Photography/ Digital Light Source; **434** (t)sciencephotos/Alamy, (b)AP Photo/Greg Campbell; **435** (t)Charles D. Winters/Photo Researchers, Inc., (b)The McGraw-Hill Companies, Inc./Jacques Cornell photographer, (t)The McGraw-Hill Companies, Inc./Stephen Frisch, photographer; **436** Park Dale/Alamy; **438** Joel Sartore/National Geographic/Getty Images; **440** Brand X Pictures/ PunchStock; **443** (tl)McGraw-Hill Companies, (tc) Brand X Pictures/PunchStock, (tr)Tetra Images/ Getty Images, (bl)The McGraw-Hill Companies, (bc) Royalty-Free/CORBIS, (br)Alexis Grattier/Getty Images; **457** The McGraw-Hill Companies Inc./Ken Cavanagh Photographer; **458-459** Keith Kapple/ SuperStock; **460** Hutchings Photography/Digital Light Source; **461** Visual&Written SL/Alamy; **462** (tl) PHOTOTAKE Inc./Alamy, (tr)Royalty-Free/CORBIS, (cl)Reuters /Landov, (cr)Flirt/SuperStock, (bl)Don Farrall/Getty Images, (br)Royalty-Free/CORBIS; **463** (tl)liquidlibrary/PictureQues, (tcl)Mark Steinmetz, (tcr)Michael Maes/Getty Images, (tr)C Squared Studios/Getty Images, (bl)Dennis Kunkel Microscopy, Inc.; **464** Hutchings Photography/ Digital Light Source; **466** (tl)Royalty-Free/CORBIS, (tr)C Squared Studios/Getty Images, (b)Hutchings Photography/Digital Light Source; **467** Zia Soleil; **468** Hutchings Photography/Digital Light Source; **469** (t)Susan Gottberg/Alamy; **470** (t)Ingram Publishing/Alamy, (c)Michael Maes/Getty Images, (b)SuperStock/age fotostock; **471** The Mcgraw-Hill Companies; **472 473** Hutchings Photography/Digital Light Source; **475** (t)Stock Connection Distribution/ Alamy, Hutchings Photography/Digital Light Source; **476 478** Hutchings Photography/Digital Light Source; **479** (t)Fletcher & Baylis/Photo Researchers, Inc, (b)Horizons Companies; **481** (cw from top) Charles D. Winters/SPL/Photo Researchers, (2) Steve Mason/Getty Images, (3)Yasuhide Fumoto/ Getty Images, (4)Phil Degginger/Alamy, (5) Westend61/SuperStock, (6)F. Schussler/PhotoLink/ Getty Images, (7)Richard Megna, Fundamental Photographs, NYC; **482** (t)Royalty-Free/CORBIS, (bl)Busse Yankushev/age footstock, (br)Spencer Jones/Getty Images; **483** (tl)liquidlibrary/ PictureQuest, (tr)Royalty-Free/CORBIS, (bl)The McGraw-Hill Companies, Inc./Jacques Cornell photographer, (br)Robert Manella for MMH; **498-499** Jean-Paul Ferrero/Minden Pictures; **500** Hutchings Photography/Digital Light Source; **501** (t) The Natural History Museum/Alamy, (b)Wolfgang Pölzer/Alamy; **503** (t)Image Source/Getty Images, (b) Michael Betts/Getty Images; **508** Michael Betts/ Getty Images; **509** (tr)Courtesy of Texas Instruments, (cr)Bettmann/CORBIS, (bl)Digital Art/ CORBIS, (br)Roger Ressmeyer/CORBIS, (bkgd) Digital Vision/Getty Images; **510** Hutchings Photography/Digital Light Source; **511** Kim Taylor/ Minden Pictures; **513** (t)AP Photo/Rita Beamish, (c) Franceso Tomasinelli/Photo Researchers, Inc., (b) Jamie Grill/Getty Images; **514** Image Source/Getty

Images; **516** (t to b)Andy Hibbert; Ecoscene/ CORBIS, (2)Image Source/Getty Images, (3)Leonard Lessin/Photo Researchers, Inc, (4)Glyn Thomas/ Alamy; **517** Glyn Thomas/Alamy; **519** Hutchings Photography/Digital Light Source; **520** Seb Rogers/ Alamy; **522** (l)Image Source/Getty Images, (r)Brian Hagiwara/Jupiter Images; **526** Image Source/Getty Images; **529** Jean-Paul Ferrero/Minden Pictures; **539** CORBIS; **542-543** Kaz Mori/Getty Images; **544** Hutchings Photography/Digital Light Source; **545** (t) Deco/Alamy, (b)Hutchings Photography/Digital Light Source; **550** (t)Royalty-Free/CORBIS, (c)Jason Hosking/zefa/CORBIS, (b)Mark Downey/Lucid Images/CORBIS; **551** (t) imagebroker/Alamy, (b) Edward Kinsman/Photo Researchers, Inc.; **552** Hutchings Photography/Digital Light Source; **553** (bkgd)Comstock/PunchStock, (inset)NOAA; **554** Hutchings Photography/Digital Light Source; **555** WireImage; **562** Hutchings Photography/Digital Light Source; **563** (t)Gustoimages/Photo Researchers, Inc., (b)NASA, ESA, and the Hubble Heritage (STScI/AURA)-ESA/Hubble Collaboration; **564** Car Culture/Getty Images; **565** Richard Megna, Fundamental Photographs, NYC; **566-570** Richard Megna, Fundamental Photographs, NYC; **582-583** Dan Barba/Photolibrary; **584** Hutchings Photography/Digital Light Source; **585** Mike Kemp/ Getty Images; **587** Jens Kuhfs/SeaPics.com; **588** Michael Mährlein/age fotostock; **589** (b)Wave Royalty Free/age fotostock; **590** Michael Mährlein/ age fotostock; **591** (t)Michael Newman/PhotoEdit, (b)Steve Hamblin/Alamy; **592** Hutchings Photography/Digital Light Source; **593** Stefano Cellai/age fotostock; **602** Hutchings Photography/ Digital Light Source; **603** (bkgd)NOAA, (inset) EMORY KRISTOF/National Geographic Stock; **606** (t)John Luke/age footstock, (bl)View Pictures Ltd/ SuperStock, (br)Zave Smith/Getty Images; **607** DAJ/Getty Images; **608** Zephyr/Photo Researchers, Inc.; **609** John Luke/age fotostock; **610** (t)Michael Mährlein/age footstock,(b)Zave Smith/Getty Images; **622-623** NASA; **624** Hutchings Photography/Digital Light Source; **625** Menno Boermans/Aurora Photos/ CORBIS; **628** X-ray: NASA/CXC/J.Hester (ASU); Optical: NASA/ESA/J.Hester & A.Loll (ASU); Infrared: NASA/JPL-Caltech/R.Gehrz (Univ. Minn.); **629** (t c)Bjorn Rorslett/Photo Researchers, Inc, (bl) Hutchings Photography/Digital Light Source, (br) Ralf Nau/Getty Images; **631** NASA; **632** Hutchings Photography/Digital Light Source; **633** Felix Stenson/age fotostock; **634** (t)izmostock/Alamy, (c)

PM Images/Getty Images, (b)D. Hurst/Alamy; **635** (t) Peter Cade/Getty Images, (c)Digital Vision/ PunchStock, (b)Richard T. Nowitz/Phototake; **638** Howard Sokol/age fotostock; **639** Carl Trinkle; **641** Joe Drivas/Getty Images; **643** (t)Simon Belcher/ Alamy, (c)NASA, (b)CORBIS/age fotostock; **644** (b) Digital Vision/age fotostock; **648** The McGraw-Hill Companies, Inc./Ken Cavanagh photographer; **660-661** Charles Krebs/Getty Images; **662** Hutchings Photography/Digital Light Source; **663** (t)Russell Burden/photolibrary.com, (b)Andrew Geiger/Getty Images; **664** (t)Hutchings Photography/Digital Light Source, (b)Jasper James/Getty Images; **665** Picturenet/Getty Images; **670** Jeremy Horner/Getty Images; **676** (background) NASA/ESA/M. Livio, (inset) NASA; **677** Hutchings Photography/Digital Light Source; **678** Image Source/Getty Images; **680** (l)Don Farrall;, (r)Don Farrall/Getty Images; **683** Getty Images; **684** Glow Images, Inc/photolibrary. com; **685** MIXA/Getty Images; **687** (t)Scott Smith/ photolibrary, (c)image100/CORBIS, (b)Simon Stuart-Miller/Alamy; **688** Hutchings Photography/Digital Light Source; **689** (t)Michael Kappler/AFP/Getty Images, (b)AP Photo/Thomas Kienzle; **692** Getty Images; **693** (tl)Phil Degginger/Getty Images, (tr) James Leynse/CORBIS, (br)Construction Photography/CORBIS; **694** (4)Bruce Ando/Getty Images; **708-709** Bettmann/CORBIS; **710** Hutchings Photography/Digital Light Source; **711** Rick Elkins/Getty Images; **715** Hutchings Photography/Digital Light Source; **717** vladimir zakharov/Getty Images; **718** (b)Alan R Moller/Getty, (bkgd)Ralph H Wetmore II/Getty; **720** (bkgd)Peter Arnold, Inc./Alamy, (inset)Jeff Greenberg/Alamy; **721** Hutchings Photography/Digital Light Source; **722** Construction Photography/CORBIS; **723** Hutchings Photography/Digital Light Source; **730** Hutchings Photography/Digital Light Source; **731** Troy Paiva/LostAmerica.com; **732** Hutchings Photography/Digital Light Source; **735** Tetra Images/Getty Images; **750-751** age fotostock/ SuperStock; **753** (t)imagebroker/Alamy, (cl)Joel Arem/Photo Researchers, Inc, (cr)Clive Streeter/ Getty Images, (b)Stockbyte/Getty Images; **756** Grambo/Getty Images; **762** Hutchings Photography/Digital Light Source; **763** Alexis Rosenfeld/Photo Researchers, Inc.; **770** Hutchings Photography/Digital Light Source; **771** BlueMoon Stock/Alamy; **774** (t)Lester Lefkowitz/CORBIS, (b) RIA NOVOSTI/SCIENCE PHOTO LIBRARY; **778** bobo/Alamy.

Science Benchmark Practice Test

Multiple Choice *Bubble the correct answer.*

1 One type of scientific investigation is an experiment. Which question could be investigated by experimentation? **SC.7.N.1.3**

- (A) In which part of the world are ammonite fossils most common?
- (B) Does light intensity affect the rate of cell division in a leaf cell?
- (C) Why does one side of Saturn's moon Iapetus appear lighter than the other side?
- (D) Which caterpillar species are eaten by young Florida scrub jays that live in Oscar Scherer State Park?

2 David tests a new type of skateboard wheel. He releases the skateboard at the top of a curved structure. He measures and records the distance from the starting point to where the skateboard stops. The table below lists some of David's data. **SC.7.N.1.2**

Trials	Distance Traveled by Skateboard with 50 kg Weight
Trial 1	21.4 m
Trial 2	23.5 m
Trial 3	20.9 m

Which scientific practice is represented by the data in this table?

- (F) repetition
- (G) replication
- (H) making predictions
- (I) control of variables

3 Empirical evidence is the basis for scientific explanations. Which is an example of empirical evidence? **SC.7.N.1.6**

- (A) a direct observation
- (B) a majority decision
- (C) a teacher's opinion
- (D) a testable hypothesis

4 A scientific law is different from a societal law. Which is an example of a scientific law? **SC.6.N.3.2**

- (F) For every action, there is an equal but opposite reaction.
- (G) It is illegal to do research on human beings without permission.
- (H) Within park boundaries, owners of dogs without leashes may be fined.
- (I) If fruit flies prefer apples to bananas, then they will land more on apple slices.

Copyright © Glencoe/McGraw-Hill, a division of The McGraw-Hill Companies, Inc.

NGSSS for Science Benchmark Practice *continued*

5 As Bernie swings back and forth on a swing at a regular rate, energy transformations take place. At the lowest point in his swing, he moves the fastest. As he swings upward, he slows down. At the highest point in his swing, he momentarily slows to a stop. Which statement must be true? **SC.6.P.11.1**

(A) He always has the same amounts of kinetic and potential energy.

(B) He has the least kinetic energy at the lowest point in his swing.

(C) He has the most potential energy at the highest point in his swing.

(D) His total kinetic energy and potential energy change as he swings.

6 Sodium chloride, also known as table salt, is an ionic compound made up of positive sodium ions and negative chloride ions. The diagram below models how sodium chloride dissolves in water.

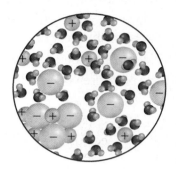

What type of force attracts the positive ends of the water molecules to the negative chloride ions? **SC.6.P.13.1**

(F) contact force

(G) electrical force

(H) gravitational force

(I) magnetic force

Copyright © Glencoe/McGraw-Hill, a division of The McGraw-Hill Companies, Inc.

NGSSS for Science Benchmark Practice continued

7 There is a greater force of gravity between objects that have a lot of mass than between objects that have very little mass. What other factor affects the gravity between two objects? **SC.6.P.13.2**

(A) the shape of the objects

(B) the volume of the objects

(C) the distance between objects

(D) the temperature of objects

8 Imagine that a play's stage crew must move a 30-kg piece of scenery on and off the stage. As they move the piece of scenery off the stage at a constant speed, the forces shown in the diagram below suddenly affect it.

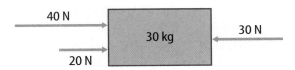

Which statement best describes what happens to the piece of scenery and why? **SC.6.P.13.3**

(F) Its motion stops because the forces acting on it are balanced.

(G) Its motion changes because the forces acting on it are balanced.

(H) Its motion changes because the forces acting on it are unbalanced.

(I) Its motion doesn't change because the forces acting on it are unbalanced.

9 Francisco has been trying different combinations of dyes for his sculpture. Tyfara is conducting an experiment that compares how long certain dyes will last under different conditions and on different materials. What makes Tyfara's work with dyes different from Francisco's work with dyes? **SC.7.N.1.3**

(A) Tyfara's work is based on facts.

(B) Tyfara's work must be reproducible.

(C) Tyfara's work cannot have mistakes or errors.

(D) Tyfara's work is easier than Francisco's work.

10 Akira makes a chart showing different types of light. At the top of the chart, she writes "Electromagnetic Waves." Under that category, she divides the chart into smaller categories. She labels one of these smaller categories "Visible Light from the Sun." Which light falls within this smaller category? **SC.7.P.10.1**

(F) black light

(G) blue light

(H) infrared light

(I) ultraviolet light

Copyright © Glencoe/McGraw-Hill, a division of The McGraw-Hill Companies, Inc.

NGSSS for Science Benchmark Practice continued

11 An architect designs an office building. She wants bright natural light to fill the front lobby. However, she does not want it to get too hot in the late afternoon. Which material would be best for the windows in the front lobby? **SC.7.P.10.2**

(A) a material that absorbs all light

(B) a material that transmits all light

(C) a material that transmits most light but reflects the rest of it

(D) a material that absorbs most light but refracts the rest of it

12 Two scientists study the effects of noise from motorboats on whales. One scientist places a detector above the surface of the water in the open air. The other places a detector underwater. Why does the underwater detector detect sounds before the open-air detector does? **SC.7.P.10.3**

(F) Sound travels faster through air than through water.

(G) Sound travels faster through water than through air.

(H) The underwater detector is farther away from the sounds.

(I) The detector located in the open air is farther away from the sounds.

13 Latisha steadily heated a pan of ice on a stove for 26 minutes while measuring its temperature. She made the graph below using the data she collected.

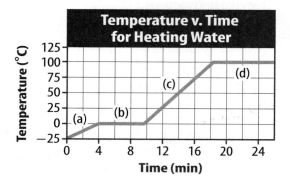

What could Latisha's friend conclude by looking at the graph? **SC.7.P.11.1**

(A) Areas **a** and **c** of the graph show when the water changed state.

(B) Areas **b** and **d** of the graph show when the water changed state.

(C) Areas **b** and **d** of the graph show when Latisha turned the stove off.

(D) Area **d** of the graph shows when there was no more water left in the pan.

Copyright © Glencoe/McGraw-Hill, a division of The McGraw-Hill Companies, Inc.

NGSSS for Science Benchmark Practice continued

14 Chris placed two fresh batteries in a flashlight. Then he switched on the flashlight and the bulb lit up. Which series of energy transformations took place? **SC.7.P.11.2**

(F) chemical energy→electrical energy→ light energy

(G) electrical energy→light energy→ chemical energy

(H) light energy→electrical energy→ light energy

(I) nuclear energy→electrical energy→ light energy

15 Hannah placed her left hand in a bowl of cold water and her right hand in a bowl of warm water. She waited a few minutes while her left hand cooled down and her right hand warmed up. Then she pulled her hands out of each bowl of water and folded them together. Which statement best describes how heat flows between her hands now? **SC.7.P.11.4**

(A) Heat flows from her left hand to her right hand until they are both the same temperature.

(B) Heat flows from her right hand to her left hand until they are both the same temperature.

(C) Heat flows from her left hand to her right hand until the right hand becomes warmer than the left hand.

(D) Heat flows from her right hand to her left hand until the left hand becomes warmer than the right hand.

16 These diagrams model the atoms of a substance in three different states.

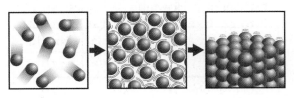

Which set of labels correctly identifies each state in the diagrams? **SC.8.P.8.1**

(F) gas → liquid → solid

(G) gas → solid → liquid

(H) liquid → gas → solid

(I) liquid → solid → gas

17 Imagine that an astronaut takes his lucky horseshoe to the Moon. What characteristic of the lucky horseshoe will be different on the Moon than on Earth? **SC.8.P.8.2**

(A) density

(B) mass

(C) volume

(D) weight

Copyright © Glencoe/McGraw-Hill, a division of The McGraw-Hill Companies, Inc.

NGSSS for Science Benchmark Practice continued

18 Antonio measures the masses and volumes of samples of four different liquids and lists them in the table below. **SC.8.P.8.3**

Liquid	Volume (mL)	Mass (g)
corn syrup	20	27.6
rubbing alcohol	30	23.5
vegetable oil	15	13.8
water	25	25

Which liquid has the greatest density?

(F) corn syrup

(G) cold water

(H) rubbing alcohol

(I) vegetable oil

19 Dr. Phung is trying to identify a substance. She knows that is a white powder at room temperature (20°C), a liquid at 750°C, and a gas at 1400°C. She knows that this means the substance melts (turns from solid to liquid) at a temperature below 750°C and boils (turns from liquid to gas) at a temperature above 1400°C. She looks up the melting and boiling points of four different white powders in a science handbook and lists them in the table below.

Substance	Melting Point (°C)	Boiling Point (°C)
Potassium bromide	734	1435
Potassium chloride	790	1420
Sodium bromide	747	1396
Sodium chloride	801	1413

What most likely is the substance based on the information in the table? **SC.8.P.8.4**

(A) potassium bromide

(B) potassium chloride

(C) sodium bromide

(D) sodium chloride

Copyright © Glencoe/McGraw-Hill, a division of The McGraw-Hill Companies, Inc.

NGSSS for Science Benchmark Practice continued

20 Janice's candy recipe states that she must stir the hot mixture constantly for 10 minutes over medium heat. She decides to stir with a wooden spoon rather than a metal spoon to avoid burning her hand. Which physical property did Janice use to choose a spoon for stirring the hot candy mixture? **SC.8.P.8.4**

(F) boiling point

(G) melting point

(H) electrical conductivity

(I) thermal conductivity

21 Hayden uses a pH meter to test an unknown liquid. He finds that it has a pH of 4. What can Hayden conclude about the unknown liquid? **SC.8.P.8.8**

(A) It is an acid.

(B) It is a base.

(C) It is a mixture.

(D) It is a solution.

22 A scientist examines a sample of a liquid under a high-powered microscope. He observes that the liquid has the same color and texture throughout. He cannot find different particles. Which term or terms could he use to describe the liquid? **SC.8.P.8.9**

(F) "substance" only

(G) "heterogeneous mixture" only

(H) "substance" or "homogeneous mixture"

(I) "substance" or "heterogeneous mixture"

23 Rhonda is a park ranger at Three Rivers State Park in northwest Florida. She prepares a yearly report on how much the park staff spends on damage repairs. She classifies each type of damage as the result of a physical change or a chemical change. Which does she describe as a physical change? **SC.8.P.9.2**

(A) burning of wooden trail markers during the spread of forest fires

(B) crumbling of roads due to water that freezes and expands in cracks

(C) rotting of wooden railings along lookouts and fishing docks

(D) rusting of steel parts in bridges and water tanks that have no paint

Copyright © Glencoe/McGraw-Hill, a division of The McGraw-Hill Companies, Inc.

24 Emily is riding in a roller-coaster car that slows down as it climbs each hill and speeds up as it rolls down the hill. How can she describe this pattern in terms of potential energy and kinetic energy? **SC.6.P.11.1**

(F) Potential energy transforms into kinetic energy as the car climbs each hill.

(G) Potential energy transforms into kinetic energy as the car rolls down each hill.

(H) Potential energy is created while kinetic energy is destroyed as the car rolls down each hill.

(I) Potential energy is destroyed while kinetic energy is created as the car rolls down each hill.

25 Babcock Ranch is a planned city that will be built in southwestern Florida. If the project is successful, it will be the first U. S. city to use only solar power during the day. Its power will come from a large photovoltaic solar plant. What kind of energy transformation occurs at a solar power plant? **SC.7.P.11.2**

(A) Chemical energy transforms to electrical energy.

(B) Electrical energy transforms to light energy.

(C) Light energy transforms to electrical energy.

(D) Nuclear energy transforms to light energy.

Copyright © Glencoe/McGraw-Hill, a division of The McGraw-Hill Companies, Inc.

PERIODIC TABLE OF THE ELEMENTS

Element — Hydrogen
Atomic number — 1
Symbol — H
Atomic mass — 1.01
State of matter

🎈 Gas
💧 Liquid
◻ Solid
⊙ Synthetic

A column in the periodic table is called a **group**.

A row in the periodic table is called a **period**.

	1	2	3	4	5	6	7	8	9
1	Hydrogen 1 H 1.01								
2	Lithium 3 Li 6.94	Beryllium 4 Be 9.01							
3	Sodium 11 Na 22.99	Magnesium 12 Mg 24.31							
4	Potassium 19 K 39.10	Calcium 20 Ca 40.08	Scandium 21 Sc 44.96	Titanium 22 Ti 47.87	Vanadium 23 V 50.94	Chromium 24 Cr 52.00	Manganese 25 Mn 54.94	Iron 26 Fe 55.85	Cobalt 27 Co 58.93
5	Rubidium 37 Rb 85.47	Strontium 38 Sr 87.62	Yttrium 39 Y 88.91	Zirconium 40 Zr 91.22	Niobium 41 Nb 92.91	Molybdenum 42 Mo 95.96	Technetium 43 Tc (98)	Ruthenium 44 Ru 101.07	Rhodium 45 Rh 102.91
6	Cesium 55 Cs 132.91	Barium 56 Ba 137.33	Lanthanum 57 La 138.91	Hafnium 72 Hf 178.49	Tantalum 73 Ta 180.95	Tungsten 74 W 183.84	Rhenium 75 Re 186.21	Osmium 76 Os 190.23	Iridium 77 Ir 192.22
7	Francium 87 Fr (223)	Radium 88 Ra (226)	Actinium 89 Ac (227)	Rutherfordium 104 Rf (267)	Dubnium 105 Db (268)	Seaborgium 106 Sg (271)	Bohrium 107 Bh (272)	Hassium 108 Hs (270)	Meitnerium 109 Mt (276)

The number in parentheses is the mass number of the longest lived isotope for that element.

Lanthanide series	Cerium 58 Ce 140.12	Praseodymium 59 Pr 140.91	Neodymium 60 Nd 144.24	Promethium 61 Pm (145)	Samarium 62 Sm 150.36	Europium 63 Eu 151.96
Actinide series	Thorium 90 Th 232.04	Protactinium 91 Pa 231.04	Uranium 92 U 238.03	Neptunium 93 Np (237)	Plutonium 94 Pu (244)	Americium 95 Am (243)

Metal

Metalloid

Nonmetal

Recently discovered

			13	**14**	**15**	**16**	**17**	**18**
								Helium 2 **He** 4.00
			Boron 5 **B** 10.81	Carbon 6 **C** 12.01	Nitrogen 7 **N** 14.01	Oxygen 8 **O** 16.00	Fluorine 9 **F** 19.00	Neon 10 **Ne** 20.18
10	**11**	**12**	Aluminum 13 **Al** 26.98	Silicon 14 **Si** 28.09	Phosphorus 15 **P** 30.97	Sulfur 16 **S** 32.07	Chlorine 17 **Cl** 35.45	Argon 18 **Ar** 39.95
Nickel 28 **Ni** 58.69	Copper 29 **Cu** 63.55	Zinc 30 **Zn** 65.38	Gallium 31 **Ga** 69.72	Germanium 32 **Ge** 72.64	Arsenic 33 **As** 74.92	Selenium 34 **Se** 78.96	Bromine 35 **Br** 79.90	Krypton 36 **Kr** 83.80
Palladium 46 **Pd** 106.42	Silver 47 **Ag** 107.87	Cadmium 48 **Cd** 112.41	Indium 49 **In** 114.82	Tin 50 **Sn** 118.71	Antimony 51 **Sb** 121.76	Tellurium 52 **Te** 127.60	Iodine 53 **I** 126.90	Xenon 54 **Xe** 131.29
Platinum 78 **Pt** 195.08	Gold 79 **Au** 196.97	Mercury 80 **Hg** 200.59	Thallium 81 **Tl** 204.38	Lead 82 **Pb** 207.20	Bismuth 83 **Bi** 208.98	Polonium 84 **Po** (209)	Astatine 85 **At** (210)	Radon 86 **Rn** (222)
Darmstadtium 110 **Ds** (281)	Roentgenium 111 **Rg** (280)	Copernicium 112 **Cn** (285)	* Ununtrium 113 **Uut** (284)	* Ununquadium 114 **Uuq** (289)	* Ununpentium 115 **Uup** (288)	* Ununhexium 116 **Uuh** (293)		* Ununoctium 118 **Uuo** (294)

* The names and symbols for elements 113-116 and 118 are temporary. Final names will be selected when the elements' discoveries are verified.

Gadolinium 64 **Gd** 157.25	Terbium 65 **Tb** 158.93	Dysprosium 66 **Dy** 162.50	Holmium 67 **Ho** 164.93	Erbium 68 **Er** 167.26	Thulium 69 **Tm** 168.93	Ytterbium 70 **Yb** 173.05	Lutetium 71 **Lu** 174.97
Curium 96 **Cm** (247)	Berkelium 97 **Bk** (247)	Californium 98 **Cf** (251)	Einsteinium 99 **Es** (252)	Fermium 100 **Fm** (257)	Mendelevium 101 **Md** (258)	Nobelium 102 **No** (259)	Lawrencium 103 **Lr** (262)